Dahlem Workshop Reports
Physical, Chemical, and Earth Sciences Research Report 9
Ocean Margin Processes in Global Change

Goal of this Dahlem Workshop:
to evaluate our understanding of the fluxes and transformation of natural and anthropogenic materials in estuaries and coastal oceans, and to examine their global consequences.

Physical, Chemical, and Earth Sciences Research Reports

Held and published on behalf of the
Freie Universität Berlin

Sponsored by:
Senat der Stadt Berlin
Marga and Kurt Möllgaard Stiftung

Ocean Margin Processes in Global Change

R.F.C. Mantoura, J.-M. Martin, and R. Wollast, Editors

Report of the Dahlem Workshop on
Ocean Margin Processes in Global Change
Berlin 1990, March 18–23

Rapporteurs: T.D. Jickells, P.S. Liss, F.T. Mackenzie, J.P. O'Kane

Program Advisory Committee:
R.F.C. Mantoura, J.-M. Martin, R. Wollast, Chairpersons
T.M. Church, J. Duinker, K.J. Hsü, I.N. McCave,
J.J. Walsh, H.L. Windom

A Wiley–Interscience Publication

John Wiley & Sons 1991
Chichester · New York · Brisbane · Toronto · Singapore

Copy Editor: J. Lupp

With 5 photographs, 113 figures, and 54 tables

Library of Congress Cataloguing-in-Publication Data

Dahlem Workshop on Ocean Margin Processes in Global Change (1990 : Berlin, Germany)
Ocean margin processes in global change : report of the Dahlem Workshop on Ocean Margin Processes in Global Change, Berlin, 1990, March 18–23 / R.F.C. Mantoura, J.-M. Martin, and R. Wollast, editors.
p. cm. — (Physical, chemical and earth sciences research report ; 9) (Dahlem workshop reports)
"Sponsored by Senat der Stadt Berlin, Marga and Kurt Möllgaard Stiftung"—Prelim. p.
"A Wiley-Interscience publication."
Includes bibliographical references and index.
ISBN 0-471-92673 6
1. Chemical oceanography—Congresses. 2. Biogeochemistry——Congresses. 3. Continental margins—Congresses. I. Mantoura, R. F.C. II. Martin, Jean-Marie. III. Wollast, R. IV. Berlin (Germany : West). Senat. V. Marga and Kurt Möllgaard Stiftung. VI. Series. VII. Series: Physical, chemical, and earth sciences research reports ; 9.
GC110.D34 1991
551.46—dc20 90–22159
CIP

British Library Cataloguing-in-Publication Data

Dahlem Workshop on Ocean Margin Processes in Global Change (1990: Berlin Germany)
Ocean margin processes in global change.
1. Oceans. Bed. Geological features
I. Title II. Mantoura R.F.C. III. Martin, J.-M.
IV. Wollast, R. V. Series
551.46088

ISBN 0-471-92673-6

Typeset by Photo·graphics, Honiton, Devon
Printed and bound in Great Britain by Biddles Ltd, Guildford, Surrey

Table of Contents

The Dahlem Konferenzen

Founders

Recognizing the need for more effective communication between scientists, the Stifterverband für die Deutsche Wissenschaft[1], in cooperation with the Deutsche Forschungsgemeinschaft (German Science Foundation), founded Dahlem Konferenzen in 1974. In January 1990 Dahlem Konferenzen became a part of the Freie Universität Berlin (Free University of Berlin). It is financed by the Berlin Senate, the Deutsche Forschungsgemeinschaft, and various private foundations.

Name

Dahlem Konferenzen was named after the district of Berlin called *Dahlem*, which has a long-standing tradition and reputation in the sciences and arts.

Aim

The task of Dahlem Konferenzen is to promote international, interdisciplinary exchange of scientific information and ideas, to stimulate international cooperation in research, and to develop and test new models conducive to more effective communication between scientists.

The Concept

The increasing orientation towards interdisciplinary approaches in scientific research demands that specialists in one field understand the needs and problems of related fields. Therefore, Dahlem Konferenzen has organized workshops, mainly in the Life Sciences and the fields of Physical, Chemical, and Earth Sciences, of an interdisciplinary nature.

The Dahlem Workshops provide a unique opportunity for posing the right questions to colleagues from different disciplines who are encouraged to state what they do not know rather than what they do know. The aim is not to solve problems or to reach a consensus of opinion, the aim is to define and discuss priorities and to indicate directions for further research.

---

[1] The Donors Association for the Promotion of Sciences and Humanities, a foundation created in 1921 in Berlin and supported by German trade and industry to fund basic research in the sciences.

THE DAHLEM WORKSHOP MODEL

MONDAY	TUESDAY	WEDNESDAY	THURSDAY	FRIDAY
A. Opening (P) B. Introduction (P) C. Selection of Problems for the Group Agendas (S) 1 2 3 4	1 2 3 4	1 3	F. Report Session 1 2 3 4	G. Distribution of the Reports H. Reading Time I. Discussion of the Group Reports (P)
D. Presentation of Group Agendas (P) E. Group Discussions (S) 1 2		2 4		J. Revision of the Reports (S) 1 2 3 4

Key: (P) = Plenary Session;
(S) = Simultaneous Sessions;
○ = one discussion group

Explanation of the Dahlem Workshop Model

A. Opening
Background information is given about Dahlem Konferenzen and the Dahlem Workshop Model.

B. Introduction
The goal and the scientific aspects of the workshop are explained.

C. Selection of Problems for the Group Agenda
Each participant is requested to define priority problems of his choice to be discussed within the framework of the workshop goal and his discussion group topic. Each group discusses these suggestions and compiles an agenda of these problems for their discussions.

D. Presentation of the Group Agenda
The agenda for each group is presented by the moderator. A plenary discussion follows to finalize these agendas.

E. Group Discussions
Two groups start their discussions simultaneously. Participants not assigned to either of these two groups attend discussions on topics of their choice.
The groups then change roles as indicated on the chart.

F. Report Session
The rapporteurs discuss the contents of their reports with their group members and write their reports, which are then typed and duplicated.

G. Distribution of Group Reports
The four group reports are distributed to all participants.

H. Reading Time
Participants read these group reports and formulate written questions/comments.

I. Discussion of Group Reports
Each rapporteur summarizes the highlights, controversies, and open problems of his group. A plenary discussion follows.

J. Groups Meet to Revise their Reports
The groups meet to decide which of the comments and issues raised during the plenary discussion should be included in the final report.

Topics

The topics are of contemporary international interest, timely, interdisciplinary in nature, and problem oriented. Dahlem Konferenzen approaches internationally recognized scientists to suggest topics fulfilling these criteria. Once a year, the topic suggestions are submitted to a scientific board for approval.

Program Advisory Committee

A special Program Advisory Committee is formed for each workshop. It is composed of 6–7 scientists representing the various scientific disciplines involved. They meet approximately one year before the workshop to decide on the scientific program and define the workshop goal, select topics for the discussion groups, formulate titles for background papers, select participants, and assign them their specific tasks. Participants are invited according to international scientific reputation alone. Exception is made for younger German scientists. Invitations are not transferable.

Dahlem Workshop Model

Since no type of scientific meeting proved effective enough, Dahlem Konferenzen had to create its own concept. This concept has been tested and varied over the years. It is internationally recognized as the *Dahlem Workshop Model*. Four workshops per year are organized according to this model. It provides the framework for the utmost possible interdisciplinary communication and cooperation between scientists in a period of $4\frac{1}{2}$ days.

At Dahlem Workshops 48 participants work in four interdisciplinary discussion groups. Lectures are not given. Instead, selected participants write background papers providing a review of the field rather than a report on individual work. These papers, reviewed by selected participants, serve as the basis for discussion and are circulated to all participants before the meeting with the request to formulate written questions and comments to them. During the workshop, each of the four groups prepares reports reflecting their insights gained through the discussion. They also provide suggestions for future research needs.

Publication

The group reports written during the workshop together with the revised background papers are published in book form as the Dahlem Workshop Reports. They are edited by the editor(s) and the Dahlem Konferenzen staff. The reports are multidisciplinary surveys by the most internationally distinguished scientists and are based on discussions of advanced new concepts, techniques, and models. Each report also reviews areas of priority interest and indicates directions for future research on a given topic.

The Dahlem Workshop Reports are published in two series:

1. Life Sciences Research Reports (LS), and
2. Physical, Chemical, and Earth Sciences Research Reports (PC).

Director
Jennifer Altman, M.A., Ph.D.

Address

Dahlem Konferenzen
Tiergartenstr. 24–27
D-1000 Berlin (West) 30

E.T. Degens (1928–1989)

Dedication

This book is dedicated to our friend Egon Degens, member of the EROS Programme Advisory Committee, who passed away at home on February 18, 1989. With his death the environmental science community mourns the loss of one of its most able and highly respected pioneers.

E.T. Degens was born on April 16, 1928 and was educated in Bonn and Wurzburg. After a stint at the Pennsylvania State University, he returned to Wurzburg to help set up one of the first organic geochemistry laboratories in the world. This laboratory was the breeding ground for some of the eminent organic geochemists at work today. Subsequently, he joined the California Institute of Technology and started his work on stable carbon isotopes, and later on biogeochemical compounds in natural waters. The book "Geochemistry of Sediments," published by Prentice Hall (1965), stems from this period and was the first geochemistry book ever to include chapters on naturally occcurring organic matter. From California he moved across the continent to the East coast, leading to yet another productive phase at the Woods Hole Oceanographic Institution. He was actively involved in the pioneering work ccarried out by the Woods Hole scientists in the Black Sea, which is the largest anoxic basin in the world, and in the Red Sea, where the hydrothermal ore deposits at the seafloor were first discovered. Further expeditions took him to the East African Rift Lakes, where his team discovered hydrothermal activity in Lake Kivu. During this period, his major research interest remained the nature of organic matter and its cycling in the ocean. Of special note is his work on stable carbon isotopes in marine plankton, its fractionation and biomineralization, and the significance of the associated organic matter in deciphering evolutionary patterns observed in certain marine organisms.

He returned to Germany to head the Department of Geology at the University of Hamburg in 1973, where within a short time he succeeded in establishing a research group in biogeochemistry dealing with the cycling of elements and its perturbation by humans. Over the years he has initiated major research efforts within the framework of SCOPE (Scientific Committee on Problems of the Environment) and UNEP (United Nations Environmental Programme) projects on biogeochemical cycling of elements in rivers, lakes, estuaries, and oceans. Throughout he has encouraged scientists from all continents to sit together and discuss environmental problems which are of concern to mankind.

He remained professionally active until the last few days of his life. Among others he edited his textbook "Perspectives in Biogeochemistry" (Springer), where he presented his comprehensive views on the evolution of Earth and Life.

Being among the best in his field, he was never elitist. He always kept in mind the words of Sir Alan Parks: "Science can be serious without being sacrosanct." He was always keen to share his charismatic view and was never condescending towards even the most modest of research groups from developing countries who turned to him for the benefit of his worldwide experience. He was a lively and generous person, both in the national and international arena, who was always genuinely concerned with the issue at hand.

He was one of the most brilliant and tireless advocates of the idea of his gathering of European scientists, which has led to the implementation of the EROS programme. As stated by Jim Lovelock, he thought that "we may find that the vital organs in the Body of Gaia are not on land surfaces but in estuaries, wetlands, and muds on the continental shelves. There, the rate of carbon adjusts automatically to regulate the concentration of oxygen and essential elements are returned to the atmosphere."

Egon was regarded as an inspiring teacher by many of us and by still more of us as a generous friend. We shall never forget our last meeting in September, 1988, on Lake Baikal, which for him represented the realization of a childhood dream.

Words cannot express our grief at his sudden death. It seems unthinkable that we shall never again hear his always constructive remarks. Although nothing can compensate for his loss, he left behind a great legacy of learning and inspirations and an active group of young geochemists at the University of Hamburg, who will undoubtedly build on his inspiration and germinate his many ideas into new discoveries on marine geochemistry.

R.F.C. Mantoura
J.-M. Martin
R. Wollast

Introduction

R.F.C. Mantoura[1], J.-M. Martin[2], and R. Wollast[3]

[1]*Plymouth Marine Laboratory*
Prospect Place, The Hoe
Plymouth PL1 3DH, U.K.

[2]*Institut de Biogéochemie Marine, Ecole Normale Supérieure*
1, rue Maurice Arnoux
92120 Montrouge, France

[3]*Laboratoire d'Océanographie Chimique*
Université Libre de Bruxelles
Campus de la Plaine CP 208, Boulevard du Triomphe
2050 Bruxelles, Belgium

Ocean margins (OMs), which comprise estuaries, coastal, shelf and shelf-edge components, are a globally critical land–ocean interface controlling the anthropogenic and terrestrial fluxes and fates of chemicals and biological production to and from the open ocean. Thus, in 1985 inspired by the late Egon Degens from the University of Hamburg, several marine scientists conceived the European River Ocean System (EROS-2000) program, funded in 1988 by the Commission of European Communities to investigate the coastal biogeochemical fluxes of natural and human-made elements in contrasting European seas. After two very successful years focusing on the influence of the Rhone discharge and delta on N.W. Mediterranean Sea (1989 Paris EROS-2000 Workshop), we felt there was a need for an international forum to evaluate not only the data base on the role of OMs in global change, but also to identify critical areas of ignorance requiring urgent research. This "brainstorming" activity is the hallmark of the Dahlem Workshop Model and its chairperson, S. Bernhard, who in consultation with its new sponsor, the Free University of Berlin, agreed last year to organize and support this meeting. We are most grateful to our colleagues in Berlin

Ocean Margin Processes in Global Change
Edited by R.F.C. Mantoura, J.-M. Martin and R. Wollast

for this support and were privileged to be in Berlin at such a momentous time of political change and reconciliation. Judging by the high quality of our background review papers and by the importance placed on this topic by several international working parties, including the Joint Global Ocean Flux Studies (JGOFS) and the International Geosphere Biosphere Programme (IGBP) currently in session, we are sure this Workshop will be critical in focusing future marine research into the key areas of OMs.

But why are OMs of global significance?

Although OMs, including estuarine, coastal, and shelf systems, comprise only ca. 10% and 0.5% of the surface area and volume of the ocean respectively, they account for up to 30% of ocean production because of the fertilizing influence of nutrient inputs from rivers and upwelling and buoyancy exchanges at the shelf edge. Over 90% of the riverine particulates and associated carbon, trace metals and pollutants are trapped in deltaic and shelf regions of OMs. The extent to which OMs can trap or export terrestrial and autochthonous phytoplankton carbon profoundly affects global carbon mass balances and their climatic consequences. Climatic feedback via radiatively active gases (CH_4, N_2O, dimethyl sulfide [DMS]) to the atmosphere are also accentuated in coastal zones. It is now known that coastal fluxes of N_2O and DMS can make significant contributions to acid rain, even in industrialized regions. Another important consideration is the impact of sea level changes associated with alternating submergence and emergence of the continental shelves on the biogeochemical mobility of trapped elements within the OMs.

Estuarine and nearshore parts of OMs are invariably the sites of major discharge of urban and industrial pollutants, many of which have very well-established effects (e.g., eutrophication, anoxia) while others (e.g., organic micropollutants) have still poorly understood impact. Human impact on these coastal ecosystem systems depends on the physical, chemical, and biological characteristics of the OMs; however, areas like the Adriatic, Southern Bight of the North Sea, New York Bight, etc. appear to have already been over exploited. There are currently major changes occurring in river discharge through damming and water diversion for irrigation, and these are having a dramatic effect (e.g., Nile) on coastal erosion, fertility, and the export of terrestrial matter to the ocean.

Coastal seas are characterized by tight biogeochemical coupling of dissolved and particulate materials within an offshore sequence of frontal systems (e.g., haloclines, riverine plumes, tidal fronts, shelf-edge upwelling) as well as vertical exchange with the shallow sediments. Since the physiography and hydrodynamics of OMs are geographically variable and since these ultimately control stratification, flushing and exchange time scales of chemicals in OMs, it was important for us at this conference to consider typological yardsticks for classifying OMs. This will allow us to test the global validity of our site-specific knowledge and to guide us in selecting globally representative regions for future studies. Already such generalized models exist for lakes (e.g., Vollenweider), rivers (deBecker), estuaries (e.g., Ketchum, Rattray) and tidal seas (Simpson-Hunter) and have proved their unifying strength in reconciling and rationalizing contradictory site-specific results.

Our workshop goal was to "evaluate our understanding of the fluxes and transformation of natural and anthropogenic materials in estuarine and coastal oceans and to examine their global consequences." The future priority research areas for OMs are clear:

1. Globally representative and socially relevant ocean margin systems now need to be typologically identified and investigated by coordinated multidisciplinary teams focusing on the transport modes and biogeochemical fluxes of carbon, nitrogen, and sulfur and their natural and human-made compounds, both within and at the landward and shelf-edge boundaries.
2. Even with the limited information available and brought together in this workshop, it is clear that OMs can no longer be ignored in global models of the elements.
3. The functional characteristics of OMs in relation to human engineering and hydrological and environmental impact must now be quantitatively evaluated if we are to predict the role and response of OMs to global change.

We hope you find that the background papers and group reports arising from the workshop discussions have indeed addressed this objective in an exciting and challenging way. We are most grateful to the authors and group rapporteurs for their incisive, exciting and timely contributions to this globally important subject.

Dissolved and Particulate Organic Matter in Rivers

A. Spitzy[1] and V. Ittekkot[2]

[1]*Max-Planck-Institut für Meterologie*
Bundesstr. 55
2000 Hamburg 13, F.R. Germany

[2]*Institut Für Biogeochemie und Meereschemie, Universität Hamburg*
Bundesstr. 55
2000 Hamburg, F.R. Germany

Abstract. Recent estimates of the transport rates of riverine dissolved and particulate organic matter are presented. Globally, 0.5 gigatons of organic carbon are transported by rivers, half of which is in the particulate form. The degradable component as estimated from detailed chemical characterization is about 30% for dissolved organic carbon and about 35% for particulate organic carbon. Refinements in these global estimates appear to be unwarranted since, for example, the river transport of organic carbon is no more than 10% of the CO_2–C released into the atmosphere from human activities. At the regional level, however, improved knowledge of the riverine organic matter's composition and fluxes, their seasonal variations, and their response to short-term hydrologic extreme events as well as to long-term anthropogenic perturbations in the catchment areas are essential, because of the controlling role that the amounts as well as the chemical nature of organic matter has on the distribution of nutrients and pollutants, and on the chemical forms in which they are introduced into the estuaries and coastal seas. Variations in quality and quantity of riverine organic matter fluxes may thus induce corresponding fluctuations in regionally specific ocean margin biogeochemical processes.

INTRODUCTION

The global cycle of water is an important carrier of carbon in its reduced state (organic carbon). In this chapter we are concerned with a major branch in this cycle, namely the transport of organic carbon by rivers from land to the sea.

Ocean Margin Processes in Global Change
Edited by R.F.C. Mantoura, J.-M. Martin and R. Wollast

The organic matter (OM) in which that carbon is contained appears in all sizes, ranging from free monomers to large particles, and is a heterogeneous mix of biochemicals, whose oxidation rates range from hours to thousands of years. It is conventionally subdivided into a dissolved phase (passing through a 0.45 micron filter) and a particulate phase (retained on a 0.45 micron filter). The terms dissolved organic carbon (DOC) and particulate organic carbon (POC) are used correspondingly. The arbitrary nature of this operational definition is evident from the fact that distinct particles have been observed down to 0.02 microns (Eisma 1988). Below, there are colloids and large organic molecules with a molecular weight greater than 1000. The fine particulate and colloidal material can be up to 15% of the DOM (Wollast and Billen 1981). The quantity of OM in rivers is well documented (Degens et al. 1990). However, so far only about 25% have been chemically characterized as carbon in carbohydrates, amino acids, fatty acids, hydrocarbons, and phenolic compounds. The former three compound classes can be viewed as largely biodegradable. The uncharacterized fraction includes a hydrophilic and a hydrophobic component, the latter has been found to be relatively stable and is termed "Gelbstoff" (Kalle 1966) or aquatic humus (Gjessing 1976). The hydrophilic fraction overlaps with part of the characterized biodegradable fraction. The hydrophobic fraction can be subdivided into fulvic acids and humic acids (precipitate at pH of 2). Much of the uncharacterized fraction has a plant and soil course. The extent to which it is refractory determines its time scale of oxidation as it moves from rivers through estuaries and coastal waters to the open ocean. A fraction of POC is oxidized, the rest is buried in estuarine and coastal sediments, though a part may bypass these areas and be directly discharged to the deep sea, such as in the Bay of Bengal (Ittekkot et al. 1986). The detailed modes and rates of oxidation of the various forms of OM are not yet sufficiently understood for assessing the role of riverine OM within the perturbed carbon cycle.

SOURCES

POC and DOC have allochthonous (external) and autochthonous (internal) sources in rivers. The main autochthonous sources for both POC and DOC are photosynthesis and degradation in the water column of rivers, lakes, and reservoirs. Dissolution of POC may contribute to DOC, while adsorption of DOC onto particles may contribute to POC. Allochthonous sources of POC are soil erosion and input of plant debris. Additional inputs of POC from floodplains during periods of high water flow occur in rivers with high sediment supply, such as the Indus, Ganges, and Brahmaputra (Ittekkot et al. 1985). Allochthonous DOC stems from plant litter and soil leachates reaching the river via overland flow, interflow, or groundwater input, predominantly during the dry season or when soils are snowcovered or frozen in winter. In addition, domestic and industrial effluents contribute both POC and DOC. Domestic sewage is largely biodegradable and respires within a short distance from the source. The biodegradability of industrial effluents depends on the type of organics and the presence of inhibiting and toxic chemicals in the effluent (Meybeck et al. 1989). An additional source of POC and DOC is resuspended riverine bottom sediments and their pore waters. Finally, OM from the sea surface microlayer may be transported landward via the atmosphere and may also contribute, though probably to a small extent, to riverine OM. The relative proportions of allochthonous and autochthonous OM is affected by a river's hydrology, such as, for example, the presence of reservoirs and lakes, which provide a strong source of phytoplankton inputs. Methods to determine sources of OM include analyses of stable carbon isotope ratios, specific bioindicators (carbohydrates, amino acids, lipids, tannin, and lignin residues) and IR- and NMR-spectroscopy on organic matter isolates (Spitzy and Leenheer 1990).

FLUXES

Past estimates of total organic carbon (TOC) flux in world rivers vary by nearly two orders of magnitude, ranging from 0.01 (Williams 1971) to 1.0 (Richey et al. 1980) gigatons of carbon (GtC) per year. Intermediate values were proposed, for example, by Garrels and MacKenzie (1971 [0.32 GtC]), Duce and Duursma (1977 [0.1–0.15 GtC]), and Schlesinger and Melack (1981 [0.4 GtC]). The first systematic attempt to measure riverine TOC fluxes on a global scale was initiated in 1982 by the SCOPE/UNEP project "Transport of carbon and minerals in major world rivers" (Degens et al. 1990), the results of which form the basis of our further discussion. POC contributes 0.5–12% to the mass of TSS (total suspended solids) (Ittekkot 1988), with concentrations decreasing with logarithmically increasing TSS concentrations. This trend is similar to that reported previously by Meybeck (1982), its reasons being (a) dilution of autochthonous organic matter by mineral matter (increased erosion [Meybeck 1982]), (b) reduced autochthonous input due to lack of light penetration at high sediment concentrations (turbidity [Thurman 1985]), and (c) differences in the sources and biogeochemical processes affecting the nature of organic matter at various stages of the hydrographic regime.

DOC concentrations in rivers range from less than 1 to more than 20 mg/l (e.g., Malcolm and Durum 1976; Naiman and Sibert 1978). DOC:POC ratios range from less than 0.5 in tropical rivers to up to 20 in temperate rivers (Meybeck 1982). Within a seasonal cycle, DOC concentrations can vary by an order of magnitude, correlating positively with runoff ("flushing effect"). The export of DOC from the major morphoclimatic zones (see

Table 1) is such that the wet tropics account for up to 65% of the global DOC transport, followed by the temperate and taiga zones. Means, ranges, and fluxes of DOC, water runoff, and area-specific DOC export are given in Table 2 for individual big rivers. The rivers listed account for 37.7% of global water discharge of 37,400 $km^3\ a^{-1}$. Extrapolation to 100% yields a global riverine DOC-flux of 218×10^{12} gCa^{-1}. The annual DOC loss per unit area of catchment correlates significantly with runoff, as shown in Fig. 1; this also holds for POC.

Annual DOC and POC export from different continental regions is shown in Fig. 2. Organic carbon fluxes calculated for selected big rivers of a region were extrapolated by relating to the region's total water and sediment discharge. The largest amounts of POC are being transported by the rivers of Asia, followed by those draining North America, Oceania, Africa, and Europe. As for DOC, South American rivers dominate the global export, followed by rivers of Asia, North America, the Arctic, Africa, Europe, and Oceania. The POC and DOC export per unit area ("yield") is shown in Fig. 3. For POC it is highest in Asia and Oceania due to the high sediment load, although the rivers draining these two regions carry POC in concentrations similar to, or less than, those in the rivers of the Arctic and Europe. South America and the Arctic have the highest DOC yield, the former due to the enormous water flux from the region, the latter due to high concentrations of DOC (around 10 mg/l) in the rivers draining that region. They are followed by Asia, North America, Africa, Europe, and Oceania. The flux data presented are minimum values because they are mainly based on samples from the surface of rivers (logistical problems in

TABLE 1. DOC and water export from major morphoclimatic zones based on discharge and "typical" DOC concentration data given by Meybeck (1988).

Morphoclimatic zone	typical DOC conc. (mg/l)	water discharge ($km^3\ a^{-1}$)	(% of total)	DOC export ($10^6 t\ a^{-1}$)	(% of total)
tundra	2	1,122	3	2.2	1
taiga	7	4,376	11.7	30.6	13
temperate	4	10,285	27.5	41.1	17.6
wet tropic	8	19,186	51.3	153.5	65.5
dry tropic	3	2,169	5.8	6.5	2.8
semi arid	1	262	0.7	0.3	0.1
total		37,400	100	234.2	100

*Mean of typical DOC—concentrations: 4.2 (2/3), discharge weighted mean global DOC concentration: 6.3 (3/3) (*would underestimate by one third).

TABLE 2. Water and DOC transport data for big rivers (the first column corresponds to the numbers plotted in Fig. 1); d.w. mean = discharge weighted mean. From Spitzy and Leenheer (1990).

	River	Discharge ($km^3 a^{-1}$)	Area ($103 km^2$)	Runoff ($mm a^{-1}$)	DOC concentration (mg/l) range	DOC concentration (mg/l) d.w. mean	DOC flux ($106t a^{-1}$)	DOC flux ($t km^{-2} a^{-1}$)	Reference
1	Amazon	5,520	6,300	876	3–5	3.6	20	3.2	Richey et al. (1985)
2	Orinoco	1,135	950	996	2–5	2.87	3.22	6.1	Paolini et al. (1987)
3	Parana	480	2,800	171	–20	6.1 (Enso:10.2)	2.8 (Enso:7.5)	1	Depetris and Paolini (1990)
4	Uruguay	158	350	451	2–8	3.2	0.5	1.4	Manosa, unpublished
5	Mississippi	439	3,267	154		8	3.5	1.1	Leenheer (1982)
6	St. Lawrence	413	1,150	359	3–5	3.7	1.6	1.4	Pocklington and Tan (1983)
7	Yukon	210	840	250		8.8	1.9	3.0	Leenheer (1982)
8	Columbia	135	670	202		2.7	0.4	0.7	Leenheer (1982)
9	Yangtsekiang	883	1,950	453	5–23	13.4	11.8	6.1	Gan Wei-Bin et al. (1983)
10	Brahmaputra	609	580	1,050	1–6	3.2	1.9	3.3	Safiullah et al. (1987)
11	Ganges	366	975	375	1–9	4.6	1.7	1.7	Safiullah et al. (1987)
12	Indus	211	950	222	2–22	16.1	0.75	0.8	Arain (1987)
13	Huanghe	44	745	59	5–25	12.3	0.54	0.72	Gan Wei-Bin et al. (1983)
14	Congo	1,237	4,000	309	3–10	8.5	10.5	2.6	Martins and Probst (1990)
15	Niger	192	1,125	171	2–6	2.9	0.55	0.5	Martins (1982)
16	Gambia	4.6	42	110	1–4	2.4	0.011	0.27	Lesack et al. (1985)
17	Lena	533	2,430	219		9.5(TOC)	5.1	2.1	Telang et al. (1990)
18	Ob	419	2,550	164		8.8(TOC)	3.7	1.5	Telang et al. (1990)
19	Yenisei	562	2,580	218		7.4(TOC)	4.1	1.6	Telang et al. (1990)
20	MacKenzie	249	1,810	138	3–6	5.2	1.3	0.7	Telang et al. (1990)
	Sum	13,635.6							

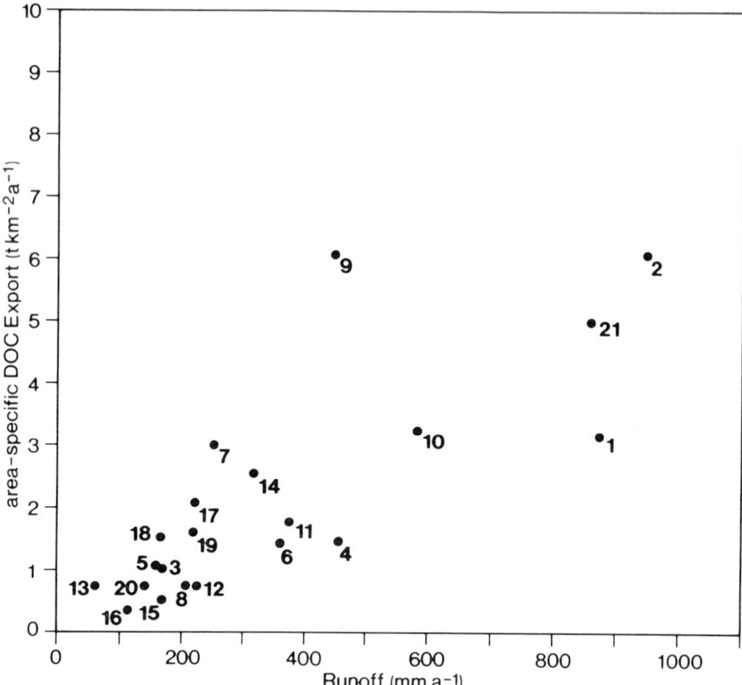

Fig. 1—Area-specific annual DOC export by rivers given in Table 2 versus their respective runoff (Spitzy and Leenheer 1990).

obtaining depth-integrating samples from most of the major rivers). In addition, the contribution of bedload is difficult to estimate. There are wide variations in the supply of sediments within this fraction. In certain rivers, for example the Zaire, the bedload equals the amount of sediment transport in the suspended fraction, whereas in others it is insignificant (Eisma 1987). Milliman and Meade (1983) contend that about 10% of the total sediment transport is in the form of bedload. However, the contribution of bedload to the organic carbon budget is probably much less than 10%, because of the relatively low POC content in silt and sand.

THE DEGRADABLE COMPONENT

For carbon cycle research it is important to know how much of the riverine OM flux to the ocean is rapidly oxidized back to the CO_2-pool of the surface ocean, and how much is buried in sediments. One approach is to compare organic carbon distribution in sediments from the deltas of major world rivers with published data on riverine transport of organic carbon. Doing

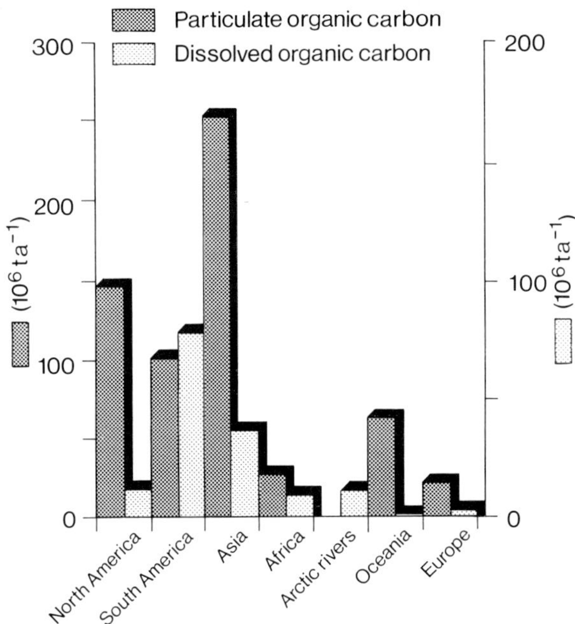

Fig. 2—Annual riverine inputs of dissolved and particulate carbon for continental regions (Degens and Ittekkot 1985).

so led Berner (1982) to suggest that a significant part of riverine carbon is decomposed in estuarine and coastal waters. Another approach is to study decomposition rates of POC from river samples. This has been done, e.g., by Eisma et al. (1985) for the Gironde river, who conclude that more than two-thirds of POC may be decomposed at the river mouth—similar to the loss of riverine OM observed in the Amazon plume by Edmond et al. (1981). Finally, one can estimate the relative proportions of labile versus stable OM components by analyzing OM for chemical compounds that are comparatively labile and will easily be degraded in the riverine or estuarine environment, such as carbohydrates and amino acids (Degens and Ittekkot 1985). Labile carbohydrates, for example, are released during disaggregation of macroaggregates at the river–sea interface, thus reducing the POC transfer from rivers to the sea (Eisma et al. 1983). During the above-mentioned SCOPE study, the carbohydrate and protein fractions of particulate and dissolved OM from several major world rivers were analyzed. These data provide lower limits of the degradable fraction since other organic compounds, e.g., fatty acids, also contribute to it.

For details of the nature of OM in riverine particulate matter we refer to Degens and Ittekkot (1985), Ittekkot and Arain (1986), and Ittekkot et

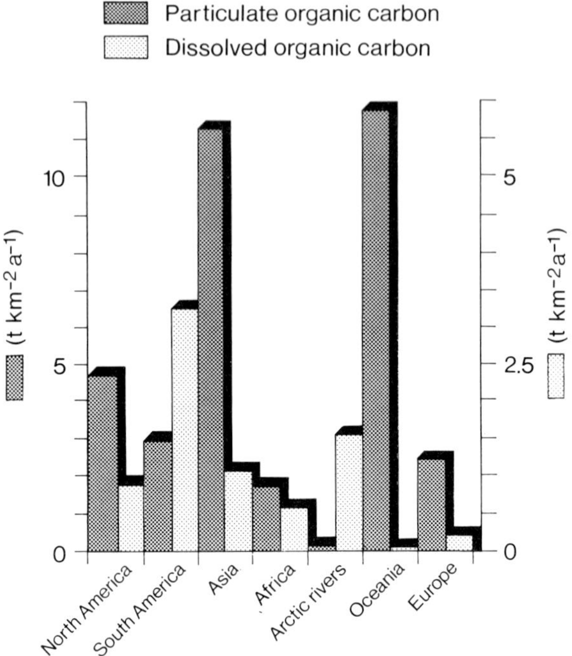

Fig. 3—Annual dissolved and particulate carbon yield for continental regions (Degens and Ittekkot 1985).

al. (1985, 1986). The percentage of total carbohydrates and amino acids in POC for individual rivers is depicted in Fig. 4. They contribute 2–11% and 5–22%, respectively, to the POC measured. The average values are 4% for carbohydrates and 11% for amino acids, which suggests that more than 15% of the POC is degradable in the surrounding environment. For individual rivers the degradable fraction of POC varies between 6.5 and 31%.

As can be seen from Figs. 4 and 5, the rivers of Southeast Asia and Oceania are relatively low in degradable POC. The reason is that most of the sediment discharge takes place within the short time interval of the high water stage, when material is flushed from the vast floodplains, where OM associated with these sediments has already been largely biodegraded. These rivers contribute most of the global sediment discharge from land to sea. Assuming that sugars and amino acids contribute only 50% of the degradable fraction, it is concluded that globally ca. 35% of POC has a potential to become decomposed (Ittekkot 1988).

For the dissolved fraction (DOC), fewer samples have been available for sugar and amino acid analysis. Amino acids comprise 3–5% of the DOC while the carbohydrates comprise 5–10% of the DOC in all analyzed rivers.

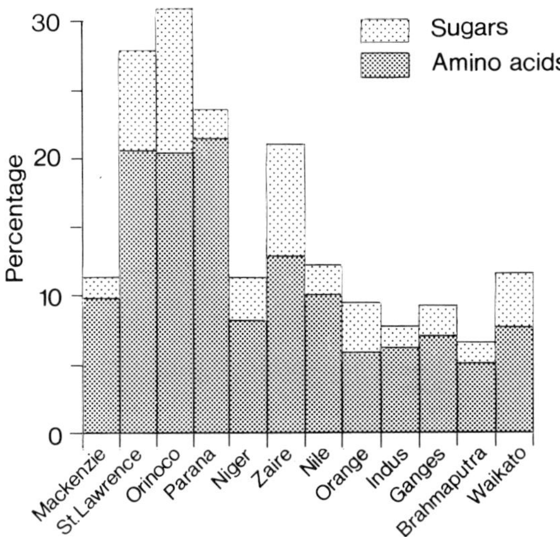

Fig. 4—The percentage of sugars and amino acid carbon associated with particulate organic carbon in major world rivers (Degens and Ittekkot 1985).

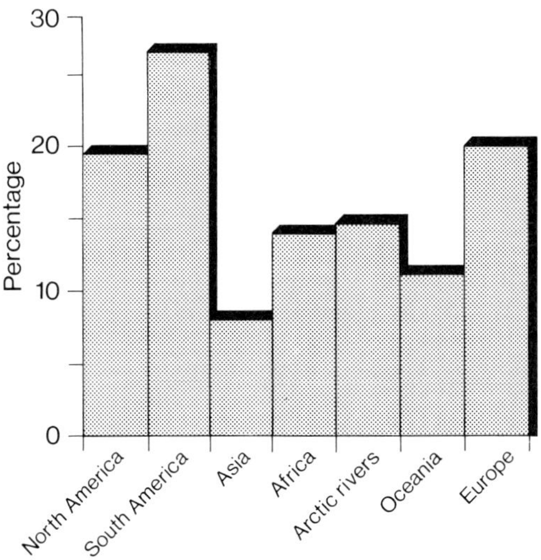

Fig. 5—The percentage of metabolizable carbon in particulate organic carbon for continental regions (Degens and Ittekkot 1985).

Using the global average, therefore, it can be estimated that about 15% of the DOC is degradable, based on sugars and amino acids only. The error introduced by this limitation is unlikely to exceed a factor of two at the most. Therefore, 30% could be considered as an upper limit for the degradable fraction of DOC.

CONCLUSIONS

Given the uncertainties in the data base (incomplete coverage of rivers, representativity of time, and space resolution of sampling) and in the water flux estimate used for extrapolation, we conclude that annual global organic carbon flux in rivers is 0.5 gigatons, of which half is DOC and half POC. This is about 10% of the annual anthropogenic CO_2–C release by burning of fossil fuels and, from the viewpoint of global carbon cycle modeling, we see no need to make further refinements in this number. Application of the DOC analysis technique, proposed recently by Sugimura and Suzuki (1988), to seawater led to a significant upward revision of estimates of the marine DOC pool. In the future it will be necessary to check the DOC data gathered by application of this technique to riverine samples. On a regional scale, OM fluxes need to be better established in order to assess their role in the biogeochemical cycles of different types of coastal environments. In the regional framework it is important to realize the potential effects of climatic/hydrologic extreme events, such as the 1983/84 El Niño/Southern Oscillation that produced a doubling of TOC fluxes in the Parana (Depetris and Paolini 1990). The close relation between runoff and TOC export should be noted in view of the enhanced hydrocycle inherent in future greenhouse warming scenarios, in particular with regard to enhanced fluxes in the tropics and subtropics (areas where dominant riverine TOC fluxes occur) and with increased release of TOC from warming arctic regions (e.g., Siberian rivers).

REFERENCES

Arain, R. 1987. Persisting trends in carbon and mineral transport monitoring of the Indus River. In: Transport of Carbon and Minerals in Major World Rivers, ed. E.T. Degens, S. Kempe, and Gan Wei-Bin, pt. 4, pp. 417–421, Sonderband 64. Hamburg: SCOPE/UNEP.

Berner, R. 1982. Burial of organic carbon and pyrite sulphur in the modern ocean and its geochemical and environmental significance. *Am. J. Science* **282**:451–473.

Degens, E.T., and V. Ittekkot. 1985. Particulate organic carbon—an overview. In: Transport of Carbon and Minerals in Major World Rivers, ed. E.T. Degens, S. Kempe and R. Herrera, pt. 3, pp. 7–27, Sonderband 58. Hamburg: SCOPE/UNEP.

Degens, E.T., S. Kempe, and J. Richey, eds. 1990. Biogeochemistry of Major World Rivers. SCOPE 42. Chichester, New York, Toronto: Wiley.

Depetris, P. and J. Paolini. 1990. Biogeochemistry of South American Rivers. In: Biogeochemistry of Major World Rivers, ed. Degens, E.T., S. Kempe, and J. Richey, SCOPE 42. Chichester, New York, Toronto: Wiley.

Duce, R.A., and E.K. Duursma. 1977. Inputs of organic matter to the ocean. *Mar. Chem.* **5**:319–339.

Edmond, J.M., E.A. Boyle, B. Grant, and R.F. Stallard. 1981. The chemical mass balance in the Amazon plume. I: The nutrients. *Deep Sea Res.* **28A**:1339–1374.

Eisma, D. 1987. Processes of nearshore accumulation of suspended material. In: Transport of Carbon and Minerals in Major World Rivers, pt. 4, eds. E.T. Degens, S. Kempe, and Gan Wei-Bin, pp. 57–70, Sonderband 64. Hamburg: SCOPE/UNEP.

Eisma, D. 1988. The terrestrial influence on tropical coastal seas. In: Transport of Carbon and Minerals in Major World Rivers, pt. 5, eds. E.T. Degens, S. Kempe, and A.S. Naidu, pp. 289–317, Sonderband 66. Hamburg: SCOPE/UNEP.

Eisma, D., P. Bernard, J.J. Boon, R. Van Grieken, J. Kalf, and W.C. Mook. 1985. Loss of particulate organic matter in estuaries as exemplified by the Ems and Gironde estuaries. In: Transport of Carbon and Minerals in Major World Rivers, pt 3, eds. E.T. Degens, S. Kempe, and R. Herrera, pp. 397–412, Sonderband 58. Hamburg: SCOPE/UNEP.

Eisma, D., J. Boon, R. Groenewegen, V. Ittekkot, J. Kalf, and W.C. Mook. 1983. Observations of macroaggregates, particle size and organic composition of suspended matter in the Ems Estuary. In: Transport of Carbon and Minerals in Major World Rivers, eds. E.T. Degens, S. Kempe and H. Soliman, pt. 2, pp. 295–314. Sonderband 55. Hamburg: SCOPE/UNEP.

Gan Wei-Bin, Chen Hui-Ming, and Han Yun-Fang. 1983. Carbon transport by the Yangtze (at Nanjing) and Huanghe (at Jinan) Rivers, People's Republic of China. In: Transport of Carbon and Minerals in Major World Rivers, pt. 2, ed. E.T. Degens, S. Kempe,and H. Soliman, pp. 459–470. Sonderband 55. Hamburg: SCOPE/UNEP.

Garrels, R.M., and F.T. MacKenzie. 1971. Evolution of Sedimentary Rocks. New York: W.W. Norton.

Gjessing, E.T. 1976. Physical and Chemical Characteristics of Aquatic Humus. Ann Arbor: Ann Arbor Science.

Ittekkot, V. 1988. Global trends in the nature of organic matter in river suspensions. *Nature* **332**:436–438.

Ittekkot, V., and R. Arain. 1986. Nature of particulate organic matter in the river Indus, Pakistan. *Geochim. Cosmochim. Acta* **50**:1643–1653.

Ittekkot, V., S. Safiullah, and R. Arain. 1986. Nature of organic matter in rivers with deep-sea connections: The Ganges, Brahmaputra and the Indus. *Sci. Tot. Env.* **58**:93–107.

Ittekkot, V., S. Safiullah, B. Mycke, and R. Seifert. 1985. Organic matter in the river Ganges, Bangladesh: seasonal variability and geochemical significance. *Nature* **317**:800–803.

Kalle, K. 1966. The problem of gelbstoff in the sea. In: Oceanographic Marine Biological Annual Review No. 4, ed. Barnes, pp. 91–104. London: Allen and Unwin.

Leenheer, J. 1982. United States geological survey data information service. In: Transport of Carbon and Minerals in Major World Rivers, pt. 1, ed. E.T. Degens, pp. 355–356. Sonderband 52. Hamburg: SCOPE/UNEP.

Lesack, L.W.F., J.M. Melack, and R.E. Hecky. 1985. Transport of organic carbon in the Gambia River, West Africa. In: Transport of Carbon and Minerals in Major World Rivers, pt. 3, ed. E.T. Degens, S. Kempe, and R. Herrera, pp. 165–178. Sonderband 58. Hamburg: SCOPE/UNEP.

Malcolm, R., and W.H. Durum. 1976. Organic carbon and nitrogen concentrations and annual organic load of six selected rivers of the United States. U.S. Geol. Survey Water Supply Paper, 1817-F.

Martins, O. 1982. Geochemistry of the Niger River. In: Transport of Carbon and Minerals in Major World Rivers, pt. 1, ed. E.T. Degens, pp. 397–418. Sonderband 52. Hamburg: SCOPE/UNEP.

Martins, O., and J.L. Probst. 1990. Biogeochemistry of major African rivers. In: Biogeochemistry of Major World Rivers, eds. Degens, E.T., S. Kempe, and J. Richey, SCOPE 42. Chichester: Wiley.

Meybeck, M. 1982. Carbon, nitrogen and phosphorous transport by major world rivers. *Am. J. Sci.* **282**:401–501.

Meybeck, M. 1988. How to establish and use world budgets of riverine materials. In: Physical and Chemical Weathering in Geochemical Cycles, ed. A. Lerman and M. Meybeck, pp. 247–272. Kluwer.

Meybeck, M., D. Chapman, and R. Helmer. 1989. Global Freshwater Quality—A First Assessment. WHO/UNEP—GEMS Report. Oxford: Blackwell.

Milliman, J.D., and R. Meade. 1983. World-wide delivery of river sediment to the oceans. *J. Geology* **91**:1–21.

Naiman, R.J., and J.R. Sibert. 1978. Transport of nutrients and carbon from the Nanaimo River to its estuary. *Limnol. Oceanogr.* **23**:1183–1193.

Paolini, J., R. Hevia, and R. Herrera. 1987. Transport of carbon and minerals in the Orinoco and Caroni Rivers during the years 1983–1984. In: Transport of Carbon and Minerals in Major World Rivers, pt. 4, eds. E.T. Degens, S. Kempe, and Gan Wei-Bin, pp. 325–328. Sonderband 64. Hamburg: SCOPE/UNEP.

Pocklington, R., and F. Tan. 1983. Organic carbon transport in the St. Lawrence River. In: Transport of Carbon and Minerals in Major World Rivers, pt. 2, eds. E.T. Degens, S. Kempe, and H. Soliman, pp. 243–252. Sonderband 55. Hamburg: SCOPE/UNEP.

Richey, J.E., J.T. Brock, R.R. Maiman, R.C. Wissmar, and R.F. Stallard. 1980. Organic carbon: oxidation and transport in Amazon River. *Science* **207**:1348–1351.

Richey, J.E., E. Salati, and U. Dos Santos. 1985. Biochemistry of the Amazon River: an update. In: Transport of Carbon and Minerals in Major World Rivers, pt. 3, eds. E.T. Degens, S. Kempe, and R. Herrera, pp. 245–258. Sonderband 58. Hamburg: SCOPE/UNEP.

Safiullah, S., M. Mofizuddin, S.M. Iqbal Ali, and S. Enamul Kabir. 1987. Biogeochemical cycles of carbon in the rivers of Bangladesh. In: Transport of Carbon and Minerals in Major World Rivers, pt. 4, eds. E.T. Degens, S. Kempe, and Gan Wei-Bin, pp. 435–442. Sonderband 64. Hamburg: SCOPE/UNEP.

Schlesinger, W.H., and J.M. Melack. 1981. Transport of organic carbon in the world's rivers: *Tellus* **33**:172–187.

Spitzy, A., and J. Leenheer. 1990. DOC in Rivers. In: Biogeochemistry of Major World Rivers, ed. E.T. Degens, S. Kempe and J. Richey, SCOPE 42. Chichester, New York, Toronto: Wiley.

Sugimura, Y., and Y. Suzuki. 1988. A high-temperature catalytic oxidation method for the determination of nonvolatile dissolved organic carbon in seawater by direct injection of a liquid sample. *Mar. Chem.* **24**:105–131.

Telang, S.A., R. Pocklington, A.S. Naidu, E.A. Romankevich, I.I. Gitelson, and M.I. Gladishev. 1990. Biogeochemistry of arctic rivers. In: Biogeochemistry of Major World Rivers, eds. Degens, E.T., S. Kempe, and J. Richey, SCOPE 42. Chichester, New York, Toronto: Wiley.

Thurman, E.M. 1985. Organic Geochemistry of Natural Waters. Boston, MA: Martinus Nijhoff/Dr. W. Junk.

Williams, P.M. 1971. The distribution of cycling of organic matter in the ocean. In: Organic Compounds in Aquatic Environments, ed. S.J. Faust and J.V. Hunter, pp. 145–163. Marcel Dekker.

Wollast, R., and G. Billen. 1981. The fate of terrestrial carbon in the coastal sea. In: Flux of Organic Carbon by Rivers to the Ocean, Rep. US-DOE Workshop, No. 016, 75–108.

N, P, and Si Retention along the Aquatic Continuum from Land to Ocean

G. Billen[1], C. Lancelot[1], and M. Meybeck[2]

[1]*Groupe de Microbiologie des Mileux Aquatiques*
Université Libre de Bruxelles,
Campus de la Plaine, CP 221, 1050 Bruxelles, Belgium

[2]*Laboratoire de Géologie Appliquée Université P & M Curie*
4, Place Jussieu, T26 5e ét.
75252 Paris cedex 05, France

Abstract. Before reaching the ocean, nutrients from land sources transit through the continuum formed by rivers, lakes, estuaries, and coastal marine areas. These systems act as successive filters, retaining a significant fraction of the nutrients transported. Retention of the aquatic continuum not only deeply affects the absolute amount of nutrients reaching the ocean, it also modifies the ratio in which N, P, and Si are transferred: P is less efficiently removed than N; Si retention is enhanced by increased N and P inputs. The removal capacity of the aquatic continuum is considerably affected by human use of land and rivers.

INTRODUCTION

Mass balance considerations show that the input of the major biogenic nutrients (N, P, Si) from terrestrial systems to the ocean plays a central role in regulating global oceanic production, hence the variations in atmospheric CO_2 (McElroy 1983). Our purpose in this paper is not to establish one more quantitative mass balance of the corresponding present, past, or future fluxes, but to present a few (sometimes speculative) generalizations based on regional observations, regarding the factors affecting these fluxes. This is of importance both for providing the basis for extrapolation when establishing global budgets of elements and for assessing the trends of variations in response to climate change or human perturbations.

Nutrients of terrestrial origin are transported to the ocean either through

Ocean Margin Processes in Global Change
Edited by R.F.C. Mantoura, J.-M. Martin and R. Wollast

atmospheric or riverine pathways. Atmospheric transport is able to bring nutrients from land over long distances to the middle of the ocean. This mode of transport, however, accounts for only a limited, albeit significant, part of the total input, especially at the scale of regional seas, close to industrialized areas (e.g., Loye Pilot et al. 1990). By far, the major input from land to sea thus occurs through river transport, which is much slower and less direct than atmospheric transport. It involves passage through very active zones where intense biological and physicochemical processes occur, leading to transformation, immobilization, or elimination of nutrients before they reach the ocean. Indeed, the aquatic terrestrial–ocean interface comprises the river system including lakes and reservoirs, estuaries, and the coastal marine areas. These form a continuum of aquatic systems which act as very effective and selective "filters" for nutrients (Fig. 1).

In this chapter we will discuss the processes of N, P, and Si immobilization or elimination affecting the transfer of these elements along the aquatic continuum from land to ocean. Particular emphasis will be devoted to the selectivity of these processes, which can deeply affect the ratio between N, P, and Si inputs to the ocean as this ratio has a strong biological effect on structure and function of the marine ecosystem.

In the first section, we will examine the sources of nutrients to continental surface waters. We will then review the basic processes involved in their transformations. These processes are the same in any aquatic environment (uptake by primary producers, mineralization, sedimentation, reactions with solid phases, etc.); however, their expression and quantitative importance depends strongly on morphological and hydrological constraints, which markedly differ between rivers, lakes, estuaries, and coastal seas. These differences, and the resulting nutrient removal role of the river–coastal water system will be discussed in the third section. Finally, we will examine some possible trends of variation of these processes in response to climate changes and human impact.

SOURCES OF NUTRIENTS TO SURFACE WATERS

Nutrients reach surface water through atmospheric deposition, soil leaching and erosion, and the direct discharge of wastewater from human activities. Table 1 summarizes available data on the magnitude of the associated fluxes. Whenever appropriate, values are expressed in mass per year and per km^2 drainage basin in order to make further comparisons more straightforward.

Atmospheric Deposition

Because of the importance of gaseous forms in the biogeochemical cycle, nitrogen is the most abundant nutrient in atmospheric deposition. Moreover,

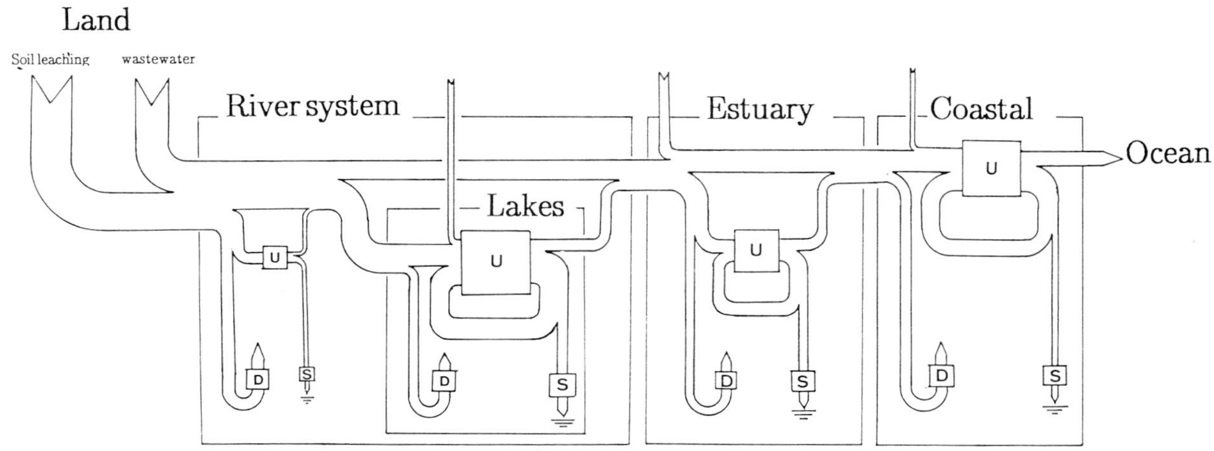

Fig. 1—Conceptual scheme of the processes leading to nutrient retention along the aquatic continuum. Abbreviations: U—uptake by primary producers; S—permanent storage in sediments; D—denitrification.

TABLE 1. Representative values of the inputs from terrestrial sources to continental surface water per unit watershed area (in kg N, P, Si km^2 yr) DON = dissolved organic nitrogen; TDN = total dissolved nitrogen; TDP = total dissolved phosphorus.

	NO_3	NH_4	DON	TDN	PO_4	TDP	Si	N:P:Si atomic ratio
Atmospheric Deposition								
Mean (exoreic) continents[1]	180	225	225	630	5	10	200	143:1:23
Industrialized regions[1,2]	1,250	225	225	1,700	20	60	200	64:1:4
Soil Leaching								
Nonagricultural watersheds								
temperate	75	7	55	135	4.6	12	3,400	25:1:323
tropical: —dry				127		5	2,000	57:1:456
—humid	60	5	168	228	36	65	10,300	8:1:180
Temperature agricultural watersheds[3]								
grassland: —loam/clay				1,000		60	3,400	38:1:65
—sand				3,500		60	3,400	133:1:65
arable land: —loam/clay				2,500		60	3,400	95:1:65
—sand				7,200		60	3,400	273:1:65

[1]Meybeck 1982, calculated with average precipitation of 1 m/yr.
[2]Brimblecombe and Stedman 1982.
[3]Billen 1990.

emission of nitrous oxides by automobile and industrial combustion has resulted in a five- to tenfold increase in the nitrate content of rainwater in industrialized countries (Brimblecombe and Stedman 1982).

Soil Leaching and Erosion

While dissolved N and P concentration in drainage water is mainly regulated by the interactions between water, soil minerals, and biota, Si content results from rock alteration and depends on the nature of the bedrock and the extent of its weathering (Meybeck 1984). Accordingly, Si in surface water is not significantly affected by agriculture, while drainage of N, and to a lesser extent of P, is considerably enhanced by agricultural practices. The occurrence of a period during which soil is devoid of vegetal cover is as important in determining N leaching as the amount of fertilizer applied; thus, at similar fertilizer application rates, grassland is subject to much lower N losses than arable land.

Wastewater Discharge

Per capita domestic loads in industrialized countries with a high living standard are shown in Table 2. These figures represent an actual input to surface water only when a sewer system serves the population. Collection of urban wastewater started in the large cities of Western Europe and the U.S. as early as the mid 19th century; however, even in industrialized countries, they do not currently serve the whole population. Industrial discharges of nutrients are mainly due to the food processing industries and

TABLE 2. Annual load of nutrients from domestic discharges in industrialized countries (yearly per capita rate in kg of N, P, and Si).

	N		P		Si
Excrements	3–4.6		0.4		
Housework wastewater	0.4		0.1		
Washing powders			0.7–1.1		
Total	3.4–4		1.3–1.7		0.5–1
Atomic N:P:Si ratio	5–9	:	1	:	0.9–0.3

The ban of polyphosphates from washing powders in Switzerland from 1986 led to a reduction of domestic load from 1.3 to 0.62 kg P/cap/yr. Sources: Verbanck et al. (1989); Officer and Ryther (1980).

the chemical sector. In industrialized countries, the N and P load discharged by industry is of the same order of magnitude as that resulting from domestic wastewater release.

Overall Inputs from Watersheds

The values present in Tables 1 and 2 along with the data gathered in Fig. 2, on N and P inputs from a number of watersheds of different type in temperate latitudes, reveal the following:

1. Inputs from forested watersheds are considerably lower in P than in both N and Si, with respect to the requirements of phytoplankton (see below). Deforestation leads to considerable enhancement of N—and to a much lower extent P—leaching.
2. Agriculture increases both N and P inputs to surface waters without a

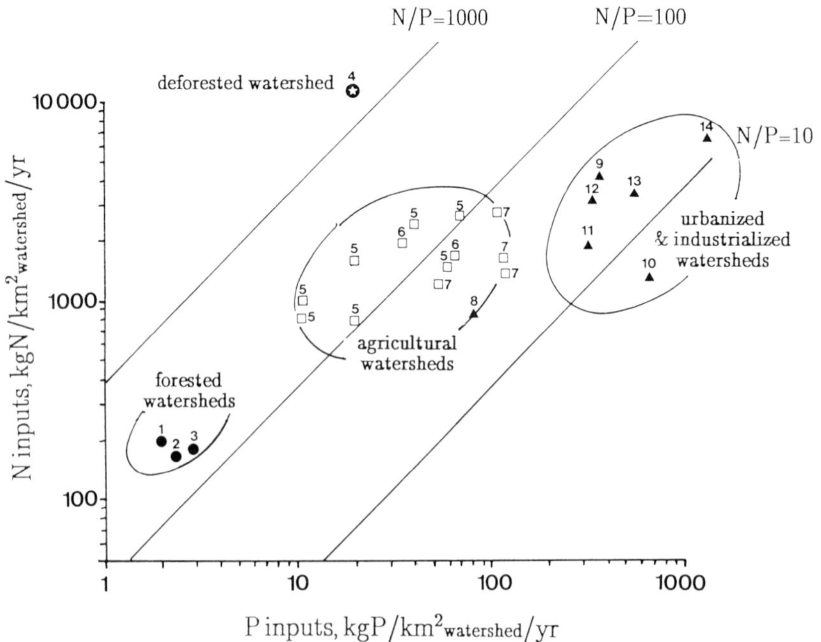

Fig. 2—N and P inputs to surface water from different types of watershed. Forested watersheds: (1) Hubbard Brook, New Hampshire, (2) Sweden, (3) New Zealand, (4) Hubbard Brook deforested. Agricultural watersheds: (5) Belgium, (6) Switzerland, (7) Ireland. Agricultural, urbanized and industrialized watersheds: (8) Chesapeake Bay, U.S.A., (9) Lake Erie, Canada, U.S.A., (10) Lake Washington, U.S.A., (11) Narragansett Bay, U.S.A., (12) Meuse, Belgium, (13) Yser, Belgium, (14) Scheldt, Belgium.

corresponding increase in Si input. N and P remain in approximately the same ratio as in forested systems (25–200 by atoms).

3. In densely populated areas domestic and industrial pollution leads to still higher inputs of P and N, again without greatly affecting Si inputs. P is released in relatively higher amounts so that the N/P atomic ratio drops close to 10.

Biological Fixation of Atmospheric Nitrogen

Microbial fixation of atmospheric N_2 constitutes an additional nitrogen source that must be examined. The current evidence, however, is that this process is much more intense in terrestrial than in aquatic systems. Except in eutrophic lakes, where planktonic N_2-fixing blue-green algae and cyanobacteria can sometimes contribute significantly to total N inputs, benthic N_2-fixation is more important. Representative rates of nitrogen fixation directly determined in a number of aquatic systems are presented in Table 3. When extrapolated to annual budgets, these data show that the local biological N_2 fixation rarely accounts for more than a few percent of N inputs from the watershed.

PROCESSES AFFECTING NUTRIENT TRANSFER

Uptake by Primary Producers

Uptake of inorganic dissolved nutrients is a basic process associated with growth (i.e., net primary production) of phytoplankton or benthic algae and macrophytes. Being essential elements in the composition of basic functional

TABLE 3. Representative values of nitrogen fixation rates in various aquatic systems (synthesis of data cited by Seitzinger 1988; Capone 1988).

	$mgN/m^2/h$
Lakes	
oligo-mesotrophic	0–0.05
eutrophic	0.07–0.3
Estuaries	0.004–0.07
Salt Marshes	0.2–2.4
Coastal Marine Areas	
coral reefs	0–0.18
other	0–0.07

and structural compounds of algae, C, N, and P are generally taken up on the long term in a constant ratio (106:16:1 by atoms, according to Redfield, but ranging between 120:20:1 and 90:5:1), even if temporary storage in the intracellular pools can occasionally occur. Silica on the other hand is only required for the growth of those organisms characterized by the presence of a frustule (diatoms) or by scales (chrosophyceas) made of biogenic opal. The molar C:Si ratio in diatoms varies between 0.1 and 0.8, the lower range corresponding generally to marine species, the higher to freshwater diatoms, which are characterized by more heavily silicified frustures (Conley et al. 1989).

Diatoms are often observed to dominate the phytoplankton community in the early stage of seasonal succession in temperate and polar latitudes, both in rivers and lakes and in marine systems. Flagellates take over later in the succession, when diatoms are reduced in numbers due to either nutrient limitation or zooplankton grazing. Lancelot (in preparation) suggested that this general competitive advantage of diatoms over flagellates, in light and temperature conditions of the early season, lies both in a better photosynthetic capacity and in lower maintenance energy requirements (Fig. 3).

Planktonic Regeneration of Nutrients

Regeneration of dissolved N, P, and Si nutrients from phytoplanktonic biomass or detritus is often viewed as the stoichiometric release of these elements resulting from organic matter degradation, according to the Redfield ratios. In fact, the basic processes leading to N, P, and Si

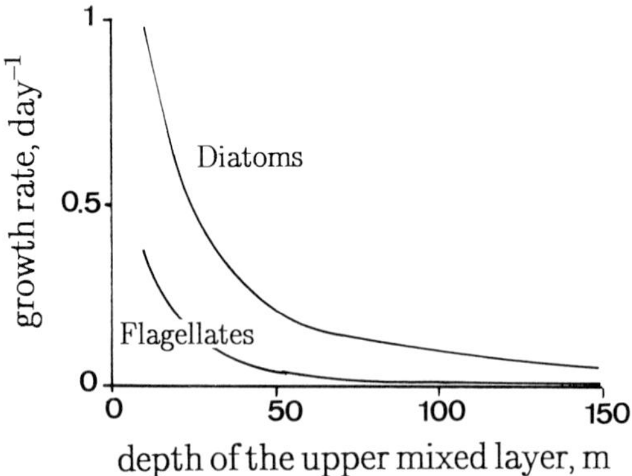

Fig. 3—Depth-integrated specific growth rate of marine diatoms and flagellates in early spring conditions of light and temperature, calculated for different mixing depths according to the physiological model of Lancelot (in prep.).

regeneration are quite different from each other and follow distinct kinetics. Bacterial utilization of organic phosphorus compounds, which cannot be transported as such across cell membranes, involves extracellular hydrolysis, with direct release of the phosphate ion, as a first obligatory step. Organic nitrogen compounds, like amino acids, on the other hand, are directly transported inside the cell where they are either incorporated as such in the biomass or intracellularly deaminated and excreted. Dissolution of biogenic opal is a slow, nonbiological process, that first requires the biological destruction of a protective organic coating after the death of the diatom.

The rate of mineralization of phytoplankton detritus has often been approximated to first-order kinetics. Table 4 presents a summary of the corresponding rate constants, showing the faster mineralization of P with respect to N and especially to Si.

Benthic Recycling of Nutrients

The fraction of organic matter not degraded in the planktonic phase eventually reaches the sediments, where mineralization continues. The presence of a mineral solid phase and the possible reducing conditions that may be established at depth, when benthic heterotrophic activity is high, deeply modifies the resulting process of nutrient regeneration.

Phosphates are adsorbed very effectively on ferric oxyhydroxide phases present in virtually all oxic sediments (dimensionless *in situ* adsorption coefficient (K):25–5000). They are much less adsorbed on reduced iron compounds such as siderite or ferrous sulfides (K:1–5 [Krom and Berner

TABLE 4. Apparent first-order constants for N and P mineralization and Si solubilization during decomposition of various fractions of phytoplankton material. Rate constants in day^{-1}. In parentheses: approximate fraction of total material undergoing decomposition at the corresponding rate, in %.

		Fraction of Phytoplankton Material		
Nutrient	Reference	Autolytic fraction	Labile fraction	Refractory fraction
P	Garber (1984)	~3 (20%)	0.02–0.2 (40%)	0.003–0.02 (40%)
N	Garber (1984)	—	0.02–0.2 (60%)	0.003–0.02 (40%)
Si	Lewin (1961)	—	0.005–0.02 (100%)	—

1980]). This observation led to the oxygen control model of benthic P regeneration, which stated that oxic conditions (or at least the presence of NO_3 [Andersen 1982]) prevent P recycling from sediments, while previously immobilized phosphates are released into solution when reducing conditions are established. Recent data (Caraco et al. 1989) shows, however, that sulfate also interferes with phosphate adsorption. In brackish and saline water with a high sulfate concentration, immobilization of phosphorus is generally not large. In freshwater systems with a very low sulfate concentration (≤ 60 μmol/l), immobilization of P occurs both under oxic and anoxic conditions. Only freshwater systems with intermediate sulfate concentrations (100–300 μmol/l) conform to the oxygen control model of P immobilization and release. Authigenic apatite or fluoroapatite precipitation is another process leading to P burial in sediments.

Burial of organic nitrogen in sediments may be significant, particularly in transient situations (Billen et al. 1989). Adsorption of ammonium on clay minerals is not important (K:0.4–1.9 [Mackin and Aller 1984]). The process of denitrification, however, plays a very significant role in nitrogen cycling of aquatic environments. The factors controlling the magnitude of this process in the benthos are summarized in Fig. 4, showing the results of Billen's (1982) theoretical model of N diagenesis. At low organic input to the benthos, nitrogen release to the water column is proportional to this input and occurs mainly as nitrate, ammonium being quantitatively nitrified. No denitrification occurs as the sediment is entirely oxic. Above a critical rate of organic input to the sediment, an anaerobic layer develops in the sediment, and the contribution of nitrate to the total mineral nitrogen recycled to the water column decreases severely. Denitrification takes place in the anaerobic layer and reduces part of the nitrate produced by nitrification in the upper oxic layer. Up to a certain point, increasing organic inputs thus result in increasing denitrification rates. With a further increase of organic input, the upper oxic layer might become so reduced that nitrate production through nitrification decreases and limits denitrification in the lower layer. Such a limitation does not occur when high nitrate concentrations are present in the overlying water, as is often the case in river or estuarine systems, because nitrate can then diffuse downwards across the sediment–water interface to be reduced in the anaerobic layer.

Silica dissolution rate is insensitive to redox conditions in sediments, and dissolved silica effectively diffuses from interstitial water to the water column. Preservation of a small fraction of biogenic opal can also occur, either by secondary organic coating or by diagenetic transformation into cristobalite or quartz. In marine or estuarine sediments, a part of dissolved silica in pore water may react with terrigenous clay minerals to form siliceous marine clays (Wollast 1974).

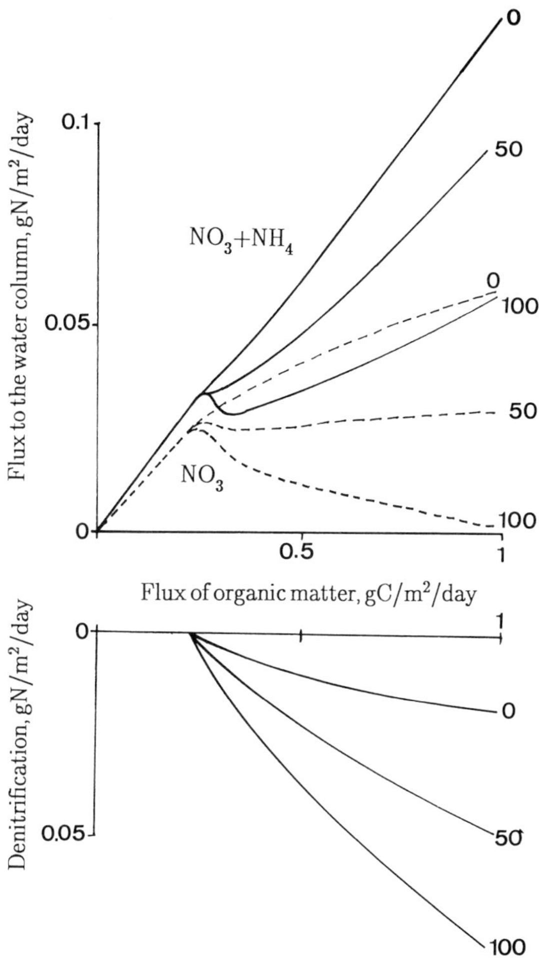

Fig. 4—Benthic nitrogen recycling as a function of the flux of particulate organic matter to the sediments (J_1+J_2). Fluxes are calculated according to the model of Billen (1982). *Top*: Fluxes of nitrate (dotted lines) and total inorganic nitrogen (solid lines) across the sediment–water interface for a concentration of nitrate in the overlying water of 0.50 and 100 μmol/l. *Bottom*: Depth-integrated denitrification rates for 0.50 and 100 μmol/l of nitrate in the overlying water.

NUTRIENT RETENTION CAPACITY OF THE AQUATIC CONTINUUM

River Systems

A major constraint imposed on planktonic microbial life by the hydrology of rivers is the high dilution rate resulting from lateral discharges. Planktonic microorganisms can only develop in rivers when their growth rate (μ) exceeds this dilution rate, i.e. when

$$\left[\mu > \frac{1}{A} \cdot \frac{dQ}{dx}\right]$$

where A is the wetted section and dQ/dx the longitudinal gradient of discharge along the river. Accordingly, in most NW European watersheds, phytoplankton can only develop in tributaries of stream-order 5 or higher (DeBecker 1986). Only benthic micro- or macrophytes develop in lower order streams, resulting in a much more limited biogeochemical impact.

Similarly, benthic processes are likely to play a more significant role with increasing stream order, both because total streambed area increases with respect to watershed area (from about 1% in first- and second-order streams to 3–5% for large watersheds) and because sediment deposition is favored in downstream large rivers with lower slopes compared with headwater streams.

Another characteristic of rivers is the significance of riparian systems in regulating fluxes of nutrients between land or aquifers and surface water. Wetlands, swamps, and riparian woods in temperate latitudes (Pinay and Decamps 1988), but also floodplain during inundation phases in tropical rivers (Saunders and Lewis 1988), have been shown to play a significant biogeochemical role in immobilizing and/or eliminating nutrients, particularly nitrogen.

A number of input–output nitrogen budgets in river systems (mostly at temperature latitudes) have been published. The results are summarized in Fig. 5. They show that a considerable retention of nitrogen occurs in highly polluted river systems with high N loads, and to a much lesser extent in streams draining intensive agricultural watersheds. Total nitrogen behaves almost conservatively in forested river systems. In most instances, the retention could be mainly attributed to benthic denitrification. A summary of direct determinations of denitrification rates in river sediments published in the literature is presented in Table 5. The differences observed between forested, agricultural, and industrialized rivers agrees well with the trends revealed by Fig. 5. Denitrification rates are particularly high in ponds, small reservoirs, or storage basins, where conditions of stagnancy allow sedimentation of fine organic material.

No data have been found in the literature concerning phosphorus input–output budgets in river systems. The dominant role of solid transport of this element renders such budgets difficult to establish accurately.

The low primary production in lower order streams implies that silica is generally not significantly taken up from water. In larger rivers, considerable development of diatoms may occur, leading to silica depletion. This depletion

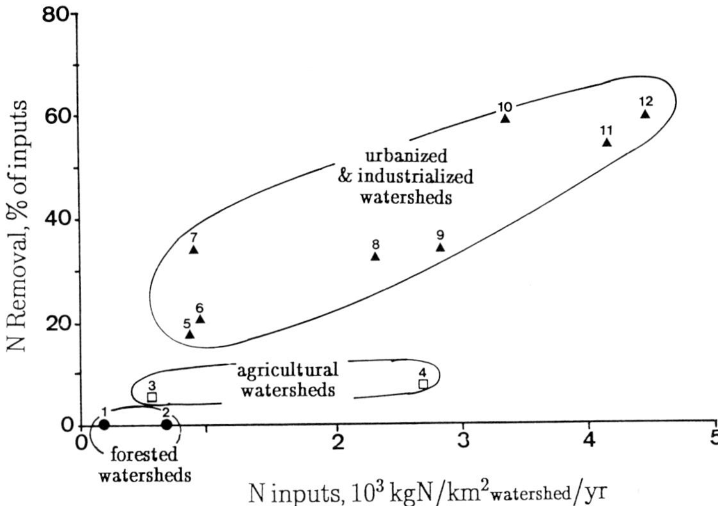

Fig. 5—Nitrogen budgets of river systems: N elimination in % of total inputs from the watershed. Forested watersheds: (1) Hubbard Brook, New Hampshire, (2) Viroin, Belgium. Agricultural watersheds: (3) Duffin Creek, Canada, (4) Dyle, Belgium. Urbanized and industrialized watersheds: (5) Aare, Switzerland, (6) Delaware River, U.S.A., (7) Potomac River, U.S.A., (8) Meuse, Belgium, (9) Rhine, Germany and the Netherlands, (10) Elbe, Germany, (11) Weser, Germany, (12) Scheldt, Belgium.

TABLE 5. Representative values of denitrification rates in the sediments of various aquatic systems (synthesis of data cited by Seitzinger 1988; Billen 1990).

	$mgN/m^2/h$
Rivers	
forested watershed	0–2
agricultural watershed	1–30
urbanized and industrialized watershed	4–60
Lakes	
oligo-mesotrophic	0.3–0.8
moderately eutrophic	0.1–0.3
eutrophic	0.6–1.5
Estuaries	60–300
coastal marine areas	1–500

is particularly evident in large eutrophicated rivers such as the lower Loire (Meybeck et al. 1988) or Rhine (Admiraal et al. 1990). In the latter instance, biogenic particulate silica dominates over the dissolved form from May to September. During this period a net sedimentation of biogenic particulate silica, representing 20–60% of the total Si transport, was observed in the lowest 150 km reaches of the three artificial branches of this strongly modified river.

Lake Systems

Lakes systems obviously differ from rivers by the magnitude of the residence time of the water masses, the depth of the water column, the importance of sedimentation, and the lack of important effects of riparine processes in the overall budget of elements. From spring to fall, phytoplankton in the upper layers often develops until nutrient limitation halts their growth. Uptake by primary production followed by sedimentation and immobilization in the benthos (or elimination in the case of denitrification) are the major processes by which nutrients are retained in lakes. Owing to their longer residence times and the resulting higher biological activity, the presence of lakes and reservoirs in a watershed strongly increases the overall nutrient removal of the river system and is therefore important to take into account when evaluating the transfer of nutrients from land to sea. On the other hand, input–output budgets are relatively easy to establish in lakes (when compared to rivers, estuaries, or coastal sea areas), so that our knowledge of nutrient retention in lakes is much more advanced. Lakes, therefore, constitute useful models for understanding retention properties of other aquatic systems.

As far as nitrogen is concerned, eliminated varies between 20 and 70% of the total inputs (Fig. 6). Nitrogen storage in sediments generally represents only a small part of this removal (Seitzinger 1988), while benthic denitrification is most often the dominant process. Comparison of the data from Table 5 with those of Table 3 also shows that denitrification largely exceeds the N_2 fixation rate.

Phosphorus retention, on the other hand, occurs only by benthic burial. Retention often ranges between 30 and 80% of the total inputs for oligotrophic or moderately eutrophic lakes (Fig. 7). In eutrophic lakes developing reduced bottom waters, phosphorus release from sediments occurs to a sometimes very large extent, and retention is strongly reduced. The lakes in Fig. 7 with no P retention are those developing strongly reducing bottom water (oxygen and nitrate depletion, with moderate sulfate concentration [Anderson 1982]).

Silica retention in lakes is always very high, ranging from 40% in oligotrophic lakes to 90% in eutrophic lakes. The case of the Great

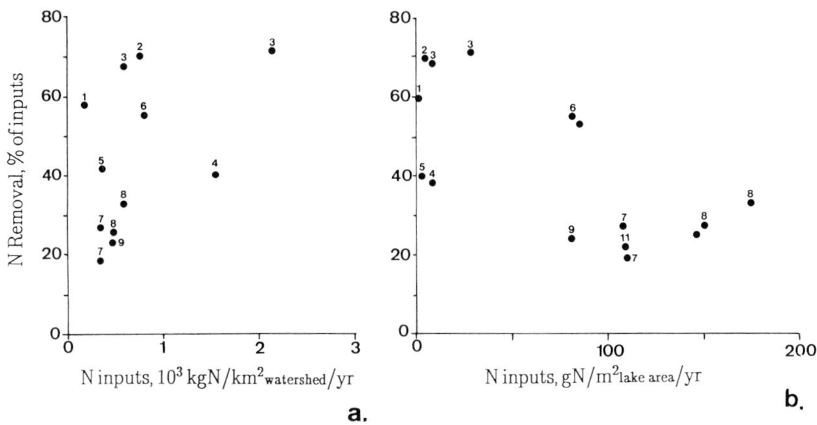

Fig. 6—Nitrogen budgets in lakes: retention in % of total inputs as a function of (a) total annual nitrogen inputs per unit watershed area and (b) of total annual nitrogen inputs per unit lake area. (1) Rawson Lake, Canada, (2) Ockechobee, Florida, (3) Kinneret, Israel, (4) Mergozzo, Italy, (5) Gardsjon, Sweden, (6) Bryrup, Denmark, (7) Kul, Denmark, (8) Kvindt, Denmark, (9) Stigsholm, Denmark, (10) Biel, Switzerland, (11) Sulejow, Poland.

Laurentian Lakes has been intensively studied in relation to eutrophication. In the pristine state, phosphorus was limiting the growth of diatoms, then the dominant component of phytoplankton, preventing silica depletion. Progressive increase of the phosphorus load, with unchanged silica inputs, first enhanced diatoms production. This altered the previously equilibrated silica mass balance and led to a progressive decrease of dissolved silica concentration in the whole water column and an increase in silica storage in sediments. After some time, depending on the residence time of the lake, silica became limiting for diatoms growth. Blue-green or green algae then replaced diatoms as the dominant component of phytoplankton. A new steady state was established for silica in which diatom development consumed all silica inputs during the algal growing season. This sequence of events can be traced back in the sediment record of the lakes by the presence of a peak of biogenic silica corresponding to the transition period between P and Si limitation of the diatoms (Schelske et al. 1983).

Estuaries

Regarding residence time and morphological characteristics, estuaries are somehow intermediate between rivers and lakes. The mixing of fresh and seawater, however, is responsible for distinct hydrological and physicochemical features which directly and indirectly affect nutrients transfer.

Flocculation of river colloidal material by seawater electrolytes in the

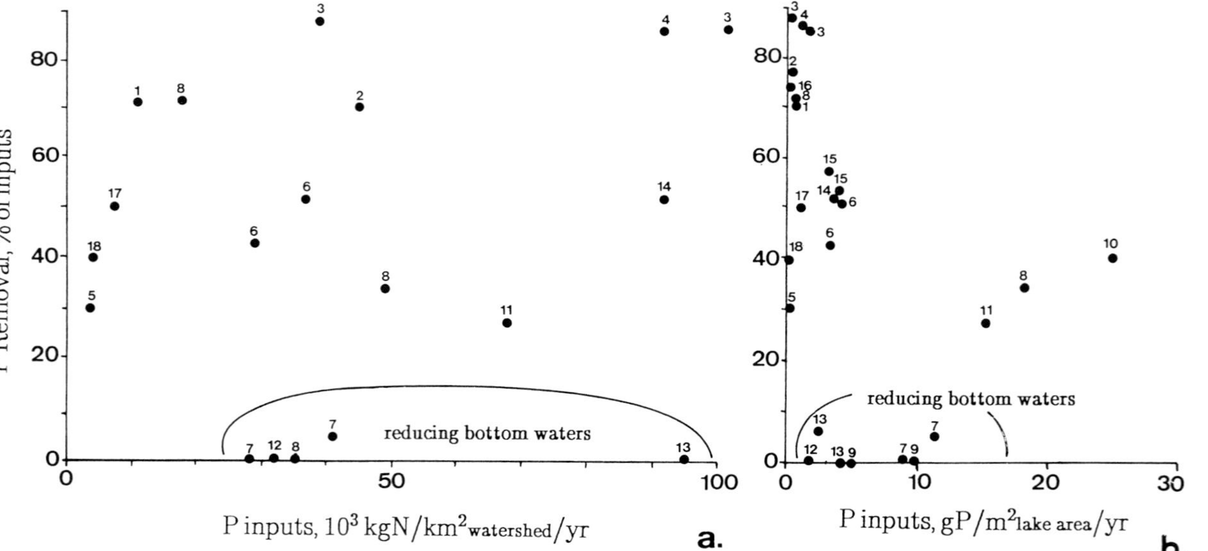

Fig. 7—Phosphorus budgets in lakes: retention in % of total inputs as a function of (a) total annual phosphorus inputs per unit watershed area and (b) total annual phosphorus inputs per unit lake area. (1) Rawson Lake, Canada, (2) Ockechobee, Florida, (3) Kinneret, Israel, (4) Mergozzo, Italy, (5) Gardsjon, Sweden, (6) Bryrup, Denmark, (7) Kul, Denmark, (8) Kvindt, Denmark, (9) Stigsholm, Denmark, (10) Biel, Switzerland, (11) Sulejow, Poland, (12) Skanderborg, Denmark, (13) Halle, Denmark, (14) Maggiore, Italy, (15) Salten, Denmark, (16) Mirror, U.S.A., (17) Karl, Denmark, (18) Fussing, Denmark.

0–5‰ salinity range is a common process to all estuaries. Furthermore, in moderately stratified estuaries, where there is both vertical mixing and a net landward flow at the bottom, suspended matter is trapped in the estuary, resulting in maximum turbidity and sediment accumulation. Part of these sediments, however, might be transported to the sea during flood episodes. In salt-wedge estuaries, where river flow is dominant over tidal forces and little exchange occurs from the fresh seaward flowing river water to the inward flowing lower seawater, sediments are also transported to the sea without accumulation within the estuary.

When it is not limited by the high turbidity, primary production in estuaries can be very high, particularly if the incoming nutrient load is high. Heterotrophic activities, both planktonic and benthic, are also very active, however; overall, many estuaries are heterotrophic systems where mineralization of organic matter produced internally and imported from the river largely dominates over local primary production.

This heterotrophic character of estuaries, and the frequent occurrence of large deposits of organic rich sediments explains the importance of benthic denitrification (see Table 5). A number of studies on nitrogen budgets in estuaries have shown that 30–60% of the total nitrogen input is eliminated through this process, irrespective of the magnitude of this input (Fig. 8). As was found for lakes, sediment accumulation of nitrogen accounts for only a small part (6–15%) of total N elimination.

In contrast to this very effective nitrogen elimination capacity, phosphorus retention by estuaries seems much more limited. Removal of dissolved

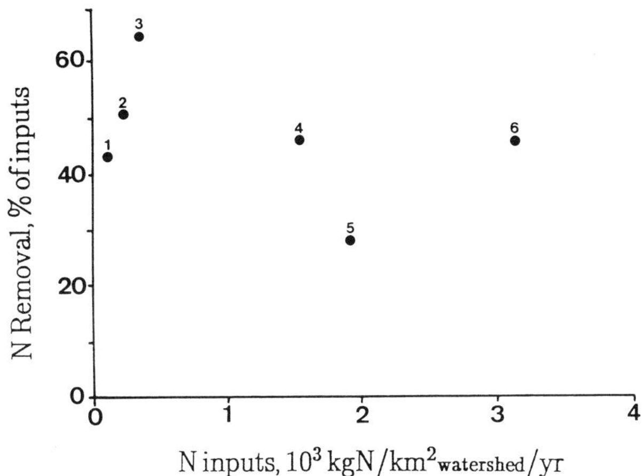

Fig. 8—Nitrogen budgets of estuaries: retention in % of total inputs per unit watershed area. (1) Ochlockonee, U.S.A., (2) Amazon, Brazil, (3) Baltic Sea, (4) Delaware Bay, U.S.A., (5) Narragansett Bay, U.S.A., (6) Scheldt, Netherlands.

orthophosphate from solution is often observed in the low salinity range of estuaries, probably by adsorption on iron oxyhydroxides precipitated there from dissolved or colloidal riverine iron. However desorption of phosphate from suspended sediments further occurs in the higher salinity range (Liss 1976). In fact, this reversible adsorption process acts as a buffering mechanism maintaining dissolved phosphate near 0.6–1.25 μmol/l. On the other hand, rapid mineralization of organic phosphorus prevents important storage in sediments. Figure 9 summarizes a number of P budgets in estuaries, showing that their retention capacity is much smaller than that of most lakes.

Although a biological removal of silica at an early stage of freshwater mixing has been described, the behavior of silica in estuaries is mostly determined through the uptake by diatoms and subsequent redissolution or burial of biogenic opal. Uptake of Si by diatoms in estuaries is often estimated from a silica versus salinity plot of data obtained in surface waters. As discussed by DeMaster (1981), this can lead to overestimation of the net amount of Si removed, because redissolution of biogenic silica in bottom waters can escape observation. DeMaster and others estimated the immobilization of silica within estuaries by directly determining the accumulation of biogenic opal in the sediments. Such estimations, as gathered in Fig. 10, show that a higher removal of silica (50–60%) occurs in those estuaries receiving more N and P inputs, i.e., as a result of eutrophication.

Coastal Marine Zones

The very last part of the water continuum that constitutes the land–ocean boundary is constituted by coastal shelf areas enriched in terrigenous nutrients by the outflow of estuaries. These zones often differ from offshore, less enriched shelf areas and from upwelling zones by the N:P:Si ratio of

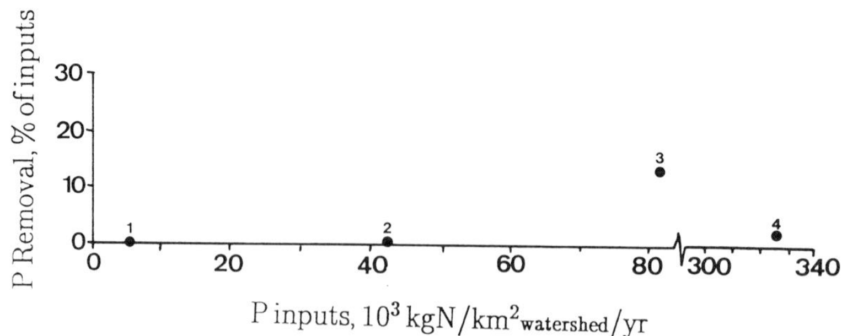

Fig. 9—Phosphorus budgets of estuaries: retention in % of total inputs per unit watershed area. (1) Ochlockonee, U.S.A., (2) Amazon, Brazil, (3) Chesapeake Bay, U.S.A., (4) Narragansett Bay, U.S.A.

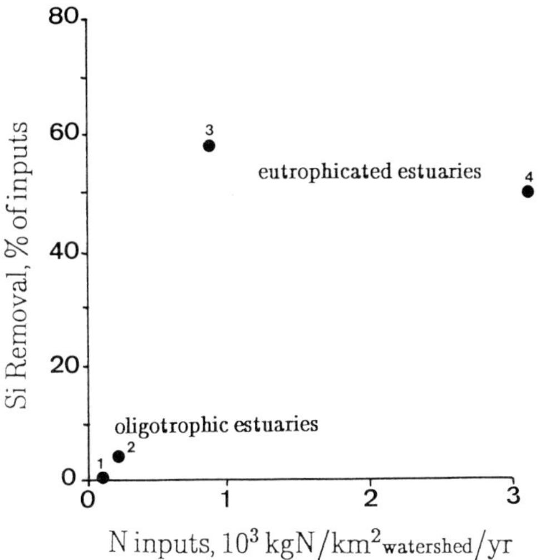

Fig. 10—Silica budget in estuaries. Silica retention in % of total river inputs as a function of total nitrogen inputs, taken here as an index of the degree of eutrophication of the estuary. (1) Ochlockonee, U.S.A., (2) Amazon, Brazil, (3) Chesapeake Bay, U.S.A., (4) Scheldt, Netherlands.

the nutrients available for new production. In offshore areas, the sequence of events observed from the beginning of the vegetative period (or from the beginning of the upwelling event) consists of the following: diatoms develop first and, in the absence of important population of grazers, form a bloom, exhausting N, P, and Si nearly simultaneously (Fig. 11a). When nutrient depletion is complete, diatoms settle down and the *new production community* is replaced by a *regenerated system*, dominated by flagellates and macro- or micro-grazers, which accumulate much less biomass and recycle N and P very efficiently. A distinct characteristic of the river-enriched coastal system is that during the course of the increasing phase of the bloom, inorganic nitrogen remains in the water column after complete silica depletion (Fig. 11b). This allows flagellate development which is not supported exclusively by regenerated nutrients. Most coastal eutrophication problems result from this *new* production of—often undesirable—flagellates. The species developing under these circumstances are often not suitable for ingestion by indigenous zooplankton and much of their primary production is exported.

Indeed, contrary to a recent statement by Smith and Hollibaugh (1989), nutrient-enriched coastal zones are most often net *autotrophic* systems, characterized by oxygen oversaturation and net exportation of autochthonous organic matter. Supporting evidence for the significance of this export comes

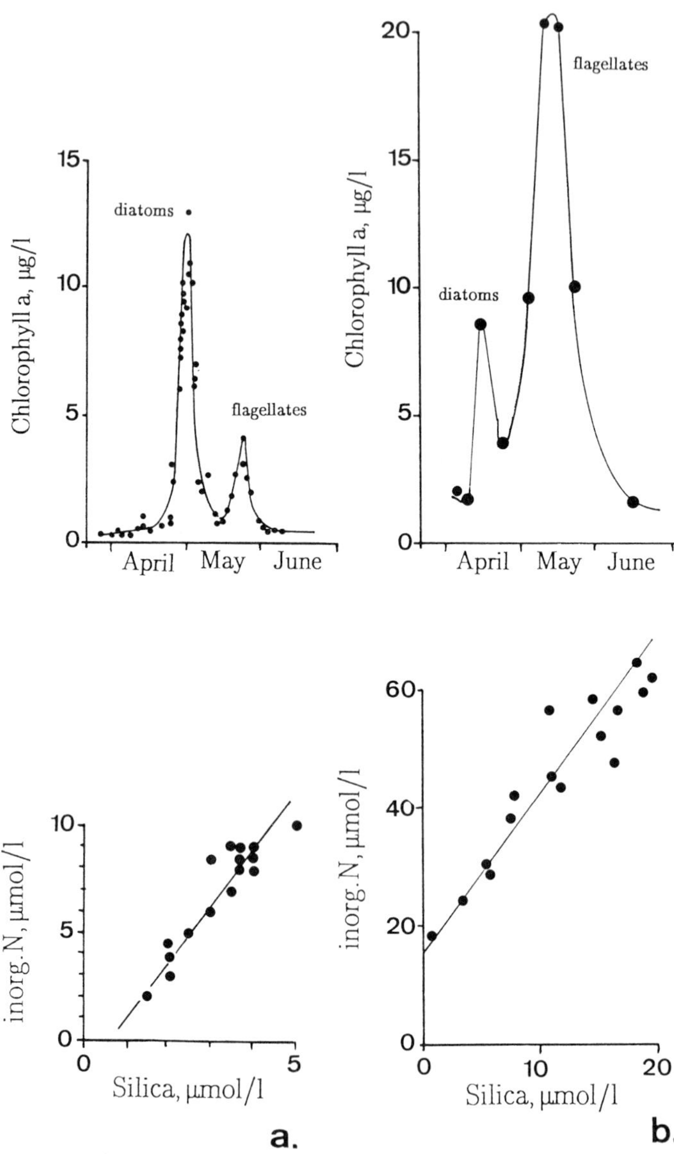

Fig. 11—Comparison of N/Si uptake and diatoms/flagellates development during the spring bloom (a) in an offshore shelf area (Fladden Ground, North Sea) and (b) in a nutrient enriched coastal area (North Sea Southern Bight).

from observations by Walsh et al. (1981): the amount of *fresh* organic material from marine origin deposited in the sediments of the upper continental slope (400–1800 m) is higher than on the shelf itself. Accumulation of this material constitutes a significant sink for river nutrients.

In addition to this storage, nitrogen is actively eliminated by denitrification in marine sediments (Table 5). Denitrification often eliminates 10 to 25% of total N benthic mineralization in nearshore sediments (Billen and Lancelot 1985).

Overall Retention along the Water Continuum: Summarizing Comments

With respect to the other nutrients, an excess of nitrogen is present in the direct inputs from land to surface water. However, it is also very efficiently eliminated through denitrification and burial, during its transfer along the aquatic continuum: up to 50% elimination may occur in polluted river systems, 50% of the inputs are retained in lakes, and a further reduction of 50% is observed in estuaries. Phosphorus is less abundant in the inputs from soil leaching and erosion. Its presence in surface water is mainly the result of direct contamination with human sewage or animal wastes. It is also less efficiently retained by river systems and estuaries. Phosphorus lake retention might be high but only under certain conditions. Put together, these observations clearly explain why phosphorus is generally the limiting nutrient in freshwater, upstream systems, while nitrogen plays this role in downstream brackish or marine waters.

Silica is generally present in excess with respect to P and N in water draining forested areas. An increase in P and N inputs resulting from agriculture and urbanization does not lead to a parallel increase of silica inputs. On the contrary, N and P enrichment increases silica retention by rivers, lakes, and estuaries up to 20, 80, and 50% of inputs, respectively. Depletion of silica with respect to P and N has major consequences in lake and coastal marine eutrophication, leading to a shift from diatom to flagellate dominance of the phytoplankton community.

PAST AND FUTURE TRENDS IN NUTRIENT TRANSFER FROM LAND TO OCEAN

On the basis of the above conclusions, a few hypothetical statements can be made concerning the possible variations in nutrient retention capacity of the aquatic continuum in response to climate change and human impact.

The rising of sea level occurring in the warming phase of the glaciation cycle has considerably increased the global area of coastal and estuarine zones. This should have enhanced the retention of nitrogen from terrestrial sources and reduced its flux to the ocean, while phosphorus and silica (in the absence of eutrophication) should have been much less significantly affected. If nitrogen inputs are indeed controlling ocean *new* phytoplankton production, the resulting lower atmospheric CO_2 uptake should have acted as a positive feedback mechanism further enhancing the warming up (McElroy 1983).

From the end of the last glaciation period, human activity has become a significant biogeochemical factor at the global scale. Its impact on the functioning of the global aquatic system is far-reaching and can result in opposing effects, depending on time and place; its net global effect still awaits quantitative evaluation. For the temperate system of NW Europe and North America, both historical records and hydrobiological data are available to reconstitute the major trends of variations of nutrient transfer along the aquatic continuum in response to human impact.

Deforestation is known to lead to a very high N and P release from soil, resulting in nitrate levels in running water higher than the drinking water standard. This must have occurred on a very large scale during the periods of first agriculture settlement. A signal of increased biogenic silica storage in sediments corresponding to enhancement of diatom production associated with early settlement and forest clearance in North America (1800–1850) is clearly detectable in the sediment record of Lake Ontario and Lake Erie (Schelske et al. 1983).

Traditional agricultural systems, which by the end of the tenth century had replaced the climax forest ecosystems in the greater part of NW Europe, relied entirely on an efficient nutrient cycling economy within the territory of the small rural communities. Lots of reglementations of this time, niggling in appearance, can be viewed as organizing an internal cycling of nutrients and limiting exports outside the system. Leaching of agricultural soils was, however, inevitably higher than in forested ecosystems. On the other hand, as hydraulic power at that time was intensively utilized for many purposes, small dams were systematically built up on nearly every stream close to human settlements, creating ponds and wetlands. The enhanced retention of such river systems probably compensated for the increased N and P output from soil leaching.

The rather recent transition to the industrial civilization has had more significant impact on nutrient transfer to the ocean. Opening the biogeochemical cycles linked to the human food system, made possible by the industrial production of fertilizers, led to increased soil leaching, on one hand, and to increased organic pollution of the river system by human and cattle wastes, no longer applied as amendments to farmlands, on the other. As far as nitrogen is concerned, these two effects probably balanced each other for years, because organically polluted river systems have a higher capacity for N elimination. Only phosphorus inputs to the ocean were increased, particularly when polyphosphates associated with synthetic detergents became an additional source. Two factors, however, led to a gradual reduction of the retention capacity of the river system. The first was the hydraulic management of rivers, characterizing the first phase of industrial age. With the purpose of extending arable land area, wetlands were drained and small rivers were channelized, strongly reducing both the residence time of water in the river system and the area with high nutrient retention capacities. For navigational purposes, large rivers were also channelized and meander loops rectified. As an example of this, rectification of the "Wild Rhine" between 1818 and 1880 resulted in reducing the bedstream area from 350 to100 m^2/km. The second factor, involving the recent development of biological wastewater purification, has led to a considerable decrease in organic contamination of river systems. Nutrient elimination by these treatments is limited: a maximum of 30–40% in the case of nitrogen unless specific, very expensive tertiary treatments are applied. Therefore, the reduction of nutrient retention capacity of the receiving water, resulting from the improvement of their oxygen status, might well counterbalance the slight reduction of nutrient load. Billen et al. (1986) estimated that secondary wastewater purification, reducing point source organic pollution of the Scheldt watershed to 90% of their present value, should result in significantly increasing nitrogen output to coastal waters.

The recent increase of marine coastal water eutrophication problems is thus not only a result of an increase in the inputs of nutrients to surface water but also of a decrease in the N and P retention capacity of the water system, with a parallel increase of Si retention in eutrophicated lakes and major rivers.

It is difficult to assess how these trends, probably applicable to both NW Europe and the eastern coast of North America, can be extrapolated to other regions in the world. Very little is known, for instance, about how tropical systems react to human impact, namely to deforestation. Another major impact of man on nutrient transfer from land to sea is the impoundments of large rivers, either for flood control, establishment of reservoirs for water supply and irrigation, or hydroelectric power generation. Man-made reservoirs presently represent an area of about 400,000 km^2 and increase at a rate of 3.5% per year. Storage and elimination of nutrients in the rapidly accumulating sediments of these reservoirs might represent a significant part of N, P, and Si transfer from land to ocean. The observed reduction in fish production in the Mediterranean off of Egypt since the impoundment of the Nile River dramatically illustrates this point.

CONCLUSIONS AND OPEN QUESTIONS

The data presented demonstrate the global significance of coupled processes affecting N, P and Si during their transfer along the water continuum to the ultimate input into the ocean. They show that it would be quite unrealistic to establish global budgets of past, present, or future nutrient inputs to the sea without taking into account these complex processes. Some trends are known, mostly for temperate systems, and models have been established for predicting the behavior of single nutrients in local situations. Much remains to be learned, however, concerning a number of questions. A few examples are:

1. What is the role of riparine processes in regulating nutrient exchanges between surface and groundwater?
2. How great is the accummulation of P in terrestrial systems?
3. What regulates nitrogen fixation and why is it so much more active in terrestrial than in aquatic environments?
4. What regulates P and Si transfer in estuaries?

The data presented also clearly shows the interdependance of the dynamics of the three nutrients and carbon. Coupled models of these elements must be established. Also, much more information is required before we can create even simplified models of these processes at the global level, particularly concerning the behavior of nutrients in tropical rivers, lakes, and estuaries.

Acknolwedgements. Part of this work was funded by the Commission of European Communities (DG XII, contract EV4V-012-B) and by the Ministry of Science Policy, Belgium (contract ENV2). G. Billen is a Research Associate of the National Funds for Scientific Research, Belgium.

REFERENCES

Admiraal, W., P. Breugem, D.M. Jacobs, and E.D. de Ruyter van Steveninck. 1990. Fixation of dissolved silicate and sedimentation of biogenic silicate in the lower river Rhine during diatom blooms. *Biogeochem.* **9**: 175–185.

Andersen, J.M. 1982. Effect of NO_3 concentration in lake water on phosphate release from the sediments. *Water Res.* **16**:1119–1126.

Billen, G. 1982. An idealized model of nitrogen recycling in marine sediments. *Am. J. Sci.* **282**:512–541.

Billen, G. 1990. N-budget of the major rivers discharging into the continental coastal zone of the North Sea: the nitrogen paradox. In: EEC Workshop on eutrophication and algal blooms in North Sea coastal zones, the Baltic and adjacent areas: Prediction and assessment of prevention actions, ed. C. Lancelot, G. Billen, and

H. Barth. Water Pollution Research Reports No. 12, Brussels: Commission of the European Communities (EUR 1219 OEN).

Billen, G., S. Dessery, C. Lancelot, and M. Meybeck. 1989. Seasonal and interannual variations of nitrogen diagenesis in the sediments of a recently impounded basin. *Biogeochem.* **8**:73–100.

Billen, G., and C. Lancelot. 1985. Carbon–nitrogen relationships in nutrient metabolism of coastal marine ecosystems. *Adv. Aq. Microbiology* **3**:263–322.

Billen, G., C. Lancelot, E. De Becker, and P. Servais. 1986. The terrestrial interface: modelling nitrogen transformations during its transfer through the Schelde river system and its estuarine zone. In: Marine Interfaces Ecohydrodynamics, ed. J.C.J. Nihoul, pp. 429–452. Amsterdam: Elsevier.

Brimblecombe, P., and D.H. Stedman. 1982. Historical evidence for a dramatic increase in the nitrate component of acid rain. *Nature* **298**:460–462.

Capone, D.G. 1988. Benthic nitrogen fixation. In: Nitrogen Cycling in Coastal Marine Environments, ed. T.H. Blackburn and J. Sorensen, Scope 33, pp. 85–114. New York: Wiley.

Caraco, N.F., J.J. Cole, and G.E. Likens. 1989. Evidence for sulfate-controlled phosphorus release from sediments of aquatic systems. *Nature* **341**:316–318.

Conley, D.J., S.S. Kilham, and E. Theriot. 1989. Difference in silica content between marine and freshwater diatoms. *Limnol. Ocean.* **34**:205–213.

DeBecker, E. 1986. Apports, transferts et transformations de l'azote dans les réseaux hydrographiques. Développement d'une méthodologie générale et applications au réseau belge. Thèse. Faculté des Sciences, Université Libre de Bruxelles, Belgium.

DeMaster, D.J. 1981. The supply and accumulation of silica in the marine environment. *Geochim. Cosmochim. Acta* **45**:1715–1732.

Garber, J.H. 1984. Laboratory study of N and P remineralization during the decomposition of coastal plankton and seston. *Est. Coast. Shelf Sci.* **18**:685–702.

Krom, M.D., and R.A. Berner. 1980. Adsorption of phosphate in anoxic marine sediments. *Limnol. Ocean.* **25**:797–806.

Lewin, J. 1961. The dissolution of silica from diatom walls. *Geochim. Cosmochim. Acta.* **21**:182–198.

Liss, P.S. 1976. Conservative and non-conservative behavior or dissolved constituents during estuarine mixing. In: Estuarine Chemistry, ed. J.D. Burton and P.S. Liss, pp. 93–130. New York: Academic.

Loye Pilot, M.D., J.M. Martin, and J. Morelli. 1990. Atmospheric input of inorganic nitrogen to the Western Mediterranean. *Biogeochem.* **9**: 117–134..

Mackin, J.E., and R.C. Aller. 1984. Ammonium adsorption in marine sediments. *Limnol. Ocean.* **29**:250–257.

McElroy, M.B. 1983. Marine biological controls on atmospheric CO_2 and climate. *Nature* **302**:328–329.

Meybeck, M. 1982. C, N and P transport by world rivers. *Am. J. Sci.* **282**:401–450.

Meybeck, M. 1984. Les fleuves et le cycle géochimique des éléments. Thesis University of Paris VI, France.

Meybeck, M., G. Cauwet, M. Somville, D.T. Gouleau, and G. Billen. 1988. Nutrients (organic C, P, N, Si) in the Eutrophic River Loire (France) and its estuary. *Est. Coast. Shelf Sci.* **27**:1–30.

Officer, C.B., and J.H. Ryther. 1980. The possible importance of silicon in marine eutrophication. *Mar. Ecol. Prog. Ser.* **3**:83–91.

Pinay, G., and H. Decamps. 1988. The role of riparian woods in regulating N fluxes between the alluvial aquifer and surface water: A conceptual model. *Regulated Rivers: Res. and Management* **2**:507–516.

Saunders, J.F., and W.M. Lewis. 1988. Transport of P, N and C by the Apure River, Venezuela. *Biogeochem.* **5**:323–342.

Schelske, C.L., E.F. Stoermer, D.J. Conley, J.A. Robbins, and R.M. Glover. 1983. Early eutrophication in the Lower Great Lakes: new evidence from biogenic silica in sediments. *Science* **222**:320–322.

Seitzinger, S.P. 1988. Denitrification in freshwater and coastal marine ecosystems: Ecological and geochemical significance. *Limnol. Ocean.* **33**:702–724.

Smith, S.V., and J.T. Hollibaugh. 1989. Carbon-controlled nitrogen cycling in a marine macrocosm: an ecosystem-scale model for managing cultural eutrophication. *Mar. Ecol. Prog. Ser.* **52**:103–109.

Verbanck, M., J.P. Vanderborght, and R. Wollast. 1989. Major ion content of urban wastewater: assessment of per capita loading. *Res. J. Water Pollution Control Fed.* **61**:1722–1728.

Walsh, J.J., E.T. Premuzic, T.E. Whiteledge. 1981. Fate of nutrient enrichment on continental shelves as indicated by the C/N content of bottom sediments. In: Ecohydrodynamics, ed. J.C.J. Nihoul, pp. 13–49. Amsterdam: Elsevier.

Wollast, R. 1974. The silica problem. In: The Sea, ed. E.D. Goldberg, vol. V, pp. 359–392. New York: Wiley.

Present and Future Roles of Ocean Margins in Regulating Marine Biogeochemical Cycles of Trace Elements

J.-M. Martin[1] and H.L. Windom[2]

[1]*Institut de Biogéochimie Marine Unité Associée au C.N.R.S. no. 386*
Ecole Normale Supérieure, 1 rue Maurice Arnoux
92120 Moutrouge, France

[2]*Skidaway Institute of Oceanography, P.O. Box 13687*
Savannah, Georgia 31416, U.S.A.

Abstract. Ocean margins act as efficient filters of dissolved and particulate trace element fluxes to the open ocean by rivers. Much of the removal of the dissolved flux is due to physicochemical processes acting in estuaries and coastal waters; however, removal through incorporation in biogenic particles is apparently also important. Trace elements cycled from the deep ocean are also removed in ocean margins due to these processes. Thus, 90–95% of the input of trace elements to the marine environment accumulates in ocean margins, having significant effects on their oceanic residence times.

Predicted future inputs of nutrients to ocean margins are estimated to increase significantly their efficiency as sinks for trace elements of riverine and oceanic origin. This, along with predicted global changes, will lead to an increasing role of ocean margins in the marine biogeochemical cycles of trace elements.

INTRODUCTION

From a geochemical perspective, ocean margins are regions adjacent to the continents, including inland seas, estuaries, and continental shelves and slopes, where the major portion of continental detritus delivered to the oceans is deposited. In these regions, chemical, biological, and physical processes combine to scavenge elements from the water column to the seafloor at rates orders of magnitude greater than those occurring in the deep ocean. Thus, ocean margins act to filter the influx of materials transported from the continents to the ocean.

Ocean Margin Processes in Global Change
Edited by R.F.C. Mantoura, J.-M. Martin and R. Wollast

Surface ocean currents circulate seawater through continental margins (i.e., continental shelf and slope regions) at rates that are rapid relative to the residence time of most elements in the sea. Ocean margins, therefore, may also act as a continual filter for open ocean waters: a process often referred to as "boundary scavenging" (Anderson et al. 1990).

Ocean margins are heterogeneous in their chemical, biological, and physical characteristics in time and space. Many studies have been conducted in specific ocean margin regions, but they cannot be considered to be representative. Although the lack of information makes it difficult to make a detailed evaluation, we will nonetheless attempt to assess the role of ocean margins in the marine biogeochemical cycles of trace elements.

In this chapter we consider specific trace elements that serve as representatives of groups having different biogeochemical characteristics. The first group, metalloids, is represented by As, which is the element within this group for which most information exists. A group referred to as nutrient-type elements is represented by Cd, Cu, Ni, and Zn. These represent those trace elements in the ocean that exhibit strong relationships to nutrients (Bruland 1983), suggesting that their cycles in the ocean are influenced predominantly by biological processes. Geochemically controlled elements are those trace elements whose fates are dominantly influenced by inorganic processes. We select Al, Co, Fe, Mn, Pb, and U as examples. Finally, the last group we consider are the rare earth elements and Ce, Eu, La, and Sm are taken as representatives.

For each group we attempt to estimate fluxes from the continents to ocean margins, their removal there,and their net fluxes to the deep ocean. We then use a two-box model to assess the role of ocean margins in the biogeochemical cycles and residence time of these trace elements. Finally, we predict how anthropogenic influences and global changes affect this role.

The data used in our assessment are from various sources and are therefore of varying quality. The amount of data available for a given element, compartment, or flux also varies and, in some cases, does not exist so that extrapolations are made. When extrapolations are made they are indicated, but we do not attempt to otherwise assess in any detail the quality of the data used.

TRACE ELEMENT FLUXES TO OCEAN MARGINS FROM LAND

Trace elements are delivered to ocean margins primarily by river runoff, atmospheric transport from land, and by inputs from the open ocean. Estimates of fluxes from land can be made relatively easily based on existing data. Although there is clear evidence for the accumulation of trace elements in ocean margins that are due to inputs from the open ocean, this flux is not easily quantified. We will, however, attempt to make some estimates of this for specific trace elements in a later section. Our intention here is to summarize the existing information on fluxes from land and to estimate their rates.

Dissolved Riverine Fluxes

The most recent reviews of credible data on the concentrations of dissolved trace elements in rivers generally agree within narrow limits (GESAMP 1987). There is still a need for additional reliable data from river systems draining geologically, climatologically, and physiologically diverse watersheds. With such data, a more accurate estimate of riverine flux could be made. Nonetheless, the existing credible data can be used to estimate the average composition of world river runoff (Table 1). Using this average

TABLE 1. Riverine trace element concentrations and fluxes to ocean margins.

	Concentrations		Fluxes	
	Dissolved μg/l	Particulate μg/g	Dissolved 10^6 kg/y	Particulate 10^6 kg/y
Metalloids				
As	1.7*	5††	65	75
Nutrient-Type				
Cd	0.01	1.2	0.4	18
Cu	1.5	100	58	1,500
Ni	0.5	90	19	1,350
Zn	0.6*	250	23	3,750
Geochemically Controlled				
Al	50†	94,000	1,925	1,410,000
Co	0.1*	20	3.8	300
Fe	40	48,000	1,540	720,000
Mn	8.2	1,050	320	15,750
Pb	0.03*	35*	1.1	525
U	0.24	3	9.2	45
Rare Earth Elements				
Ce	0.08	80**	3.1	1,200
Eu	0.001	1.5	0.04	23
La	0.05	45	1.9	675
Sm	0.008	7	0.31	105

*All values for concentrations were taken from Martin and Gordeev (1982); recent data from GESAMP (1987).
**From Goldstein and Jacobson (1988).
†More recent data suggests a value closer to 20 μg/l (R. Wollast, pers. comm.; H. Windom, unpublished data).
††Probably closer to 7–9 μg/g (Huang et al. 1988).

composition and an estimated global river discharge of ca. 38.5 × 10^3 km^3/yr (Baumgartner and Reichel 1975), the gross riverine flux to the ocean margin can be estimated (Table 1).

Particulate Riverine Fluxes

The major contribution of continentally derived particles to the marine environment is through rivers. Although various estimates have been made during the past 30 to 40 years, Milliman and Meade (1983), using more recent data, have calculated the annual total sediment discharge to the oceans from world rivers to be about 15 × 10^{12} kg. This includes suspended sediment and bedload and probably represents the best estimate of total sediment discharge at the present time (for a new estimate see Milliman, this volume). The average composition of riverine particulates is fairly well known (Martin and Gordeev 1982) and can be coupled with riverine sediment flux to yield estimates of particulate trace element fluxes (Table 1).

Atmospheric Trace Element Fluxes

A recent review (GESAMP 1989) provides estimates for the atmospheric inputs to the world ocean for some of the trace elements listed in Table 1. For example, the maximum total fluxes of As, Cd, Cu, Ni, Pb, and Zn are estimated to be 8.5, 4.0, 52, 28, 88, and 228 × 10^6 kg/yr, respectively, while their soluble fluxes (i.e., dissolvable in seawater) are 3.6, 2.6, 30, 10, 80, and 100, respectively. The soluble fluxes of Al and Fe derived from mineral aerosols are 3600 and 3200 × 10^6 kg/yr. These values can be used to derive estimates of the atmospheric fluxes to ocean margins in the following way. A reasonable estimate of the portion of atmospheric transport to the ocean that is delivered to the margins is 50% (R. Duce, pers. comm.). The trace metals bound on particles (i.e., not dissolvable) would be insignificant when compared to riverine particulates (Table 1). Thus, the important atmospheric contribution to ocean margins is that which is dissolvable, estimates of which are given in Table 2.

There are exceptional ocean margin regions, such as the Mediterranean Sea (Martin et al. 1989), where atmospheric fluxes represent the most significant contributions to the total trace element inputs. These regions, however, are generally limited to ocean margins adjacent to arid and/or highly developed continental regions with low river inputs.

TABLE 2. Soluble atmospheric trace element flux to ocean margins.

	10^6 kg/y
Metalloids	
As	1.8
Nutrient-Type	
Cd	1.3
Cu	15
Ni	5
Zn	50
Geochemically Controlled	
Al	3600
Co	0.4*
Fe	1600
Mn	51*
Pb	40
U	0.1*
Rare Earth Elements	
Ce	4.5*
Eu	0.06*
La	2.1*
Sm	0.25*

*Values with asterisk based on the assumption that these elements in atmospheric inputs are solubilized as efficiently as aluminum and that their abundance relate to aluminium is the same as average crustal material.

OCEAN MARGINS AS A FILTER FOR CONTINENTAL FLUXES

Removal of Continentally Derived Particulate Trace Elements

Meade (1972) pointed out the importance of ocean margins in trapping continental detritus and reducing its transport to the deep ocean. Recent studies suggest that generally less than 10% and probably closer to 5% of the suspended sediments delivered by rivers escapes the ocean margins (i.e., estuaries, continental shelves, and slopes) (Table 3). Direct transport to the deep sea takes place mainly when a river mouth is located near to or at the shelf edge (Niger, Mississippi), when the shelf is very narrow (San Francisco), or where a canyon funnels suspended matter downward (Ganges–Brahmaputra) (Eisma 1987).

Based on the above, it can be conservatively estimated that about 95% of the trace element flux associated with riverine particulates (Table 1) is removed and deposited in ocean margins. Of course, this assumes that an

TABLE 3. Removal of riverine sediment inputs in ocean margins.

Area	Removal, in % From Total Suspended Transport	Reference
Mississippi	90	Trefry and Presley (1976)
St. Lawrence	93	Bewers and Yeats (1977)
Zaire	95	Eisma et al. (1978)
Rivers of the Black and Azov Seas	83	Demina et al. (1978)
Kura	90–95	Ivanov (1955)
S.E. U.S. Rivers	>90	Windom and Gross (1989)
N.E. U.S. Rivers (Cape Cod to Cape Lookout)	≈90	Meade (1972)
Rhone	≈90	Guieu, Martin, Thomas, and Elbaz-Poulichet (1990)
Yellow (Huang He)	≈95	Martin et al. (submitted)
Yangtze (Chang Jiang)	≈80	Chen Ji Yu (pers. comm.)

insignificant fraction of the trace elements associated with riverine particulates is remobilized (i.e., dissolved) in ocean margins.

Removal of Continentally Derived Dissolved Trace Elements

The efficiency of filtering or reducing the dissolved trace element fluxes from the continents to the ocean depends on processes that convert dissolved trace elements to particulate phases. Two types of processes can be considered. The first type includes all the physicochemical processes such as precipitation, flocculation, adsorption, redox reactions, and ion exchange. These dominate in the estuarine region of ocean margins where chemical and particulate gradients are larger. The second type consists of biologically mediated particle production. Primary production and formation of fecal material are the dominant processes. They occur throughout ocean margins but are most pronounced in coastal regions and upwelling areas adjacent to surface ocean current fronts on outer continental shelves and slopes (e.g., the Gulf Stream).

In the following, we estimate the significance of estuarine and biological removal to evaluate the relative efficiency of ocean margins in scavenging, and thus decreasing net fluxes of the different classes of dissolved trace elements to the deep ocean.

Estuarine Removal

Metalloids. With the exception of arsenic, very little information exists on the estuarine behavior of metalloids. In the case of As, behavior has been observed to be conservative in southeastern U.S. estuaries (Waslenchuk 1977; Waslenchuk and Windom 1978; Froelich et al. 1985), in British estuaries (Langston 1983; Howard et al. 1984), French (Seyler 1985), and in Chinese estuaries (Huang et al. 1988). Other studies of seasonal cycles in estuarine and coastal environments, however, have shown nonconservative distributions involving removal and/or remobilization processes. It has been demonstrated that As may be removed during one part of the year, due to incorporation with chemically similar phosphorus in phytoplankton, and released later in reduced forms such as organo-arsenic compounds (Johnson and Burke 1978; Byrd 1988). Nevertheless, there are insufficient data to suggest that there is a net estuarine removal of As and, more generally, of metalloids (Table 4).

Nutrient-type trace elements. In the deep ocean it is clear that certain trace metals are involved in biogeochemical cycles similar to nutrients. The evidence for this is the covariance in concentrations of metals, such as Cd, Cu, Ni, and Zn with PO_4, NO_3, or SiO_2 in deep ocean water column profiles. This indicates that metals are incorporated into phytoplankton in the surface euphotic zone, along with nutrients, and released at depth due to microbial degradation of organic detritus.

A few studies have suggested that a similar biological control on trace metals concentrations may be active in estuaries (Eaton 1979a; Church 1986). Others, however, (e.g., Windom et al. 1990) suggest that biological processes cannot explain the behavior of these nutrient-type trace metals in most cases. It is more likely that inorganic processes exert the major influence on the estuarine behavior of these metals.

A few studies of the nutrient-type trace metals have indicated estuarine removal. For example, Windom and Smith (1985) have reported that up to 50% of the zinc transported by southeastern U.S. river systems is removed during estuarine transport. Others, however, have indicated that Zn is conservative or enriched in estuarine waters during transport (Klinkhammer and Bender 1981). Enrichment in estuarine waters, due to release from particles during estuarine transport, is more typical of the behavior of the other nutrient-type metals. The best example of this is cadmium. It is generally accepted that mobilization processes, occurring when river water mixes with seawater, lead to the release of Cd from particles and formation of soluble Cd–chloro complexes. This has been observed in the lower part of the Scheldt Estuary (Salomons and Kerdijk 1986; Duinker et al. 1981). Evidence for some Cd remobilization has also been reported by Boyle et

TABLE 4. Total natural removal of dissolved trace elements in ocean margins.

	Estuarine Removal (10^6 kg/y)	Biological Removal* (10^6 kg/y)	Total Removal (10^6 kg/y)	Percent of Input**
Metalloid				
As	0	1.5	1.5	2
Nutrient-Type				
Cd	−2	4	2	120
Cu	0	3	3	74
Ni	0	3	3	13
Zn	0	27	27	37
Geochemically Controlled				
Al	580	52	630	11
Co	1.9	0.45	2.4	57
Fe	1,390	68	1450	47
Mn	160	2.1	162	44
Pb	0	2.1	2.1	5
U	0.92	1	1.9	20
Rare Earth Elements				
Ce	1.86	0.84	2.7	35
Eu	0.024	0.015	0.04	40
La	1.38	0.6	2.0	50
Sm	0.17	0.12	0.3	53

*Using concentrations of metals in phytoplankton reported by Martin and Knauer (1973), Collier and Edmond (1984), Knauss and Ku (1983), and Sanders and Windom (1980). The latter reference was for As concentration in biogenic sediment in experimental microcosms.
**Combined dissolved riverine and atmospheric fluxes given in Tables 1 and 2.

al. (1982) in the Amazon plume. More recently, Edmond et al. (1985) reported similar Cd mobilization in the Changjiang, Yellow, and Orinoco rivers, and Elbaz-Poulichet et al. (1987) estimated that the Cd input to the ocean might be increased by a factor of from 2 to 30 due to estuarine release from particles.

For the discussions presented here we assume, as a first approximation, that Cu, Ni, and Zn are conservatively transported through estuaries, although variable behavior has been reported for each (Boyle et al. 1982; Duinker et al. 1981; Edmond et al. 1985; Windom and Smith 1985). Thus we assume no estuarine removal of these trace metals in Table 4. For Cd we estimate that an additional flux of five times the dissolved riverine flux results from estuarine biogeochemical processes. This we indicate as a negative estuarine removal in Table 4.

Geochemically controlled trace elements. With the exception of lead, studies of the estuarine chemistry of representative elements in this group (i.e., Al, Co, Fe, Mn, and possibly U) indicate that they are removed during estuarine mixing. Studies of Pb in French (Elbaz-Poulichet et al. 1984), southestern U.S. (Windom et al. 1985), and Asian (Windom et al. 1988) estuaries all suggest that it is not significantly removed from solution during estuarine mixing. The low levels in rivers imply that particle scavenging of Pb is efficient in the fresh aquatic environment.

The nonconservative behavior of iron in estuaries is well documented and an almost universal feature of estuarine mixing (Windom et al. 1971; Coonley et al. 1971; Bewers et al. 1974; Boyle et al. 1977; Figueres et al. 1978; Sholkovitz et al. 1978). While various estimates of the net transport vary, removal of Fe is always reported as being high, and various mechanisms are proposed to accomplish this, including precipitation of hydrated oxides and flocculation due to destabilization of colloids. We use an estimate of 90% removal in Table 4.

For manganese, the low solubility of MnO_2 and comparatively high solubility of Mn(II) explains the observed contrasting behavior of this element in different estuaries. Examples are given by Eaton (1979b) for Chesapeake Bay, Ackroyd et al. (1986) in the Tamar, Wollast et al. (1979) in the Scheldt, and Windom and Smith (1985) for estuaries of the southeastern U.S.

Although removal of dissolved Mn in coastal areas is often observed to be as high 90% (Windom and Smith 1985), subsequent remobilization from reduced sediments of ocean margins certainly decreases the consequences of estuarine removal. Also, photoreductive processes may solubilize particulate Mn in coastal waters. Nevertheless, for the purpose of the following discussions, we assume an estuarine removal of 50% of the dissolved riverine input.

Although few studies of the transport of cobalt through estuaries have been made, data from dissolved Co-salinity relationships in coastal waters can be used to extrapolate back to an apparent zero salinity concentration. Such data exist for Hudson Bay, the South Atlantic Bight (U.S.), and the Gulf of Thailand (GESAMP 1987). These values (0.06, 0.03, and 0.04 μg/kg), when compared to the mean concentration of Co in rivers (Table 1), suggest that about 50%of this metal is removed in estuaries (Table 4).

Martin and Gordeev (1982) estimate that about 30% of the dissolved Al delivered by rivers is removed in estuaries. This estimate, based on the work of Bewers and Yeats (1977) on the St. Lawrence River, is used here as well (Table 4).

It has been generally accepted that uranium isotopes behave conservatively in estuaries (Figueres et al. 1982; Borole et al. 1982); however, there is some evidence that significant removal may occur by sorption onto particles or coagulation in the low salinity area (Martin, Meybeck, and Pusset 1978; Martin, Nijampurkar, and Salvadori, 1978; Maeda and Windom 1982; Church et al. 1981). Moreover, it has long been speculated that a significant fraction of isotopes introduced through rivers may deposit in nearshore sediments rich in organic matter (Koczy 1954). Such sediments are in contact with bottom waters low in oxygen, which may induce a reducing environment where insoluble U^{4+} precipitates. Examples are given by Veeh (1967) and Figueres et al. (1982), and most recently by Anderson (1987). Such processes would probably account for the removal of ca. 10–20% of the river input or more. More recent data obtained by McKee et al. (1987) on Amazon shelf sediments indicate that the frequent physical disturbance, the input of extensively weathered lateritic material, and the Fe-dominated redox chemistry combine to support conditions conducive to release of U to the sea in that specific area. They computed that this process increases the river supply of U by a factor of 5. We, however, still choose 15% estuarine removal for Table 4.

Rare Earth Elements (REE). The short residence time of REE in the ocean implies that their ocean margin chemistry is likely to be controlled by particulate removal reactions, such as coagulation and flocculation processes, involving humic substances and/or hydrous oxides as well as adsorption processes. Few papers discuss the fate of REE in ocean margins. Martin et al. (1976) observed that uptake of dissolved REE onto particles reached 70% for the light REE, as river water mixes with seawater. Further work by Sholkovitz and Elderfield (1988) in the eastern U.S. and Goldstein and Jacobsen (1988) in Canada confirms the importance of estuarine removal of REE. Goldstein and Jacobsen (1988) estimated that the percent removal of La, Ce, Sm, and Eu is 73, 60, 54, and 61, respectively (used in Table 4). They also showed that REE and Fe behave differently during estuarine mixing in the Great Whale river estuary. Fe removal is almost complete at low salinity whereas a significant fraction of REE still remains in solution. The shape of the Fe and REE vs. salinity curves is not consistent with a simple model of coagulation of Fe- and REE-bearing materials. They considered that the observed linear relationship between the activity of free ion REE^{3+} and pH is consistent with a simple ion exchange model for REE removal.

Biological Removal

The incorporation of trace elements into particles due to biological uptake is clearly important in removing them from seawater. It is well known that primary production is enhanced in or near ocean margins (Berger et al.

1987). Organic carbon burial rates (Romankevich 1984) suggest that nearly 90% of the removal of biogenic particles occurs at the continental margin. Berner (1982) estimates that 5.7×10^{12} gC/y is buried in pelagic sediments. If the nitrogen associated with this carbon burial yields a C:N molar ratio of 15, then only about 1% of the total N delivered by rivers reaches these sediments. The rest must accumulate in ocean margins.

Liu (1979) has estimated that N burial in the ocean is between $9–38 \times 10^{12}$ g/y, most of which must accumulate in ocean margins. More recent N budgets assume organic N burial rates of 21×10^{12} g/y (Christensen et al. 1987; Codispoti and Christensen 1985).

If we assume that all of this N is buried in the ocean margins as organic detritus having the approximate composition of plankton, then literature values of trace element concentrations in such material can be used to estimate their biological removal. The results (Table 4) indicate that biological removal exceeds dissolved inputs for Cd. This suggests that ocean margins act as sinks for some oceanic trace elements. Alternatively, uncertainties in the composition of plankton may account for the apparent excess removal. For example, the Cd:N ratios for plankton range from 2 to 620×10^{-6}, assuming the N content of plankton is ca. 9% (Collier and Edmond 1984). We used a mean value of 250×10^{-6}.

SIGNIFICANCE OF OCEAN MARGINS IN THE GLOBAL BIOGEOCHEMICAL CYCLING OF TRACE ELEMENTS

The significance of ocean margins in the biogeochemical cycling of trace elements can be evaluated by considering a two-box model which divides the ocean into two lateral compartments as shown in Fig. 1. As specified above, we consider the ocean margin to be that portion of the ocean shallower than 1 km (i.e., 5% of the surface area). In this model we also assume that the average depth of the ocean margin is 250 m. Due to the difficulty in estimating the fraction of the atmospheric delivery to ocean margins which ultimately reach the open ocean, their contribution has not been considered in the model. The remaining assumptions are specified in the figure.

Ocean Margin Control of Trace Element Budgets of the Open Ocean

We will first consider the mass balance of this two-box model to assess whether the removal in ocean margins, estimated above, results in a net export to the deep ocean that balances with other inputs and outputs.

We have assumed that only 5% of the particle-bound trace elements delivered by rivers reaches the deep oceans. This, added to the net soluble trace element transport through ocean margins (i.e., input given in Table 1 minus removal given in Table 4) yields the total output from the ocean margins to the deep ocean (Table 5). The flux from the ocean margins, as

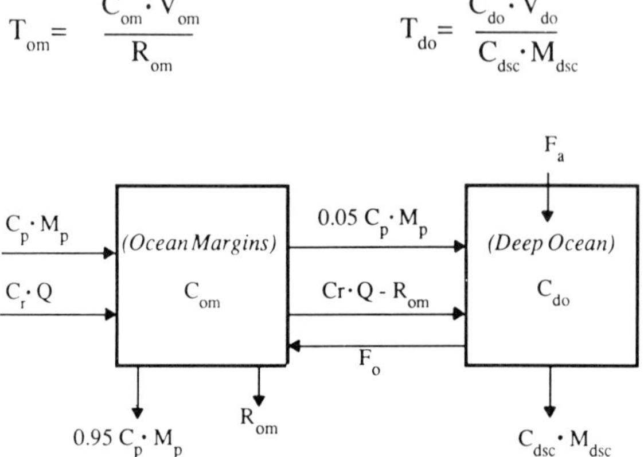

Fig. 1—Two-box model of oceanic trace element input and removal.
C_p—Concentration of trace element in riverine particles (Table 1)
C_r—Concentration of dissolved trace element in river water (Table 1)
M_p—Mass of sediment transported by world rivers ($15\ 10^{12}$ kg y^{-1})
Q—Discharge of world rivers ($38.5\ 10^3$ km^3 y^{-1})
C_{om}—Concentration of dissolved trace element in ocean margin waters (Table 6)
R_{om}—Removal of dissolved trace elements in ocean margins (Table 3)
C_{do}—Concentration of dissolved trace elements in deep ocean (Table 3)
C—Average concentration of trace element in deep-sea clays (Martin and Whitfield 1983)
M_{dsc}—Deep-sea clay accumulation rate ($1.1\ 10^{12}$ kg y^{-1})
F_a—Atmospheric flux of trace elements to the ocean (GESAMP 1989)
F_o—Flux from the deep ocean to ocean margins
T_{om}—Residence time of trace element in ocean margin
T_{do}—Residence time of trace element in the deep ocean
V_{om}—Volume of ocean margin waters ($4.5\ 10^6$ km^3)
V_{do}—Volume of deep ocean waters ($1.37\ 10^9$ km^3)

estimated, along with atmospheric trace element flux appears to balance, reasonably well, the estimated removal to deep-sea clays (Table 5). The factor of two or more difference between removal and input rates for Co and Mn may reflect significant hydrothermal sources. The imbalance between the greater input of Cd and Zn relative to removal has been recognized by others (Yeats and Bewers 1983; Windom 1990) and has been suggested as being due to anthropogenic mobilization of these metals, since this additional input would not yet be recorded in deep-sea clays. Alternatively, these metals may be underestimated in deep-sea clays. Uranium imbalance has also been recognized by others (Veeh 1967); however, this imbalance is probably due to the underestimation of removal of anoxic ocean margin sediments, which have been estimated to be much higher (Anderson 1987).

More recently, Barnes and Cochran (1990) reevaluated the uranium mass-

TABLE 5. Open ocean mass balance for trace elements using natural net output from ocean margins (all values are in 10^6 kg/y).

	Output from Ocean Margins	Atmos. Input to Global Ocean†	Total Input	Total Removal††
Metalloids				
As	65.3	1.8	67.1	14
Nutrient-Type				
Cd	0.6	1.1	1.7	(0.3)
Cu	145	7	152	220
Ni	89	11	100	220
Zn	230	23	253	(130)
Geochemically Controlled				
Al	76,000	18,000	94,000	104,000
Co	17	4*	21	60
Fe	37,000	16,000	53,000	66,000
Mn	1,000	255*	1,255	6,600
Pb	65	45	110	90
U	9.6	0.5	10	(2)
Rare Earth Elements				
Ce	65	23*	88	110
Eu	1.2	0.3*	1.5	1.6
La	36	11*	47	50
Sm	5.5	1.3*	6.8	7.7

†Data taken from GESAMP (1989); *indicates that value was obtained by using crustal ratios of element to aluminum and multiplying by the atmospheric Al flux.
††Calculated using the average composition of deep-sea clay (Martin and Whitfield 1983) and a deep-sea clay accumulation rate of 1.1×10^{12} kg/y.

balance problem. They pointed out that suboxic hemipelagic sediments, which are of significantly greater areal extent than anoxic sediments, may account for about twice the removal of U occurring in anoxic sediments. According to these authors, U removal to suboxic and anoxic sediments would account for about 40% of the river input. The total sedimentary removal including calcium carbonate would average 66% of the river input. A similar process might explain the As imbalance.

The role of ocean margins as a filter for fluxes from the continents to the deep ocean appears to be consistent with the mass balance presented in Table 5. The significance of ocean margins in the cycling of trace elements in the ocean can be appreciated by comparing the difference between the total removal to deep-sea clays (Table 5) and the riverine fluxes, both soluble and particulate (Table 1). Generally, over 80–90% of the total trace element input to the oceans accumulates in ocean margins.

Residence Times of Trace Metals in Ocean Margins Compared to those in the Open Ocean

Assuming steady state, the two-box model (Fig. 1) can also be used to estimate residence times of trace elements. Residence times for trace elements in ocean margins are based on total dissolved content and removal from solution (Table 4). This assumes no significant mobilization of trace metals transported to this box in particles. Residence times for trace elements in the deep ocean are based on total water column content (which generally includes both dissolved and particulate) and total removal (Table 6).

For As, Cd, Zn, and U, residence times are much longer in the deep ocean. This suggests that the scavenging efficiency is considerably greater in ocean margins for elements that have overall longer residence times in

TABLE 6. Comparison of trace element residence times in ocean margin and the deep ocean.

	Concentration (10^{-6} kg/m^3)		Residence Time (years)	
	Ocean Margin*	Deep Ocean†	Ocean Margin	Deep Ocean
Metalloid				
As	1.1	1.7	$3.3 \cdot 10^3$	$1.7 \cdot 10^5$
Nutrient-Type				
Cd	0.02	0.078	$4.5 \cdot 10^1$	$3.3 \cdot 10^5$
Cu	0.4	0.25	$6 \cdot 10^2$	$1.5 \cdot 10^3$
Ni	0.6	0.47	$9 \cdot 10^2$	$2.9 \cdot 10^3$
Zn	0.4	0.39	$7 \cdot 10^1$	$4.1 \cdot 10^3$
Geochemically Controlled				
Al	8.3	0.81	$5.9 \cdot 10^1$	$1.1 \cdot 10^1$
Co	0.03	0.0018	$5.6 \cdot 10^1$	$4.2 \cdot 10^1$
Fe	0.4	0.056	$1.2 \cdot 10^0$	$1.2 \cdot 10^0$
Mn	0.6	0.27	$1.7 \cdot 10^2$	$5.6 \cdot 10^1$
Pb	0.025	0.0021	$5.3 \cdot 10^1$	$3.2 \cdot 10^1$
U	3.0	2.86	$7 \cdot 10^3$	$1.9 \cdot 10^6$
Rare Earth Elements				
Ce	0.017	0.0028	$2.8 \cdot 10^1$	$3.4 \cdot 10^1$
Eu	0.0004	0.00014	$4.5 \cdot 10^1$	$1.2 \cdot 10^2$
La	0.008	0.0042	$1.8 \cdot 10^1$	$1.1 \cdot 10^2$
Sm	0.0002	0.0006	$3.0 \cdot 10^0$	$1.1 \cdot 10^2$

*Values taken from: Byrd (1988); Windom et al. (1988); Windom and Smith (1985); Kremling (1985); Kremling (1983); Maeda and Windom (1982).
†From Bruland (1983).

marine waters. Residence times of the other geochemically controlled elements and REE are similarly short in both compartments. The calculated residence times of from one to a few hundred years for these elements are considerably shorter than previous estimates based on a one-box model ocean using river inputs (Broecker and Peng 1982).

Four elements, As, Cu, Ni, and U, have residence times in ocean margins that are significantly longer than the residence time of water (ca. 100 years). This may be due to the underestimation of their removal (particularly U). Alternatively, removal of these materials in ocean margins requires their cycling between the two boxes of the model.

Ocean Margins as a Sink for Trace Elements of Oceanic Origin

The 21×10^{12} g/y of nitrogen removed in ocean margins discussed above (see section on Biological Removal) compares to an annual natural riverine flux of 21×10^{12} g particulate N and 14×10^{12} g dissolved N (GESAMP 1987). Most of the particulate N is probably trapped in estuaries and bays where denitrification may account for half of this input (Seitzinger et al. 1984). This would leave a reasonably well-balanced N budget for ocean margins. Recent studies, however, suggest that denitrification of shelf sediments provides a N loss averaging about 50×10^{12} g/y (Christensen et al. 1987). This clearly indicates a large imbalance of N in ocean margins. The total oceanic N budget of Christensen et al. (1987) appears to be imbalanced by 60 to 90×10^{12} g/y of losses over inputs; their budget, however, uses an atmospheric input of 40×10^{12} g/y rather than the 100×10^{12} g/y atmospheric input estimated by Wollast (1983).

Christensen et al. (1987) explain their imbalanced N budget in the following way: During the present interglacial period, the greatest losses of N occur on the shelf due to denitrification. During the maximum glacial extent, shelf regions were 25% of their present area; thus, assuming similar denitrification rates, N loss was then about 25% of today's rate, or about 12×10^{12} g/y. At the same time, river input of N would be more efficiently delivered to the deep ocean and primary production may have been enhanced, therefore increasing the input of organic matter to pelagic and hemipelagic sediments. This suggests that during glacial periods, the deep ocean accumulates N due to smaller losses caused by denitrification. During the transition from glacial to interglacial, deep ocean N is cycled more and more efficiently through ocean margins, where it is processed (i.e., converted to biogenic particles) and ultimately lost by denitrification and burial. The present imbalance may therefore be explained as a lag in the equilibration of N to interglacial conditions.

Regardless of the scenario, recent estimates of N budgets clearly suggest that ocean margins are a sink for trace elements transported from the deep ocean along with the "excess" N accumulated there. The enhanced biogenic particle production in ocean margins, driven by the deep ocean N, provides for a more efficient transfer of trace elements from the water column to sediments. Some portion of the increased trace element flux to ocean margin sediments is probably not released back to the water column during denitrification.

The particle-rich environment of ocean margins may also lead to enhanced water column scavenging. Removal of plutonium has been shown to occur extensively in various European and North American estuaries and coastal waters (Jeandel et al. 1981; Duursma et al. 1983; Sholkovitz and Mann 1987). Mass-balance studies carried out in the Gironde estuary (Elbaz-Poulichet et al. 1982) and in the Savannah estuary (Olsen et al. 1989) compared observed removal to the increased Pu in suspended sediments. They both concluded that the excess particulate Pu must originate from the scavenging of ocean water. Similar processes could represent a sink for other chemical elements which may also be scavenged from oceanic waters by sorption in turbid estuarine/coastal areas and subsequently transported to sediments.

THE INFLUENCE OF GLOBAL CHANGES ON TRACE ELEMENT FLUXES TO AND THROUGH OCEAN MARGINS

Atmospheric CO_2 Increase

Based on experimental evidence and theoretical considerations it appears that an increase in level of CO_2 may result in increased terrestrial photosynthetic production (cf. Lerman et al. 1977), provided other variables are unchanged. It is difficult, however, to translate this into quantitative terms on a global scale owing to changes in agricultural practices, species distribution, increased deforestation, and desertification. Thus it is difficult to speculate about potential effects upon trace element cycles. Changes in species may have some effect upon trace element speciation. Hekstra (1986) speculated about the possible change in the C_4/C_3 plant ratio as a function of CO_2 increase. He considered that if nutrients do not keep pace, the plants will produce more carbohydrates and less protein. This obviously would have some effect upon dissolved trace metal speciation. This, however, may not be significant when compared to predicted changes due to EDTA use (F. Morel, pers. comm.).

Increased Nutrient Mobilization

Wollast (1983) has estimated an anthropogenic mobilized dissolved nitrogen flux in rivers of 21×10^9 kg/y. As of 1980, this extrapolated to an annual per capita load of 7 kg N/yr. The per capita loading was estimated on an average world population basis, since highly developed societies mobilize N at a higher rate (e.g., in Europe 10 kg/y) than less developed ones.

Assuming this additional amount of mobilized N is removed in ocean margin sediments, an additional biological removal of trace elements would be expected. This additional removal, above that removed naturally, can be estimated following the approach described above (see section on Biological Removal and Fig. 2).

We can predict the future increase in N mobilization by assuming that as world population increases, N mobilization would increase at the same per capita rate (i.e., 7 kg N/y) as at the present time. This is clearly a rough estimate of how N mobilization would increase because the degree of development of a country determines its per capita rate of N mobilization.

If we assume this increased N mobilization with population growth is reasonable, we can predict the increased efficiency of trace element removal in ocean margins by, again, assuming the new N is totally removed in this region. Demeny (1987) has predicted world population growth through the year 2050. We multiply this increased population by the present per capita N mobilization to predict future trends in trace element removal in ocean margins (Fig. 2). From this prediction it can be seen that the present efficiency of ocean margins in removing trace elements may be significantly increased in the future.

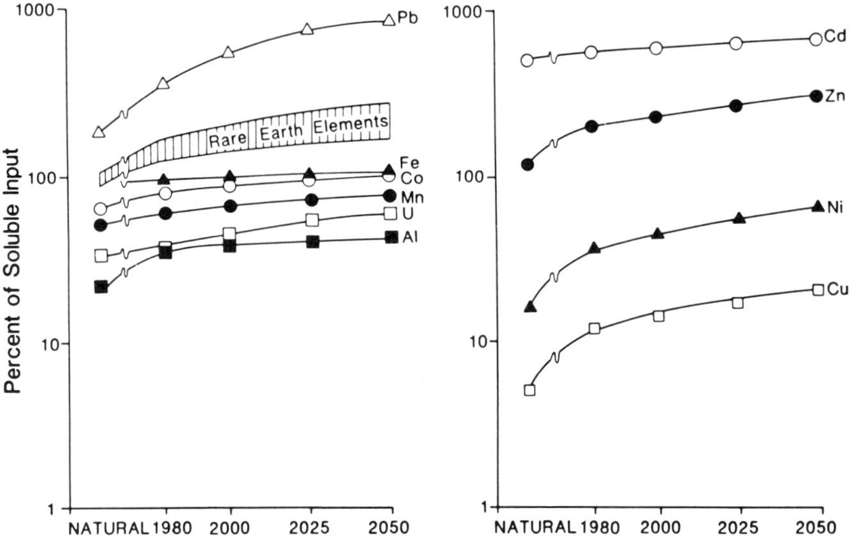

Fig. 2—Predicted future changes in dissolved trace element removal in ocean margins. Values are percent of dissolved input by rivers.

Importance of River Runoff

Impacts of rising global temperature on the hydrologic cycle will include changes in rainfall patterns and evapotranspiration regimes around the globe. Glantz and Wigley (1987) argued that if average water resources are not significantly modified, the variability, frequency of extreme events, and seasonal cycles may be altered. However, they considered that the most sophisticated models developed for case studies are difficult to extrapolate on a global basis. L'vovich (1987) considered the natural water cycle and its link with economy. He estimated that a maximum 10% decrease might occur in river runoff by 2050 but that damming and channeling of rivers might have a more important effect. The effect will primarily be on sediment discharge and associated trace element delivery on particles.

Climate Change

With global warming, sea level rise will increase the area of continental shelves. For example, a rise in a sea level of 1 m corresponds to an increase in a continental shelf area of ca. 10^6 km^2. Following the scenario described above (Ocean Margins as a Sink for Trace Elements of Oceanic Origin) it would appear likely that ocean margins will become an even more important sink for trace elements as well as other materials in the future.

REFERENCES

Ackroyd, D.R., A.J. Bale, J.M. Howland, S. Knox, G.E. Millward, and A.W. Morris. 1986. Distributions and behaviour of dissolved Cu, Zn and Mn in the Tamar estuary. *Est. Coast. Shelf Sci.* **23**:621–640.

Anderson, R.F. 1987. Redox behavior of uranium in anoxic marine basins. *Uranium* **3**:145–164.

Anderson, R.F., Y. Lao, W.S. Broecker, S.E. Trumbore, H.J. Hofman, and W. Wolfli. 1990. Boundary scavenging in the Pacific Ocean: a comparison of ^{10}Be and ^{231}Pa. *Earth Planet. Sci. Lett.* **96**:287–304.

Barnes, C.E. and Cochran, J.K. 1990. Uranium removal in oceanic sediments and the oceanic U balance. *Earth Planet. Sci. Lett.* **97**:94–101.

Baumgartner, A. and E. Reichel. 1975. The World Water Balance. Amsterdam, Elsevier.

Berger, W.H., K. Fischer and C. Lai, and G. Wu. 1987. Ocean productivity and organic carbon flux. *Scripps Inst. Ocean. Ref. Ser.* 87–30.

Berner, R.A. 1982. Burial of organic carbon and pyrite sulfur in the modern ocean. *Am. J. Sci.* **282**: 451–473.

Bewers, J.M., I.D. McCauley, and B. Sundby. 1974. Trace metals in the waters of the Gulf of St. Lawrence. *Can. J. Earth Sci.* *11*:939–950.

Bewers, J.M. and P.A. Yeats. 1977. Oceanic residence times of trace metals. *Nature* **268**:595–598.

Borole, D.V., S. Krishnaswamy, and B.L.K. Somayajulu. 1982. Uranium isotopes in rivers, estuaries and adjacent coastal sediments of western India: their weathering, transport and oceanic budget. *Geochim. Cosmochim. Acta* **46**:125–137.

Boyle, E.A., J.M. Edmond, and E.R. Sholkovitz. 1977. The mechanism of iron removal in estuaries. *Geochim. Cosmochim. Acta* **41**:1313–1324.

Boyle, E.A., S.S. Huested, and B. Grant. 1982. The chemical mass balance of the Amazon plume II. Copper, nickel and cadmium. *Deep-Sea Res.* **29**:1355–1364.

Broecker, W.S. and T.-H. Peng. 1982. Tracers in the sea. Palisades, N.Y.: Eldigio Press, Columbia Univ.

Bruland, K.W. 1983. Trace elements in sea-water. In: Chemical Oceanography, ed. J.P. Riley and R. Chester, vol. 8, p. 157–220. London: Academic Press.

Byrd, J.T. 1988. The seasonal cycle of arsenic in estuarine and nearshore waters of the South Atlantic Bight. *Marine Chem.* **25**:383–394.

Christensen, J.P., J.W. Murray, A.H. Devol and L.A. Codispoti. 1987. Denitrification in continental shelf sediments has major impact on the ocean nitrogen budget. *Glob. Biogeochim. Cyc.* **1**:97–116.

Church, J.M., C.J. Lord, III and B.L.K. Somayajulu. 1981. Uranium, thorium and lead nuclides in a Delaware salt marsh sediment. *Est. Coast. Shelf Sci.* **13**:267–275.

Church, T. 1986. Biogeochemical factors influencing the residence time of micronutrients in a large tidal estuary, Delaware Bay. *Marine Chem.* **18**:393–406.

Codispoti, L.A. and J.P. Christensen. 1985. Nitrification, denitrification and nitrous oxide cycling in the eastern tropical South Pacific Ocean. *Marine Chem.* **16**:277–300.

Collier, R. and J.M. Edmond. 1984. The trace element geochemistry of marine biogenic particulate matter. *Prog. Ocean.* **13**:113–199.

Coonley, S., E.B. Baker, and H.D. Holland. 1971. Iron in the Mullica River and Great Bay, New Jersey. *Chem. Geol.* **7**:51–63.

Demeny, P. 1987. Population Change: Global Trends and Implications. In: Resources and World Development, ed. D.J. McLaren and B.J. Skinner, Dahlem Konferenzen PC6, pp. 29–48. Chichester: Wiley.

Demina, L.L., V.V. Gordeev, and L.S. Fumina. 1978. Dissolved and particulate Fe, Mn, Zn and Cu in rivers and estuaries of the Black Sea, Sea of Azov and Caspain Sea. (In Russian) *Geochimica* **8**:1211–1229.

Duinker, J.C., R.F. Nolting, and D. Michel. 1981. Effects of salinity, pH, and redox conditions on the behavior of Cd, Zn, Ni and Mn in the Scheldt Estuary. Thalassia Jugosl. **18**:191–202.

Duursma, E.K., M.J. Frissel, J.C. Guary, J.M. Martin, J. Nieuwenhuizel, R.M.J. Pennders, and A.J. Thomas. 1983. Plutonium in sediments and mussels of the Rhine Meuse-Scheldt estuary. In: Proc. Int. Seminar on the Behaviour of Radionuclides in Estuaries, Rept. No. XII/380/85-EN, pp. 71–106. Luxemburg: CEC.

Eaton, A. 1979a. Observations on the geochemistry of soluble copper, iron, nickel and zinc in the San Francisco Bay Estuary. *Envir. Sci. Tech.* **13**:425–432.

Eaton, A. 1979b. Removal of "soluble" iron in the Potomac river estuary. *Est. Coast. Marine Sci.* **9**:41–49.

Edmond, J.M., A. Spivack, B.C. Grant, Hu Ming-Hui, Chen Zexiam, Chen Sung, and Zeng Xiushau. 1985. Chemical dynamics of the Changjiang Estuary. *Cont. Shelf Res.* **4**:17–36.

Eisma, D. 1987. Processes of nearshore accumulation of suspended material. *Mitt. Geol. Paleontol. Inst. Univ. Hamburg* **64**:57–69.

Eisma, D., J. Kalf and S.J. Van der Gaast. 1978. Suspended matter in the Zaire estuary and adjacent Atlantic Ocean. *Neth. J. Sea Res.* **12**(3/4):382–406.

Elbaz-Poulichet, F., P. Holliger, W.W. Huang, and J.M. Martin. 1984. Lead cycling in estuaries, illustrated by the Gironde estuary, France. *Nature* **308**(5958):409–414.

Elbaz-Poulichet, F., W.W. Huang, J. Jednacak-Biscan, J.M. Martin, and A.J. Thomas. 1982. Trace metals behaviour in the Gironde estuary: the problem revisited. *Thalassa Jugoslavica* **18**(1–4):61–95.

Elbaz-Poulichet, F., W.W. Huang, J.M. Martin, and J.X. Zhu. 1987. Dissolved Cd behaviour in some selected French and Chinese estuaries. Consequences on Cd supply to the Ocean. *Marine Chem.* **22**:125–136.

Figueres G., J.M. Martin, and M. Meybeck. 1978. Iron behaviour in the Zaire estuary. *Neth. J. Sea Res.* **12**(3/4):329–337.

Figueres G., J.M. Martin, and A.J. Thomas. 1982. Apports par les fleuves d'uranium dissous à l'océan, exemple du Zaire. *Oceanol. Acta* **5**(2):141–147.

Froelich, P.N., G.A. Hambrick, M.O. Andreae, R.A. Mortlock, and J.M. Edmond. 1985. The geochemistry of inorganic germanium in natural waters. *J. Geophys. Res.* **90**:1133–1141.

GESAMP. 1987. Land/sea boundary flux of contaminants: contributions from rivers. Reports and Studies No. 32, Joint Group of Experts on the Scientific Aspects of Marine Pollution, 172 p. Paris: UNESCO.

GESAMP. 1989. Atmospheric input of trace species to the world ocean. Reports and Studies No. 38, Joint Group of Experts on the Scientific Aspects of Marine Pollution, 111 p. Geneva: World Meteorological Organization.

Glantz, M.H. and T.M.L. Wigley. 1987. Climatic variations and their effects on water resources. In: Resources and World Development, ed. D.J. McLaren and B.J. Skinner, pp. 625–641. Dahlem Konferenzen PC 6. Chichester: Wiley.

Goldstein, S.J., and S.B. Jacobsen. 1988. REE in the Great Whale River estuary, Northwest Quebec, *Earth Planet. Sci. Lett.* **88**:241–252.

Guieu C., J.-M. Martin, A. J. Thomas, and F. Elbaz-Poulichet. 1990. Atmospheric versus river input of metals to the Gulf of Lions: total concentration, partitioning and fluxes. *Mar. Poll. Bull.* in press.

Hekstra, G.P. 1986. Will climatic changes flood the Netherlands? Effects on agriculture, land use and well-being. *Ambio.* **15**(4):316–326.

Howard, A.G., M.H. Arbad-Zavar, and S. Apte. 1984. The behaviour of dissolved arsenic in the estuary of the river Beaulieu. *Est. Coast. Shelf Sci.* **19**:493–504.

Huang, W.W., J.M. Martin, P. Seyler, J. Zhang, and X.M. Zhong. 1988. Distribution and behaviour of arsenic in the Huang He (Yellow River) Estuary and Bohai Sea. *Marine Chem.* **25**:75–91.

Ivanov, K.K. 1955. Sedimentation of suspended matter near the mouth of the River Kura (in Russian). *Proc. Gov. Ocean. Inst.* **28**(40):131–136.

Jeandel, C., J.M. Martin, and A.J. Thomas. 1981. Les radionucléides artificiels dans les estuaires francais. In: Impacts of radionuclide releases into the marine environment, SM 248/123, 15–32. Vienna: I.A.E.A.

Johnson, D.L., and R.M. Burke. 1978. Biological mediation of chemical speciation. II: arsenate reduction during phytoplankton blooms. *Chemosphere* **8**:645–648.

Klinkhammer, G.P., and M.L. Bender. 1981. Trace metal distributions in the Hudson River Estuary. *Est. Coast. Shelf Sci.* **12**:629–643.

Knauss, K., and T.L. Ku. 1983. The elemental composition and decay series radionuclide content of plankton from the East Pacific. *Chem. Geol.* **39**:125–145.

Koczy, F.F. 1954. Geochemical balance of uranium and thorium in the hydrosphere. In: Nuclear Geology, ed. H. Faul, pp. 120–127. New York: Wiley.

Kremling, K. 1983. Trace metal fronts in European shelf waters. *Nature* **303**:225–227.

Kremling, K. 1985. The distribution of cadmium, copper, nickel, manganese, and aluminum in surface waters of the open Atlantic and European shelf area. *Deep-Sea Res.* **32**:531–555.

Langston, W.J. 1983. The behaviour of arsenic in selected United Kingdom estuaries. *Can. J. Fish. Aquat. Sci.* **40**:143–150.

Lerman, A., M. Bernhard, B. Bolin, C.C. Delwiche, D.H. Ehhalt, et al. 1977. Fossil Fuel Burning: its effects on the biosphere and biogeochemical cycles. In: Global Chemical Cycles and their Alteration by Man, ed. W. Stumm, pp. 275–289. Dahlem Konferenzen. Berlin: Abakon Verlag.

Liu, K.K. 1979. Geochemistry of inorganic nitrogen compounds in two marine environments: the Santa Barbara Basin and the ocean off Peru. Ph.D. diss., University of California, Los Angeles.

L'vovich, M.I. 1987. Ecological foundations of global water resources conservation. In: Resources and World Development, ed. D.J. McLaren and B.J. Skinner, pp. 831–839. Dahlem Konferenzen PC6. Chichester: Wiley.

McKee, B.D., B.J. DeMaster, and C.A. Nittrouer. 1987. Uranium geochemistry on the Amazon Shelf: evidence for uranium release from bottom sediments. *Geochim. Cosmochim. Acta* **151**:2779–2786.

Maeda, M., and H.L. Windom. 1982. Behavior of uranium in two estuaries of the southeastern United States. *Marine Chem.* **11**:427–436.

Martin, J.H., and G.A. Knauer. 1973. The elemental composition of plankton. *Geochim. Cosmochim. Acta* **37**:1639–1653.

Martin, J.-M., F. Elbaz-Poulichet, C. Guieu, M.D. Loye-Pilot, and G. Han. 1989. River versus atmospheric input of material to the Mediterranean Sea: an overview. *Marine Chem.* **28**:159–182.

Martin, J.-M., and V.V. Gordeev. 1982. River input to ocean systems: a reassessment. In: Estuarine Processes: An Application to the Tagus Estuary. Proc. UNESCO/IOC/CNA Workshop, pp. 203–240. Paris: UNESCO.

Martin, J.-M., O. Hogdahl, and J.C. Philippot. 1976. Rare earth element supply to the ocean. *J. Geophys. Res.* **81**(18):3119–3124.

Martin, J.-M., M. Meybeck, and M. Pusset, 1978. Uranium behaviour in Zaire estuary. *Neth. J. Sea Res.* **12**:338–344.

Martin, J.-M., V. Nijampurkar, and F. Salvadori. 1978. Uranium and thorium isotope behaviour in estuarine systems. In: Biogeochemistry of Estuarine Sediments. Proc.UNESCO/SCOR Workshop, pp. 111–122. Melreux, Belgium: UNESCO.

Martin, J.-M., and M. Whitfield. 1983. The significance of the river input of chemical elements to the ocean. In: Trace Metals in Seawater, ed. C.S. Wong, E. Boyle, K.W. Bruland, J.D. Burton, and E.D. Goldberg. New York and London: Plenum.

Meade, R.H. 1972. Sources and sinks of suspended matter on continental shelves. In: Shelf Sediment Transport: Process and Pattern, ed. D.J. Swift, D.B. Duane, and O.H. Pilkey, pp. 249–262. Stroudsburg, PA: Hutchinson and Ross, Inc.

Milliman, J.D., and R.H. Meade. 1983. Worldwide delivery of river sediment to the oceans. *J. Geol.* **91**:1–21.

Olsen, C.R., M. Thein, I.L. Larsen, P.D. Lowry, P.J. Mulholland, N.H. Cutshall, J.T. Byrd, and H. Windom. 1989. Plutonium, lead-210 and carbon isotopes in the Savannah estuary: riverborne versus marine sources, *Envir. Sci. Tech.*, **23**(12):1475–1481.

Romankevich, E.A. 1984. Geochemistry of Organic Matter in the Ocean. Berlin, Springer.

Salomons, W. and H.N. Kerdijk. 1986. Cadmium in fresh and estuarine waters. In: Cadmium in the Environment, ed. H. Mislin, and O. Ravera, 144 pp. Stuttgart: Birkhauser Verlag.

Sanders, J.G., and H.L. Windom. 1980. The uptake and reduction of arsenic species by marine algae. *Est. Coast. Marine Sci.* **10**:555–567.

Seitzinger, S.P., S.W. Nixon, and M.E.Q. Pilson. 1984. Denitrification and nitrous oxide production in a coastal marine ecosystem. *Limnol. Ocean* **29**:73–83.

Seyler, P. 1985. Formes chimiques et comportement de l'arsenic en milieu estuarien. Ph.D. diss. University of Paris.

Sholkovitz, E.R., E.A. Boyle, and N.B. Price. 1978. The removal of dissolved humic acids and iron during estuarine mixing. *Earth Planet. Sci. Lett.* **40**:130–136.

Sholkovitz, E.R. and H. Elderfield. 1988. Geochemical cycling of dissolved rare earth elements in Chesapeake Bay. *Glob. Biogeochem. Cyc.* **2**(2):157–176.

Sholkovitz, E.R., and D.R. Mann. 1987. 239, 240 Pu in estuarine and shelf waters of the North Eastern United States. *East. Coast Shelf Sci.* **25**:413–434.

Trefry, J.H. and B.J. Presley. 1976. Trace metal transport from the Mississippi River to the Gulf of Mexico. In: Marine Pollutant Transfer, ed. H.L. Windom and R.A. Duce, pp. 39–76. Lexington, MA: Lexington Books.

Veeh, H.H. 1967. Deposition of uranium from the ocean. *Earth Planet. Sci. Lett.* **3**:145–150.

Waslenchuk, D.G. 1977. The geochemistry of arsenic in the continental shelf environment. Ph.D. diss., Georgia Inst. of Tech., Atlanta, GA.

Waslenchuk, D.G. and H.L. Windom. 1978. Factors controlling the estuarine chemistry of arsenic. *Est. Coast. Marine Sci.* **7**:455–462.

Windom, H.L. 1990. Flux of particulate metals between east coast North American rivers and the North Atlantic Ocean. *Sci. Total Envir.*, in press.

Windom, H.L., K. Beck, and R. Smith. 1971. Transport of trace metals to the Atlantic Ocean by three southeastern rivers. *Southeast Geol.* **12**:169–181.

Windom, H.L., J. Byrd, R. Smith, Jr., M. Hungspreugs, S. Dhamvanij, W. Thumtrakul, and P. Yeats. 1990. Trace metal-nutrient relationships in estuaries. *Marine Chem.*, in press.

Windom, H.L. and T.F. Gross. 1989. Flux of particulate aluminum across the southeastern U.S. continental shelf. *Est. Coast. Shelf. Sci.* **28**:327–338.

Windom, H.L., and R.G. Smith. 1985. Factors influencing the concentration and distribution of trace metals in the South Atlantic Bight. In: Oceanography of the Southeastern U.S.Continental Shelf, ed. L.P. Atkinson, D.W. Menzel, and K.A. Bush, pp. 141–152. Washington, D.C.: AGU.

Windom, H.L., R.G. Smith, and M. Maeda. 1985. The geochemistry of lead in rivers, estuaries and the continental shelf of the southeastern United States. *Marine Chem.* **17**:43–56.

Windom, H.L., R. Smith, Jr., C. Rawlinson, M. Hungspreugs, S. Dharmvanij, and G. Wattayakorn. 1988. Trace metal transport in a tropical estuary. *Marine Chem.* **24**:293–305.

Wollast, R.G. 1983. Interaction in estuaries and coastal waters. In: The Major Biogeochemical Cycles and their Interactions, ed. B. Bolin, and R.B. Cook, pp. 385–410. New York: Wiley.

Wollast, R., G. Billen, and J.C. Duinker. 1979. Behaviour of manganese in the Rhine and Scheldt estuaries: I. Physicochemical aspects. *Est. Coast. Marine Sci.* **9**:161–169.

Yeats, P.A., and J.M. Bewers. 1983. Potential anthropogenic influences on the trace metal distributions in the North Atlantic. *Can. J. Fish. Aquat. Sci.* **40**:124–131.

Flux and Fate of Fluvial Sediment and Water in Coastal Seas

J.D. Milliman

Woods Hole Oceanographic Institution
Woods Hole, Massachusetts 02543, U.S.A.

Abstract. Rivers presently discharge about 35×10^3 km^3 of freshwater and 15×10^9 tons of sediment into the world oceans annually. About 65% of the water and 80% of the sediment comes from southern Asia, Oceania and northeastern South America. Since sea level reached its present position about 6,000 years ago, most fluvial sediment has remained trapped in coastal and inner shelf areas; most riverborne sediment escapes to the deep sea during lower stands of sea level.

Because discharge values can vary dramatically with changes in climate and human land-use, these data are not applicable for forecasting future trends. In the near future, erosion rates probably will continue to increase (particularly in Asia). However, as more dams and barrages are built and utilized, the transfer of both water and sediment to the sea almost certainly will decrease. Uncompensated (or even accelerated) subsidence of low-lying deltas will be one effect of river/sediment diversion, but accentuated coastal erosion may lag considerably. In contrast, effects from the decreased flux of fresh water, such as shrinking mangrove forests and decreased coastal fisheries, probably will be felt immediately.

INTRODUCTION

River discharge is a prime source of nutrients and fresh water to the sea, balancing most of the evaporative loss of water from the sea surface. In terms of sediment input, rivers presently account for approximately 70% of the world's total; this number, of course, can change with changes in climate (e.g., glaciations), volcanic activity, or terrestrial erosion. Rivers also serve as a major conduit for the introduction of pollutants to the ocean. Maintaining or changing the quantity or quality of river discharge, therefore, becomes critically important as human beings try to cope with existing and future climatic and environmental changes.

Ocean Margin Processes in Global Change
Edited by R.F.C. Mantoura, J.-M. Martin and R. Wollast

Predicting changes in fluvial processes and impacts, however, assumes knowledge of river flow, transport and discharge to the sea as well as defining the fate of the discharged products. While data for water discharge of many mid-sized and large rivers are available (largely in the grey literature), sediment loads are poorly documented except for moderate-size rivers in Europe and North America and a few very large rivers in Asia and South America. As most readers are aware, the fate of fluvially derived nutrients and sediments in the marine environment is even less documented—not a very good information base for those who manage river basins and coastal areas.

In this chapter, four major topics will be discussed. First, river discharge of sediment and water to the sea is reviewed. While earlier papers (e.g., Milliman and Meade 1983; Milliman 1989) emphasized sediment input to the oceans, the storage, evaporation or discharge of river-derived fresh water to the sea is important to marine scientists.

Second, these numbers are subject to both short-term changes (such as annual variation in weather patterns) and longer-term changes, related to human activities (e.g., increased erosion due to farming versus decreased sediment flux to the ocean due to river damming and diversion) and climate change.

Third, it is worthwhile to consider the fate of sediment once it reaches the sea: where does it go and when? The decoupling of river sediment from river water at the land–sea interface means that the pathway and rates at which fluvial sediment reaches the deep sea are quite different from those for river water itself.

Finally, it is worth considering possible future problems as fluvial pollution (particularly in southern Asia rivers) increases, river flow is increasingly diverted, and sea level rises. Coupled with our ignorance about fluvial discharge, a preliminary list of research priorities can be compiled.

WATER FLUX FROM RIVERS

The amount of water discharged by the world rivers to the ocean is estimated to be between 32 (Livingstone 1963) and 37 (Unesco 1978) $\times 10^3$ km^3/yr. I estimated a volume of 33×10^3 km^3/yr by extrapolating known river discharge for specific areas over larger areas (Milliman 1989). For the purposes of this discussion, a value of 35×10^3 km^3 seems a realistic estimate.

Meteorologists often calculate the term "runoff" differently: as [precipitation−evaporation]. Baumgartner and Reichel (1975), for example, calculated the continental runoff to be 39.7×10^3 km^3/yr, balanced by an equal net evaporation from the oceans. It is perhaps not surprising that this runoff value of Baumgartner and Reichel is somewhat higher than values derived from river measurements alone: meteorological data include precipitation and evaporation from the Arctic and Antarctic, areas not used in calculating river discharge. Perhaps more important, at least some of Baumgartner and Reichel's runoff consists of groundwater, much of which would not be reflected in river discharge measurements. Finally, the river discharge estimates cited here are derived by integrating measured discharge from moderate and large rivers. However, although few small rivers on mountainous islands have been documented, they may have larger discharge rates per unit basin area because they discharge directly to the sea (i.e., little loss to evaporation, lake storage or groundwater); if so, the runoff numbers presented here and in other river papers may be too conservative.

Nevertheless, the Baumgartner and Reichel data are interesting to compare with the data from rivers (Table 1). For instance, the strong zonal influence of precipitation and therefore runoff can be seen, with greatest runoff occurring in tropical and subtropical areas. However, this is clearly an oversimplification, as much of the 30°N–30°S zone in Africa (for example) has very low or negative runoff (i.e., evaporation equals or surpasses precipitation). Runoff is also influenced by the monsoonal interaction with mountainous southeast Asia and the adjacent high-standing oceanic islands. In these areas, runoff locally exceeds 3 m, nearly an order of magnitude greater than the worldwide average calculated by Baumgartner and Reichel.

Nowhere is the effect of latitudinal (and therefore climatologic) variations in the water cycle on river discharge better seen than in Asia. Five of the world's largest rivers (in terms of discharge) drain Asia: two in the Eurasian Arctic (Lena and Yenissei) and three in southern Asia (Changjiang, Ganges-Brahmaputra, and Mekong) (Fig. 1). However, in terms of specific discharge (discharge/drainage basin area), the northern rivers are small, generally less

TABLE 1. Comparison of zonal distribution of runoff (1) (precipitation minus evaporation) as calculated by Baumgartner and Reichel (1975) and zonal runoff of major rivers to the sea (2) (data derived from Milliman and Meade 1983).

	Vol. runoff (km^3)	% Land area[1]	% Runoff[1]	Runoff[1] (mm/yr)	River Discharge[2] (mm/yr)
60–90°N	3551	11.6	8.9	207	197
30–60	8252	31.4	20.8	251	399
0–30	12597	24.5	31.8	525	516
0–30	11746	19.6	29.6	503	563
30–60	1567	9.4	3.9	365	
60–90°S	1987	9.4	5.0	141	

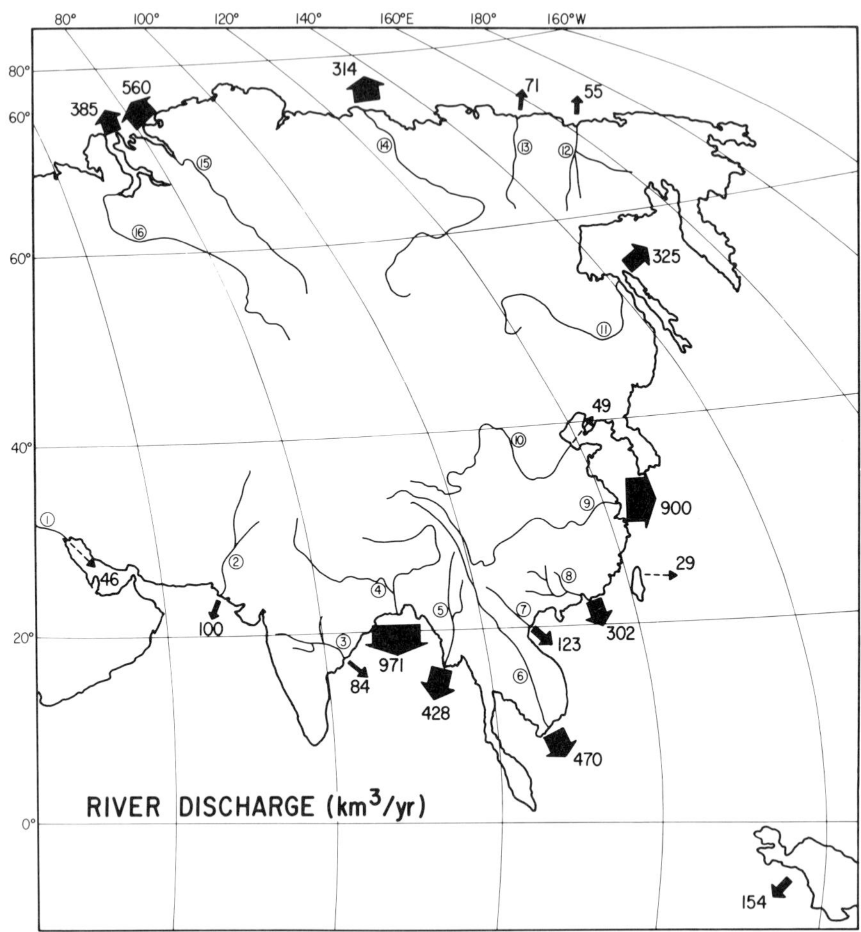

Fig. 1—Annual river water discharge of the 16 largest rivers draining Eurasia. Values are in km^3/yr; arrows are proportional to these values. Most data are from Milliman and Meade (1983). Indus value represents flow prior to damming in the early 1960s (data from Milliman et al. 1984). Numbered rivers are as follows: (1) Tigris-Euphrates, (2) Indus, (3) Godavari, (4) Ganges-Brahmaputra, (5) Irrawaddy, (6) Mekong, (7) Hunghe (Red River), (8) Pearl River, (9) Changjiang (Yangtze River), (10) Haunghe (Yellow River), (11) Amur, (12) Kolyma, (13) Indagirka, (14) Lena, (15) Yenissei, (16) Ob. Also shown are combined river discharges of several rivers from the islands of Taiwan and Papua New Guinea; the reason for showing these numbers becomes obvious in Fig. 2.

than 0.2 m/yr, in accordance with the low net precipitation (Fig. 2). In contrast, most rivers in southern Asia drain smaller basins but experience very high rainfall. Several Asian rivers have specific discharge values in excess of 1 m/yr, and rivers in Taiwan (whose individual discharges are small) exceed 2 m/yr, the result of high rainfall, mountainous terrain, and small river-basin area.

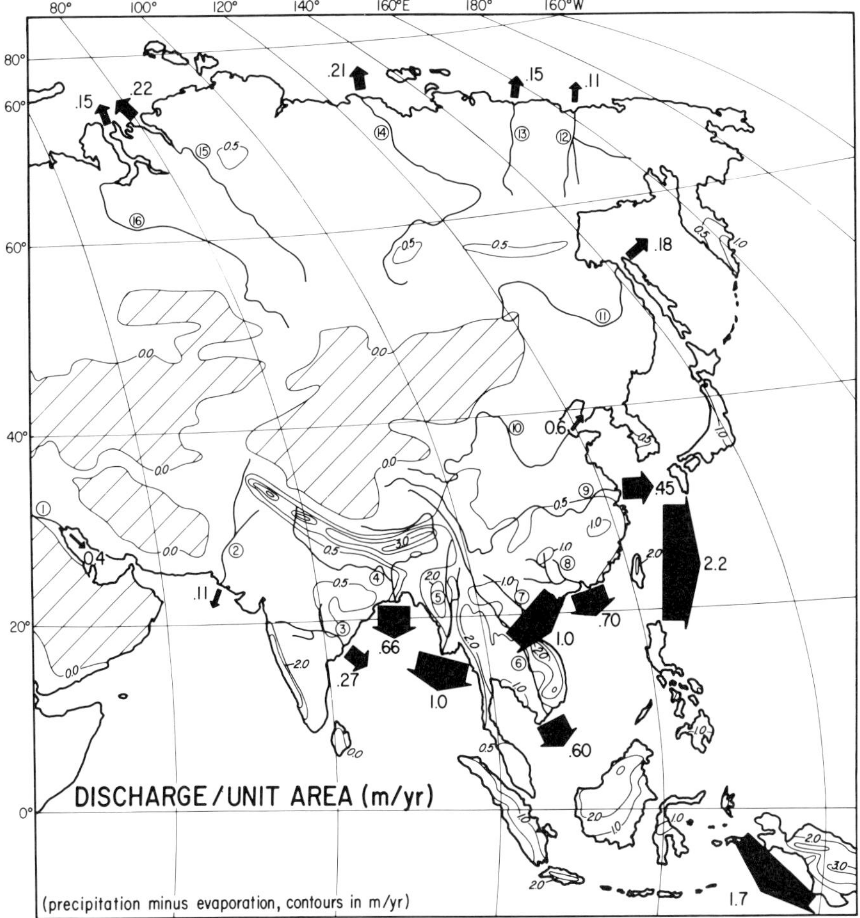

Fig. 2—Discharge of river water to the oceans normalized for drainage basin area. Also shown are (precipitation minus evaporation) values (after Baumgartner and Reichel 1975). Both parameters are expressed in m/yr. Note that the normalized water discharge for southern Asian rivers is much higher than for the large areas in the Eurasian Arctic and that the Taiwan and New Guinea rivers have even higher values, about equal to the amount of net precipitation.

In terms of worldwide water discharge, the world's ten largest rivers account for about 38% of the total fluvial water entering the ocean, slightly greater than their combined percentage of drainage basin area (Table 2). The Amazon River alone contributes more than 15% of the world total (about 6300 km^3/yr), more than the combined total of the next 8 largest rivers! Not surprisingly, tropical areas with heavy rainfall—specifically southern Asia, Oceania and northeastern South America—are the prime contributors, about 65% of the global total (Fig. 3). In contrast, with the exception of the Zaire and Niger Rivers, Africa contributes practically no fluvial water to the oceans (the Nile River being effectively dammed).

SEDIMENT FLUX FROM RIVERS

Data pertaining to fluvial sediment loads provide rates of terrestrial denudation and flux to the ocean. To obtain denudation rates in a river basin, geologists usually use values from tributary rivers, where sediment loads generally are highest. Some riverborne sediment settles out (i.e., is "stored") downstream in the form of fans, floodplains or lake/reservoir deposits (see Meade 1988). Holeman (1981), for instance, estimated that only about 5% of the soil eroded in the continental U.S. presently reaches the ocean, meaning a 20-fold difference between denudation and sediment delivery. For this reason, it is not surprising that fluvial values used to estimate denudation rates are higher (32 and 51 $\times$ 10^9 t/yr; [Gilluly 1955 and Fournier 1960]) compared to estimates of sediment flux to the sea (12 to 18 $\times$ 10^9 t/yr [Lisitzin 1972; Holeman 1968; Milliman and Meade 1983]).

Early estimates of sediment discharge by Strakov (e.g., Lisitzin 1972) and Holeman (1968) contain outdated and inaccurate data. A more recent compilation by Milliman and Meade (1983) updates the earlier estimates, and Meade (1989) lists new estimates of sediment loads for some of the larger rivers. The present fluvial suspended sediment flux to the world oceans is approximately 13.5 $\times$ 10^9 t/yr, with an additional 1–2 $\times$ 10^9 t/yr coming from flood discharge and bed load sediment (Milliman and Meade 1983); 15 $\times$ 10^9 t/yr is a reasonable estimate. With a total drainage basin area accessible to the ocean of about 10^8 km^2, the average sediment yield (sediment discharge/unit area of drainage basin) is 150 t/km^2/yr; the significance of this number becomes apparent in the following paragraphs.

Southern Asia and Oceania contribute about 70% of the world flux, although they account for only about 15% of the land area draining into the oceans (Figs. 4 and 5; Table 2); northeastern South America (i.e., the Amazon, Orinoco, and Magdalena rivers) contributes another 11%. In contrast, rivers draining the Eurasian Arctic, a basin area similar in size to southern Asia and Oceania (combined), contributes about two orders of magnitude less sediment. As with water discharge, the tropics exhibit the

TABLE 2. Ranking of the world's major rivers in terms of drainage area, discharge and sediment load. Data from Milliman and Meade (1983), modified slightly with new compilations by J.C. Kammerer (written communication). The Nile's drainage basin is listed even though it presently is nearly completely dammed.

A. Drainage Basin Area

	River	$\times 10^6$ km^2
1.	Amazon	6.15
2.	Zaire	3.72
3.	Mississippi	3.27
4.	(Nile)	(3.03)
5.	La Plata	2.83
6.	Yenissei	2.58
7.	Ob	2.99
8.	Lena	2.49
9.	Changjiang	1.81
10.	Amur	1.86
11.	MacKenzie	1.81
	Total	29.24 (29% of world total)

B. Average Annual Water Discharge

	River	km^3/yr
1.	Amazon	6300
2.	Zaire	1250
3.	Orinoco	1100
4.	Ganges/Brahmaputra	971
5.	Changjiang	921
6.	Mississippi	580
7.	Yenissei	560
8.	Lena	514
9.	Plata	470
10.	Mekong	470
	Total	13115 (40% of world total)

C. Average Annual Suspended Load

	River	$\times 10^6$ t/yr
1.	Ganges/Brah.	1670
2.	Huanghe	1080
3.	Amazon	900
4.	Changjiang	478
5.	Irrawaddy	285
6.	Magdalena	220
7.	Mississippi	210
8.	Orinoco	210
9.	Hunghe	160
10.	Mekong	160
	Total	5373 (40% of world total)

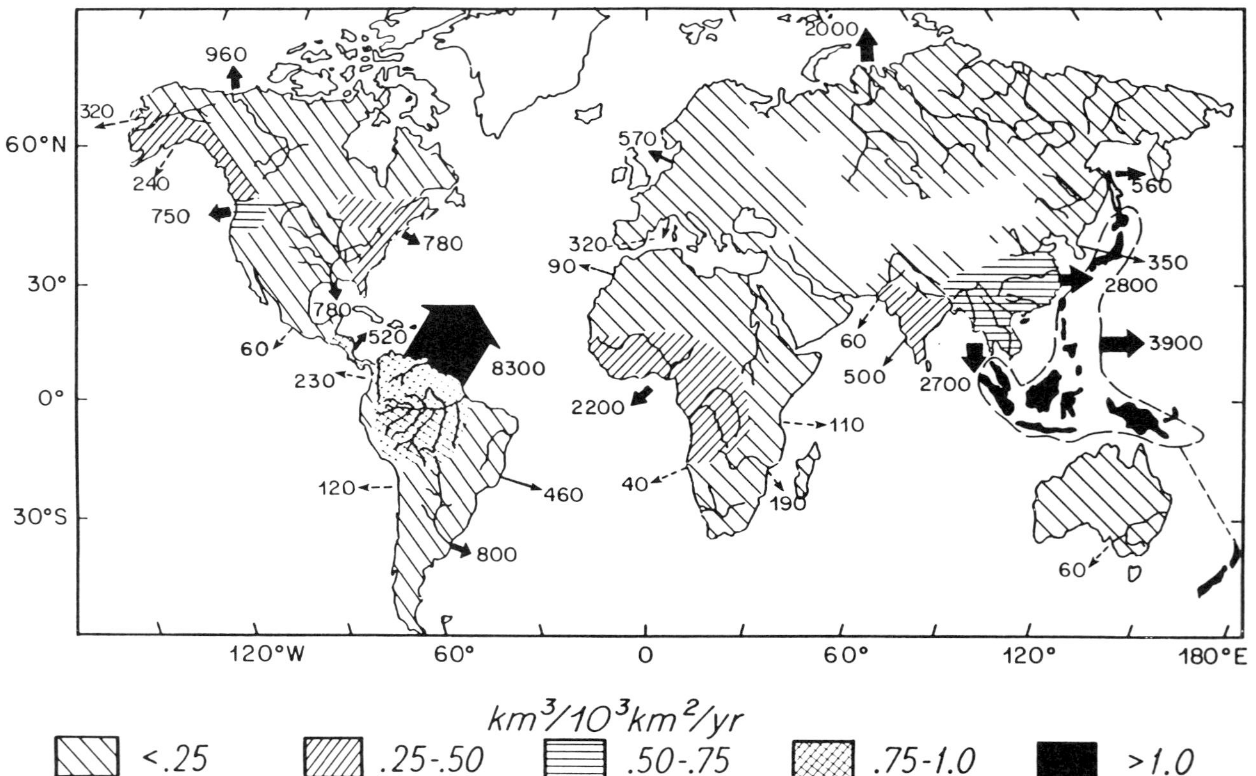

Fig. 3—Annual discharge of fluvial water to the oceans. Numbers in km^3/yr; arrows proportional to the numbers. Most data from Milliman and Meade (1983).

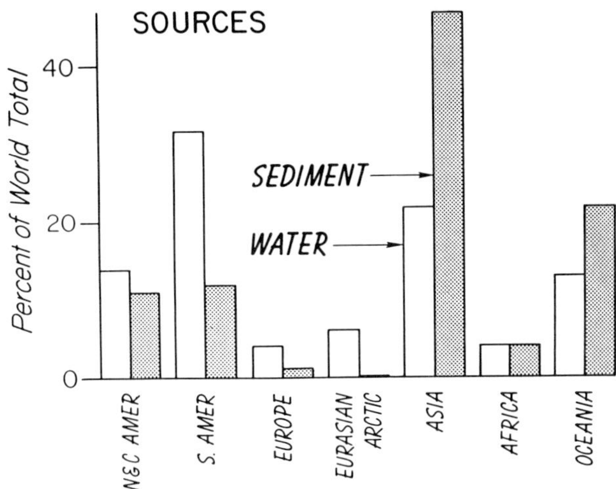

Fig. 4—Land sources for fluvial sediment and water discharged into the ocean.

highest sediment yields, although wet-based glaciers (e.g., southern Alaska and Spitsbergen) have yields in excess of 1000 t/km^2/yr. Although rivers draining semi-arid and arid areas exhibit discontinuous flow patterns, periodic storms can result in yields as great as or greater than those in the tropics.

Smaller drainage basins have less area to store sediments; as a result, the sediment yield of smaller basins increases by as much as 7-fold for each order of magnitude decrease in basin area (Schumm and Hadley 1961; Lopatin 1962). The result is that many rivers draining smaller basins can discharge far more sediment to the sea than a single river draining a much larger basin. In part this explains the very high sediment input from islands in Oceania; for example, as cited in Milliman and Meade (1983), the small rivers draining both Taiwan and the South Island of New Zealand discharge comparable quantities of sediment to the sea as the Mississippi River, even though they have much smaller drainage areas. Not surprisingly, then, rivers draining only 10% of the world's drainage basins account for more than 60% of the sediment discharge to the ocean (Table 3).

Again, it is interesting to note the variation in sediment discharge in the Asian continent, in relation both to rainfall (see Fig. 2) and topography (Fig. 6). With the exception of the Kalyma River, sediment yield (river load/drainage basin area) in Eurasian Arctic rivers is extremely low: 4–9 t/yr, the result of low rainfall and flat topography. In contrast, rivers draining the high mountains and plateaus in southern Asia (with high rainfalls and easily erodable sediments—see below) have sediment yields considerably higher than 100 and locally (Ganges-Brahmaputra) approaching 1000. The

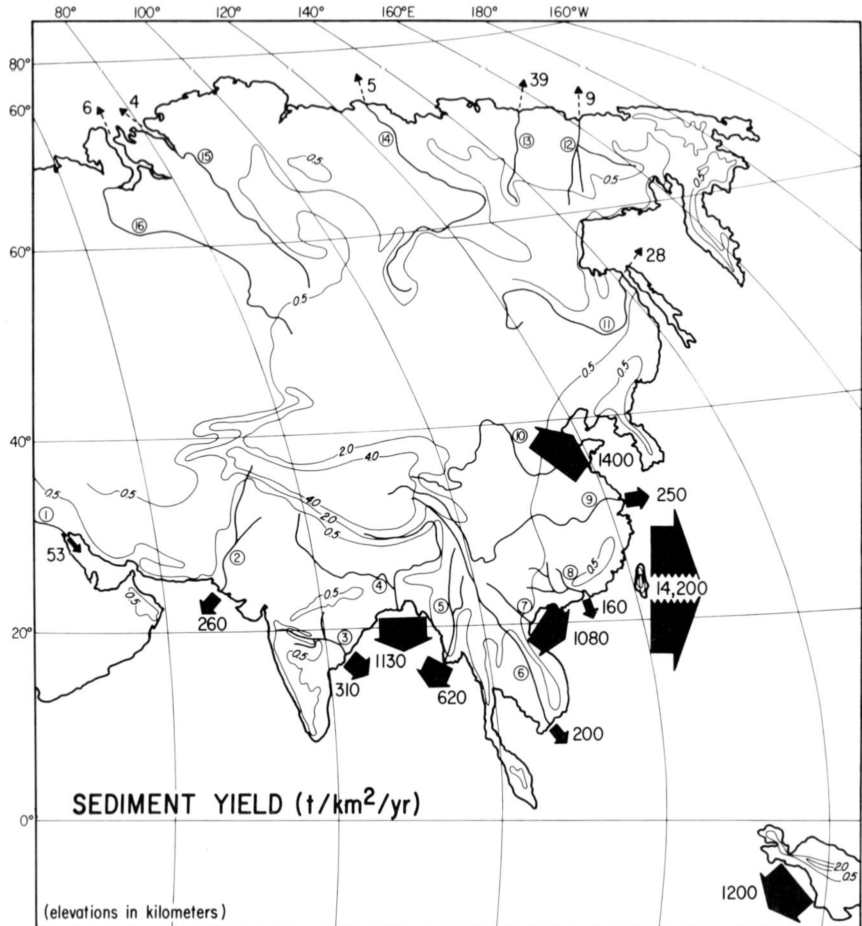

Fig. 5—Sediment yield (annual river sediment load normalized for drainage basin area, expressed as t/km^2/yr) for the 16 largest rivers in Eurasia. Arrows are proportional to the values. Note that the sediment yield for Taiwan is so large (14, 200) that the arrow is broken. Also shown are topographic contours (in km) to illustrate the close relationship between high topography and high erosion rates/sediment yield.

Huanghe (Yellow River) drains relatively low-lying terrain with low rainfall, but the Pleistocene loess hills in northern China provide most of the sediment; sediment yield is 1400 t/yr. High values have been documented in rivers in New Guinea, and similarly high values are assumed for many of the high-standing islands in Oceania. Taiwan is perhaps the ultimate example of high sediment yield (greater than 14,000 t/km^2/yr, two orders of magnitude greater than the world mean) due to its intensely cultivated lands.

TABLE 3. Distribution of sediment yields for all river basins draining into the ocean. Data from Milliman and Meade (1983).

Sediment Yield ($t/km^2/yr$)	% Land Area	% Annual Sediment Discharge
5–10	23	1
10–50	31	5
50–100	15	7
100–500	21	24
500–1000	6	34
> 1000	4	30

TABLE 4. Annual sediment loads and highest daily loads of the Santa Clara River (at Montalvo, CA). Annual loads over the 18-year interval in which the river was gauged varied by three orders of magnitude, indicating the difficulty in discussing "mean annual sediment loads" for many small rivers. Note also that the highest daily load, usually the result of a flood, averages about half the annual load, and in some years (e.g. 1974, 1977) reaches or exceeds 70%. Data courtesy of R.H. Meade (1989).

Year	Annual Load ($\times 10^6$ t/y)	Highest Daily Load ($\times 10^6$ t/y)	Percentage
1968	0.076	0.034	45
1969	50.5	20.4	40
1970	0.66	0.28	42
1971	2.4	1.58	70
1972	0.48	0.23	48
1973	4.3	2.67	62
1974	0.49	0.33	70
1975	0.54	0.26	48
1976	0.068	0.044	66
1977	0.62	0.47	77
1978	29.2	11.8	40
1979	2.26	0.89	39
1980	9.8	3.2	33
1981	0.12	0.035	30
1982	0.18		
1983	17.9	7.1	40
1984	0.22	0.10	48
1985	0.089	0.058	65
mean	6.02		51

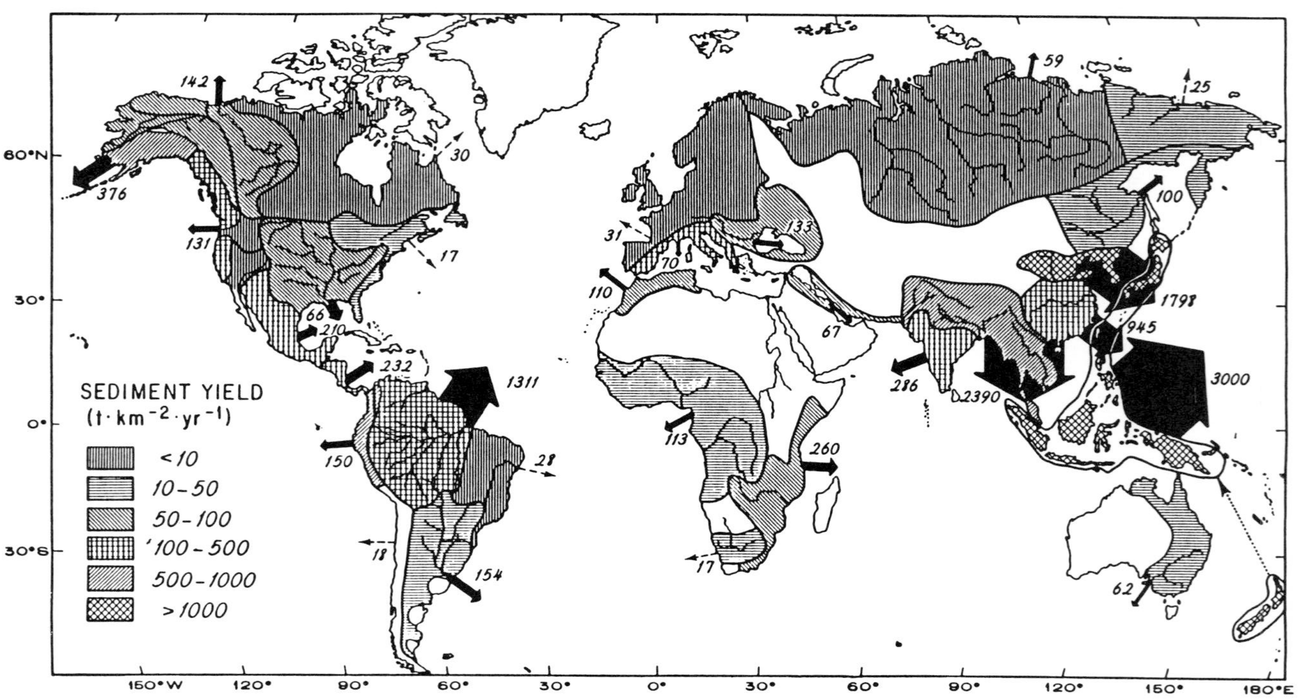

Fig. 6—Annual fluvial sediment flux from large drainage basin areas to the oceans. Numbers in millions of tons (per year); arrows proportional to the numbers. From Milliman and Meade (1983).

The reason for these very high yields relates directly to the mountainous terrain, heavy rainfall and very small drainage basins, which deny the accumulation of riverborne sediment.

VERACITY OF DATA AND FUTURE TRENDS

(Much of this discussion comes from Milliman 1990.)

With the exception of the Chinese rivers, most large rivers in Asia and South America are poorly documented whereas many rivers in North America and Europe are well studied but have small sediment loads. Some large rivers (the Ganges-Brahmaputra, Irrawaddy and la Plata being obvious examples) have scant data basis that may not be seasonally averaged. As a result, estimates of the sediment loads for many large rivers (as well as most small ones) are exactly that—estimates.

The very high sediment loads in modern Asian rivers reflect a considerable influence from human activities, particularly poor agricultural practices in conserving soil. The Yellow River, for example, transports an order of magnitude more sediment than it did prior to widespread cultivation of the loess plateau in northern China (Milliman et al. 1987). Assuming that present-day Asian and Oceania river loads are five times greater than before humans began deforestation and farming, the worldwide quantity of fluvial sediment reaching the ocean 2500 years ago might have been less than 7×10^9 t/yr, and the percentage from Asia and Oceania would have accounted for a correspondingly smaller fraction of the world's total than they do at present.

Moreover, sediment loads for rivers are changing continually. On one hand, erosion rates in many areas of the world have increased in response to poor soil conservation. Deforestation of Nepal has increased dramatically the sediment loads in the upper reaches of the Ganges River, for example. On the other hand, more rivers are being diverted or dammed for irrigation, flood control, and hydroelectric power. Several examples of damming are well known, e.g., the complete cessation of sediment flux from the Colorado and Nile rivers. The Rhone carries only about 5% of the load it did in the 19th century (Corre et al. 1990), and the Indus River discharges less than 20% of the load it did before construction of barrages in the late 1940s (Milliman et al. 1984). Values for many other Asian and South American rivers may be similarly misleading, either because of sparse or erroneous data or because new river diversions have changed the loads dramatically since the cited data were obtained.

While most long-term data come from larger rivers, the potential impact of sediment flux from smaller rivers should not be minimized. The best examples are the many rivers draining oceanic islands as well as rivers draining glaciated mountains in southern Alaska. In Prince William Sound, for example, the many small rivers (individual basins no larger than 100 to 500 km^2) collectively may discharge as much or more sediment as the adjacent Copper River (drainage basin 60,000 km^2). The fact that these smaller basins also are more likely to feel the effect of a cataclysmic event (such as an earthquake, mudslide or flood) only accentuates the potential impact of small rivers. If these rivers were studied in sufficient detail, it might require a drastic upward revision of sediment escaping to the sea.

Catastrophic events in small drainage basins, particularly large floods, can increase river sediment loads by one to several orders of magnitude. One of the best examples is the Santa Clara River (southern California), in which three floods (representing a total of 6 days) in the span of 18 years accounted for nearly 60% of the total sediment transport measured during that period (Meade et al. 1989; Table 4). The impact of peak events and interannual variation, however, is not necessarily restricted to small rivers. Over the past 40 years the Changjiang (Yangtze River), with a very large basin area that modulates local weather patterns, has seen only a twofold range in annual sediment load, whereas the Huanghe, with its smaller drainage area and more changeable weather patterns, has experienced annual variations that exceed an order of magnitude (Fig. 7). In most large rivers, the drainage basin may be sufficiently large to modulate events such that a major flood in one part of

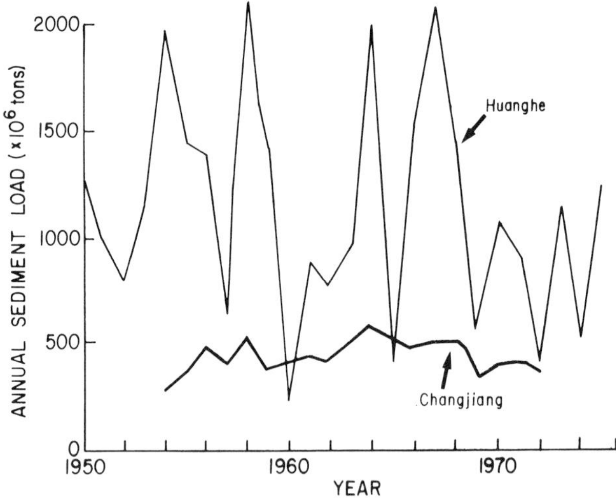

Fig. 7—Contrast of sediment discharge from the Huanghe and Changjiang. Variable annual weather patterns (i.e., precipitation) in large part explain the large annual variation in sediment load discharging from the smaller Huanghe basin. In contrast, the Changjiang's basin is more than twice as large as the Huanghe, such that weather changes are modulated; as a result, sediment loads vary only slightly year to year.

the basin would have relatively little effect on the sediment discharge of the entire river; therefore, very long records may not be required, although, ironically, it is these rivers for which we usually have the longest records.

The numbers presented here therefore represent the world condition as of about 1980 and, in some instances (where data are poor or old) they don't even do that. The irony is that while erosion may be increasing worldwide, sediment delivery may be decreasing. Thus, data for some rivers are unrealistically high and other estimates are low. One can assume that within the next 50 years many large rivers will have gone the way of the Colorado, Nile and Indus.

FATE OF FLUVIAL SEDIMENT AND NUTRIENTS

There has been considerable debate about the fate of fluvial sediment once it reaches the sea. Aggregation (a term I favor rather than "flocculation," which conveys an electrochemical process and negates biological influence) of particles in estuarine waters results in a rapid decoupling of fresh water from the fluvial sediment. Fresh water moves with surface currents controlled mostly by wind and drift, mixing with oceanic waters as it moves. Sediment remains mixed within the water column only as long as turbulent mixing remains active; otherwise it quickly settles through the water column, where it can follow an entirely different transport path, often moving onshore in response to estuarine circulation.

Once sufficient sediment has settled, primary producers (mainly diatoms in coastal waters) respond to increased light levels and the nutrient-rich medium to create biological blooms. In the Amazon River, this happens when terrigenous sediment concentrations fall below about 10 mg/l (Milliman and Boyle 1975; DeMaster et al. 1983). The turbid plumes seen in countless satellite images that many scientists have touted to represent sediment escaping from a river mouth, in fact, often represent (phyto)plankton blooms, the sediment long since having settled to near-bottom waters.

Meade (1972) pointed out that little sediment carried by rivers draining the eastern U.S. actually reaches the ocean, most of it being trapped within large estuaries, most notably Chesapeake and Delaware bays. In fact, the eastern U.S. estuaries actually serve as sinks for sediment moving landward from the continental shelf (Meade 1972). Thus, according to still-accepted belief, not only is river sediment not escaping (except, perhaps, locally during catastrophic storms), but the estuary is filling in at both ends: from rivers on the landward end and from the continental shelf on the seaward end.

McCave (1972) urged caution in extrapolating Meade's observations to other parts of the world, suggesting that more than half of the riverborne sediment discharged to the ocean makes it at least as far as the continental slope, the bulk of it being deposited on the fans and cones off large rivers. McCave offered a number of examples in which sediment clearly escaped estuaries, but a more recent review of sediment transfer from major rivers (Milliman 1990) indicates that for the ten largest rivers (which collectively contribute 36% of the fluvial load reaching the oceans; see Table 1C), probably less than 25–30% escapes the shelf break; rather most remains in coastal areas. For example, both Huanghe and Changjiang sediments remain in the delta and coastal areas in the Yellow and East China Seas (Milliman et al. 1987; Bornhold et al. 1986); only about 5–15% apparently escapes beyond the inner shelf, and much less escapes the shelf (Milliman et al. 1987).

Of the ten rivers, only the Mississippi and the Ganges-Brahmaputra at present appear to discharge significant percentages of their sediments beyond the shelf break. The case of the Ganges-Brahmaputra sediment, however, is problematical: it may bypass the shelf and be deposited directly onto the adjacent slope, rise and fan (e.g., Kuehl et al. 1989), but this conflicts with the observations of Emmel and Curray (1985), who concluded that the deep-sea fan is presently sediment-starved. If little sediment does escape the shelf break, then another possible sediment sink could be the outer (subaerial) Bengal Delta, particularly if subsidence rates are sufficiently high.

As a general rule, then, we can speculate that much of the sediment carried by rivers discharging onto narrow shelves may make its way to the deep sea, whereas fluvial sediment discharging onto broad shelves probably remains trapped within the coastal zone and inner shelf. Following this scenario, much of the riverborne sediment draining into the Pacific and northwestern Indian oceans may bypass the shelf; the obvious exceptions are the Malaysia and Indonesia rivers that empty into the shallow Sunda Sea (water depths generally less than 100 m). In contrast, most of the sediment carried by rivers draining into the Atlantic Ocean presently remains in the large estuaries or on the broad shelves that characterize this ocean.

It is only during lower stands of sea level, when most continental shelves are subaerially exposed, that terrigenous sediment ubiquitously escapes directly to the deep sea (Milliman 1990). Working at much longer time scales and with a much more extensive data base, seismic stratigraphers have reached similar conclusions (e.g., Vail et al. 1977). Sediment flux to and beyond the shelf break, of course, was not continuous during lows stands of sea level, at least not at higher latitudes. Rather, it probably happened in short bursts, e.g., rapid melting events. The sediment wedge on the outer shelf and uppermost slope off northern New Jersey (on the Hudson Apron), for instance, is over 60 m thick; it is thought to have been deposited during one or several late Pleistocene deglaciation events (Milliman et al. 1990), each lasting perhaps only a few tens or hundreds of years.

FUTURE CHANGES

Rivers and coastal areas are under increasing environmental stress at a time when engineering projects and possible climatic change from greenhouse warming threaten even greater stress. With the projected climatic warming during the next century, higher rates of evapotranspiration and (locally) altered rainfall could lead to higher rates of erosion and thus greater input of fluvial sediment to the sea (LeHouérou 1990). As developing countries (wherein lie most of the major rivers; see Table 1) continue their economic development, however, there will be an increased demand for "river control and utilization"—that is, diversion projects in the form of dams and barrages. Increasing hydroelectric power, maximizing agricultural use of the waters (by irrigation) and minimizing the impact of floods or droughts are obvious reasons to consider such projects. Concommittent with these trends, however, are (a) escalating population densities along river courses and (particularly) at river mouths, (b) accelerating upland erosion rates due to poor agricultural practices and increasing deforestation, and (c) partly a result of points (a) and (b) increasing levels of pollution within the water.

Perhaps nowhere will these environmental impacts be greater than in the deltaic countries of southern Asia: distributaries for some of the world's largest rivers (Indus, Ganges-Brahmaputra, Irrawaddy, Mekong, Pearl, Changjiang) and sites for some of the large urban areas (Calcutta, Dhaka, Rangoon, Bangkok, Guangzhou, Shanghai). Although clearly oversimplified, one can predict a gloomy yet logical sequence of events: increased river diversion; decreased flow (particularly during floods); less overtopping of banks; less sediment accumulating on the (subsiding) delta; increased shoreline erosion and decreased mangrove (and coastal biological) productivity; greater local sea-level rise because of eustatic sea-level rise and the resulting uncompensated subsidence; downstream waters become increasingly polluted and saltwater intrusion contaminates shallower water supplies; greater need for groundwater as rivers are dammed upstream; accelerated subsidence rates as groundwater removal causes accelerated soil compaction.

Perhaps the most rapid and drastic example of river diversion was the Huanghe, whose river path changed during floods in 1854, the mouth shifting from the Yellow Sea south of the Shandong Peninsula to the Gulf of Bohai, 300 km to the north. While the delta front immediately around the river mouth eroded quickly, most of the coastline continued to accrete. Erosion has become widespread only in recent years. Presumably, this lag in delta-wide erosion reflects the time for nearshore waters to reach a new wave-base equilibrium. Evidence of this is seen in the deepening of offshore waters (by 1–3 m) between the mid 1800s and the mid 1900s.

In contrast, diverting or damming a river can have drastic and immediate consequences on the biological environment. In the Mediterranean waters of the Nile Delta, the anchovie industry essentially died the year (Wahby and Bishara 1981) following the completion of the Aswan Dam and the effective cessation of nutrient flux from the Nile. Even a slight increase in the salinity of coastal waters can seriously affect mangrove forests, an incredibly productive ecosystem that serves also as a fisheries hatchery and nursery as well as offering protection against coastal erosion (see, for example, Ormond 1985).

It is not too far fetched to say that within the next 20 years environmental conditions on the Nile Delta will have followed the sequential scenario proposed above. Already the Aswan Dam (which has greatly increased hydroelectric power and most assuredly mitigated the worst effects of the recent North African drought) has decreased freshwater discharge to the Mediterranean Sea to a trickle (discharge at less than 3 km^3/yr, compared to more than 100 km^3/yr prior to construction of the dam). Offshore productivity has been reduced and locally, coastal erosion has been as great as hundreds of meters per year (Frihy 1988; Smith and Abdel-Kadar 1988). The future need to pump groundwaters in the Nile Delta, whose natural subsidence rates are as high as 5 mm/yr (Stanley 1988)—about three times the current rate of sea-level rise—could increase subsidence rates considerably. Using the most pessimistic scenario (maximum sea-level rise, maximum subsidence rates), we (Milliman et al. 1989) calculated that by the year 2100, more than 25% of Egypt's presently habitable land could be inundated by rising sea level. With a national population that is doubling every 20–30 years, such a prediction is gloomy indeed.

Egypt is not alone, however. Nearly half the area of Bangladesh lies at elevations of less than 5 m above sea level. The international cry to stop the disastrous effects of river and coastal flooding may result in the diking of both river banks and coastal areas as well as the damming of the Brahmaputra River itself. Decreased river flow almost certainly would have a deleterious effect on the Sundarban mangrove forest, upon which no less than 30% of the Bangladeshi population presently depends (UN/ESCAP 1986). But as we have little idea of the transfer or fate of fluvial sediment in this delta area, nor do we know rates of subsidence, we are at a loss to predict the extent to which engineering projects would alter the environment, other than to say almost certainly the effects would be negative and quite probably (at least locally) disasterous.

FUTURE RESEARCH DIRECTIONS

A strong argument can be made for new studies that address basic questions for which there are almost immediate environmental, economic, social and political ramifications. These questions include:

1. How is river sediment transferred to the marine environment?
2. Once reaching the estuary/coastal area, what is the fate of this sediment?
3. With question (2) in mind, how much of the sediment budget can be explained by accumulation within subsiding subaerial environments, and what would be the effect of river diversion upon local sea-level rise?
4. How will the coastal ecosystem react to changes in the flow/discharge of the adjacent river system?
5. Is it possible to place monetary values on the advantages and disadvantages of river diversion; if so, can these be translated to social and political implications?

While the transfer and fate of fluvial water, nutrients and sediment continue to be subjects of intense academic interest to a wide diversity of research scientists, practical implications to these studies should not be minimized. For those altruistic marine scientists, the capacity to relate basic science to real human needs is clear. For those of us less altruistic, these economic, social and political needs ultimately might translate into an effective rationale for funding the study of interesting natural systems.

Acknowledgements. Many of the ideas expressed in this paper have resulted from collaborative research and discussions with colleagues at WHOI and other institutes, particularly in the People's Republic of China. Support for much of the data collection and contemplation of their significance was provided by the Office of Naval Research and the National Science Foundation. R.H. Meade gave his usual insightful views that helped considerably in preparing the final version of this paper. D.A. Ross, F. Gable and J. Blanton also offered valuable comments. WHOI Contribution No. 7266.

REFERENCES

Baumgartner, A., and E. Reichel. 1975. The World Water Balance. Elsevier Publ., Amsterdam, 179 p.

Bornhold, B.D. et al. 1986. Sedimentary framework of the modern Huanghe (Yellow River) delta. *Geo-Marine Letters* **6**:77–83.

Corre, J.J. et al. 1990. Implication des changements climatiques. Etude de case le Golfe du Lion (France). In: Implications of Expected Climate Changes in the Mediterranean Region, ed. J.D. Milliman, in press. Oxford: Pergamon Press.

DeMaster, D.J., G.B. Knapp, and C.A. Nittrouer. 1983. Biological uptake and accumulation of silica on the Amazon continental shelf. *Geochim. Cosmochim. Acta* **47**:1713–1724.

Emmel, F.J., and J.R. Curray. 1985. Bengal Fan, Indian Ocean. In: Submarine Fans and Related Turbite Systems, ed. A.H. Bouma, W.R. Normark, and N.E. Barnes, pp. 107–112. New York: Springer-Verlag.

Fournier, F. 1960. Climat et erosion. Paris, Presses Univ. de France.

Frihy, O.E. 1988. Nile Delta shoreline changes: Aerial photographic study of a 28-year period. *J. Coast. Res.* **4**:597–606.

Gilluly, J. 1955. Geologic contrasts between continents and ocean basins. *Geol. Soc. Amer. Spec. Paper 62*, 7–18.

Holeman, J.N. 1968. Sediment yield of major rivers of the world. *Water Resources Res.* **4**:737–747.

Holeman, J.N. 1981. The national erosion inventory of the Soil Conservation Service, U.S. Dept. Agriculture, 1977–79. In: Erosion and sediment transport measurement. *Int. Assoc. Hydrol. Serv.* **133**: 315–319.

Kuehl, S.A., Hariu, T.M., and Moore, W.S. 1989. Shelf sedimentation off the Ganges-Brahmaputra River system: Evidence for sediment bypassing to the Bengal Fan. *Geology* **17**:1132–1135.

LeHouérou, H.N. 1990. Vegetation and land-use in the Mediterranean Basin by the year 2050: A prospective study. In: Implications of Expected Climate Changes in the Mediterranean Region, ed. J.D. Milliman, in press. Oxford: Pergamon Press.

Lisitzin, A.P. 1972. Sedimentation in the world ocean. Soc. Econ. Paleont. Mineral Spec. Publ. 17, 218 p.

Livingstone, D.A. 1963. Chemical composition of rivers and lakes. US Geol Survey Prof. Paper 440-G, 64 p.

Lopatin, G.W. 1962. Intensität der Wassererosion auf dem Gebiet der Ud.S.S.R. *Int. Assoc. Sci. Hydrology* **59**:43–51.

McCave, I.N. 1972. Transport and escape of fine-grained sediment from shelf areas. In: Shelf Sediment Transport: Processes and Patterns, ed. D.J.P. Swift, D.B. Duane, and O.H. Pilkey, pp. 225–247. Stroudsburg, PA: Dowden, Hutchinson and Ross Inc.

Meade, R.H. 1972. Sources and sinks of suspended matter on continental shelves. In: Shelf Sediment Transport: Processes and Patterns, ed. D.J.P. Swift, D.B. Duane, and O.H. Pilkey, pp. 249–262. Stroudsburg, USA: Dowden, Hutchinson and Ross. Inc.

Meade, R.H. 1988. Movement and storage of sediment in river systems. In: Physical and Chemical Weathering in Geochemical Cycles, ed. A. Lehrman, and M. Meybeck, pp. 165–179. Dordrecht, Holland: Kluwer Acad. Publ.

Meade, R.H. 1989. River-sediment inputs to major deltas. In: Rising Sea Level and Subsiding Coasts, ed. J.D. Milliman, and S. Sabhasri, in press. SCOPE series.

Meade, R.H., T.R. Yuzuk, and D.J. Day. 1989. Movement and storage of sediment in rivers of the United States and Canada. In: The Geology of North America, Surface Water Hydrogeology. ed. W.G. Wolman and H.C. Riggs, in press. Boulder (CO): Geol. Soc. Amer.

Milliman, J.D. 1989. River discharge of water and sediment to the oceans: Variations in space and time. In: Facets of Modern Biogeochemistry, ed. V. Ittekkot, S. Kempe, W. Michaelis, and A. Spitzy, pp. 83–90. Heidelberg: Springer-Verlag.

Milliman, J.D. 1990. Discharge of fluvial sediment to the oceans: Global, temporal and anthropogenic implications, in press. Washington, DC: NAS/NRC Press.

Milliman, J.D., and E.D. Boyle. 1975. Biological uptake of dissolved silica in the Amazon River estuary. *Science* **189**:995–997.

Milliman, J.D., J.M. Broadus, and F. Gable. 1989. Environmental and economic implications of rising sea level and subsiding deltas: The Nile and Bengal examples. *Ambio* **18**:340–345.

Milliman, J.D., J.Z. Zhuang, A.C. Li, and J. Ewing. 1990. Late Quaternary sedimentation on the outer and middle New Jersey continental shelf: Result of two local deglaciations? *J. Geol.* **98**: 966–976.

Milliman, J.D., and R.H. Meade. 1983. World-wide delivery of river sediment to the oceans. *J. Geol.* **91**:1–21.

Milliman, J.D., Y.S. Qin, M.E. Ren, and Y. Saito. 1987. Man's influence on the erosion and transport of sediment by Asian rivers: The Yellow River (Huanghe) example. *J. Geol.* **95**:751–762.

Milliman, J.D., G.S. Quraishee, and M.N.A. Beg. 1984. Sediment discharge from the Indus River to the ocean: Past, present and future. In: Marine Geology and Oceanography of the Arabian sea, ed. B.U. Haq, and J.D. Milliman, pp. 65–70. New York: van Nostrand Reinhold Co.

Ormond, R. 1985. Status of critical marine habits in the Indian Ocean. IOC/Unesco Workshop on Regional Cooperation in Marine Science in the Central Indian Ocean and Adjacent Seas and Gulfs, IOC (Paris) Workshop Rept. 37 (supplement), 167–193.

Schumm, S.A., and R.F. Hadley. 1961. Progress in the application of landform analysis in studies of semiarid erosion. U.S. Geol. Survey Circular 437, 14 p.

Smith, S.E., and A. Abdel-Kadar. 1988. Coastal erosion along the Egyptian Delta. *J. Coast Res.* **4**:245–255.

Stanley, D.J. 1988. Subsidence in the northeastern Nile Delta: Rapid rates, possible causes and consequences. *Science* **240**:497–500.

UN/ESCAP 1986. Coastal environmental management plan for Bangladesh. Unpubl. Draft Manuscript. Bangkok.

Unesco 1978. World register of rivers discharging into the oceans. Unpubl. ms.

Vail, P.R. et al. 1977. Seismic stratigraphy and glocal changes in sea level (a series of papers). Amer. Assoc. Petroleum Geol. Memoir 26.

Wahby, S.D., and N.F. Bishara 1981. The effects of the River Nile of Mediterranean water before and after the construction of the High Dam at Aswan. In Proceedings of a Review Workshop on River Input to Ocean Systems, Unesco (Paris), 311–318.

Chemical Exchange at the Air–Coastal Sea Interface

R.A. Duce

Graduate School of Oceanography
University of Rhode Island
Narragansett, RI 02882, U.S.A.

Abstract. There is increasing recognition that the atmosphere can be an effective transport path for chemicals entering the coastal zone. Both wet and dry deposition processes can be significant, although our ability to measure directly dry deposition fluxes of either particles or gases is primitive at best. Estimates from some of the semi-enclosed seas in Europe as well as along the east coast of North America indicate that in some regions atmospheric input can be similar to or greater than that from rivers, sewage, dumping, etc. for certain trace metals, and particularly for a number of synthetic organic chemicals. Atmospheric input of nitrogen compounds can also be significant, particularly when atmospheric deposition to an entire watershed, followed by runoff to the coastal zone, is considered. At the present time, atmospheric flux estimates have considerable uncertainty, however, with estimates probably only reliable to within a factor of 2 to 4 at best. Improvements in these estimates are unlikely until adequate methods are developed for the direct measurement of both dry and wet fluxes.

INTRODUCTION

Marine chemistry and marine pollution research have long been concerned with understanding the budgets and processes associated with natural and pollution-derived chemical species in the ocean. Research in this area initially focused on the most obvious input flux to the ocean—that carried by rivers. Because of the many studies of rivers and estuaries, much has been learned over the years about this source of chemicals in the oceans and about the chemical and biological processes associated with these inputs. However, over the past decade it has become apparent that the atmosphere is also a significant pathway for the transport of many natural and pollutant materials from the continents to the oceans, not only far from land in mid-ocean

Ocean Margin Processes in Global Change
Edited by R.F.C. Mantoura, J.-M. Martin and R. Wollast

regions but also to the coastal environment. Many of these substances are natural materials, such as mineral dust and plant residues. Others are pollution derived, including vanadium from heavy fuel oil combustion, lead from the burning of gasoline containing additives, nitrogen species from combustion processes, fertilizers and pesticides from agricultural use, and synthetic organic compounds from a myriad of industrial uses.

Some atmospherically derived substances, such as lead and certain chlorinated hydrocarbons, are potentially detrimental to marine biological systems and water quality. Other substances may actually enhance marine productivity since they are nutrient materials; these include nitrogen species and iron. Atmospheric input may also be important in the natural marine chemical cycles of some substances, such as aluminum and certain rare earth elements. Atmospheric inputs can thus have a variety of impacts on marine systems, and it is necessary to begin to develop an understanding of the magnitude and geographical distribution of the atmospheric fluxes of these materials to the oceanic system.

The Group of Experts on the Scientific Aspects of Marine Pollution (GESAMP 1989) has recently calculated the atmospheric input of several trace species to the global ocean. Substances evaluated included a number of trace metals, mineral matter, nitrogen species, phosphorus, and several synthetic organic species. For the soluble form of metals entering the global ocean, GESAMP estimated that atmospheric inputs are approximately equal to those from rivers for Fe, P, Cu, Ni, and As, while for Zn, Cd, and Pb atmospheric input dominates that from rivers. In the case of nitrogen, it is estimated that the atmosphere is the primary source of new nitrogen to the global ocean. For synthetic organic species, the atmospheric pathway apparently dominates for such species as PCBs, chlordane, DDT, dieldrin, HCB, and HCH.

One might expect the impact of atmospheric materials to be most important in open ocean regions because the atmosphere provides an efficient and rapid transport pathway to these areas. For example, average transport times for Asian substances to reach the central North Pacific gyre range from 5 to 10 days. In contrast, in the coastal marine environment there are many strong local sources for pollutants, including rivers and streams, sewage outfalls, dumping, etc. The magnitude and relative importance of atmospheric input of materials to the coastal zone will be explored in this chapter.

ATMOSPHERIC DEPOSITION PROCESSES

Aerosol Dry Removal

At present there are no satisfactory methods to measure directly aerosol dry deposition to a water surface. Dry removal of aerosol particles to a water surface is often estimated using the dry deposition velocity, v_d. For dry deposition, the flux is given by:

$$F_d = v_d \cdot C_a \tag{1}$$

where F_d is the dry deposition flux (e.g., mg m^{-2} sec^{-1}), v_d is the dry deposition velocity (e.g., m sec^{-1}), and C_a is the concentration of the aerosol substance in air (e.g., mg m^{-3}). In this formulation, v_d incorporates all processes of dry deposition, including diffusion, impaction, and gravitational settling of the particles to the water surface. It is very difficult to parameterize accurately the dry deposition velocity since each of these processes is acting on a particle population, and they are dependent upon a number of factors, including wind speed, particle size, relative humidity, roughness of the ocean surface, near-surface atmospheric stability, etc. Utilizing a variety of information, GESAMP (1989) assumed the following dry deposition velocities during their considerations of aerosol fluxes to the ocean:

Submicrometer aerosol particles:	0.1 cm sec^{-1} $\pm$ a factor of three
Supermicrometer crustal particles, not associated with sea-salt	1.0 cm sec^{-1} $\pm$ a factor of three
Giant sea-salt particles and materials carried by them	3.0 cm sec^{-1} $\pm$ a factor of two

Proper use of Eq. (1) requires that information be available on the particle size distribution of the aerosol particles and the material present in them.

Aerosol Wet Removal

Aerosol particles and their associated chemicals are also transported to the sea surface via precipitation. Direct measurements of rain concentrations should be the best approach for determining wet deposition, but problems with sampling, contamination, and the natural variability of the concentration of trace substances in precipitation often make representative flux estimates difficult. Studies have shown that the concentration of a substance in rain is dependent upon the concentration of that substance in the atmosphere. This relationship can be expressed in terms of a scavenging ratio, S:

$$S = C_r \cdot \rho \cdot C_a^{-1} \tag{2}$$

where C_r is the concentration of the substance in rain (e.g., mg kg^{-1}), ρ is the density of air (about 1.2 kg m^{-3}), and S is dimensionless. Values for S for species present in aerosol particles range from a few hundred to a few thousand, which roughly means that one gram (or one cm^3) of rain scavenges about one m^3 of air. S is, of course, dependent upon a number of factors, including particle size, chemical composition, vertical concentration distribution, vertical extent of the precipitating cloud, etc. Because of these factors, the use of scavenging ratios requires great care. However, if the aerosol concentration is known, the scavenging ratio approach can be used to estimate wet deposition fluxes as follows:

$$F_r = P \cdot C_r = P \cdot S \cdot C_a \cdot \rho^{-1} \quad (3)$$

where F_r is the wet deposition flux (e.g., in mg m^{-2} yr^{-1}) and P is the precipitation rate (e.g., in m yr^{-1}), with appropriate conversion factors to translate rainfall depth to mass of water per unit area. Note that $P \cdot S \cdot \rho^{-1}$ is equivalent to a wet deposition velocity.

Gas Removal

Direct measurement of gas phase fluxes to the water surface is only possible for one or two species at the present time. Modeling of the flux of gaseous species to the sea surface or to rain droplets requires consideration of the Henry's Law constants and exchange coefficients as well as atmospheric and oceanic concentrations of the species of interest. For many species, appropriate Henry's Law constants and exchange coefficients are poorly known. Models for such calculations are described in detail in GESAMP (1989) and in many other references and will not be discussed further here.

In the sections that follow the information available on the atmospheric input of certain trace species to coastal waters will be reviewed. The estimates presented will generally be for the direct input to the water surface in the coastal regions unless indicated otherwise. The atmospheric input will also be compared with the input via other pathways when possible. Note that there are very large uncertainties in all the numbers given here, both for atmospheric fluxes and for inputs from other sources. Unless otherwise indicated, it should be assumed that the atmospheric input rates have uncertainties ranging from a factor of 2 to at least 4.

TRACE METALS

There is a growing body of information available on the atmospheric input of trace metals to the coastal ocean. Some information is from the U.S. east coast, but most is from the enclosed and semi-enclosed seas around Europe, including the North, Mediterranean, and Baltic Seas.

North Sea

There has probably been more research into the evaluation of atmospheric fluxes to the North Sea than any other coastal or near-shore region. A number of valuable aerosol and precipitation concentration data sets are available, many covering extensive periods of time. Particularly useful data and calculations are reported by Pattenden and Branson (1987), Dedeuwaerder (1988), Cambray et al. (1975, 1979), Van Aalst et al. (1982), Krell and Rockner (1988), Yaaqub (1989), Pacyna et al. (1984), and Balls (1989). Again, GESAMP (1989) has evaluated all these data and estimated the total input of these trace metals to the North Sea. It should be noted that Pattenden and Branson (1987) have found a decline in the atmospheric Pb concentration in Britain, which they attributed to the decreasing use of alkyl leads in gasoline. GESAMP corrected for these decreasing lead concentrations with time. GESAMP (1989) utilized scavenging ratios of 200–1000 for Cd, Cu, Ni, Zn, Mn, Pb, V, and As and 500–2000 for elements with mass particle-size distributions skewed toward larger particles, such as Al and Fe. These are somewhat larger values for S than are found in open ocean regions. A precipitation rate of 68 cm yr^{-1} was used. 0.1 cm sec^{-1} was used for v_d for As, Cd, Cu, Ni, Pb, and Zn, while 1.0 cm sec^{-1} was used for Al, Fe, Mn, and V. From their calculations, dry deposition appears to account for less than 10% of the total flux in this region. Calculations were made for both the southern and northern areas of the North Sea, and the total calculated fluxes are presented in Table 1 along with the estimated total atmospheric emissions of these metals from Europe and the corresponding fraction of the emissions entering the North Sea. These fractions range from 1 to 15%. Table 2 presents a compilation of estimates of the atmospheric input of trace metals to the entire North Sea basin made by other authors. Note that the agreement among these estimates is reasonably good but the estimates do vary by over an order of magnitude for many species. In general, the more recent estimates are likely more reliable because of the increased availability of data during the past five years.

The North Sea Conference (1987) estimated the percentage of the total input to the North Sea of these trace metals that is due to atmospheric transport, utilizing their own estimates for both atmospheric fluxes and those via other pathways (Table 3). Note that atmospheric input clearly dominates for Pb and Cd and that it is also very significant for Hg, Zn, and Ni. Except for As, these estimates indicate that atmospheric input is similar to or greater than riverine input for the metals listed.

TABLE 1. Total atmospheric fluxes and deposition of trace metals to European coastal seas.

Metal	Total Atmospheric Flux ($mg\ m^{-2}\ yr^{-1}$)	Total Atmospheric Deposition ($10^9\ g\ yr^{-1}$)	Estimated European Anthropogenic Emissions* ($10^9\ g\ yr^{-1}$)	% of European Emissions Entering the Sea (%)
		North Sea (GESAMP 1989)		
Pb	4–23	2.3–12	120	2–10
Cd	0.1–0.5	0.05–0.25	2.7	2–9
Zn	5–23	2.7–12	80	3–15
Cu	1–4.4	0.5–2.3	16	3–14
As	0.2–1.1	0.13–0.58	6.5	2–9
V	0.4–2.1	0.23–1.1	35	0.7–3
Ni	0.4–2.1	0.23–1.1	16	1–7
Mn	1–5	0.5–2.7	18	3–15
Fe	38–150	20–81	—	—
Al	38–150	20–81	—	—
		Northwestern Mediterranean Sea (Bergametti 1987)		
Pb	29†††	15		13
Cd**	1	0.5		19
Zn**	34	17		21
Cu**	4.2	2.1		13
As**	1	0.5		7
V**	25	13		—
Mn	22	11		—
Fe†	720	360		—
Al	970	440		—
Si	3100	1500		—
P	32	16		—
		Baltic Sea (GESAMP 1989)		
Pb	2.4	1.0		1
Cd	0.14	0.06		2
Zn	11	4.7		6
Cu	2.9	1.2		8
As	0.46	0.19		3
V	1.1	0.44		1
Cr	0.53	0.22		—
Mn	2.4	1.0		6
Fe	87	36		—
Hg††	0.014	0.006		—

*Pacyna et al. (1984); **derived from aerosol data and scaled to measured Pb flux; †same approach as ** but scaled to Al; ††wet flux only; †††see comment in text.

TABLE 2. Various estimates of the atmospheric deposition of trace metals to the North Sea.

Metal	Reference (1) (10^9 g yr^{-1})	(2) (10^9 g yr^{-1})	(3) (10^9 g yr^{-1})	(4) (10^9 g yr^{-1})	(5) (10^9 g yr^{-1})	(6) (10^9 g yr^{-1})	(7) (10^9 g yr^{-1})
Pb	2.3–12	2.6–7.4	3.6–13	5.8	8.1	3.6	5.6
Cd	0.05–0.25	0.05–0.24	0.11–0.43	—	—	0.53	0.53
Zn	2.7–12	4.9–11	7.2–58	16	38	7.0–14	15
Cu	0.5–2.3	0.4–1.6	1.4–100	5.6	6.9	2.0–4.0	4.9
As	0.13–0.58	0.04–0.12	0.22–0.72	<0.5	1.5	—	—
V	0.34–1.1	—	—	—	—	—	—
Cr	—	0.3–0.9	0.07–1.4	0.74	0.84	0.7	0.7
Ni	0.23–1.1	0.3–1.0	0.36–3.6	<1.9	1.4	1.7	1.7
Mn	0.5–2.7	—	—	—	—	—	—
Fe	20–81	—	—	—	—	—	—
Hg	—	0.01–0.03	<0.04	—	<0.02	0.006	0.006

(1) GESAMP (1989)
(2) North Sea Conference (1987)
(3) van Aalst et al. (1982)
(4) Cambray et al. (1979)
(5) Cambray et al. (1975)
(6) RSU (1980)
(7) ICES (1978)

TABLE 3. Estimates of the percentage contribution of various sources of trace metals entering the North Sea (North Sea Conference 1987).

Metal	Atmosphere (%)	Rivers (%)	Direct Discharges (%)	Dumping (%)
Pb	43–67	10–15	2–3	21–39
Cd	34–74	15–34	6–15	7–17
Zn	22–40	26–34	4–5	30–39
Cu	12–35	30–40	7–10	28–38
As	5–13	38–39	23–27	26–30
Cr	7–18	14–15	10–11	58–68
Ni	21–45	13–17	5–8	37–54
Hg	20–40	28–40	8–9	24–31

Mediterranean Sea

There have been several studies of the atmospheric input of trace metals to the Mediterranean (Bergametti 1987; Bergametti et al. 1989; Arnold 1985; Chester et al. 1981; Loye-Pilot et al. 1986; Dulac et al. 1987). Bergametti (1987) provided estimates of the atmospheric flux to the northwest Mediterranean using data from a station in northwestern Corsica (Table 1). It was noted that the temporal variability of measured total (wet plus dry) deposition fluxes of metals in this area was not related linearly to the measured atmospheric concentrations. Bergametti (1987) pointed out, for example, that the scavenging ratio was much higher in the summer, when most of the precipitation was from convective storms, than in winter, when stratiform precipitation systems predominated. Because the source regions were relatively close compared with open-ocean situations, the particle-size distribution still showed a significant number of large particles. This resulted in dry deposition fluxes being more important in this area than for mid-ocean regions. For the northwestern Mediterranean, Bergametti (1987) estimated that 30 to 50% of the total deposition was attributed to dry removal. Note in Table 1 that the input of several of these trace metals to the northwestern Mediterranean represent between roughly 5 and 20% of the human-derived emissions of these metals from Europe. Recent results from the EROS (European River Ocean System) 2000 program suggest that the flux of lead to the northwestern Mediterranean may be a factor of 4 to 8 lower than that reported in Table 1 (Guieu et al., submitted).

Baltic Sea

Similar estimates of trace metal input have been made for the Baltic Sea, and many of these studies are summarized in GESAMP (1989). Most of the data in this region were obtained from a monitoring network developed by the Baltic Marine Environment Protection Commission, also known as HELCOM. Studies of the atmospheric input of metals to the Baltic have been presented by Schneider (1987, 1989); Iverfeldt and Rodhe (1988); Oblad and Selin (1986); Kemp (1984); Petersen et al. (1989); and Rodhe et al. (1980). The concentrations of four elements, Pb, Cd, Zn, and Cu, have been measured at five coastal stations around the Baltic since 1984; Schneider (1989) has used these data to estimate the fluxes into five sub-basins. Atmospheric concentrations of other elements, including As, V, Cr, Mn, and Fe, have been determined in southern regions of the Baltic; GESAMP (1989) used these data to calculate dry deposition fluxes using Eq. (1) and the indicated dry deposition velocities. For wet deposition, the precipitation data of Rodhe et al. (1980) at coastal sites were utilized. To estimate deposition gradients between the coastal measuring sites on the open Baltic Sea, use was made of the EMEP model results of Petersen et al. (1989) for Pb and Cd. Simulations of the long-range transport of these substances over Europe were used to evaluate gradients over the Baltic. These gradients could then be applied to the other metals of interest to obtain estimates to the entire Baltic (Table 1). Note that in general these fluxes to the Baltic are considerably lower than those to the northwestern Mediterranean and also represent a smaller fraction of the total atmospheric emissions from Europe.

Coastal North America

A study was undertaken in the area of the New York Bight by Duce et al. (1976). The atmospheric input of Pb, Cd, Zn, and Fe was estimated for an area of 10^4 km^2 in the New York Bight region; this was compared with the total input of trace metals to this region by barge dumping, runoff, sewage, and rivers (Table 4). Note that the flux to this region is in general higher than fluxes reported for the European seas. This is because the New York Bight is a relatively small area located directly downwind from one of the major population and industrial regions on earth, so this is not surprising. Note also that the percentage of the total input of these metals from the atmosphere is considerably less than those for the North Sea (Table 3), reflecting the intense dumping and river input from this major population center. The atmospheric input becomes a more important source for material in the water at increasing distances from the coast since runoff, sewage, and river inputs enter directly on or very near the coast. The atmospheric input is spread more uniformly over the entire area of interest and beyond. For

TABLE 4. Total atmospheric fluxes and deposition of trace metals to coastal U.S. locations.

Metal	Total Atmospheric Flux (mg m^{-2} yr^{-1})	Total Atmospheric Deposition (10^9 g yr^{-1})	Input from Rivers, Runoff Dumping, etc. (10^9 g yr^{-1})	Percentage Atmospheric (%)
		New York Bight (Duce et al. 1976)		
Pb	60	0.60	4.1	13
Cd	1.2	0.012	0.86	1.4
Zn	81	0.81	9.6	7.8
Fe	370	3.7	78	4.5
		South Atlantic Bight (Windom 1981)		
Pb	10	0.62	0.07	90
Cd	0.13	0.008	0.003	73
Zn	12	0.71	0.31	70
Cu	3.5	0.21	0.11	64
As	0.7	0.04	0.02	67
Ni	6.1	0.37	0.22	67
Mn	1.0	0.06	0.09	40
Fe	92	5.5	0.95	85

example, considering the total atmospheric transport of these metals off the coastline around New York City, Duce et al. (1976) calculated that only 6% of the atmospheric Pb, 6% of the Cd, 24% of the Zn, and 30% of the Fe would be deposited within the first 100 km of the coast, assuming no rain. The remainder would be transported further out to sea.

Another study along the U.S. east coast was undertaken by Windom (1981) in the area of the South Atlantic Bight off the coasts of South Carolina, Georgia, and northern Florida (Table 4). Note that the atmospheric fluxes for the anthropogenically derived metals Pb, Cd, and Zn are one-fifth to one-tenth of those to the New York Bight, reflecting the lower source strengths for air pollutants in this region. Note also, however, that while the source is weaker and the atmospheric flux less, the percentage of the total flux to this region contributed by the atmosphere is apparently greater than in the New York Bight area and may be the dominant source for all metals investigated except Mn. This may be typical for areas not directly downwind of major industrial and population centers. It also obviously depends upon the size of the area over which the calculation is made. The greater the water area, the greater the impact of atmospheric deposition relative to the total input.

SYNTHETIC ORGANIC COMPOUNDS

A wide variety of high molecular weight synthetic organic compounds are released to the atmosphere, and a significant fraction of these releases make their way ultimately to the coastal zone and the open ocean. Like the trace metals, most of the studies of atmospheric input of organic materials to coastal waters have been undertaken in the European regional seas areas. Much of this work has been summarized by GESAMP (1989); this publication also provides new estimates for some of these regions. The calculation of atmospheric fluxes of these compounds to the ocean surface is complicated by the fact that many of them are found primarily in the gas phase. In most cases, the wet and dry removal rates for the gas phase dominated that for the particulate phase. The numbers presented refer to the total atmospheric fluxes, both gaseous and particulate.

Since the atmospheric residence times of most of the gas-phase compounds considered are longer than those of aerosol particles, the potential source areas for these species in a particular coastal region may be quite widely dispersed, in some cases as far as a thousand km or more from the coast. Calculations of fluxes are also complicated by the fact that the gas/particle partitioning of these species can be affected by the number, size, and composition of the aerosol particles as well as by the temperature. Data are few, and the estimates of fluxes are consequently quite uncertain.

Estimates of the atmospheric fluxes of a number of organic species to the North, Baltic, and Mediterranean Seas are presented in Table 5. Much of the data is from GESAMP (1989), with other information given where available. Note that the ranges of fluxes determined by GESAMP for these species (mg m^{-2} yr^{-1}) are identical for all three seas. With the small data base and the same large European source region for these substances, using different estimates for the mean *concentrations* of these species over the various seas was deemed to be unrealistic. There is a range of fluxes calculated from these common mean concentrations for the different seas because of the temperature dependence of precipitation scavenging of the vapor phase and because of differences in precipitation amounts in the various regions. In general, GESAMP (1989) believes that the high end of the flux range is more typical of the North and Baltic Seas. The lower end of the flux range is more typical of the Mediterranean because that area has higher temperatures and lower rainfall, leading to an expected lower gas phase flux to the water surface.

GESAMP's (1989) calculations indicated that direct gas exchange may account for one-half to two-thirds of the total deposition of most of these organic species. This assumes that the water is 90% saturated with respect to these organic substances. Obviously there are significant uncertainties in the gas phase calculations. For example, if significant quantities of organic

TABLE 5. Total atmospheric fluxes and deposition of synthetic organics to European coastal seas.

Organic Compound	Total Atmospheric Flux (μg m^{-2} yr^{-1})	Total Atmospheric Deposition (10^6 g yr^{-1})	Riverine Input* (10^6 g yr^{-1})	Percentage Atmospheric (%)	Other Atmospheric Deposition Estimates (10^6 g yr^{-1})
		North Sea (GESAMP 1989)			
α-HCH	8.5–20	10	1	91	4–40*
γ-HCH	10–54	18	1.5	92	6–60*
Σ HCH	18–74	28	2.5	82	10–100*
Σ DDT	0.6–1.7	0.9	—	—	1.2–12*
Σ PCB	3.3–6.8	3.7	—	—	10–160*
Dieldrin	1.2–2.0	1.1	—	—	1.2–12*
Chlordane	0.08–0.15	0.1	—	—	—
HCB	0.3–0.9	0.5	4	11	2–17*
		Northwestern Mediterranean Sea (GESAMP 1989)			
α-HCH	8.5–20	4.3	—	—	1.5**
γ-HCH	10–54	5.6	—	—	8.4**
Σ HCH	18–74	9.9	—	—	9.9**
Σ DDT	0.6–1.7	0.3	—	—	1.3**
Σ PCB	3.3–6.8	1.7	3.8†	31–85	6.8 (22)†
Dieldrin	1.2–2.0	0.6	—	—	—
Chlordane	0.08–0.15	0.05	—	—	—
HCB	0.3–0.9	0.2	—	—	0.20**
		Baltic Sea (GESAMP 1989)			
α-HCH	8.5–20	7.7	—	—	5††
γ-HCH	10–54	13	—	—	4††
Σ HCH	18–74	21	—	—	9††
Σ DDT	0.6–1.7	0.7	—	—	14††
Σ PCB	3.3–6.8	2.8	—	—	—
Dieldrin	1.2–2.0	0.9	—	—	—
Chlordane	0.08–0.15	0.1	—	—	—
HCB	0.3–0.9	0.4	—	—	—

substances enter the coastal waters in precipitation and this material cannot be removed rapidly to subsurface waters, a potentially important fraction of the wet flux could be revolatilized to the atmosphere. This possibility has been discussed in detail by Swackhamer et al. (1988) for lake systems and by Rodhe et al. (1980) for the Baltic Sea.

From the data in Table 5 it is difficult to assess the relative importance of atmospheric and riverine input of these organic materials to the North and Mediterranean Seas. It appears that atmospheric input may dominate riverine input for HCH, but the opposite may occur for HCB and PCB. It is interesting to note that GESAMP (1989) has calculated that atmospheric input dominates (80–99%) riverine input of these species to the global ocean.

NITROGEN SPECIES

There has been relatively little research, until very recently, on the input of nitrogen species to the coastal ocean. Paerl (1985) has suggested that in certain coastal and near-urban regions, atmospheric nitrogen input may be quite important for biological activity. Paerl investigated the impact of rainfall on chlorophyll *a* production in the waters of Bogue Sound, North Carolina. Figure 1 presents the results he obtained following the addition of 10 and 20% v/v distilled water, 10 and 20% v/v rainwater (each from two separate rainfalls), 100 and 200 ppb nitrate, and 50 and 100 ppb phosphate to samples of Bogue Sound water. The pH of Rain A was 5.85, which Paerl suggests is typical of marine rain. The pH of Rain B was 4.05, which Paerl suggests is representative of continental rain with considerable anthropogenic components. As Fig. 1 shows, the strongly acidic rainfall resulted in increased chlorophyll *a* production relative to the more neutral rain. Paerl related this increased biological production to the higher nitrate/nitrite concentration in the strongly acidic rain. Parallel nitrate and phosphate additions (Fig. 1) showed exclusive nitrate stimulation of productivity. Both rains stimulated productivity for several days, but the more acidic rain resulted in stimulation lasting 6–7 days, compared with 2–3 days for the more neutral rain. Note, however, that the 10–20% additions of rain to seawater in the laboratory experiments may be rather large relative to the effect one might expect from several cm of rain falling on the ambient ocean surface. Other tests indicated that the acidity itself was not a factor. Paerl (1985) pointed out that acidic rainfall with its associated high nitrate concentrations may well be important in affecting both the patterns and magnitudes of primary productivity along coastal regions in proximity to extensive urban areas or industrial activities. The eastern coasts of continents in mid-latitudes may be particularly susceptible to the input of atmospheric anthropogenic nitrate from the conversion of emitted NO_x. Paerl's paper

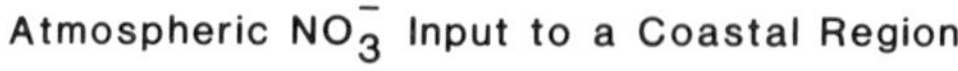

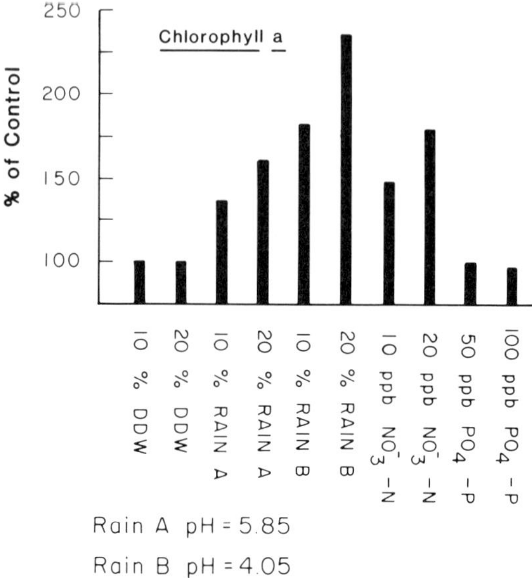

Fig. 1—Effect of rainwater on chlorophyll *a* production in Bogue Sound, North Carolina. All treatments were compared with "no addition" controls. Adapted from Paerl (1985).

highlights the importance of the episodic input and potential impact of trace species from the atmosphere to the coastal ocean. This is particularly important for substances such as nitrogen, which may be limiting productivity in some regions.

Perhaps the publication that has most effectively galvanized interest and research on the atmospheric input of nitrogen species to coastal waters was the report by the Environmental Defense Fund (Fisher et al., 1988) on the role of acid rain in polluted coastal waters of the northeastern United States. This report indicated that a major source of nitrogen to Chesapeake Bay and other Atlantic coastal waters is atmospheric deposition of nitrate in acid rain. Fisher et al. (1988) indicated that approximately 40% of all the nitrogen contributed by human activity to the Chesapeake Bay enters via acid rain falling directly on the Bay or on its watershed (see Table 6). Note that this is different from the other numbers we have discussed, which only considered deposition directly on the water surface. For Chesapeake Bay rain, input of nitrate exceeded that from animal waste runoff and was approximately the same as input from point sources. Only fertilizer runoff exceeded the atmospheric input of nitrogen. Fourteen percent of the total nitrogen input

TABLE 6. Input of nitrogen to Chesapeake Bay (Fisher et al. 1988).

Source	Total Input (10^9 g yr^{-1})	Areal Input Rate (g m^{-2} y^{-1})	Percent of the Total Input
Precipitation:			
nitrate	35	3.1	25
ammonium	19	1.7	14
Animal waste	5	0.4	4
Fertilizers	48	4.2	34
Point sources	33	2.9	24
TOTAL	140	12.3	100

was contributed by ammonium in precipitation. Of the 35×10^9 grams of nitrate–nitrogen entering the Bay from the atmosphere each year, about 8×10^9 g fall directly on the Bay, with the rest falling on the watershed. It is interesting to note that the total nitrogen fertilizer applied to croplands in the Chesapeake Bay watershed is about 5.4 g m^{-2} yr^{-1}, while the atmospheric input of nitrate nitrogen is about 3.1 g m^{-2} yr^{-1}. It is clear that the Chesapeake Bay is almost as heavily polluted from the atmosphere as croplands are fertilized in that watershed. Fisher et al. (1988) pointed out that in the future, atmospheric input of nitrate may become even more important since point sources and agricultural runoff are the current targets for reduction of nitrate input to coastal waters. In the absence of future nitrogen oxide emission controls, they estimate that direct atmospheric nitrate could contribute as much as 42% of the total nitrogen input to the Bay by 2030.

Other estimates have been made for nitrogen inputs to the coastal zone. A number of these are summarized in Table 7. Sinderman and Swanson (1979) estimated that direct atmospheric deposition contributed 13% to the overall annual input to the New York Bight, while another 29% came from runoff, a significant fraction of which originates as atmospheric deposition on land. Fisher et al. (1988) reported that 23% of the total input of nitrogen to Long Island Sound comes from the atmosphere, with roughly 7% of that from direct deposition on the Sound and the remaining 16% from fallout on the watershed that ultimately reaches the Sound. North Carolina Division of Environmental Management (1985) has estimated that 18% of the nitrogen input to the lower Neuse River comes from atmospheric nitrogen falling on the water itself. Roenner (1985) has estimated that 25% of the input of

TABLE 7. Input of nitrogen to coastal areas.

Region	Total Atmospheric Input* (10^9 g yr^{-1})	Total Input All Sources (10^9 g yr^{-1})	Percent Atmospheric Input (%)	References
New York Bight	—	—	13†	(1)
Chesapeake Bay	54	140	39	(2)
Long Island Sound	11	49	23	(2, 3)
Neuse River Estuary, NC	1.7	7.5	23	(2, 4)
North Sea	400†	1500	27†	(5)
Western Mediterranean Sea	400†	403††	50†	(6)
Baltic Sea	—	—	25†	(7)

*Total from direct atmospheric deposition and runoff of atmospheric material from the watershed.
†Direct atmospheric deposition to the water only.
††Total from atmospheric and riverine input only.
(1) Sinderman and Swanson (1979).
(2) Fisher et al. (1988).
(3) Farro et al. (1986).
(4) North Carolina Div. of Environmental Management (1985).
(5) North Sea Conference (1987).
(6) Martin et al. (1989).
(7) Roenner (1985).

nitrogen (as nitrate and ammonium) to the Baltic Sea is derived from direct input to these waters. North Sea Conference (1987) found that about 27% of the total input of nitrogen to the North Sea is from the atmosphere. In the western Mediterranean, Martin et al. (1989) found that 50% of the combined input from the atmosphere and rivers comes via the atmosphere. The data in Table 7 are relatively consistent and indicate that when atmospheric input to a watershed (with subsequent input to the water body of some faction of that total) and directly to the coastal water are considered, roughly 20–30% of the total can be traced to an atmospheric pathway. When only riverine and atmospheric input are compared, the atmospheric percentage is somewhat higher, when only direct deposition to the water surface is considered, the atmospheric percentage is probably in the 10–25% range.

CONCLUSIONS

From the considerations in this chapter it is apparent that the atmosphere can provide an effective transport path for chemical substances entering the waters of the coastal zone. In many areas and for certain substances, atmospheric input may even dominate that from more commonly investigated sources such as rivers, sewage outfalls, dumping, etc. Nevertheless, estimates of atmospheric deposition are still very uncertain, and in general the estimates are only reliable to within a factor of 2 to 4. Significant improvement in these estimates will only come with increased temporal and geographical measurement of the atmospheric and precipitation concentrations of these substances, and from the development of accurate techniques for measuring directly the particle and gas exchange with water surfaces. Improvements in our ability to model these exchange processes will also require better information on the kinetic parameters (e.g., transfer and deposition velocities). Atmospheric input is typically highly episodic in nature, with the input rate depending on pollutant source strengths and a variety of meteorological conditions. Particularly important is the amount and character of precipitation, since wet deposition is quite important for many species. Other rapidly changing meteorological factors, such as wind transport patterns from source regions and atmospheric stability, will always make modeling of atmospheric input to coastal regions quite uncertain. Significant annual variability of the atmospheric input of pollutants to any coastal region should be expected.

REFERENCES

Arnold, M. 1985. Géochimie et transport des aérosols métalliques au-dessus de la Méditérranée Occidentale. Ph.D. diss., Univ. of Paris VII, France.

Balls, P.W. 1989. Trace metal and major ion composition of precipitation at a North Sea coastal site. *Atmos. Envir.* **23**:2751–2759.

Bergametti, G. 1987. Apports de matière par voie atmosphérique à la Méditerranée Occidentale: aspects géochimiques et météorologiques. Ph.D. diss., Univ. of Paris VII, France.

Bergmaetti, G., A.L. Dutot, P. Buat-Ménard, R. Losno, and E. Remoudaki. 1989. Seasonal variability of the elemental composition of atmospheric aerosol particles over the northwestern Mediterranean. *Tellus* **41B**:353–361.

Burns, K.A., and J.P. Villeneuve. 1987. Chlorinated hydrocarbons in the open Mediterranean ecosystem and implications for mass balance calculations. *Marine Chem.* **20**:337–359.

Cambray, R.S., D.F. Jefferies, and G. Topping. 1975. An estimate of the input of atmospheric trace elements into the North Sea and Clyde (1972–1973), AERE Rep. R7733. Harwell, U.K.: AERE.

Cambray, R.S., D.F. Jefferies, and G. Topping. 1979. The atmospheric input of trace elements to the North Sea. *Mar. Sci. Comm.* **5**:175–194.

Chester, R., A.C. Saydam, and E.J. Sharples. 1981. An approach to the assessment of local trace metal pollution in the Mediterranean atmosphere. *Mar. Poll. Bull.* **12**:426–431.

Dedeuwaerder, H. 1988. Study of the dynamic transport and of the fallout of some ecotoxicological heavy metals in the troposphere of the southern Bight of the North Sea. Ph.D. diss., Vrije Univ., Brussels, Belgium.

Duce, R.A., G.T. Wallace, and B.J. Ray. 1976. Atmospheric trace metals over the New York Bight. NOAA Technical Report ERL 361–MESA 4. Boulder, CO: US Dept. of Commerce.

Dulac, F., P. Buat-Menard, M. Arnold, and U. Ezat. 1987. Atmospheric input of trace metals to the western Mediterranean Sea. 1. Factors controlling the variability of atmospheric concentrations. *J. Geophys. Res.* **92**:8437–8453.

Farro, D.R.G. et al. 1986. The National Coastal Pollutant Discharge Inventory: Boundaries for Long Island Sound. Rockville, MD: Natl. Oceanic and Atmospheric Administration.

Fisher, D., T. Ceroso, T. Mathew, and M. Oppenheimer. 1988. Polluted Coastal Waters: The Role of Acid Rain. New York: Environmental Defense Fund.

GESAMP. 1989. The Atmospheric Input of Trace Species to the World Ocean. Reports and Studies No. 38. Geneva: World Meteorological Organization.

ICES. 1978. Input of Pollutants to the Oslo Commission Area. Cooperative Res. Rep. 17. Charlottenlund, Denmark: Intl. Council for the Exploration of the Sea.

Iverfeldt, A., and H. Rodhe. 1988. Atmospheric transport and deposition of mercury in the Nordic countries. Progress report prepared for the Nordic Council of Ministers.

Kemp, K. 1984. Multivariate analysis of elements and SO_2 measured at Danish EMEP stations. Tech. Rep. of Air Pollution Laboratory (MST LUFT-A88). Danish National Agency of Environmental Protection.

Korolev, S.M. 1984. Input of organochlorine pesticides from the atmosphere into the Baltic Sea water area. Baltic Marine Environment Protection Committee. Meeting, Stockholm, Sweden.

Krell, U. and E. Rockner. 1988. Model simulations of the atmospheric input of lead and cadmium into the North Sea. *Atmos. Envir.* **22**:375–381.

Loye-Pilot, M.D., J.M. Martin, and J. Morelli. 1986. Saharan dust: influence on the rain acidity and significance for atmospheric input to the Mediterranean. *Nature* **321**:427–428.

Martin, J.-M., F. Elbaz-Poulichet, C. Guieu, M.-D. Loye-Pilot, and G. Han. 1989. River versus atmospheric input of material to the Mediterranean Sea: an overview. *Marine Chem.* **28**:159–182.

North Carolina Division of Environmental Management. 1985. Nutrient Management in the Neuse Basin: an Update. Raleigh, NC.

North Sea Conference. 1987. Quality Status of the North Sea. Second Intl. Conference on the Protection of the North Sea. London: Dept. of the Environment.

Oblad, M., and E. Selin. 1986. Measurements of elemental composition in the background aerosol on the west coast of Sweden. *Atmos. Envir.* **20**:1419–1432.

Pacyna, J.M., A. Semb, and J.E. Hanssen. 1984. Emission and long-range transport of trace elements in Europe. *Tellus* **36B**:163–174.

Paerl, H.W. 1985. Enhancement of marine primary production by nitrogen enriched acid rain. *Nature* **315**:747–749.

Pattenden, N.J., and J.R. Branson. 1987. Relation between lead in air and in petrol in two urban areas of Britain. *Atmos. Envir.* **21**:2481–2483.

Petersen, G., H. Weber, and H. Grassl. 1989. Modeling the atmospheric transport of trace metals from Europe to the North Sea and Baltic Sea. NATO Workshop on Control and Fate of Atmospheric Heavy Metals, Oslo, Norway.

Rodhe, H., R. Soederlund, and J. Ekstedt. 1980. Deposition of airborne pollutants on the Baltic. *Ambio* **9**:168–173.

Roenner, U. 1985. Nitrogen transformations in the Baltic proper: denitrification counteracts eutrophication. *Ambio* **14**:134–138.

RSU. 1980. Umweltprobleme der Nordsee. Stuttgart and Mainz: Verlag Kohlhammer.

Schneider, B. 1987. Source characterization for atmospheric trace metals over Kiel Bight. *Atmos. Envir.* **21**:1275–1283.

Schneider, B. 1989. Input of atmospheric trace metals to the Baltic Sea area—an estimate based on the EGAP (HELCOM) monitoring data. Baltic Environment Protection Commission, HELCOM, in press.

Sinderman, C.J., and R.L. Swanson. 1979. Historical and regional perspectives. In: Oxygen Depletion and Associated Benthic Mortalities in the New York Bight, ed. R.L. Swanson and C.J. Sinderman. NOAA Professional Paper 11, pp. 1–16. Washington, D.C.: U.S. Dept. of Commerce.

Swackhamer, D.L., B.D. McVeety, and R.A. Hites. 1988. Deposition and evaporation of polychlorobiphenyl congeners to and from Siskiwit Lake, Isle Royale, Lake Superior. *Envir. Sci. Tech.* **22**:664–672.

van Aalst, R.M., R.A.M. Van Ardenne, J.F. de Kreuk, Jr., and Th. Lems. 1982. Pollution of the North Sea from the atmosphere. TNO, Report CL 82/152. Delft: Netherlands Organization for Applied Scientific Research.

Villeneuve, J.-P., and C. Cattini. 1986. Input of chlorinated hydrocarbons through wet and dry deposition to the western Mediterranean. *Chemosphere* **15**:115–120.

Windom, H.L. 1981. Comparison of atmospheric and riverine transport of trace elements to the continental shelf environment. In: River Inputs to Ocean Systems, ed. J.M. Martin, J.D. Burton, and D. Eisma, pp. 360–369. Paris: UNEP/UNESCO.

Yaaqub, R. 1989. Composition of atmospheric aerosols in rural East Anglia and meteorological controls. Ph.D. diss., Univ. of East Anglia, Norwich, U.K.

Standing, left to right:
Sandro Spitzy, Viatcheslav Gordeev, Marie-Alexandrine Sicre, Robert Duce, Jean-Marie Martin, Gilles Billen, Jens Meincke
Seated, left to right:
Nick McCave, Peter Liss, Herb Windom, John Milliman

Group Report: What Regulates Boundary Fluxes at Ocean Margins?

P.S. Liss, Rapporteur
G. Billen
R.A. Duce
V.V. Gordeev
J.-M. Martin
I.N. McCave
J. Meincke
J.D. Milliman
M.-A. Sicre
A. Spitzy
H.L. Windom

INTRODUCTION

Boundary fluxes to ocean margins include inputs by river runoff, exchanges through the water–atmosphere interface, exchanges with the adjacent slope and open ocean areas, and permanent burial in margin sediments.

The group assigned itself the task of discussing the regulation of the processes responsible for these fluxes, their temporal and geographical variations, and their global significance. Some specific aspects of these questions have been dealt with in the background papers associated with the Group 1 Report, as well as in some others presented in other sections, particularly those of Blackburn and Wollast.

WHAT PROCESSES, NATURAL AND ANTHROPOGENIC, REGULATE BOUNDARY FLUXES?

There are clearly a large number of processes that regulate the fluxes of matter into and out of ocean margins. Since to deal with them all systematically would require a very large effort, the group decided to concentrate on certain important understudied aspects, in particular (a) exceptional/extreme events, (b) chemical "bombs," and (c) anthropogenic effects on fluxes.

Ocean Margin Processes in Global Change
Edited by R.F.C. Mantoura, J.-M. Martin and R. Wollast

Exceptional/Extreme Events

The frequency/magnitude curve for many environmental processes follows a Gaussian distribution. This means that there is a symmetrical decrease in frequency of occurrence about a most common value, with larger and larger events becoming progressively less likely. However, for some processes the frequency of occurrence has a non-Gaussian form, with no limit on the size of the maximum event, albeit with a low (but nonzero) frequency of occurrence. Since in terms of mass flow it is the *product* of size and frequency which is important, the open-ended distribution may lead to rare events, but of very large magnitude, dominating the process. Such events can have very large effects on the environment and catastrophic impacts on human activities.

By their very nature rare events are difficult, if not impossible, to predict, and hence observe and quantify. Because of this they have received less attention than more frequent events. Before we suggest how this might be rectified, some examples of extreme events that may be important in ocean margins are worth mentioning. The most obvious example is exceptional river flows: they not only carry large amounts of water to coastal areas but can transport orders-of-magnitude more sediment in suspension than is usually the case. Milliman (this volume) cites the example of the Santa Clara River (U.S.A.), in which six days of flow accounted for nearly 60% of the sediment transported by the river over an 18-year time span. Another example is the melting of large *grounded* ice sheets, which will lead to a substantial rise in sea level globally, with clear impacts in coastal zones. This must have occurred in the past and could occur in the relatively near future (probably several hundred years from now) if human-induced greenhouse warming of the atmosphere becomes a dominant climate-regulating process. Other, but more local, examples are the flushing of fjord and marginal sea bottom waters by periodic large-scale spillage of seawater over the sill, hurricanes and tsunamis, red dust events from desert storms, and inputs to the atmosphere from nuclear accidents, such as happened at Chernobyl. The last two are certainly observed in coastal seas where they provide reactive particles and radioactivity.

To better comprehend and therefore predict extreme events, we need to improve our basic understanding of nonlinear interactions between processes of different time and space scales resulting in such events. Further, it is necessary to have a better knowledge of the form of the flux magnitude/frequency distribution for a variety of substances in a range of contrasting environments. We note that with respect to riverborne material, small drainage basins are more susceptible to exceptional events, since larger basins have a greater capacity to buffer against external and internal changes. Although further mathematical development of the form of magnitude/frequency curves is probably required, the most important need is for monitoring programs to establish event frequency in the environment. Since exceptional events, by definition, do not occur frequently, there will be in many cases a mis-match between the time scales of the process of interest and the funding horizon of the observational program. To deal with this problem, full use will have to be made of proxy data records with annual resolution from ice cores, tree rings, coral reefs, and sediments from lakes, bogs, salt marshes, and coastal marine environments. There is a great need for existing long-term observing stations to be maintained and for the establishment of new ones, especially at environmentally sensitive locations.

Chemical "Bombs"

By this phrase we refer to chemical/biological processes that involve a critical instability and can lead to a step function change in one or several properties of the system. An obvious example in coastal environments impacted by human activities is anoxia in bottom waters. Input of nutrients or organic matter leads to consumption of oxygen, particularly in bottom waters and sediments away from the atmosphere. At a certain loading (which will vary between situations) the oxygen concentration drops below a critical value and the biology and chemistry change drastically, with generally deleterious consequences on the environment. Such situations are common in lakes and are becoming well documented in coastal marine environments (e.g., in Walvis Bay and in parts of the Baltic, north Adriatic and southern North Seas). Another example, which is much less understood, is the occurrence of toxic algal blooms in coastal waters. Clearly some factor or combination of factors triggers a particular species of algae to bloom, but at the present we can only guess at which factors are critical. Another, although hypothetical, example is that a storm in, e.g., the North Sea might lead to a release and dispersal of a toxic chemical previously dumped and thought to be immobilized in the sediments.

The coastal seas are the obvious places to examine such chemical "bombs" since compared to the open oceans they receive greater natural and man-made inputs and have a much smaller capacity to dilute them. Also, natural "bombs," such as red tides and mass marine mortalities, have a much greater potential impact on human populations when they occur in coastal ocean areas. Other potential "bomb" situations need to be discovered and the critical step or factor which induces them to "flip" between states identified.

Anthropogenic Effects on Fluxes

Here we deal with anthropogenic effects on fluxes in marine margins, first in terms of how humans can enhance/diminish existing fluxes and second from the point of view of the inputs of synthetic (industrially produced) chemicals new to the environment or where the chemical form (speciation) of the inputs is altered by human activities.

Enhancement/diminution of fluxes. A clear change in riverine fluxes to the coastal zone is brought about by alterations to drainage basins resulting from anthropogenic activities. The clearing of forests and construction of dams are obvious examples of such activities, but there are many others. The impact of the change is on both the physical and chemical properties of the hydrology. For example, deforestation and agriculture generally increase river water concentrations of plant nutrients, organic matter, and trace metals. At the same time, they also change the water flow and the amount of sediment it carries. Such alterations in freshwater inputs to coastal zones will affect the buoyancy and hence the stability of the water column, as well as the sediment dynamics (see also Jickells et al., this volume).

Another case where humans affect fluxes arises in industrial areas from the increased acidity of rain due to sulfur and nitrogen oxides from fossil fuel combustion. Although somewhat regionalized in its impact, this is an effect of global magnitude; for example, it is estimated that worldwide riverine flow of sulfate to the oceans has increased by about a factor of two since industrialization. It seems certain that river flows of other substances (nutrients, metals) must have increased by a similar, or possibly greater, amount.

Synthetic chemicals and changes in form. The chemical industry has synthesized a very large number of chemicals never previously seen on Earth. A significant fraction of these compounds are made in commercial quantities and ultimately some portion escapes to the environment. Over 90% of these substances are not adequately characterized in the environment. DDT and tri-butyltin are obvious examples where such release has led to damage to the marine environment. Since there are so many compounds and there are great difficulties in their analysis, it behoves us to treat them with great caution, particularly in near-shore marine areas.

An example of where man has altered the form of a preexisting substance is lead in the atmosphere. Not only have we raised lead levels in the air by emissions from smelters and car exhausts, the form has also been changed from dominantly coarse aerosol particles before industrialization to mainly a fine aerosol fraction at the present. This is because industrial and automobile lead is emitted in a gaseous form which condenses into or onto fine particles in the atmosphere.

In making suggestions for future research we note that there are few natural (i.e., unaffected by human intervention) river systems left in the world. If atmospheric inputs are taken into account, then there are essentially no pristine basins at all. We see considerable advantages in studying anthropogenic effects on small river catchments. Apart from the fact that large systems are inherently difficult to sample, small systems offer a larger range of river types and are also likely to be those most affected by anthropogenic activities and global climatic and other changes. Further, although small rivers by definition carry only small volumes of water, they supply a disproportionately large amount of particulate and dissolved material to the coastal oceans. Some effort should be devoted to identifying a representative set of such small catchments from differing climatic and tectonic regimes.

WHAT IS THE RELATIVE IMPORTANCE OF DIFFERENT INPUT/OUTPUT ROUTES?

Inputs

Ocean margins receive inputs of material, including water and energy, from the following main sources: rivers, the atmosphere, bottom sediments, the ocean beyond the continental shelf/slope, and groundwater. Estimates of these various inputs are uncertain both on a local basis and when integrated or estimated globally. We consider that very few if any of the fluxes are known to be better than a factor of two, and many have orders of magnitude uncertainties associated with them. The fluxes vary both temporally and spatially; changes of a hundredfold are not uncommon with, for example, season and wind direction. Over the last two decades considerable efforts have been mounted to improve these flux estimates, and significant improvements are now being realized.

We asked ourselves the question, why it is important to be able to compare the magnitudes of the different sources of material to near-shore areas? Apart from the intellectual element of curiosity, planning future work and making accurate predictions requires such comparisons. A further and very pragmatic reason is from the point of view of giving advice to governments on how to regulate pollutant inputs. For example, if atmospheric inputs are dominant, as they are for lead in the North Sea, then any attempt to control input of this element should concentrate on sources to the atmosphere, rather than those discharging to rivers or other transport paths.

In view of the large uncertainties in the various flux estimates, it is generally possible to make only semiquantitative statements about the relative importance of the various routes. If a comparison is made between riverine and atmospheric sources, it is clear that the ratio of magnitudes of the two will vary with distance from the continents, with the relative importance of the atmospheric route increasing with distance from land. However, this statement ignores other sources and must be applied site-specifically and with caution. For example, the ratio of atmospheric to riverine supply of nitrogen certainly increases with distance away from land; however, in many shelf seas, the supply of fixed nitrogen (mainly nitrate) from upwelling[1] of deep ocean waters dominates the total supply. Another general statement is that groundwater contributes less than 5% to the total inflow of water to the coastal zone. However, in some areas the groundwater may be very enriched in, for example, nitrate (where it is derived from fertilized agricultural catchments) and methane gas (Gulf of Mexico). A further point is that many of the sources of material to coastal regions are very variable with time. This is especially so for atmospheric and riverine inputs, probably less so for upwelling, with groundwater probably exhibiting a much greater constancy of composition.

An important conclusion of our discussion was the importance of upwelling of ocean water as a source material to many coastal marine environments. For example, Wollast (this volume) estimates that global upwelling of nitrate-rich deep seawater contributes about ten times more nitrogen to shelf zones than river inflow. However, such statements are made as a result of budget calculations based on few direct measurements. In view of the apparently large size of the upwelling source, there is a clear need for greater study by direct approaches, and further assessments should be made of how its magnitude may alter in the future due to climatic and other changes.

Outputs

So far the discussion has been concerned with inputs of material to the ocean margins; the other side of the coin is how material is lost to the other reservoirs. Mass balance ensures that there is an outflow of water to the open oceans, but chemical budgets for some important elements (e.g., nitrogen) imply that this outflow is more than balanced by inflow of upwelled deep ocean water. There are also important losses to the atmosphere of volatile forms of several elements. These include CH_4, N_2O, and dimethyl sulphide (DMS), some of which are considered further in a later section. Transfer of material to the atmosphere occurs via sea spray. Losses from the system in particulate form are discussed below.

The massive shifts in sediment flux from continent to deep sea between glacials and interglacials make the 10^5 year time scale appropriate when considering the process. At low stand of sea level, sediment is delivered to the present location of the outer shelf and upper slope and discharged into the head of submarine canyons. From these unstable positions,mass movement could take material to the deep sea in individual events of great

[1] In this report, "upwelling" is used in the sense of a geochemical transport route that is facilitated by physical processes such as coastal and shelf-edge upwelling, boundary-layer mixing, and advection.

magnitude (e.g., the Storegga slide of 6,000 km^3 or the Saharan Slide of 600 km^3) during both glacial and interglacial times. At high stands of sea level, as at the present, it is apparent that virtually all the sediment coming down rivers discharging onto wide shelves and embayments and a significant fraction of that on narrower shelves is trapped in estuaries and the coastal zone. In a few cases, canyons intercept coastal sediment flux but this is uncommon. Thus, it is probable that 60–90% is trapped in shelf and near-shore areas at present (Milliman, and Martin & Windom, this volume).

Rivers, although dominant, are not the only source of particles. There is a small input from coastal and shoreline erosion (which may have an abnormally high POC content) and from the atmosphere and seabed erosion. These inputs are not well quantified in global terms, at present. Although shelf circulations tend to trap material, outer shelf and upper slope currents and internal waves can result in the resuspension of sediment and its dispersal over the continental margin and down canyons in intermediate and bottom nepheloid layers. Of course, if the margin is defined to include all of the continental slope, this might not result in much escape to the open pelagic realm. These processes are likely to be common on all margins of the world at all time scales down to the seasonal (3 months).

HOW DO THE REGULATING PROCESSES CHANGE WITH TIME: PAST AND FUTURE?

In discussing temporal changes in the regulating processes, it is useful to classify the time scales involved in terms of events and processes (see Table 1). Our understanding of how the processes change decreases (exponentially?) as we go back in time. Human intervention is well documented compared to glacial/interglacial changes, which have to be interpreted from the

TABLE 1. Classification of time scales.

Time (years)	Event	Process/Effect
10^4–10^6	Glacial-Interglacial	Climate change Altered erosion rate Sea level rise/fall
10^2–10^4	Holocene climate change Rapid deglaciation	Altered rainfall/ evaporation
10^1–10^3	Medieval optimum Little Ice Age Anthropogenic intervention	Land-use change Input of pollutants Changed river use

geological record (sediments, ice cores). In between these two extremes, proxy records may reveal mesoscale climatic changes, such as the Little Ice Age and the mid-Holocene climatic optimum.

River Sediment Loads

Although for most river systems we are largely ignorant of changes in their hydrology in the past, for a few, well-studied basins a tentative assessment of how, for example, sediment load varies can be attempted, as shown in Fig. 1. It should be noted that the sediment load is plotted on a logarithmic scale!

River and aeolian inputs to the ocean system have varied greatly in the past and presumably will experience even greater variations in the future. In the past 2,500 years (in Asia) inefficient farming techniques have resulted

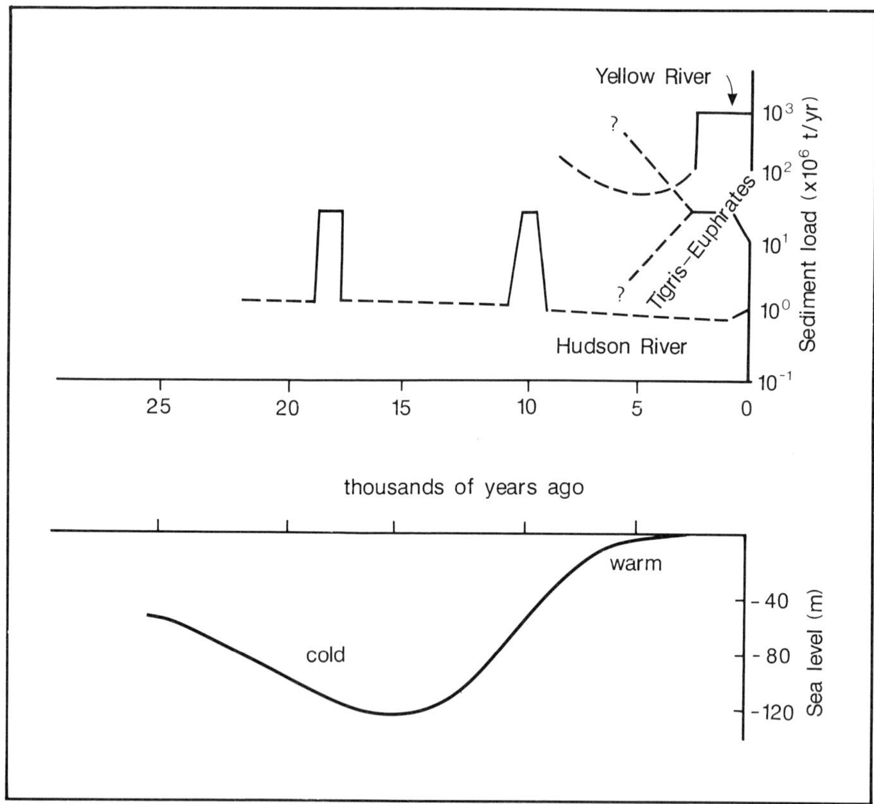

Fig. 1—Sediment load as a function of time for several river systems. Also shown is a curve of sea level on the same time axis.

in accelerated erosion, the Yellow River, e.g., experiencing an order-of-magnitude increase in its suspended load following widespread farming of the North China Loess plateau. In contrast, the Hudson River probably experienced a slight increase in sediment load several hundred years ago as European settlers began to farm eastern North America, but the major sediment fluxes from the river seem to have been related to two major glacial-ice melts 18 and 12 thousand years ago. It is entirely possible that during peak melt events (which may have lasted only a few years each) the sediment load of the Hudson exceeded that of the Yellow River! The Tigris-Euphrates' sediment load has been even more complex (and less well documented). Recent damming upstream has decreased the river's load substantially. Before the increasing aridity in the Middle East, beginning 3 to 4 thousand years ago, it is not clear whether the load was higher or lower. Higher river flow (because of higher precipitation and lower evaporation) may have carried higher sediment loads, but greater vegetation cover (as opposed to at present) almost certainly would have resulted in decreased sediment erosion. One certain result of increasing aridity, however, was the increased aeolian flux of sediment. For example, increased sediment accumulation over the past 4,000 years in Kuwait Bay appears to have exceeded 20 meters!

Glacial Conditions

There appears to be a concensus that in the last glacial period ocean margin sediments contained more organic carbon (mainly in slope sediments) than at present because upwelling was intensified over the shelf. We have already stressed the present-day importance of coastal upwelling in ocean margin productivity. A caveat to this is that during lower sea level in glacial periods shelf zones were considerably reduced in area. Thus, although the organic carbon content of margin sediments was probably greater at that time, their areal extent was likely to have been less.

A useful approach to estimating organic carbon in near-shore waters in the past may be to use the finding that at the present time there is a reasonably fixed relationship between dissolved organic carbon, particulate organic carbon and total sediment load. It is probably reasonable to assume that these properties exhibited similar ratios in the past.

There is considerable evidence that atmospheric dust loadings during the last glacial period were one or two orders of magnitude greater than at present. This may well have contributed to greater ocean productivity due to increased inputs of nutrients such as N, P, and Fe from the atmosphere. These higher dust levels are generally ascribed to stronger winds and greater aridity on land at that time. However, an examination of what currently controls dust levels in air indicates that wind is only one of the controlling factors. Measurements over many years of atmospheric dust in Barbados and for a more limited period over the North Pacific show that the Saharan and Asian deserts, respectively, vary by factors of severalfold in their apparent source strengths as a function of time. The explanation of this variability is that the dust concentration observed at a site remote from the source region depends on the complex interplay of three factors, namely the degree of drought in the desert source, the direction of wind transporting the dust, and the rainfall along the air trajectory since deposition in precipitation is a powerful removal mechanism. Thus, in order to make more than very general statements about dust levels in the past, a much improved understanding of paleometeorology will be needed.

Future Warmer Climate

Many of the things said about present-to-glacial changes will be applicable, but probably in the opposite sense, to changes in the future. We note the following points particularly. A warmer atmosphere will lead to greater evaporation from the ocean surface. This in turn implies an intensification of the hydrological cycle, with changes in the type and intensity of rainfall. We might predict heavier rains and less snow in temperature zones. However, because air temperatures in polar regions will still be well below zero, the greater vapor amounts in the atmosphere may lead to enhanced snowfall at high latitudes. It should be noted that such predictions must be treated with caution since changes in type and amount of cloud cover have thus far been ignored.

It is difficult to predict how ocean margin productivity may change in a warmer future. The predicted decrease in mean wind strength might argue for decreased coastal upwelling and hence nutrient supply from the oceans. However, since such upwelling is a local/regional phenomenon and is dependent on wind direction as well as strength, more sophisticated atmospheric circulation models than are currently available will be needed in order to resolve the matter.

Decreased winds almost certainly will lead to a decrease in the rate of air–sea gas exchange. It is clear that this process is strongly coupled to wind speed but in a nonlinear manner, at least for most of the biogeochemically active gases of interest here.

On the basis that the past is probably the best guide we have to predict the future, we recommend that efforts should be made to exploit the potential of sediments in anoxic basins (e.g., Black Sea, S. California borderlands, Cariaco Trench) to provide information on changes over a 10^2–10^3 year time scale. Areas with particularly high rates of sediment accumulation (many meters per thousand years) are also useful because of their potentially high resolution, which is enhanced by a lack of bioturbation that minimizes vertical mixing. Salt marshes and mangrove swamps may also be useful in this context. Ice cores should also be exploited, particularly where they can be linked to historical records, such as for the Little Ice Age in Iceland and Greenland.

HOW MUCH GEOGRAPHICAL GENERALITY IS THERE IN BOUNDARY FLUXES?

In assessing the role that ocean margins play in the global system, it is highly desirable that a minimum set of geohydrological variables be found to characterize such regions. The aim is to simplify and focus interdisciplinary discussions, and to develop more general schemes that couple ocean margin processes with global-scale models for the fluxes of water, gases, and particulate and dissolved matter. Present attempts are restricted to specific regional ecosystem models. It seems likely that 100-fold modifications of such models would be needed to describe the bulk of worldwide ocean margin situations, including their extension into adjacent ocean areas. The following approaches to the problem were discussed.

End-member Concept Based on Hydrological Volumes and Fluxes

The controls of an ocean margin volume can be expressed in terms of the ratio of riverine to ocean inputs versus flushing times. Then, from appropriate global maps, ocean margins can be classified by their position in a rectangular plane defined by the following end members: river- or ocean-dominated margins, both regimes with either slow or fast flushing (see Fig. 2). This concept, in its simplest application, is valid for uniformly distributed, dissolved substances and for fixed time averages of the controlling processes.

Hydrodynamical Boundary Flux Controls

Physical processes controlling fluxes across the ocean boundary of a marginal ocean area are of primary importance in understanding the coupling between the margins and the open ocean. A first-order distinction between different types of margins is made by grouping them into "narrow and steep," where fluxes are controlled by ocean dynamics under the influence of coastal bathymetry, and "wide and flat" (shelf seas), where in addition to ocean dynamics acting on shelf edge/slope bathymetry, the dynamics internal to the shelf sea have also to be considered.

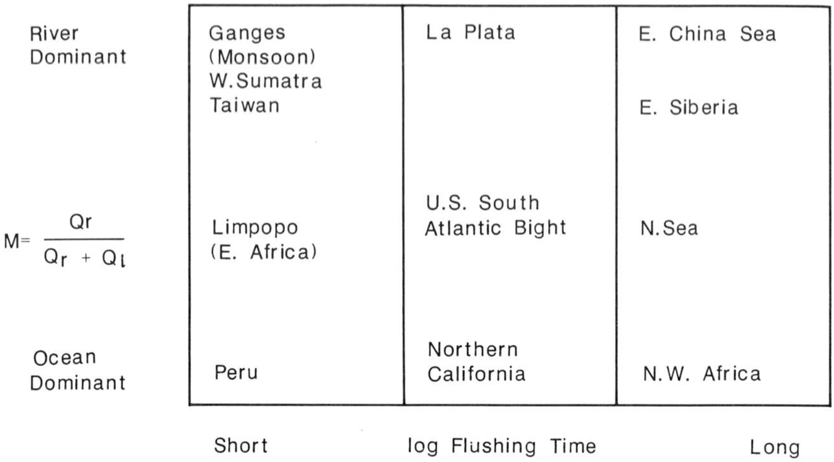

$M = \frac{Q_r}{Q_r + Q_l}$	Short	log Flushing Time	Long
River Dominant	Ganges (Monsoon) W.Sumatra Taiwan	La Plata	E. China Sea E. Siberia
	Limpopo (E. Africa)	U.S. South Atlantic Bight	N.Sea
Ocean Dominant	Peru	Northern California	N.W. Africa

Fig. 2—The Dahlem marginal water flushing diagram with examples of particular regions applied to it. Q_r = river discharge, Q_l = Oceanic water flux into the margin.

End-member Concept Based on Accumulation Erosion Rates

A third classification considers ocean margin areas as sources, sinks, or bypasses. These characteristics are end members that reflect the net result of boundary fluxes. For a given component, does a shelf act as a sink, a source, or does a significant amount escape from the shelf; and how does this balance vary with time?

Discussion of further general descriptors, such as climatic or tectonic regimes, showed them to be covered by one or more of the schemes discussed above. Further, it was noted that these three concepts each have sufficient potential to cover more special characteristics and have enough overlap to be used in combination according to the processes to be studied.

Synthesis of Geographical Variability

The foregoing flux end members are well expressed in the fluvial hydrology/oceanographic hydrography diagram proposed by Group 2 (see Jickells et al., this volume). There is a strong geomorphological control, of tectonic origin, in the x-axis of this diagram, and a climatic control on the y-axis (Fig. 2). Low values of M correspond to arid areas, whereas high values occur in regions of high river flow, at least episodically. The flushing time of the shelf has a first-order control by shelf width and here there is a primary contrast between narrow steep shelves on tectonically active margins (e.g., the Pacific margin) and wider gently sloping shelves of passive margins (e.g., Atlantic and western Indian Ocean; Fig. 3).

WHAT IS THE RELATIONSHIP BETWEEN BOUNDARY FLUXES AND GLOBAL CHANGE?

The group considered several geochemical balance models of the role of the ocean margins in the global cycles of carbon and nitrogen. In view of the fact that this exercise was a central charge of Group 4 (Mackenzie et al., this volume), our group decided to confine its attention to certain specific aspects of these two cycles. We also made some remarks on the shortcomings of geochemical models, the role of the coastal oceans in the global sulfur cycle, and the global implications of seasonal ice formation over the Arctic shelves.

Carbon Cycle

Ocean margins have a large *potential* for sequestering man-made carbon dioxide from the atmosphere. It has been estimated that up to 0.75 gigatons (10^{15} g) of anthropogenic carbon may be buried in shelf areas each year. This amount corresponds to about 15% of the total emissions from fossil fuel combustion. However, examination of how the above figure is determined reveals several important uncertainties. For example, the removal to coastal sediments is partly due to deposition of organic carbon brought in by rivers. However, it is not known how much of this organic carbon is new terrestrial production, and hence a real additional sink, and how much is old fixed carbon, which is not. The other major component is due to increased *in situ* carbon fixation in estuaries arising from additional inputs of anthropogenic nutrients. There is uncertainty concerning this process also, since, although it is thought that fertilized shelf areas are net sinks for atmospheric CO_2, this does not invariably seem to be the case.

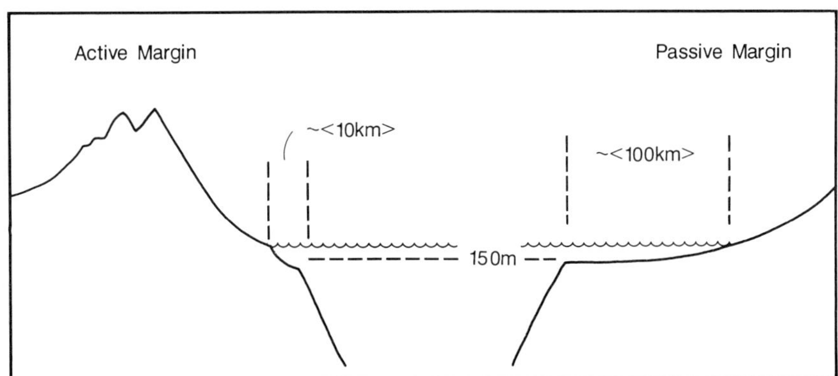

Fig. 3—Principal physiographic contrasts between active and passive (tectonically) margins.

Methane gas is another important component of the carbon cycle, particularly in the atmosphere due to its strong radiative properties. Although high methane levels have been observed in certain coastal areas, these can generally be traced back to riverine inputs (e.g., in the southern North Sea to the Rhine and Scheldt rivers). It does not appear that coastal sediments are sufficiently reducing for methane production to occur.

Nitrogen Cycle

In terms of biogeochemical importance, the role of nitrogen is second only to that of carbon. However, its cycle is significantly more complex than that of carbon due to the variety of redox states in which it can exist in the environment. For this and other reasons, such as lack of estimates of denitrification and nitrogen fixation rates, the nitrogen cycle is very poorly quantified. The importance of upwelling in controlling coastal productivity has already been stressed, but its magnitude is poorly known.

Geochemical budgets often ignore the atmosphere as a source/sink of nitrogen compounds. However, recent estimates indicate that atmospheric inputs to coastal waters may be 10–30% of riverine inputs. In the opposite sense, coastal seawaters may be significant suppliers of the nitrogen-containing gases ammonia and methylamines to the atmosphere. As far as we are aware, no measurements exist and so no estimates of the fluxes of these gases have been attempted for coastal waters. Open ocean data indicate the sea as a significant source at least for atmospheric ammonia and some of the methylamines. Since these compounds are essentially the only alkaline atmospheric gases, this lack of knowledge is in need of urgent rectification.

Geochemical Models

In our consideration of the carbon and nitrogen cycles we noted the large errors associated with many of the fluxes. Often the budgets (assumed to be at steady-state) do not balance, although the discrepancy is generally within the combined errors of the flux estimates. Such closure problems, however, may also indicate the omission of some significant pathway(s) in the computation. We recommend that all geochemical flux balance models should show the range of each estimate, as well as the central value. Then unwarranted confidence in the predictions of the models will be avoided and presently unknown pathways should be more easily predicted.

Sulfur Cycle

Only certain aspects of sulfur cycling in coastal waters were considered. Particular attention was paid to dimethyl sulfide (DMS). This gas is produced by certain algae in marine waters.

On evasion to the atmosphere, where it is oxidized to SO_2 *inter alia*, it plays important roles in controlling the acidity of rain and aerosol particles, and may be involved in climate regulation. It is well established that productive coastal waters can show up to an order of magnitude greater concentration of DMS than open ocean waters and so are potentially powerful sources. The proximity of coastal areas to land means that the resulting atmospheric acidity is more likely to be deposited there rather than redeposited on the sea, as can happen to a more significant fraction over the open oceans.

The production of DMS by algae is very seasonal, with highest values in spring and summer. Its formation is also species specific, for example, diatoms appear to make very little whereas phaeocystis and coccolithophores are powerful producers. This means that change in species abundance brought about by, for example, climatic change or enhanced nutrient inputs, can have a large effect on DMS formation. Only a fraction of the DMS (and its precursor compound, dimethylsulfoniopropionate [DMSP]) formed by the phytoplankton actually escapes to the atmosphere; the rest is either broken down photochemically or by bacteria, or sinks to the bottom with dead organism remains (as discussed by Wolfe et al., this volume). Thus, the concentration of DMS in the surface water, which provides the driving force for evasion to the atmosphere, is the result of the significantly larger internal cycle within the water column. The implication of this is that only a small change in one of the other production or destruction processes has the potential to lead to a substantial alteration in the concentration of DMS, and hence of its flux to the atmosphere.

Ice Formation in the Arctic

The Arctic ice shelves constitute a large but mostly unstudied area of ocean margin. We indicate here a process in the Arctic with profound global implications and which is pivotal in global climate change studies.

Seasonal ice formation over the Arctic shelves results in seasonal brine release to the bottom shelf waters. From here the extremely dense water spreads into the Arctic Ocean, driving the thermohaline circulation of the deep Arctic/Subarctic basin. This circulation is a precondition for the Greenland and Iceland Seas to operate convective gyres, which are the source for the North Atlantic overflows and thus for the formation of North Atlantic deep water. Its characteristics and global spreading are central components in the oceans' climate role. Similar arguments, of course, apply to deep water formation around Antarctica.

SELECTED REFERENCES

Blackburn, T.H., and J. Sorensen, eds. 1988. Nitrogen Cycling in Coastal Marine Environments. SCOPE Report 33. Chichester: J. Wiley.

Davies, A.M., ed. 1990. Modelling Marine Systems. Boca Raton, Florida: CRC Press.

Degens, E.T., S. Kempe, and J. Richey, eds. 1990. Biogeochemistry of Major World Rivers. SCOPE Report 42. Chichester: J. Wiley.

GESAMP 1987. Land/Sea Boundary Flux of Contaminants: Contribution from Rivers. Reports and Studies No. 32, Joint Group of Experts on the Scientific Aspects of Marine Pollution. Paris: UNESCO.

GESAMP 1989. The Atmospheric Input of Trace Species to the World Ocean. Reports and Studies No. 38, Joint Group of Experts on The Scientific Aspects of Marine Pollution. Geneva: World Meteorological Organization.

Gumbel, E.J. 1958. Statistics of Extremes. New York: Columbia Univ. Press.

Jamalov, R.G., I.S. Zectker, and A.V. Meshetell. 1977. Underground Water Discharge to the Seas and Oceans. Moscow: Nauka (in Russian).

Lerman, A., and M. Meybeck, eds. 1988. Physical and Chemical Weathering in Geochemical Cycles. Dordrecht: Kluwer.

Lisitzyn, A.P. 1987. Avalanche Sedimentation and Interrupted Accumulation in Seas and Oceans. Moscow: Nauka (in Russian).

L'vovich, M.I., N.L. Bratseva, G. Ya Karasik, G.P. Medvedeva, and A.V. Meleshko. 1989. The map of modern land erosion of the earth. *Proc. USSR Acad. Sci. Geogr. Ser.* **3**:17–30 (in Russian).

Martin, J.-M., F. Elbaz-Poulichet, C. Guieu, M.-D. Loye-Pilot, and G. Han. 1989. River versus atmospheric input of material to the Mediterranean Sea: an overview. *Marine Chem.* **28**:159–182.

Meade, R.H., T.R. Yuzuk, and T.J. Day. 1989. Movement and storage of sediment in rivers of the United States and Canada. In: The Geology of North America, vol. 0–1, Surface Water Hydrogeology. Boulder: Geol. Soc. America.

Seitzinger, S.P. 1988. Denitrification in freshwater and coastal marine ecosystems: ecological and geochemical significance. *Limnol. Oceanogr.* **33**:702–724.

Walsh, J.J. 1988. On the Nature of Continental Shelves. San Diego: Academic.

Biologically Mediated Removal, Transformation, and Regeneration of Dissolved Elements and Compounds

S.W. Fowler

International Laboratory of Marine Radioactivity, IAEA
19, Avenue des Castellans
98000 Monaco

Abstract. Evidence is reviewed to assess the importance of biologically mediated processes in effecting removal, transformation, and regeneration of elements and compounds in the sea, with the general aim of better understanding fluxes and transport of natural and anthropogenic substances in ocean margin systems. Element scavenging and removal, organic compound transformations, and biogenic particle regeneration of these materials are closely coupled to biological activity, particularly in the epipelagic and mesopelagic zones where the bulk of oceanic biomass resides. Because our knowledge of biogenic particle formation rates and variability therein is sparse on a global scale, it is at present difficult to evaluate quantitatively the relative importance of certain biological processes compared to competing physical and geochemical mechanisms. Some suggestions are made on the type of data needed to test and refine existing marine biogeochemical models for element and compound transport.

INTRODUCTION

Physical, geochemical, and biological processes act in concert to control the distribution and biogeochemical cycles of elements and organic compounds in the sea. The relative importance of biologically mediated removal, transformation, and regeneration mechanisms of these materials is a function of the production of oceanic biomass at any given location. Most of the oceanic biomass resides in the upper few hundred meters. Thus, the zone of most active biogenic particle production, particularly near oceanic margins, is in the surface layers where primary production occurs. Therefore, to understand fluxes and transformations of natural and anthropogenic elements

Ocean Margin Processes in Global Change
Edited by R.F.C. Mantoura, J.-M. Martin and R. Wollast

and compounds in ocean margins, it is important to evaluate the biologically mediated processes which regulate their biogeochemistries of uptake, removal, and regeneration. In this brief review selected examples of biological mechanisms of bioaccumulation, excretion, food chain transfer, and vertical transport are discussed as they pertain to element/compound scavenging, transformation, and regeneration. Emphasis is placed on the importance of small species and biogenic particles, and how they relate to the removal and transport of anthropogenic materials introduced at the sea surface.

BIOACCUMULATION

Element uptake by organisms occurs by adsorption onto external surfaces, absorption across surfaces such as integument, gill and gut walls, or a combination of both. For small organisms such as phyto-, nano-, and pico-plankton, most evidence suggests that uptake of trace elements and radionuclides is very rapid (hours to days) and is a surface sorption phenomenon involving multiple binding sites (Fisher and Fowler 1987; Lion and Rochlin 1989). In the case of hydrophobic organic compounds, concentration takes place by direct partitioning between seawater and the lipid pool. For heterotrophic species, the alternative mode of uptake is by assimilation from food.

Concentrations of these materials in organisms are in a state of dynamic equilibrium and are the net result of both uptake and elimination occurring simultaneously. Rates of these processes are controlled by exposure time, the physicochemical form of the element, salinity, temperature, competitive effects with other substances, life cycle of the organism, physiology, feeding habits, etc. Therefore, element concentrations in organisms can be expected to vary with time and will be a function of both uptake rates and the source term (e.g., riverine input, atmospheric input, oceanic water mass). For these reasons, concentrations and concentration factors are best viewed in terms of ranges rather than absolute values. Concentration factors for several trace metals and radionuclides in phytoplankton and zooplankton are given in Table 1. They vary widely in plankton from approximately 1 to nearly 10^6 depending on the species as well as the element involved. The highest concentration factors are noted for elements which are "particle reactive" in seawater, whereas low values are typical for elements which behave conservatively in seawater. Equally high concentration factors of 10^5–10^6 have been reported for DDT and PCBs in planktonic species (Fowler 1982). Such high enrichment factors alone make small species potentially important in affecting the subsequent redistribution of these materials throughout the water column.

Besides accumulating elements and compounds directly from water,

TABLE 1. Element and radionuclide concentration factors† for phytoplankton and crustacean zooplankton (from Fowler 1990).

Element or RN	Phytoplankton	Microzooplankton††	Macrozooplankton*
**Hg	5×10^4	1×10^4	2×10^4
**Ni	2×10^3	4×10^3	6×10^2
**Co	1.5×10^3	1×10^4	6×10^3
**Cd	2×10^4	4×10^4	2×10^4
**Cu	1×10^4	5×10^4	1×10^5
**Fe	5×10^5	3×10^5	2×10^5
**Zn	5×10^5	2×10^5	1×10^5
Tc	$\approx 10^0$	10^0–10^1	$\approx 10^0$
$^{239+240}Pu$	$9 \times 10^4 - 1 \times 10^5$	5×10^3	1×10^2
^{241}Am	$2 \times 10^4 - 1 \times 10^5$	3×10^3	1×10^3
^{144}Ce	9×10^4	1×10^3	—
^{106}Ru	1×10^5	3×10^3	—
^{238}U	1×10^1	5×10^0	—
^{232}Th	2×10^4	2×10^4	—
^{230}Th	8×10^3	4×10^3	—
^{228}Th	2×10^4	6×10^3	—
^{226}Ra	2×10^3	1×10^2	—
^{210}Po	1×10^4	2×10^4	1×10^4

†Defined as element/g wet animal divided by element/g water.
††Mainly copepods.
*Euphausiids.
**Computed using recent values for element concentration in seawater.

heterotrophs also assimilate these materials from the food chain. In the case of nutrients like carbon, nitrogen, phosphorus, and some essential trace elements, ingestion is undoubtedly the main vector. However, for many trace elements and microcontaminants the situation is far from clear, and there is much controversy in the literature as to the relative importance of the food and water pathways in uptake processes (Fowler 1982). This, in turn, has led to much confusion about contaminant "biomagnification" in marine food chains. In general, most evidence from the field demonstrates that trace element and radionuclide concentrations in marine organisms tend to decrease with each increasing step in the food chain. The only exceptions appear to be cesium and mercury, which have been found to increase slightly in some plankton–nekton food chains. However, for mercury it is not known which form (inorganic or organic) is taken up. Furthermore, the question is open as to whether in smaller heterotrophs mercury is primarily accumulated from water or biomagnified through the food chain. Studies on organochlorine compounds in similar species have also produced conflicting results; however, it appears that while biomagnification of PCBs and DDT occurs at the higher end of marine food chains in large organisms with high lipid contents, the opposite is true for smaller planktonic species in which uptake (partitioning) from water is more important (Fowler 1982).

ELIMINATION

Regardless of how elements and compounds are accumulated, their excretion is important in maintaining elemental balance within the organism as well as affecting the biogeochemical cycles of these materials. Release occurs by passive desorption, active excretion of soluble element, and particulate excretion in the form of feces, molts, and reproductive products. Most often elements are lost from marine species more slowly than they are accumulated. Loss rates are rarely constant and are subject to control by the same parameters that affect accumulation rates. Half-times for turnover of trace elements, radionuclides, and chlorinated hydrocarbons vary greatly and range from a few hours to several days for small microorganisms (Fowler 1982; Fisher and Fowler 1987). It has also been noted that the shorter the previous exposure to anthropogenic contaminants, the more rapid the subsequent release rate, a finding which has important implications for contaminant biogeochemical cycles. Generally, turnover time for elimination increases with increasing organism size. Excretion rates are also highly species dependent; for example, microcrustaceans which molt lose a large fraction of their element content at frequent intervals (Fowler 1982). Furthermore, molting rates increase with temperature, therefore material elimination via molting may be enhanced in the warmer, upper water layers.

Another important mode of excretion is defecation. Particularly for nonbiologically essential elements and compounds, or those which are poorly assimilated, excretion via feces becomes increasingly important. Preferential assimilation of organic matter over other ingested material leaves the latter in a more concentrated state (unit weight basis) in fecal material. An example of the fractionation of element elimination by zooplankton is given in Table 2. It is evident that for nonessential elements like plutonium, most of the radionuclide taken up is excreted with the feces. In contrast, elements like Se and Hg, which bind to thiol groups, are excreted mainly in soluble form. Furthermore, it is apparent from Table 2 that with planktonic crustaceans, defecation is a more important contributor to particulate element flux, principally because plankton defecation rates are higher than rates of molt production.

TABLE 2. Relative distribution of element elimination in euphasiids by molting, defecation, and soluble excretion (from Fowler 1982).

Element	Molting (%)	Defecation (%)	Soluble excretion (%)
Zn	1.1	92.6	6.3
Cd	3.3	84.5	12.2
Se	2.4	54.4	43.2
Hg (inorg.)	2.5	29.1	68.4
$^{239+240}$Pu	0.8	98.6	0.6

TRANSPORT MECHANISMS

Passive Transport

Biological production in the upper water layers is the principal process which fixes dissolved inorganic carbon into marine particles. Since these primary particles constitute the initial biological surfaces for the uptake of natural and anthropogenic materials, it is of paramount importance to have accurate measurements of their rate of formation in the euphotic zone. Because of the large spatial and temporal variability in the oceanic distribution of phytoplankton ("patchiness"), such estimates are difficult to obtain on a global scale by traditional methods. Formation of biogenic particles also occurs through the secondary production of heterotrophic organisms such as bacteria, zooplankton, and nekton. The rate of organic carbon formation by secondary production processes is limited by the rate of primary production and generally does not exceed approximately 20–25% of the rate of photosynthesis. However, under certain conditions (e.g., after a phytoplankton bloom) or in certain habitats (e.g., mesopelagic and bathypelagic zones) secondary production processes may be very important for biogenic particle formation. Furthermore, production of bacteria can occur by utilizing dissolved substrates and thus represents an important pathway for the introduction of soluble organic matter into particulate pools.

Although soluble elements and compounds initially become associated with primary particulates, small biogenic particles such as individual phytoplankton cells sink very slowly (≤ 1 m d^{-1}) and thus have minimal effect on the downward flux of incorporated materials. Through various aggregation processes, however, such small particles are "packaged" into larger detrital particles which rapidly sediment at sinking speeds of 10s to 100s m d^{-1}. One common aggregate is fecal material. It occurs as relatively loose, amorphous material (fecal matter) that readily disintegrates or as dense, tightly packed fecal pellets encased in membranes which maintain their integrity (Fowler and Knauer 1986). Fecal material is somewhat adhesive, which increases its probability to agglutinate with smaller particles.

Another important large particle aggregate is "marine snow": an amorphous, flocculent detrital material which is at least partly biogenic (Alldredge and Silver 1988). It is one of the most common components of particulate material caught in sediment traps and is a reactive site for material scavenging. Other biogenic particles of varying importance in element flux are plankton hard parts (pteropod shells, crustacean molts, diatom frustules, etc.) and dead organisms, many of which, depending upon their chemical composition, undergo dissolution before reaching great depths (Fowler and Knauer 1986; Alldredge and Gotschalk 1989). Whatever their origin, the sinking of such large particles has been identified as the primary mechanism determining the vertical fluxes of carbon and other materials in marine waters. The few analyses that have been made to date show that biogenic aggregates are highly enriched in many natural and anthropogenic substances (Table 3). These particles are often substantially more enriched than the organisms producing them and thus have greater potential for removing elements and compounds from the upper water column.

Active Transport

The horizontal and vertical movements of living organisms can affect the transport and eventual removal of elements and compounds associated with these species. Small species like phytoplankton and bacteria cells move little and are largely controlled by water mass movements. However, many larger species of zooplankton and nekton undertake diel vertical migrations over several hundred meters and thus have the potential of rapidly transporting materials vertically within the same depth range. Nevertheless, model studies point out that migratory species only constitute a relatively small fraction of the total biomass and that migration is restricted to roughly the upper 1000 m; therefore, large biogenic particles produced by these organisms are more important vectors for material transport through the entire water column (Lowman et al. 1971). In the case of Fe, for example, computations indicate that in the open ocean, sinking planktonic debris (molts, fecal pellets, dead organisms) would remove Fe from the upper mixed layer in about eight years whereas removal would take 90 years by downward vertical migration alone (Lowman et al. 1971). However, in coastal regions where biomass is greater, the biological transport rate by migration becomes a larger proportion of the total transport rate. In subsequent assessments (for review, see Fowler 1982) the fraction of the elements removed annually by biological processes from the upper layers of the ocean were calculated and

TABLE 3. Levels (μg g^{-1} dry) of trace elements, radionuclides, and PCBs in zooplankton and biogenic particles (from Fowler 1982, 1989).

Element	Macro-zooplankton	Fecal Pellets	Molts (%)*		Food (microplankton)	Marine snow
Ag	0.7	2.1	2.9	(31)	0.7	
Cd	0.7	9.6	2.1	(22)	2.1	3.4
Co	0.2	3.5	0.8	(34)	0.9	
Cr	0.9	38	5.3	(48)	4.9	
Cu	48	226	35	(6)	39	10
Fe	64	24,000	232	(28)	570	12,800
Mn	4	243	12	(21)	18	148
Ni	0.7	20	6.7	(78)	8.1	25
Pb	1	34	22	(≈100)	11	9
Zn	62	950	146	(18)	483	40
Hg	0.3	0.4	0.2	(4)	0.1	
Se	4	7	2	(3)	3	
Ce	0.2	200	1.2	(44)	0.3	
Eu	0.002	0.66	0.008	(26)	0.013	
Cs	0.062	6.0	0.019	(2)	0.08	
Sb	0.071	71	0.8	(87)	0.22	
Sc	0.009	2.8	0.03	(26)	0.13	
Sr	117	78	350	(23)	520	
$^{239+240}Pu$**	0.4	98	4.8	(90)	4.0	19
^{241}Am**		72				10
^{210}Po**	1100	24,500	360	(2.5)	3400	
^{210}Pb**		10,400				
^{232}Th**	0.35	250	2.6	(57)	17	
^{238}U**	21	520	245	(90)	340	
PCBs	0.62	16	1.4	(17)	4.5	

*Percent total body burden in molt.
**pCi/kg dry. 1 pCi = 37 mBq.

compared with fractions of the same elements lost each year by geochemical sedimentation. These comparisons suggested that Mn, Fe, Zn, Co, Zr, and Pb were removed from the upper mixed layer more by sinking biodetritus than by either vertical migration or geochemical processes. Lowman et al. (1971) have also evaluated the relative effectiveness of vertical transport of elements upward from deep waters by both biological and physical processes. Their calculations show that concentration factors in deep-dwelling zooplankton would have to be in the range of 10^5 to 10^6 for biological transport upward to equal physical transport. As concentration factors rarely exceed 10^4–10^5 for most elements (see Table 1), it was concluded that physical transport is generally more effective than biological transport in moving trace elements from depth to the surface.

Although such model studies underscore the importance of biogenic detritus in effecting the downward vertical transport of elements, vertically migrating species could considerably bias transport rates measured by sediment traps in the upper 1000 m. Whereas sediment traps in deep waters measure integrated particle flux consisting of all biogenic particles generated in the overlying waters, traps located in epipelagic or mesopelagic waters will miss all excretory material carried down in the stomachs of vertical migrants that fed in the upper layers and defecated at a depth beneath the traps (Angel 1984). Such a short-circuit mechanism leads to an underestimate of biogenic particle flux to the deep sea and will be particularly important for larger species, whose migration rates are commensurate with or faster than the settling rates of larger fecal pellets, and for those species whose gut retention times are long enough for the animal to reach maximum depth before defecation, but short enough to assure that the organism will not transport ingested material back up in the water column during its next ascent. Furthermore, downward seasonal migrations performed by many organisms could result in the same effect, i.e., gross changes in the depth of input of biodetrital particles and of the resultant pattern of vertical particulate flux (e.g., see Youngbluth et al. 1989). Unfortunately, hard data with which to test these hypotheses are scarce; however, the limited information that exists for vertically migrating zooplankton suggests that the fraction of total defecation that occurs at depth during the day is very small, i.e., < 1% (Dagg et al. 1989).

BIOLOGICALLY MEDIATED REMOVAL

Nutrients and Organic Compounds

The major nutrients carbon, nitrogen, and phosphorus are removed from surface waters when organisms die or are eaten and sink in the form of detritus. Fractions of the particulate organic carbon (POC) and nitrogen (PON) produced in the euphotic zone which are exported to deeper waters as "new production" vary from a few percent in oligotrophic regions to several tens of percent in eutrophic waters. Generally, POC and PON flux decreases with depth but some increases have been noted in the mesopelagic zone suggesting *in situ* production of organic matter in the aphotic zone (Karl and Knauer 1984). Anomalous increases in particle flux at depth, however have also been shown to result from horizontal transport of particles from the shelf and slope to deeper waters, either as resuspended bottom sediments (e.g., nepheloid layers [Monaco et al. 1990]) or as layers of fresh biogenic aggregates flowing along isopycnal surfaces (e.g., fecal pellets [Youngbluth et al. 1989]). At present our understanding of such transport processes is very limited but they are likely of far more importance in ocean margins than in the open ocean.

A large variety of lipids, amino acids, natural hydrocarbons, sugars, sterols, wax esters, and pigments have been measured in fecal pellets and other biogenic detritus (Lee and Wakeham 1988), and are effectively removed from the upper water layers by large particle sinking. Since many of these compounds are labile and are transformed and/or regenerated during descent (see below), the organic composition of the particles will be dramatically altered upon reaching great depth.

Anthropogenic compounds such as PCB, DDT, HCB, lindane, and petroleum hydrocarbons are common components of biogenic particles and appear to be more stable. Relatively high PCB concentrations (1–16 $\mu g\ g^{-1}$ dry) have been measured in zooplankton fecal pellets, and it has been proposed that zooplankton defecation is an important mechanism for removing PCBs from surface waters (Fowler 1982; Burns et al. 1985). Evidence from sediment trap studies bears this out and shows that this removal mechanism is at least partially responsible for the decrease in PCB concentrations over time noted in Mediterranean coastal waters. PCB concentrations in biogenic particles with depth as well as comparison with sedimentation data indicate that PCBs are relatively stable in sinking particles (Fowler et al. 1990). On the other hand, more soluble chlorinated hydrocarbon compounds like lindane and HCB are partially leached from the particles before they reach the sediment. Likewise, petroleum hydrocarbons entering surface waters are also removed by fecal pellet sedimentation; however, they are more labile in these particles than are the chlorinated hydrocarbons (Burns et al. 1985).

Trace Elements and Radionuclides

As shown in Table 3, biogenic particles are highly enriched in many elements and radionuclides. This enrichment occurs both on external surfaces and within the organic matrix (Collier and Edmond 1984), hence, the sinking of both live and detrital particles will remove those elements from the water layers where they were accumulated. Furthermore, the sinking particles may scavenge additional dissolved elements from the water by adsorption or bacterial-related uptake processes (Jannasch et al. 1988). Some evidence for these processes has been derived from examining dissolved trace element profiles which show various degrees of element depletion or enrichment with depth (Bruland 1983). For example, Cd and Zn are two trace elements whose profiles show a marked surface depletion and enrichment at depth. As such, these elements are closely correlated with phosphate, nitrate, and silica concentrations at all depths, suggesting a nutrient-type regeneration pattern. In contrast, Cu profiles in the open Pacific are characterized by a near-linear increase with depth. This pattern is typical when *in situ* scavenging onto sinking particles occurs.

The best evidence of element removal comes from sediment trap studies, which allow describing the rates of change in particulate trace metal concentrations with depth and how the sinking particles lead to either element enrichment or depletion in the water column (e.g., Masuzawa et al. 1989). Close linkage to biologically mediated processes is suggested by long-term sediment trap observations showing that fluxes of Al, Cd, Cu, Fe, Mn, Ni, P, Pb, V, and Zn in deep waters (3200 m) vary temporally and are closely correlated to the seasonal cycle of primary production in the surface waters (Jickells et al. 1984). It is more probable that such deep water trace element flux cycles are, in fact, closely coupled to the seasonal cycle of "new production" or particulate carbon export from the euphotic zone; however, the necessary data to support this hypothesis are still forthcoming.

Similar studies of both natural and artificial radionuclides have furnished some of the most striking evidence to date of the biological control on element removal and vertical flux (for review, see Fowler and Knauer 1986). In Japanese coastal waters, inventories of the particle-reactive, natural radionuclides ^{234}Th, ^{210}Po, and ^{210}Pb have been found to be minimal during the spring phytoplankton bloom. In the highly productive waters off the California coast, Coale and Bruland (1985) measured the rate of removal from dissolved to particulate form (scavenging rate) and the removal rate of particulate ^{234}Th. They found that the residence time of dissolved Th with respect to the first-order scavenging rate constant varied from 6–50 days and was a function of primary productivity or particle abundance, while the very short particle residence times for ^{234}Th (2–20 days) in the upper 50 m were a function of zooplankton (salp) grazing and fecal pellet production. Simultaneous measurements with sediment traps confirmed that large biogenic particles such as fecal pellets were removing ^{234}Th from the overlying waters.

The surface-introduced transuranium nuclides, $^{239+240}$Pu and ^{241}Am, are strongly subject to biological scavenging and removal from surface waters. Seasonal sediment trap studies in the deep north Atlantic have shown that $^{239+240}$Pu fluxes at 3200 m co-vary with primary productivity cycles in the overlying waters (Bacon et al. 1985). In the N.E. Pacific, very low $^{239+240}$Pu concentrations in surface seawater and concomitant high fractions (32%) of filterable particulate $^{239+240}$Pu have been measured during phytoplankton blooms (Fowler et al. 1983). Furthermore, copepod fecal pellets from surface waters were found to be greatly enriched in $^{239+240}$Pu and ^{241}Am most probably due to inefficient radionuclide assimilation. Similar biogenic particles caught in sediment traps at these sites show increasing transuranic concentrations with depth to approximately 1500 m. Such concentration increases most likely result from radionuclide adsorption to the particles as they sink through a subsurface zone of relatively high soluble transuranic concentrations in the water (Fowler 1990). These particles also show an increase in the Am/Pu ratio with depth indicating that ^{241}Am is being removed from the upper water column more rapidly than $^{239+240}$Pu.

The most convincing data come from similar experiments carried out prior to and following the Chernobyl accident. In the Mediterranean, high resolution, time-series sediment trap measurements demonstrated that fission products arriving at the sea surface as a single pulse were rapidly transported to 200 m in approximately seven days (Table 4). Examination of the samples containing the bulk of the radioactivity indicated they were composed mainly of copepod fecal pellets. Fresh pellets collected from copepods grazing above the traps were found to contain similar radionuclide concentrations and ratios as those in the trap materials, confirming that such pellets were responsible for removing the radioactivity to depth. Furthermore, isotopic ratios in pellets, air, water, and copepods indicated that particle-reactive radionuclides like ^{144}Ce, ^{141}Ce, ^{106}Ru, and ^{103}Ru were scavenged to a far greater extent by sinking fecal pellets than the cesium nuclides. This and other Chernobyl studies in the North Sea, Black Sea, and Pacific have clearly demonstrated the rapidity (18–200 m d^{-1}) by which sinking biogenic particles can transfer surface-introduced, particle-reactive elements to depth (Buesseler et al. 1990). By inference one would expect enhanced biogenic particle flux of trace elements during periods of high atmospheric input of these elements, but there are few data available to test this hypothesis (e.g., see Buat-Menard et al. 1989).

Model studies are now beginning to examine the important concepts of how element flux is related to spatial and temporal variations in phytoplankton and zooplankton biomass and, moreover, quantify the coupling to organic carbon export from the euphotic zone (e.g., Fisher and Fowler 1987; Fisher et al. 1988). To verify these models, simultaneous measurements of element flux and biological parameters such as particle formation rates and new production are urgently needed.

TRANSFORMATION AND REGENERATION

Organic Compounds

The classic example of biologically mediated remineralization is the oxidative regeneration of nutrients from dead organisms and organic detritus. For example, the downward vertical flux of POC, which exhibits a nonlinear decrease with increasing depth, is believed to be due to microbial decomposition. This hypothesis, however, has recently been questioned in view of new data, which show that biogenic particles sinking beneath the

TABLE 4. Chernobyl fallout radionuclides in particles collected at 200 m depth in a 2200 m water column and in copepod fecal pellets from surface waters (from Fowler et al. 1987).

	Sediment Trap Samples						Fecal Pellets
Date 1986	13–20 April	20–26 April	26 April–2 May	2–8 May	8–15 May	15–21 May	6 May
			(Bq g^{-1} dry wt.)				
^{95}Zr	<0.07	<0.2	<0.3	<0.2	24.5 ± 1.4	<0.2	1.4 ± 0.8
^{95}Nb	<0.03	<0.1	<0.2	<0.1	31.8 ± 1.1	<0.2	<0.2
^{103}Ru	<0.06	<0.1	<0.2	3.7 ± 0.2	23.6 ± 1.0	14.0 ± 0.4	16.0 ± 1.9
^{106}Ru	<0.2	<0.4	<0.8	1.1 ± 0.5	5.4 ± 1.8	3.5 ± 0.7	5.8 ± 2.9
^{134}Cs	<0.05	<0.05	<0.05	0.41 ± 0.05	2.1 ± 0.2	1.9 ± 0.1	3.4 ± 0.6
^{137}Cs	<0.05	<0.05	0.15 ± 0.08	0.85 ± 0.08	3.8 ± 0.3	4.0 ± 0.1	6.3 ± 1.0
^{141}Ce	<0.2	<0.2	<0.3	1.3 ± 0.7	12.6 ± 0.6	1.1 ± 0.5	0.9 ± 0.4
^{144}Ce	<0.06	<0.3	<0.3	<0.2	13.6 ± 0.7	<0.4	2.5 ± 1.3

Values are decay-corrected for midpoint of sampling period.

euphotic zone are, in general, poor habitats for bacterial growth and are unlikely sites for the active remineralization of organic matter (Karl et al. 1988). If this is the case, biogenic particle fragmentation by physical means and subsequent solubilization of smaller, nonsinking particles may be the predominant processes in controlling the loss of organic carbon from the upper water column. A final problem that can not be ignored is whether the rapid organic carbon regeneration rates inferred from sediment trap-derived, carbon flux profiles are real. Many sediment trap studies, particularly the earlier ones, did not carefully consider the inclusion of microscopic zooplankton "swimmers" that enter traps and contaminate particle samples. It has recently been demonstrated that even a few, cryptic swimmers and/or their detrital products can add considerably to the organic carbon content of the sample. Where organic carbon content has been corrected for swimmers, the resultant POC flux profiles are much more uniform, implying a greatly reduced "regeneration" rate compared to those reported to date (Michaels et al. 1990). The proper assessment of such artifacts is of fundamental importance in quantifying global carbon cycles in ocean margins.

In contrast to the situation in the open ocean, difficulties can arise in estimating regeneration rates of POC and other elements from sediment trap experiments in ocean margins. Shelves and slopes are areas with highly complex hydrography, which leads to sediment resuspension and lateral advection of particles to deeper waters. Such processes often result in an apparent increase in vertical particle flux with depth, which greatly complicates interpretation of element regeneration rates (Monaco et al. 1990).

The grazing and metabolic activities of zooplankton result in soluble and fecal excretion of carbon and nitrogen compounds with markedly different biochemical composition than the ingested food (Lee and Wakeham 1988; Altabet 1988). In addition, composition of excreta will depend upon whether herbivory or carnivory is involved. Hydrolysis products of phytoplankton carotenoids and intact carotenoids have been found in zooplankton fecal pellets and have been used to differentiate feeding activities of different animal types. In contrast, phaeopigments in fecal pellets undergo little change over depth suggesting no significant regeneration of these pigments during descent. Sediment trap studies have resulted in a myriad of examples of the transformation of lipids, amino acids, natural hydrocarbon sterols, wax esters, etc. in organisms and biogenic particles (for review, see Lee and Wakeham 1988). These works have contributed significantly to interpreting the distribution of organic compounds in deep-sea sediments as well as delineating zones in the water column where particle reworking activities occur.

Far less effort has gone into studying the transformation and regeneration of the more refractory anthropogenic hydrocarbons. Work in the Mediterranean coastal zone shows that PCB flux associated primarily with biogenic particles leaving the euphotic zone agrees well with PCB deposition rates in sediments at 200 m, indicating little or no PCB regeneration from large particles within this depth range (Burns et al. 1985). However, in deeper areas of the Mediterranean, sinking particles trapped near bottom and benthic flocs contain PCB concentrations two to three orders of magnitude higher than levels in the surface layer of sediments (Fowler et al. 1990). This suggests that some regeneration of these compounds is taking place at the benthic boundary layer. For the more water soluble compounds, such as lindane and HCB, there is strong evidence that they are lost from sinking particles before reaching great depth (Burns et al. 1985). These differences in chlorinated hydrocarbon regeneration behavior in sinking particles should result in quite different soluble concentration profiles of these compounds in the water column; however, to date very few studies have examined this relationship.

Elements

Besides particle fragmentation and subsequent solubilization, element remineralization is a function of the relative reactivity of the element with the organism or biogenic particle (for review, see Fowler and Knauer 1986). Analyzing sediment trap particles at several depths, Masuzawa et al. (1989) have classified various trace elements according to changes in element concentration and element/Al ratio with depth. Both concentration and element/Al ratio decreased with depth for I, Ba, Ca, and Sr indicating rapid remineralization of these "biogenic elements" which are associated with sinking particles. While the ratios of As, Sb, Se, and Ag in the same samples also decreased with depth, indicating element remineralization from the biogenic phase, their element concentrations remained nearly constant suggesting that uptake and scavenging was taking place simultaneously. In contrast, Ce and Mn concentration ratios increased with depth indicating strong scavenging by particles. It is clear from radionuclide studies that Ce deposited in surface waters is rapidly scavenged by biogenic particles (Fowler et al. 1987); however, recent evidence indicates that Ce and other rare earth elements are also regenerated from the same types of particles at depth (Buesseler et al. 1990).

In the case of transuranic radionuclides, there is strong evidence from sediment trap studies and sediment core data that Pu and Am are also remineralized from biogenic particles in deep waters (Fowler and Knauer 1986; Fowler 1990). Such remineralization rates for elements and radionuclides are difficult, if not impossible, to measure *in situ*; however, laboratory radiotracer studies have shown that Pu and Am are not irreversibly bound to fecal pellets and other biodetritus but are regenerated with half-times on the order of one to three weeks (Fowler 1982; Fisher and Fowler 1987). These rates will vary depending upon the type of biogenic particle and whether the element/radionuclide is internally incorporated into the particle matrix (e.g., fecal pellets from grazing zooplankton) or adsorbed to the external surface via element scavenging by the particle. Carefully controlled radiotracer experiments (Fisher and Fowler 1987; Jannasch et al. 1988) and trace element leaching studies (Collier and Edmond 1984) with biogenic particles offer much promise in furnishing rate information on element transfer between the soluble and particulate pools.

Finally, element scavenging and regeneration processes are strongly controlled by the redox chemistry of the surrounding waters. For example, dissolved Mn is scavenged and removed from surface water by biogenic particles; however, as the particles sink through the oxygen minimum, Mn is released only to be readsorbed onto particles below the oxygen minimum (Martin and Knauer 1985). For Se in reducing waters, a depletion of selenite and selenate and concomitant increase of organic selenide has been observed in the water column. Regeneration experiments with biogenic particles have demonstrated that on short-time scales (hrs.), organic selenide is, in fact, the species primarily regenerated with very little selenite and no selenate emanating from the decomposing detritus (Cutter 1982).

CONCLUSIONS

Evidence coming from a variety of sources strongly implicates biological processes in the removal, transformation, and regeneration of elements and compounds in the sea. Because our knowledge of particle formation rates is scanty on a global scale, it is difficult to evaluate quantitatively the coupling between biogenic particle production and export from the overlying waters, and element/compound scavenging, transformation, and regeneration observed at various depths in the water column. Nevertheless, in ocean margins where biomass is often the greatest, such processes become very important, and obtaining quantitative rate information will allow a direct comparison of the relative importance of biologically mediated processes with competing physical and geochemical mechanisms. Furthermore, deriving aggregation/disaggregation rates for biogenic particles at various depths as well as information on element uptake and release by these particle types under different conditions will greatly enhance our current understanding of the overall importance of the processes involved.

Acknowledgements. The International Laboratory of Marine Radioactivity operates under an agreement between the International Atomic Energy Agency and the Government of the Principality of Monaco. I thank Tom Church for critically reviewing the manuscript.

REFERENCES

Alldredge, A.L., and C.C. Gotschalk. 1989. Direct observations of the mass flocculation of diatom blooms: characteristics, settling velocities and formation of diatom aggregates. *Deep-Sea Res.* **36**:159–171.

Alldredge, A.L., and M.W. Silver. 1988. Characteristics, dynamics and significance of marine snow. *Prog. Ocean.* **20**:41–82.

Altabet, M.A. 1988. Variations in nitrogen isotopic composition between sinking and suspended particles: implications for nitrogen cycling and particle transformation in the open ocean. *Deep-Sea Res.* **35**:535–554.

Angel, M.V. 1984. Detrital organic fluxes through pelagic ecosystems. In: Flows of Energy and Materials in Marine Ecosystems, ed. M.J.R. Fasham, pp. 475–516. New York: Plenum.

Bacon, M.P., C.-A. Huh, A.P. Fleer, and W.G. Deuser. 1985. Seasonality in the flux of natural radionuclides and plutonium in the deep Sargasso Sea. *Deep-Sea Res.* **32**:273–286.

Bruland, K.W. 1983. Trace elements in sea-water. In: Chemical Oceanography, ed. J.P. Riley and R. Chester, vol. 8, pp. 157–220. London: Academic.

Buat-Menard, P., J. Davies, E. Remoudaki, J.C. Miquel, G. Bergametti, C.E. Lambert, U. Ezat, C. Quetel, J. La Rosa, and S.W. Fowler. 1989. Non steady-state biological removal of atmospheric particles from Mediterranean surface waters. *Nature* **340**:131–134.

Buesseler, K.O., H.D. Livingston, S. Honjo, B.J. Hay, T. Konuk, and S. Kempe. 1990. Scavenging and particle deposition in the southwestern Black Sea—evidence from Chernobyl radiotracers. *Deep-Sea Res.* **37**: 413–430.

Burns, K.A., J.-P. Villeneuve, and S.W. Fowler. 1985. Fluxes and residence times of hydrocarbons in the coastal Mediterranean: how important are the biota? *Est. Coast. Shelf Sci.* **20**:313–330.

Coale, K.H., and K.W. Bruland. 1985. ^{234}Th:^{238}U disequilibria within the California Current. *Limnol. Ocean.* **30**:22–33.

Collier, R., and J. Edmond. 1984. The trace element geochemistry of marine biogenic particulate matter. *Prog. Ocean.* **13**:113–199.

Cutter, G.A. 1982. Selenium in reducing waters. *Science* **217**:829–831.

Dagg, M.J., B.W. Frost, and W.E. Walser, Jr. 1989. Copepod diel migration, feeding, and the vertical flux of phaeopigments. *Limnol. Ocean.* **34**:1062–1071.

Fisher, N.S., J.K. Cochran, S. Krishnaswami, and H.D. Livingston. 1988. Predicting the oceanic flux of radionuclides on sinking biogenic debris. *Nature* **335**:622–625.

Fisher, N.S., and S.W. Fowler. 1987. The role of biogenic debris in the vertical transport of transuranic wastes in the sea. In: Oceanic Processes in Marine Pollution, ed. T.P. O'Connor, W.V. Burt, and I.W. Duedall, vol. 2, pp. 197–207. Malabar, FL: Krieger.

Fowler, S.W. 1982. Biological transfer and transport processes. In: Pollutant Transfer and Transport in the Sea, ed. G. Kullenberg, vol. 2, pp. 1–65. Boca Raton, FL: CRC.

Fowler, S.W. 1989. Transport and redistribution of trace metals and radionuclides in the marine environment by biogenic particles. In: Proceedings of the Twenty-first European Marine Biology Symposium, ed. R.Z. Klekowski, E. Styczynska-Jurewicz and L. Falkowski, pp. 575–584. Wroclaw: Ossolineum.

Fowler, S.W. 1990. Trace element transfer and transport by marine microorganisms. In: 2nd International Workshop on XRF and PIXE Applications in Life Science, ed. R. Moro and R. Cesareo, pp. 65–78. Singapore: World Scientific.

Fowler, S.W., S. Ballestra, J. La Rosa, and R. Fukai. 1983. Vertical transport of particulate-associated plutonium and americium in the upper water column of the Northeast Pacific. *Deep-Sea Res.* **30**:1221–1233.

Fowler, S.W., S. Ballestra, and J.-P. Villeneuve. 1990. Flux of transuranium nuclides and chlorinated hydrocarbons in the northwestern Mediterranean. *Cont. Shelf Res.* **10**, in press.

Fowler, S.W., P. Buat-Menard, Y. Yokoyama, S. Ballestra, E. Holm, and H.V. Nguyen. 1987. Rapid removal of Chernobyl fallout from Mediterranean surface waters by biological activity. *Nature* **329**:56–58.

Fowler, S.W., and G.A. Knauer. 1986. Role of large particles in the transport of elements and compounds through the oceanic water column. *Prog. Ocean.* **16**:147–194.

Jannasch, H.W., B.D. Honeyman, L.S. Balistrieri, and J.W. Murray. 1988. Kinetics of trace element uptake by marine particles. *Geochim. Cosmochim. Acta* **52**:567–577.

Jickells, T.D., W.G. Deuser, and A.H. Knap. 1984. The sedimentation rates of trace elements in the Sargasso Sea measured by sediment trap. *Deep-Sea Res.* **31**:1169–1178.

Karl, D.M., and G.A. Knauer. 1984. Vertical distribution, transport, and exchange of carbon in the northeast Pacific Ocean: evidence for multiple zones of biological activity. *Deep-Sea Res.* **31**:221–243.

Karl, D.M., G.A. Knauer, and J.H. Martin. 1988. Downward flux of particulate organic matter in the ocean: a particle decomposition paradox. *Nature* **332**:438–441.

Lee, C., and S.G. Wakeham. 1988. Organic matter in seawater: biogeochemical processes. In: Chemical Oceanography, ed. J.P. Riley, vol. 9, pp. 1–51. New York: Academic.

Lion, L.W., and K.L. Rochlin. 1989. Adsorption of Pb (II) by a marine bacterium: the effect of cell concentration and pH. *Est. Coast. Shelf Sci.* **29**:11–22.

Lowman, F.G., T.R. Rice, and F.A. Richards. 1971. Accumulation and redistribution of radionuclides by marine organisms. In: Radioactivity in the Marine Environment, pp. 161–199. Washington, D.C.: National Academy of Sciences.

Martin, J.H., and G.A. Knauer. 1985. Lateral transport of Mn in the northeast Pacific Gyre oxygen minimum. *Nature* **314**:524–526.

Masuzawa, T., S. Noriki, T. Kurosaki, S. Tsunogai, and M. Koyama. 1989. Compositional change of settling particles with water depth in the Japan Sea. *Marine Chem.* **27**:61–78.

Michaels, A.F., M. W. Silver, M.M. Gowing, and G.A. Knauer. 1990. Cryptic zooplankton "swimmers" in upper ocean sediment traps. *Deep-Sea Res.* **37**: 1285–1296.

Monaco, A., T. Courp, S. Heussner, J. Carbonne, S.W. Fowler, and B. Deniaux. 1990. Seasonality and composition of particulate fluxes during ECOMARGE-I in the western Gulf of Lions. *Cont. Shelf Res.* **10**, in press.

Youngbluth, M.J., T.G. Bailey, P.J. Davoll, C.A. Jacoby, P.I. Blades-Eckelbarger, and C.A. Griswold. 1989. Fecal pellet production and diel migratory behavior by the euphausiid *Meganyctiphanes norvegica* effect benthic–pelagic coupling. *Deep-Sea Res.* **36**:1491–1501.

Circulation Processes along Oceanic Margins in Relation to Material Fluxes

J.O. Blanton

Skidaway Institute of Oceanography
P.O. Box 13687
Savannah, Georgia 31416, U.S.A.

Abstract. Wind stress, bottom friction, and buoyancy forces exert considerable influence on advection and mixing processes along coastal margins that affect the flux of material. Their interplay results in the considerable complexities in circulation encountered along continental margins. Some of these are detailed in this chapter.

Frontal zones, such as those produced by freshwater buoyancy fluxes and fronts along western boundary currents, have mixing features that are remarkably consistent in appearance. Some models appear to simulate these features, but quantifying their influence on the flux of material over long time periods remains a challenge.

The circulation in bottom boundary layers influences the transport of material deposited to the bottom. This is particularly true in shallow water where the surface boundary layer, due to wind mixing, merges with the frictional boundary layer generated at the bottom. Oceanographers have some qualitative understanding of the role of reversing wind stress in coastal margins on near-bottom currents, but the rates of material flux due to these currents is far from understood.

Circulation regimes brought about by wind stress, bottom friction, and buoyancy forces are significantly modified by site-specific features such as capes and canyons. These features change flows along margins and induce significant cross-margin currents. Localized upwellings at these sites enhance biological production. Capes, such as Cape Hatteras, may be the sites where large quantities of material are mobilized either to be transported to the open ocean or to settle and be recycled back to the margins.

Ocean Margin Processes in Global Change
Edited by R.F.C. Mantoura, J.-M. Martin and R. Wollast

INTRODUCTION

The cycling, fate, and exchange of material across oceanic margins are affected to varying degrees by biogeochemical and physical processes. Differences in input, in time scales of transport processes, in biological productivity, and in sedimentary regimes which act as sources and sinks lead to differences in the flux and fate of material transferred to and from the continents to the open ocean. This chapter will focus on circulation processes encountered along continental margins and how these processes affect the flux of material. Continental margins are defined here as regions of the oceans in which the circulation is affected by the presence of the ocean bottom and/or the continental land boundary.

The velocity and density fields of the ocean along continental margins are controlled by air–sea interaction processes that alter the balance between buoyancy input and vertical mixing. Processes that dissipate energy through internal and boundary friction are particularly important. For example, the balance between buoyancy input from freshwater discharge and vertical mixing from tidal currents and/or wind stress will determine the location and configuration of a coastal front or will determine whether an estuary has two layers or is vertically well mixed. (There is some dispute about whether local buoyancy and vertical mixing processes versus vertical mixing alone govern front locations [Simpson and Hunter 1974; Stigebrandt 1988]). This chapter does not include an exhaustive discussion of coastal processes since there are good books covering these (e.g., Csanady 1982; Fischer et al. 1979). Rather, I will select some important circulation processes that represent challenges to our understanding of how material is transported across continental margins.

After a brief general discussion of processes and the notion of advective and turbulent fluxes, processes that play a significant role in governing these fluxes will be discussed. Accordingly, this chapter describes exchange processes associated with boundary currents, frontal instabilities, currents in bottom boundary layers, cascading associated with air–sea interactions, and circulation processes associated with the blocking or deflection of along-shelf flow. Finally, some provocative issues will be addressed.

PROCESSES AFFECTING FLUXES

Advective and Diffusive Fluxes

This chapter is concerned with the physical processes that affect the concentration of material. Changes in concentration can be described by the Reynold's-averaged equation for the conservation of material (Pond and Pickard 1978) which, using the equation of continuity, is written

$$\frac{\partial \overline{Q}}{\partial t} + \frac{\partial}{\partial x}(\overline{QU}) + \frac{\partial}{\partial y}(\overline{QV}) + \frac{\partial}{\partial z}(\overline{QW}) = -\frac{\partial}{\partial x}(\overline{Q'u'}) - \frac{\partial}{\partial y}(\overline{Q'v'}) - \frac{\partial}{\partial z}(\overline{Q'w'}) \tag{1}$$

where Q is concentration, (u,v,w) are orthogonal components of velocity in the (x,y,z) directions, and t is time. $Q = \overline{Q} + Q'$, $u = \overline{u} + u'$, $v = \overline{v} + v'$, and $w = \overline{w} + w'$; the overbars represent temporally averaged quantities and primes their deviation from the average value. The quantities ($\overline{Q'u'}$, $\overline{Q'v'}$, and $\overline{Q'w'}$) are turbulent fluxes.

Simple models derived from Eq. (1) have played a pivotal role in advancing our understanding of mixing processes. For example, much of our knowledge of estuarine dynamics has evolved from the simple model that balances the downstream advection of salt by freshwater discharge with turbulent fluxes of salt upstream (Fischer et al. 1979). The models say nothing about the physical processes causing the turbulent mixing. Many models (including those concerning nonconservative tracers) are based on a Fickian diffusion analogy which relates the turbulent fluxes to a diffusion coefficient, such as

$$\overline{Q'u'} = K_x \frac{\partial Q}{\partial x}, \text{ or } \overline{Q'w'} = K_z \frac{\partial Q}{\partial z}. \tag{2}$$

Advection–diffusion models employing various assumptions about K_x, K_y, and/or K_z are useful for analyzing some data sets but they hide the processes that lie behind the diffusion coefficients. Their use can even add confusion as later discussion about shelf-break exchange processes will show. The remainder of this section will discuss the interaction of oceanographic processes along continental margins that result in some of the complexities observed there.

Processes Affecting Mixing and Stratification

The proximity of continental margins is where the circulation is modified by the presence of land boundaries and this distinguishes coastal regimes from those of the open ocean. The density of coastal seawater is often lower than open ocean density due to the presence of freshwater runoff from the nearby continents. Coastal waters are shallower than adjacent open ocean areas, and wind can more quickly transfer its momentum to the bottom where friction retards the current.

The variations of kinetic energy occur at different time scales in coastal areas than do those in open ocean areas (Walsh 1976). Weather systems input energy at frequencies of 0.1 to 0.3 cycles/day (3- to 10-day periods), and a major portion of the kinetic energy spectrum in coastal areas is found at these frequencies. Open ocean variation of kinetic energy occurs at much lower frequencies (i.e., 0.01 to 0.02 cycles/day). The kinetic energy of currents due to astronomical tides is also higher in coastal than in open ocean regimes. Thus, coastal environments contain more high-frequency variability than do open ocean environments at similar depths.

The balance between mixing and buoyancy forces is the single most important factor governing the presence or absence of coastal frontal zones (Simpson and Hunter 1974). These zones, by their very nature, inhibit the transport of contaminants across continental margins.

Vertical mixing processes. Wind stress at the air–sea interface imparts kinetic energy to the water in the form of ocean currents and surface gravity waves. Work is performed because momentum is transferred downward and toward the bottom where the energy is finally dissipated. In shallow water, the downward transfer of momentum reaches the bottom before much kinetic energy is lost, thus providing an efficient source of energy which can mix the water. The shallower the water, the more efficiently mixing occurs throughout the depth range.

Tidal currents exert a stress on the ocean bottom, which mixes water in the same manner as the wind-induced currents. Co-oscillating tides driven by oceanic forcing at the shelf edge are particularly important in producing changes in sea surface elevation and in strong tidal currents. If tidal current amplitude is sufficiently large, and water depth shallow enough, then the tidally induced turbulence at the seabed can penetrate through to the sea surface. In regions where this turbulent Ekman bottom layer reaches to the sea surface, the water column will usually be well mixed.

In deep water with small tidal currents, stratification can occur within the water column, and in some cases may impinge upon the bottom topography. Whenever stratification intersects the bottom topography and there is significant tidal forcing, internal tides can be generated. Internal tides generated at the shelf edge can propagate both onshore and out into the deep ocean, and can provide mixing energy for the water column.

Cooling by the atmosphere can increase the density of surface water which will sink. This convective process is an efficient source of mixing and can lead also to vertically homogeneous water in coastal regions. Evaporation is also a convective process. It increases the density of surface water by leaving salt behind. The convective process can become important in coastal regimes where there are large amplitude cycles in air temperature or where evaporation is high.

The intensity of surface wind stress and the rate at which its momentum is diffused downward are particularly important in determining the bed stress during a major wind event. In addition to these effects, turbulence at the bottom associated with the surface wind wave field can significantly enhance bed stress (Christoffersen and Jonsson 1985; Grant and Madsen 1986) and may cause major sediment movement in many situations.

Buoyancy fluxes. Inputs of heat through solar radiation and freshwater are major sources of buoyancy (i.e., buoyancy fluxes) which act to stratify the water. If buoyancy fluxes are greater than available sources of vertical mixing energy (the work due to surface and bottom stress), the water column is vertically stratified. The water column becomes vertically homogeneous when mixing energy overcomes the available buoyancy forces. Water depth is the important parameter, determining whether the water column will or will not be stratified under given amounts of vertical mixing energy and buoyancy flux (Phillips 1966). The turbulent processes whereby the wind's momentum can erode the pycnocline and subsequently cause it to deepen are very complex and are usually parameterized in terms of some form of entrainment velocity or Richardson number.

The input of heat through solar radiation lowers the density of surface water. A distinct upper layer is formed in the absence of vertical mixing, with a thermocline (pycnocline) at the base. As long as mixing energy is low, relative to the buoyancy forces, the water will remain stratified, although the thermocline could strengthen and weaken, and its depth could change in some relationship to time variations in mixing energy.

Input of freshwater from land can spread out across the surface of more saline (and dense) seawater. The effects of this buoyancy flux are the same as for heating, and a two-layer system is often present in the absence of vertical mixing. Estuaries become vertically stratified when a high input of freshwater forms a strong flux of buoyancy relative to the tidal energy available for mixing. Increased tidal mixing relative to a constant buoyancy flux eventually produces vertically homogeneous estuaries.

While precipitation on the sea surface is a source of buoyancy, its effect in coastal regions is usually small compared to river runoff. This is true because land areas adjacent to coastal waters gather precipitation over many hectares and concentrate the input through riverine systems.

Topographic control. Seabed topography can significantly influence ocean currents. This influence, associated with conservation of vorticity, induces flow to follow f/h contours, where f is the Coriolis parameter and h is water depth. For example, topographic steering is responsible for the major northerly outflow from the North Sea, which is intensified and constrained by the Norwegian trench, a region about 260 m deep along the west coast of Norway. The steering provided by topography can induce lateral shear in currents and thus horizontal mixing. The balance between horizontal buoyancy inputs and horizontal mixing can govern the presence or absence of horizontal density gradients in a manner analogous to the balance of vertical buoyancy fluxes and vertical mixing discussed above.

The shelf edge is an important topographic feature, forming a natural boundary between physical processes in the open ocean and those on the shelf. It can be a region of significant upwelling, associated front formation, and biological activity. It can also support certain wave forms, namely shelf waves, which are intrinsically related to the slope of topography.

The coastline itself is equally important. Coastal features such as headlands can induce horizontal gyre-type circulations within which material can be trapped. Wind-induced setup and setdown of sea level against coastlines and associated pressure gradients can induce strong bottom currents. These, combined with wind waves, can move sediment and adsorbed contaminants.

In summary, oceanic margins are battlegrounds between buoyancy sources and mixing processes whose horizontal scales range from localized river plumes to the shear zones bounded by eastern and western boundary currents. The turbulent convective fluxes of density (ρ) represented by $\overline{\rho'u'}$, $\overline{\rho'v'}$, and $\overline{\rho'w'}$ and material requested by $\overline{Q'u'}$, $\overline{Q'v'}$, and $\overline{Q'w'}$ are governed by the dynamics of mixing and buoyancy processes which alter or redistribute the flux fields of density and materials. The remainder of this chapter will describe circulation processes that play important roles in driving material fluxes and that present formidable challenges in our understanding before they can be incorporated into flux models.

EXCHANGE PROCESSES IN BOUNDARY CURRENTS

Fluxes on many continental shelves are affected to a large extent by the adjacent deep ocean. This is particularly true of upwelling regions of eastern and western boundary currents. Upwelling causes an exchange of continental margin water with the open ocean. Supplies of nutrients are transported to the shelf along isopycnals while the carbon produced is exported. Particulate matter resuspended from the bottom near the shelf break can be advected seaward along isopycnal surfaces over deeper waters of the slope (Churchill et al. 1988). Thus, mixing processes in the bottom boundary layer introduce material to deeper water along the margin. The role of diapycnal (vertical) mixing must also play a major role in transferring material from one isopycnal to another. Several processes can induce an exchange of oceanic and shelf water. These include baroclinic instabilities in shelf-edge frontal zones that physically detach parcels of shelf water (see below), eddies in the ocean that induce cross-shelf currents, and turbulent entrainment (a one-way process) by which a turbulent boundary current entrains adjacent, more quiescent, shelf water.

Disturbances in the form of meanders and eddies propagate along frontal zones in western boundary currents. An example is provided off the southeastern U.S., where frontal disturbances result in exchanges between the water masses of the open ocean and those of the adjacent continental shelf (Fig. 1). The exchanges are driven by intrusions of western boundary current water onto the shelf, thereby displacing equal volumes of shelf water. As a rule, surface intrusions occur in winter while bottom intrusions occur in summer.

The exchange process driven by propagating disturbances has been extensively studied in the Gulf Stream and adjacent shelf off the southeastern U.S. (Hofmann et al. 1981; Atkinson et al. 1987; Lee and Pietrafesa 1987). The resulting invasion of Gulf Stream water can extend halfway or more across the shelf when the disturbance is accompanied by sustained upwelling-favorable wind stress.

Despite studies of these and other western boundary current disturbances, the *net* flux of material such as nitrate, when averaged over time and along the shelf break for extended distances, is still unknown. The ability to quantify fluxes associated with frontal disturbances is confounded by the shear magnitude of measurements required over both time and space. One is faced with determining small differences in large numbers. One study

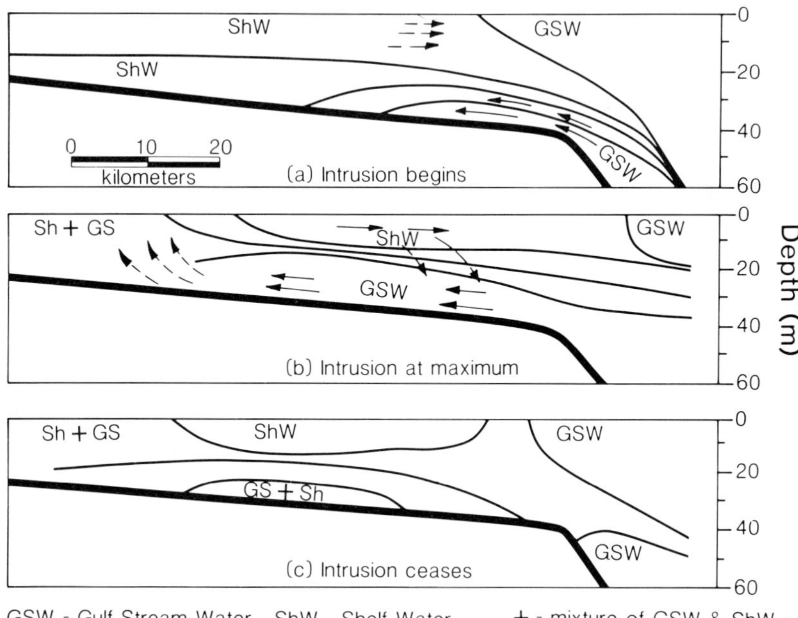

Fig. 1—A schematic sequence of cross-shelf motion illustrating a subsurface intrusion of Gulf Stream Water onto the continental shelf off the southeastern U.S. Adapted from Blanton (1971).

(Lee et al. 1981) reported an onshore nitrate flux averaged over a single 7-day eddy period of 18.6 μM m^{-2} s^{-1}. This is based on temperature (T) and cross-shelf current (u) fluctuations with a positive value of $\overline{u'T'}$, i.e., warmer/colder than average water is transported offshore/onshore. Since the temperature gradient is positive offshore, the turbulent heat flux due to eddies is against the temperature gradient (first observed by Oort 1964, in the axis of the Gulf Stream). If one resorts to an advective–diffusion model to interpret these observations, one would have to use a negative diffusion coefficient. Clearly, this would violate the Fickian diffusion premise upon which such models are based and would offer little insight into such dynamic processes.

Presently, the average value of $\overline{u'T'}$ over a long length (~1000 km) of continental margin is uncertain; however, evidence suggests that the algebraic sign may differ depending upon where along the length measurements are made (T.N. Lee, pers. comm.). It almost certainly depends upon the averaging time, i.e., whether daily, monthly, or yearly, etc. Thus the *net* effect of cross-margin exchanges is an important problem requiring much work on isopycnal and diapycnal transport processes.

FRONTAL ZONES AND INSTABILITIES

Frontal zones are common along oceanic margins. They occur when mixing processes which act to diffuse the front are overcome by buoyancy transport processes. There are three dynamic processes acting at most fronts found along continental margins (Blanton et al. 1989): (a) buoyancy is continuously supplied along one side of the front; (b) along-front wind stress induces cross-front advection of opposite sign in surface and bottom layers; and (c) vertical mixing diffuses the front through such agents as surface wind work, tidal current work at bottom, and convective overturning due to horizontal advection and atmospheric cooling.

While frontal zones, by nature, inhibit the cross-front transport of material, upwelling favorable winds induce an offshore transfer of frontal zone water in the upper layers that is mixed with water seaward of the frontal zone. Model simulations of this condition (Fig. 2) predict that lens of low density coastal water break away from the front providing a mechanism that transfers coastal water across the front (Blanton et al. 1989).

Disturbances along western boundary currents can physically transport material from one side of the front to the other (see next section). This and other examples represent a process whereby small fluctuations in the dynamic state grow in amplitude in time and space. The rate of growth is determined by dynamics associated with the interaction of the initially small perturbations with the original flow state (Pedlosky 1987).

A particularly illuminating study (Qui et al. 1988) describes a model of

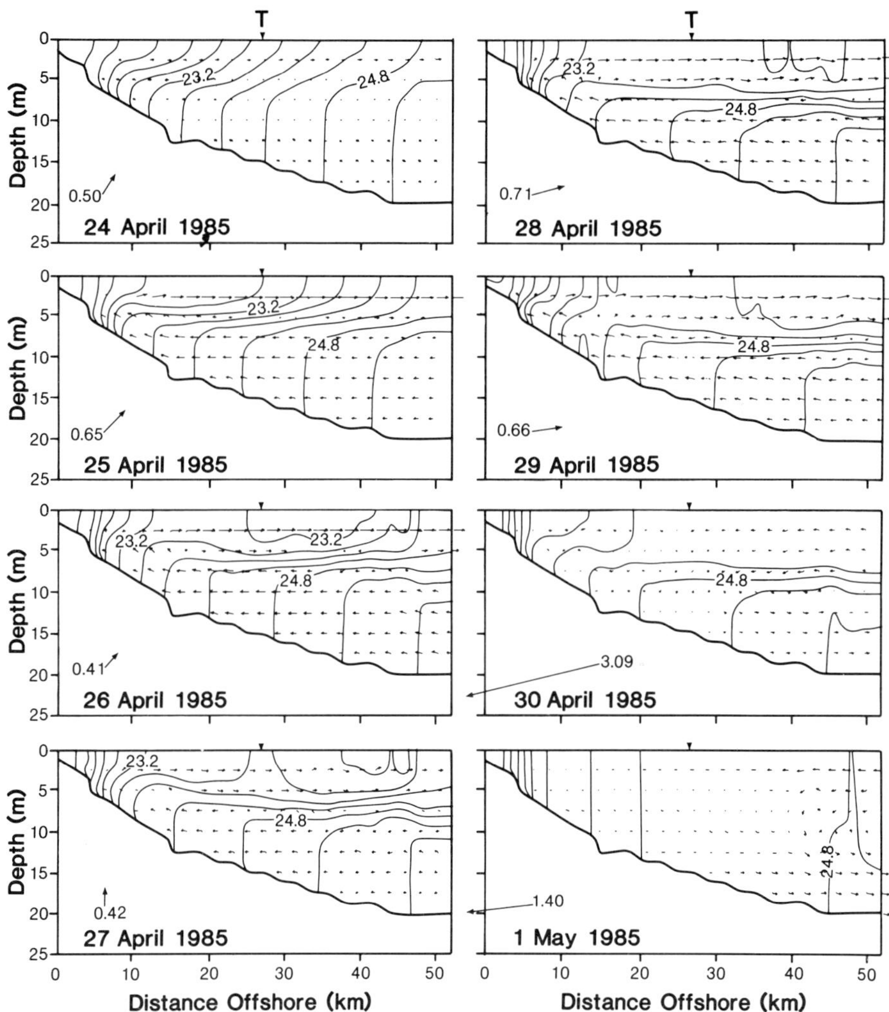

Fig. 2—Daily-averaged contours of sigma-t superimposed on cross-shelf velocity vectors for a shallow coastal front as simulated by a time-dependent x-z numerical model using wind data at location "T." Daily-averaged wind stress (Pa × 10) is plotted at lower left of each panel. Isopycnals are plotted at 0.4 intervals in sigma-t. Cited, with permission, from Blanton et al. (1989).

a strong coastal buoyancy source, such as the Yangtze River in the East China Sea which forms a frontal zone extending 100 km or so along the coast. When buoyancy input (freshwater discharge) is sufficiently large compared to horizontal eddy diffusion, wave-like disturbances that grow in time are triggered along the front. Baroclinic instability in the coastal current produces flow patterns that physically transport passive particles from one side of the front to the other and provides a mechanism that mixes coastal water with ambient ocean water (Fig. 3).

Similar patterns are found in many of the frontal zones present along coastal margins (Fig. 3). Eddies in the Norwegian Coastal Current (Fig. 3, upper right) are thought to be generated by the combined effects of topographic steering, vortex stretching, and barotropic and baroclinic instabilities (Johannessen et al. 1989). Off the west coast of the U.S. (Fig. 3, lower left) swirls, jets, and eddies are produced by the interaction of the meandering California Current with coastal waters. These mixing patterns may also be produced by baroclinic instabilities associated with vertical shear between the surface current and a poleward flowing deeper current (Ikeda and Emery 1984). Similar eddy structures occur on both flanks of the Gulf Stream (Fig. 3, lower right); the swirls on the eastern flank have received little attention. The processes that form those on the western side, while apparently associated with baroclinic instability, are still not well understood (Lee and Atkinson 1983).

The dipole structure of eddies shown and discussed above are probably generic to many coastal regions. The mechanisms which trigger these eddying motions are illusive. Accurate parameterization of this mixing process represents one of the major challenges for determining the flux of material across frontal zones.

Fig. 3—Illustrations of dipole eddy structures that develop as heavier fluid intrudes into lighter buoyant fluid. The heavier fluid is stippled dark. Upper left panel shows sequential patterns of passive particle movements at the surface of a modelled coastal frontal zone showing the onset of instability (adapted from Qui et al. 1988). Light and dark patterns here represent particles initially placed inside and outside the front, respectively. The other three panels show similar dipole patterns redrawn from satellite images. Upper right panel: Norwegian Coastal Current adapted from Figure 1 in Johannessen et al. (1989); lower left panel: Northern California Shelf adapted from an Advanced Very High Resolution Radiomemter image of July 8, 1981 of sea-surface temperature; lower right panel: adapted from the U.S. Naval Oceanographic Office Ocean Frontal Analysis of the Gulf Stream off the southeastern U.S. Gulf Stream for 23 February 1977.

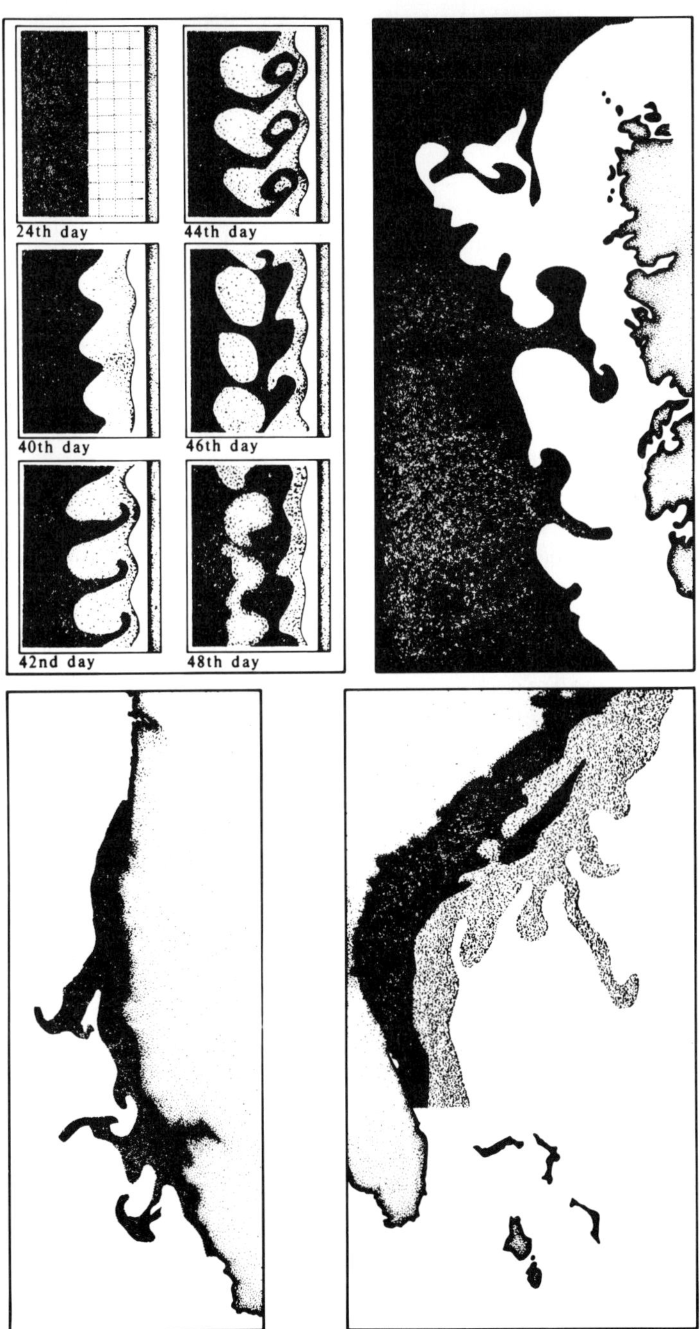
24th day
44th day
40th day
46th day
42nd day
48th day

CURRENTS IN BOTTOM BOUNDARY LAYERS

The depth of a boundary layer (d) is related to u_* (a surface or bottom friction velocity) and the frequency (ω) of u_*, by

$$d = \frac{\kappa u_*}{\omega} \tag{3}$$

where $\kappa = 0.4$ is von Karman's constant. High frequency gravity waves generate thin boundary layers that have a large influence on bottom stress (Grant and Madsen 1986). The value of d can become larger than the bottom depth from u_* induced by the relatively low frequency tidal and wind-generated currents. With a wind drag coefficient of 2×10^{-3}, a wind speed at 10 m above surface of about 20 m/s can vertically homogenize a water column 100 meters thick if there is no input of buoyancy.

If the depth of wind stress influence is on the order of 10% or less of the total water depth, the volume flux due to wind is at right angles to the surface stress. These conditions produce classical upwelling and downwelling circulation along coastal margins with associated adjustments in sea level, water mass, and longer shore currents. The cross-shelf fluxes are carried in surface and bottom frictional boundary layers. The familiar cycles of upwelling and downwelling have been extensively studied (see papers in Richards 1981) and are relatively well understood at times when the frictional boundary layers are separated by a geostrophic interior flow.

In shallow coastal regions, the surface and bottom boundary layers are not necessarily separate. The resulting currents in such a regime influence the flux of material in a complex and little-understood way, although some one-dimensional models are emerging which show promise in advancing our understanding (Mofjeld 1988; Jenter and Madsen 1989). The horizontal flux of material depends importantly on the veering of horizontal currents from surface to bottom.

Earlier (in the section on PROCESSES AFFECTING FLUXES) I described how strong winds in shallow water transmit their momentum vertically to the bottom with little alteration. The presence of stratification resulting from surface heating and from freshwater input at lateral boundaries can inhibit the vertical and horizontal fluxes of momentum and material. There are several parameterization models such as Richardson number, Monin-Obukhov length, and higher-order turbulence closure models that aim to quantify the balance between vertical mixing and the stabilizing effect of stratification. The understanding of currents in bottom boundary layers must ultimately rely on proper understanding of how energy resulting from surface and bottom friction is dissipated. The problem becomes quite complex when the bottom contains moveable material that, when suspended, further alters the balance between vertical mixing and buoyancy forces.

Given a coastal region where the bottom is covered with fine-grained material, upwelling favorable wind stress generates bottom flow with an onshore component at some distance offshore. As depth decreases near the coast, the bottom flow turns parallel to shore. If bottom currents are strong enough to mobilize the material, it would be in a convergent zone which acts to prevent its offshore movement but would allow it to advect alongshore in shallow water. The opposite case of downwelling favorable wind stress generates offshore flow near bottom. Material mobilized in this regime would presumably travel downwind with an offshore component.

The concentration of particulate material (in this case detritus as represented by total chlorophyll) can be seen in results from shallow coastal waters off Georgia in the southeastern U.S. (Fig. 4). Upwelling favorable winds were correlated with relatively high concentrations of total chlorophyll on the bottom and supports the hypothesis that this material was trapped by the onshore component of flow near the bottom. Note that chlorophyll concentrations near the surface were lower than near the bottom. During a downwelling event, chlorophyll data did not show the significant concentrations expected near surface. However, time series of chlorophyll near the area (Walsh 1988) show a clear response to downwelling circulation in which offshore bottom currents advected relatively high chlorophyll concentrations past a mooring in water 15 m deep.

These two hypothetical cases, while useful for discussion, obscure the fact that the *trajectory* taken by a suite of particles is a three-dimensional problem. This trajectory results from currents that are not only a function of space and time but also depend on bottom roughness, the presence of surface gravity waves, and the size and weight distribution of the particles (which might also be a function of space and time!). Clearly, advancements in our understanding of material fluxes in shallow-water environments resulting from simple wind-generated currents will require innovative approaches requiring interaction of chemists, biologists, and physicists.

CASCADING VIA AIR–SEA INTERACTION

Intense cooling of relatively warm shelf waters can increase the density to the point that the water will advect (cascade) off the shelf. Cold continental air masses formed in winter are frequently advected eastward over relatively warm shelf waters. The cold dry winds associated with these cold-air outbreaks remove large quantities of sensible and latent heat from shelf waters and lower the density to the point that the water will sink to greater depths over the continental slope.

Wind stress associated with cold-air outbreaks usually drives oceanic water present at the shelf break landward over the continental shelf where it forms a mid-shelf frontal zone with the colder and denser shelf waters (Oey 1986).

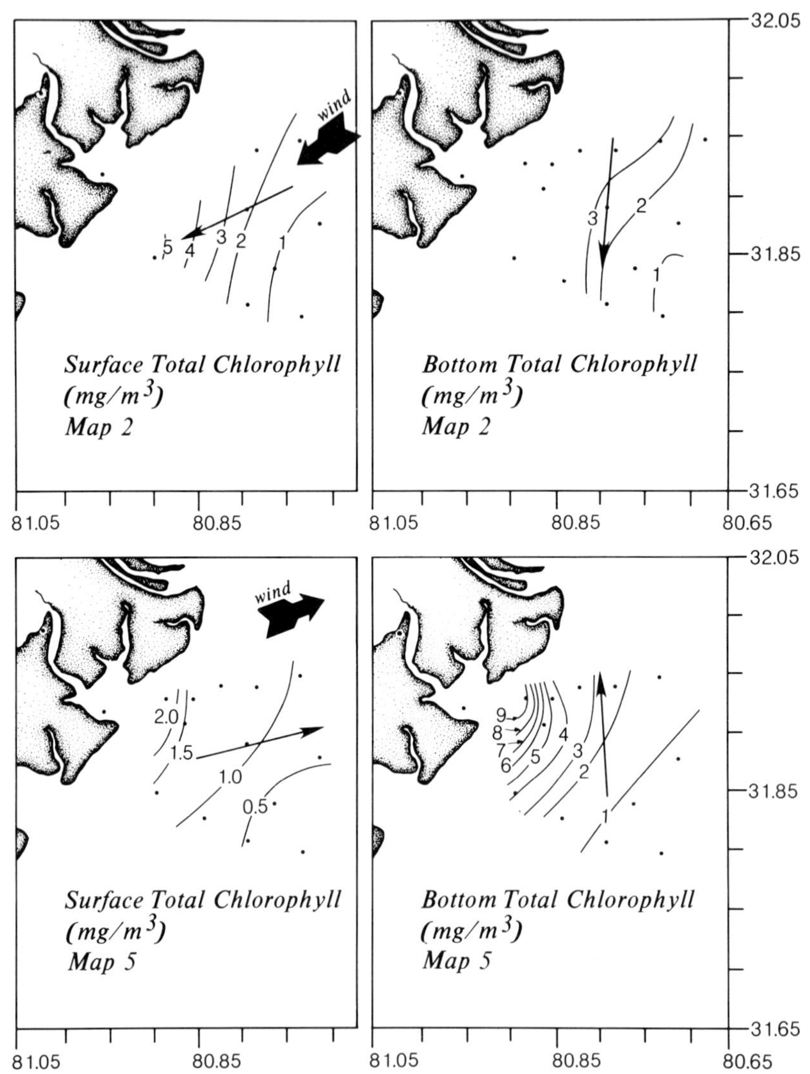

Fig. 4—Surface and bottom concentrations of total chlorophyll off the southeastern U.S. coast during downwelling (upper panels) and upwelling (lower panels) favorable winds. Water depth over the domain varies between 5 and 20 m. Long slender arrows are qualitative indicators of currents based on current meter observations.

There is enhanced vertical mixing at the front as the warm salty water is cooled by air–sea interactions (Fig. 5). When wind relaxes, water masses become unstable: relatively dense water along the front slides underneath the outer shelf waters and cascades to deeper waters beyond the shelf break (Fig. 5). Yoder and Ishimaru (1989) have suggested that cascading on warm temperate shelves could transport a significant amount of phytoplankton carbon off the shelf to the open ocean during winter.

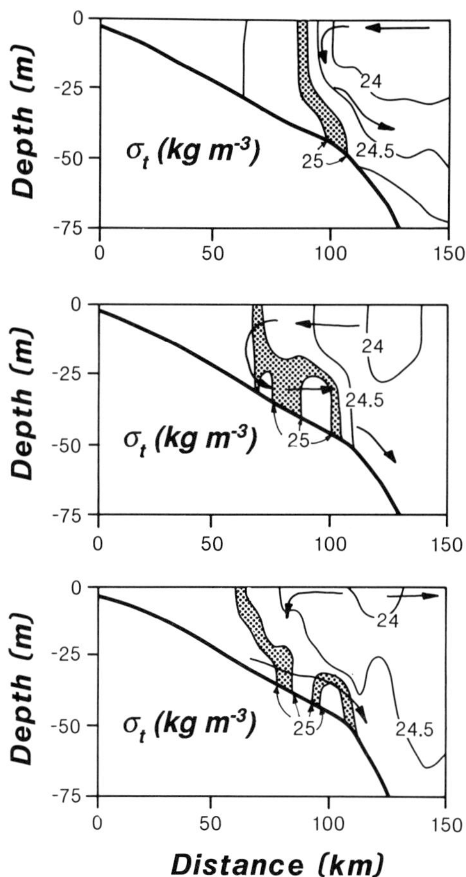

Fig. 5—Cross-shore velocity and density fields 1 day (top), and 5 days (middle) after the start of a strong downwelling-favorable windstress. Bottom panel corresponds to 1 day after wind stress ceases. Adapted from Oey (1986).

TOPOGRAPHIC BLOCKING OF CURRENTS FLOWING ALONG MARGINS

Current components directed across continental margins (hereafter called cross-shelf currents) are usually weak relative to their along-shelf components. Currents of subtidal frequencies tend to follow isobaths. However, changes of topography associated with capes, ridges, and troughs alter the flow along isobaths, and under such conditions, the cross-shelf flux of material becomes enhanced, and density and momentum is redistributed.

Capes and their associated underwater topography play two important roles that influence the flux of material. First, they deflect the currents and induce a cross-shelf component to the motion. Thus, the presence of capes can be viewed as locations where material can be injected farther seaward. Large capes, such as Cape Hatteras (U.S.) and Inubo Saki (Japan), are areas where strong boundary currents flowing poleward separate from the land while the weaker equatorward flow on the northern side is deflected seaward. Second, capes induce upwelling through the conservation of vorticity, which can enhance local biological productivity (Arthur 1965; Blanton et al. 1981). The local production of new material probably finds its way to the sediments of the continental slope and rise. There is circumstantial evidence that the zones of fine-grained organic material found off Cape Canaveral, Florida and on the topographic rise that deflects the Gulf Stream off Charleston, South Carolina (Milliman et al. 1972) are the result of the enhanced productivity associated with topographically induced upwelling.

Topographic features such as canyons and troughs that cut across the continental margin may intercept material transported along margins and convey the material up and down the canyon's axis (Freeland and Denman 1982). If the canyon is narrow, the cross-shelf sea level slope drives an ageostrophic current along the axis. Canyons may play a significant role in net long-term fluxes of material across margins.

CONCLUSIONS—PROVOCATIVE POINTS

Conceptual models of material flux depend upon the type of material and the physical environment. Further development of our ability to quantify these fluxes requires a two-front attack. First, new sampling strategies need to be developed to account for the fact that large fluxes of material occur over relatively brief events. One frequently hears that one tropical cyclone can remove an annual supply of sedimentation. New sampling gear requiring low power must be able to sample particles and process these samples *in situ*. These data need to be acquired simultaneously with data on currents and water masses. This will require effective system integration with the ability to sample energetic events and to shut down during quiescent periods.

A unique system that would yield new insight might follow a mass of water over an event sampling physical, chemical, and biological data in a Lagrangian framework.

Second, current models of the transport and fate of nonconservative materials are too idealistic; most parameterize particles as spheres with an assumed settling velocity. Yet it is becoming increasingly apparent that material consists of nonuniform particles and flocs that are perhaps transformed and even destroyed. What are relevant time scales for transformation and destruction of material? How can we enhance our ability to simulate the behavior of this material in hydrodynamic models of the flow field in a manner that would allow us to predict the trapping of these particles in frontal zones formed by convergences in a bottom boundary layer, thus bringing together several processes discussed separately in this paper.

Many of the processes discussed above are nonlinear. For example, motions across a frontal zone depend upon lateral buoyancy fluxes and upon eddy (mixing) processes, both of which interact to change continuously the cross-front velocity component. It is natural to ask to what extent such motion may have preferred bounds or modes that can occur based on even simple nonlinear interaction. The topic of "chaos" is increasingly popular, but one does not have to invoke it to explain the unstable modes in many models (e.g., Fig. 3, upper left). One can, however, legitimately ask whether even simple nonlinear models of frontal zones produce a velocity field that exhibits a strange attractor (or preferential modes of behavior), *and* do they have physical (rather than mathematical) significance?

There are instances of preferential modes in oceanic currents. Can we place time scales on such modes? For example, the Kuroshio Current off Japan has two modes of meanders (Chao 1984). The large-deflection mode and associated upwelling might produce (hypothetically) more carbon such that long-time-scale changes in boundary current behavior could influence material fluxes along margins. Even the Gulf Stream is thought to exhibit such bimodality (Bane and Dewers 1988), but the time scales have not been adequately defined.

Coastal fronts, in general, play a fundamental role in controlling the fluxes of material across continental margins. Do they exhibit the characteristics of strange attractors with preferred states of motion? The Lorenz model of convection in the atmosphere (Lorenz 1964) has the property of a strange attractor and has been extensively studied by mathematicians. In some sense, the mixing processes shown in Fig. 3 (upper left) can be modeled as convection turned on its side. The thermally induced vertical velocity is replaced by horizontal buoyancy flux. The restoring force of gravity is replaced by friction. Is this analogy useful? Can the system of nonlinear partial differential equations describing frontal dynamics be realistically truncated to a simple system of ordinary differential equations such as Lorenz (1964) did, and can one relate the results to physically meaningful parameters such as buoyancy strength, friction, and horizontal mixing?

Acknowledgements. I express my appreciation to J. Yoder and S. Bishop of Skidaway Institute of Oceanography for providing the total chlorophyll data for Fig. 4. I wish to thank Dr. Dave Menzel who offered many helpful suggestions on the original version of this paper. I also greatly appreciate support from the State of Georgia and the U.S. Department of Energy (Grant DE-FG09-85ER60351).

REFERENCES

Arthur, R.S. 1965. On the calculation of vertical motion in eastern boundary currents from the determination of horizontal motion. *J. Geophys. Res.* **70**:2799–2803.

Atkinson, L.P., T.N. Lee, J.O. Blanton, and G.-A. Paffenhofer. 1987. Summer upwelling on the southeastern continental shelf of the U.S.A. during 1981: Circulation. *Prog. Ocean.* **19**:231–266.

Bane, J.M., Jr, and W.K. Dewers. 1988. Gulf Stream bimodality and variability down-stream of the Charleston Bump. *J. Geophys. Res.* **93**:6695–6710.

Blanton, J.O. 1971. Exchange of Gulf Stream Water with North Carolina shelf water in Onslow Bay during stratified conditions. *Deep-Sea Res.* **18**:167–178.

Blanton, J.O., L.P. Atkinson, L.J. Pietrafesa, and T.N. Lee. 1981. The intrusion of Gulf Stream Water across the continental shelf due to topographically-induced upwelling. *Deep-Sea Res.* **28A**:393–405.

Blanton, J.O., L.-Y. Oey, J. Amft, and T.N. Lee. 1989. Advection of momentum and buoyancy in a coastal frontal zone. *J. Phys. Ocean.* **19**:98–115.

Chao, S.-Y. 1984. Biomodality of the Kuroshio. *J. Phys. Ocean.* **14**:92–103.

Christoffersen, J.E., and I.G. Jonsson. 1985. Bed friction and dissipation in a combined current and wave motion. *Ocean Eng* **12**:387–423.

Churchill, J.H., P.E. Biscayne, and F. Aikman III. 1988. The character and motion of suspended particulate matter over the shelf edge and upper slope off Cape Cod. *Cont. Shelf Res.* **8**:789–809.

Csanady, G.T. 1982. Circulation in the coastal ocean. Dordrecht: D. Reidel.

Fischer, H.B., E.J. List, R.C.Y. Koh, J. Imberger, and N.H. Brooks. 1979. Mixing in Inland and Coastal Waters. New York: Academic.

Freeland, H.J. and K.L. Denman. 1982. A topographically controlled upwelling center off southern Vancouver Island. *J. Marine Res.* **40**:1069–1093.

Grant, W.D., and O.S. Madsen. 1986. the continental-shelf bottom boundary layer. *Ann. Rev. Fluid Mech.* **18**:265–305.

Hofmann, E.E., L.J. Pietrafesa, and L.P. Atkinson. 1981. A bottom water intrusion in Onslow Bay. *Deep-Sea Res.* **28**:329–345.

Ikeda, M., and W.J. Emery. 1984. Satellite observations and modeling of meanders in the California Current System off Oregon and northern California. *J. Phys. Ocean.* **14**:1434–1450.

Jenter, H.L. and O.S. Madsen. 1989. Bottom stress in wind-driven depth-averaged coastal flows. *J. Phys. Ocean.* **19**:962–974.

Johannessen, J.A., E. Svensen, S. Sandven, O.M. Johannessen, and K. Lygre. 1989.

Three-dimensional structure of mesoscale eddies in the Norwegian Coastal Current. *J. Phys. Ocean.* **19**:3–19.
Lee, T.N., and L.P. Atkinson. 1983. Low-frequency current and temperature variability from Gulf Stream frontal eddies and atmospheric forcing along the southeastern U.S. outer continental shelf. *J. Geophys. Res.* **88**:4541–4567.
Lee, T.N., L.P. Atkinson, and R. Legeckis. 1981. Observations of a Gulf Stream frontal eddy on the Georgia continental shelf, April 1977. *Deep-Sea Res.* **28**:347–378.
Lee, T.N., and L.J. Pietrafesa. 1987. Summer upwelling on the southeastern continental shelf of the U.S.A. during 1981: Circulation. *Prog. Ocean.* **19**:267–312.
Lorenz, E.M. 1964. Deterministic nonperiodic flow. *J. Atmos. Sci.* **20**:130–141.
Milliman, J.D., O.H. Pilkey, and D.A. Ross. 1972. Sediment of the continental margin off the eastern United States. *Geol. Soc. Amer. Bull.* **83**:1315–1334.
Mofjeld, H.O. 1988. Depth dependence of bottom stress and quadratic drag coefficient for barotropic pressure-driven currents. *J. Phys. Ocean.* **18**:1658–1669.
Oey, L.-Y. 1986. The formation and maintenance of density fronts on the U.S. southeastern continental shelf during winter. *J. Phys. Ocean.* **16**:1121–1135.
Oort, A.H. 1964. Computations of the eddy heat and density transports across the Gulf Stream. *Tellus* **16**:55–63.
Pedlosky, J. 1987. Geophysical Fluid Dynamics (2nd edition). New York: Springer.
Phillips, O.M. 1966. The Dynamics of the Upper Ocean. London: Cambridge.
Pond, S., and G.L. Pickard. 1978. Introductory Dynamic Oceanography. Elmsford, NY: Pergamon.
Qui, B., N. Imasato, and T. Awaji. 1988. Baroclinic instability of buoyancy-driven coastal density currents. *J. Geophys. Res.* **93**:5037–5050.
Richards, F.A., ed. 1981. Coastal Upwelling. Coastal and Estuarine Sciences 1. Washington, DC: American Geophysical Union.
Simpson, J.H., and J.R. Hunter. 1974. Fronts in the Irish Sea. *Nature* **250**:404–406.
Stigebrandt, A. 1988. A note on the locus of a shelf front. *Tellus* **40A**:439–442.
Walsh, J.J. 1976. Herbivory as a factor in patterns of nutrient utilization in the sea. *Limnol. Ocean.* **21**:1–13.
Walsh, J.J. 1988. On the Nature of Continental Shelves. San Diego: Academic.
Yoder, J.A., and T. Ishimaru. 1989. Phytoplankton advection off the southeastern United States continental shelf. *Cont. Shelf Res.* **9**:547–553.

Heterogeneous Reactions in Coastal Waters

F.M.M. Morel[1], D.A. Dzombak[2], and N.M. Price[1]

[1]*Ralph M. Parsons Laboratory, Bldg. 48-425*
Massachusetts Institute of Technology
Cambridge, MA 02139, U.S.A.

[2]*Department of Civil Engineering*
Carnegie Mellon University
Pittsburgh, PA 15213-3890, U.S.A.

Abstract. In the transition from fresh water to seawater, sudden changes in the chemical environment promote a variety of reactions that control the geochemical cycle of trace elements in coastal waters and estuaries. Notable among these is the desorption of trace metals from suspended particles brought about by the increase in ionic strength and the competitive sorption of alkaline earth cations. New quantitative models of sorption allow us to analyze how dilution, changes in ionic environment, and changes in pH each contribute to sorption and desorption reactions, sometimes canceling each other. Among the other heterogeneous processes affecting geochemical cycles in coastal waters, attention has recently been brought to bear on coagulation and sedimentation, on photochemical dissolution, and on biologically mediated transformations. The interdependency of the biological processes results in a coupling of the geochemical cycles of all bioactive elements.

INTRODUCTION

When a chemical element leaves the continents and enters the coastal waters of the oceans, it is subjected to a radical change in its chemical environment: major ion concentrations and ionic strength increase suddenly while the hydrogen ion concentration usually decreases. Concomitantly, with the exception of a few major ions, the concentration of the element itself gradually decreases owing to dilution of the riverine inputs by oceanic waters. As geochemists we wish to understand what transformations, particularly what phase transitions,

Ocean Margin Processes in Global Change
Edited by R.F.C. Mantoura, J.-M. Martin and R. Wollast

befall such an element. Is it diluted conservatively by seawater? Is it precipitated and trapped in coastal sediments? Or, on the contrary, does it dissolve from the suspended load of the river?

In this chapter we address the issue of heterogeneous reactions in coastal waters, focusing particularly on sorption/desorption reactions. Our goal is to utilize the recent progress in understanding and quantifying reactions of solutes at solid interfaces to examine how the changes in chemical conditions from fresh water to seawater might affect solid/solute partitioning of elements and to compare the predictions with observations. In addition, we examine briefly the role of other reactions involving particles in the chemical transformation of riverine inputs, particularly the role of the biota. Throughout we do not intend to review all known facts regarding heterogeneous processes in coastal waters, rather we attempt to discuss aspects in which new insights have been obtained over the past few years.

SORPTION REACTIONS

The solubility of many if not most trace elements and compounds in natural waters is controlled by sorption on suspended solids rather than precipitation of a mineral phase. All suspended materials in aquatic systems exhibit some affinity for solutes, both organic and inorganic. Sorption of ions is effected chiefly by minerals, particularly oxides of iron, aluminum, and manganese which possess large numbers of surface reactive sites. Sorption of nonionic solutes involves nonspecific interactions (van der Waals forces) that depend primarily on the properties of the compound and of the sorbent. There is little experimental or theoretical chemical basis for discussing the fate of nonionic compounds in coastal waters; the processes are to a first approximation identical in fresh and marine waters. We thus focus here on the sorption and desorption of ionic compounds, particularly trace metals, which are affected by dilution, by ionic strength, by competition for sorbing sites, by complexation, and by pH.

Sorption reactions of ionic solutes in natural waters primarily involve chemical interactions between the ions and surface functional groups. These reactions are also affected by long-range coulombic interactions between all neighboring charged moieties on the surface and the sorbing ions. As a result, the equilibrium constant for a reaction such as Zn^{2+} sorbing on hydrous ferric oxide,

$$\equiv FeOH^0 + Zn^{2+} \rightleftharpoons \equiv FeOZn^+ + H^+;\ K_{ads}, \quad (1)$$

must be written as the product of an intrinsic chemical term and a variable coulombic term:

$$K_{ads} = K_{chem} \cdot K_{coul}. \quad (2)$$

The coulombic term can be derived from some version of the Gouy-Chapman-Stern theory. In the case of the Generalized Two Layer sorption model we use later (Dzombak and Morel 1990), the coulombic term is based strictly on the Gouy-Chapman formula; surface charge is considered to result exclusively from surface acid-base and coordination reactions. The mass law equation for sorption reactions at equilibrium is then written

$$\frac{(\equiv FeOZn^{+})\{H^{+}\}}{(\equiv FeOH)\{Zn^{2+}\}} = K_{chem} \cdot K_{coul} \tag{3}$$

where the surface activities have to be expressed with coherent units such as mol/l^{-1} (= mol $kg^{-1} \times kg\ l^{-1}$). According to Eq. 3, the partitioning of the sorbate is affected by the pH, the concentration of available sorbing sites, and factors that influence the coulombic interactions.

Physical and Chemical Effects on Sorption

Dilution. Consider the situation of a trace metal (say Zn) that is chiefly in particulate form (e.g., $[Zn]_T \approx [\equiv FeOZn^+] \gg [Zn^{2+}]$) sorbed on a solid whose sorption sites are far from saturated (e.g., $[\equiv FeOH^0] \approx [\equiv FeOH]_T \gg [\equiv FeOZn^+]$). The dissolved zinc concentration is then given by the equation:

$$(Zn^{2+}) = \frac{1}{\gamma_{Zn}} \frac{1}{K_{chem}} \frac{1}{K_{coul}} \frac{(Zn)_T\{H^+\}}{(\equiv FeOH)_T} \tag{4}$$

Under conditions where the pH, the activity coefficient, and the coulombic term are constant (i.e., constant ionic strength), and the total metal and sorbent are diluted simultaneously, the dissolved metal concentration (Zn^{2+}) remains constant. Thus, upon dilution, the dissolved/particulate fraction increases proportionally to the dilution factor (Fig. 1a) while the dissolved concentration is perfectly buffered by desorption until the dilution is sufficient to bring the dissolved/particulate ratio near unity (Fig. 1b). At higher dilutions the sorbate is chiefly dissolved and diluted conservatively. Thus, unless there are cancelling effects, dissolved concentration versus salinity plots for trace elements should follow conservative mixing lines only if the sorbed fraction is small. In general one cannot expect to obtain the river endmember concentration by extrapolating dissolved concentration data obtained at high salinity (i.e., high dilution by seawater) to zero salinity. Nonetheless, as shown by Boyle et al. (1974), such extrapolation is a convenient way to calculate a flux across the given salinity surface.

Solid removal. Besides dilution, the particulate load of estuarine and coastal waters can be decreased by removal of the suspended solids to the sediments. In such a case both the total sorbent concentration $(\equiv FeOH)_T$

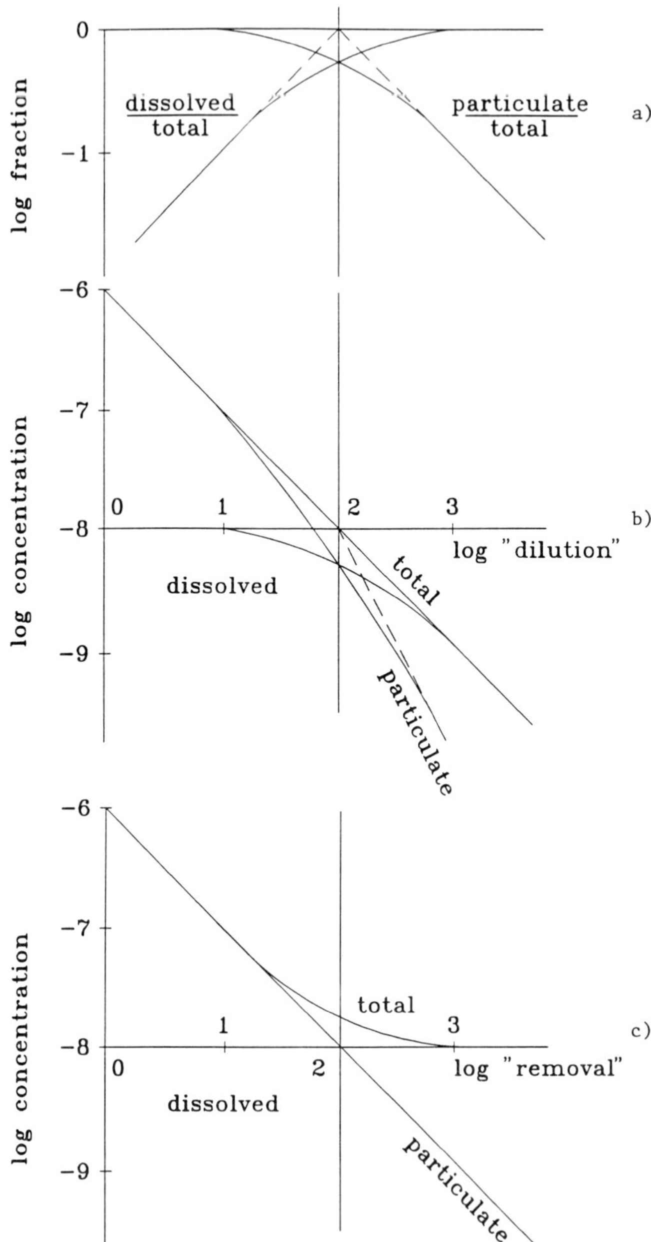

Fig. 1—Effects of dilution and solid removal on the partitioning of a sorbing solute. (a) Fraction of particulate and dissolved sorbate as a function of either "dilution" (seawater/(seawater + freshwater) = salinity/36) or "solid removal" (solid in suspension/original solid concentration). (b) Concentrations of particulate and dissolved sorbate as a function of dilution. Initial undiluted concentrations of 10^{-6} M and 10^{-8} M are assumed for the particulate and dissolved species. (c) Concentration of particulate and dissolved sorbate as a function of solid removal.

and the sorbed concentration ($\equiv FeOZn^+$) are decreased by the same factor, α. The dissolved metal (Zn^{2+}) is thus again constant:

$$(Zn^{2+}) = \frac{1}{\gamma_{Zn}} \frac{1}{K_{chem}} \frac{1}{K_{coul}} \frac{\alpha(\equiv FeOZn^+)\{H^+\}}{\alpha(\equiv FeOH)_T} . \qquad (5)$$

This equality obtains under all conditions; unlike Eq. 4 it does not require the sorbate to be mostly sorbed. The dissolved/particulate fraction then decreases proportionally with α as seen in Fig. 1c. Because our primary interest is in the behavior of the dissolved concentrations of elements, in the following model calculations, we do not examine such possible effects of solid removal.

Ionic strength. It has been many times observed that sorption of cations on oxides is little affected by ionic strength. This puzzling result is explained by a near perfect balance between the decrease in the activity of the cation in solution (e.g., $\{Zn^{2+}\}$), and hence in its tendency to react with surface groups, and the decrease in the coulombic repulsive energy through shielding of the electrostatic forces between the positively charged surface and the cation (Dzombak and Morel 1990). In situations where sorption does not occur against electrostatic repulsion, however, such cancellation of ionic strength effects is neither expected nor observed, as evidenced by data on anion sorption on positively charged oxides. Sorbed organic matter imparts a net negative charge to suspended particles (Hunter 1980; Beckett and Le 1990), such that increased ionic strength should decrease both the coulombic attraction for cations and the activity coefficient of the ions. The situation is further complicated by the effect on surface charge of the sorption of major divalent cations (Mg^{2+}, Ca^{2+}) and anions (SO_4^{2-}), as discussed below.

Surface competition. Competition among sorbates for surface binding sites could in principle be very important when fresh water particles are introduced into seawater where the concentrations of divalent cations and anions are high. Such competition has been difficult to demonstrate, however, which is not surprising since site saturation has also been difficult to document. Effective competition among sorbates can occur only if sorbing sites are saturated. Nonetheless, both saturation and competition are evident in some data sets when one of the sorbing ions is sufficiently concentrated. This is clear for competition among anions (e.g., Balistrieri and Murray 1987; Hingston et al. 1971). Few experiments with sufficiently large concentrations of cations, especially Ca^{2+} and Mg^{2+}, have been reported, and thus competition among cations has not been clearly demonstrated. Dempsey and Singer (1980) observed essentially no competition between Ca^{2+} and Zn^{2+} for sorption on hydrous ferric oxide even with $Ca_T \gg (\equiv FeOH)_T$. Additional investigation of surface site competition among ionic solutes is needed, but owing to the very large concentrations of reactive divalent ions in seawater, Mg^{2+}, Ca^{2+}, SO_4^{2-} we may expect them to be as effective in promoting competitive desorption of trace elements in estuaries and coastal waters as they are in modifying the surface charge of suspended particles (Beckett and Le 1990).

Medium complexation. Complexation of trace cations by the major anions of seawater should be effective in desorbing metals that have sufficient affinity for chloride or sulfate. Thus, cadmium, lead, or mercury, which form stable chlorocomplexes in seawater, are expected to desorb in this manner (Comans and van Dijk 1988). More complicated is the possible effect of organic complexation. Fresh water may contain higher rather than lower concentrations of organic complexing ligands. Further, organic ligands themselves may promote sorption (rather than desorption) through a bridging mechanism involving a surface–ligand–metal complex. Such a bridging mechanism, however, is probably important only at low pH, where cations tend to desorb and organic ligands to adsorb. At circumneutral pH's, it is the competition between surface sites and dissolved organic ligands that is likely to be of primary interest. Thus the metals that tend to be strongly bound to natural organic ligands are most likely to be affected by such a process. In particular we expect copper to be efficiently solubilized by binding to humic material.

pH. The effect of varying pH on sorption is of course the most dramatic and complicated. As is well known, cation sorption generally increases sharply with pH while anion sorption decreases. The sorption of cations with increasing pH is effected by decreased competition from surface binding by H^+ and by decreased repulsion from gradually neutralized surface charge. For weak acid anions, the opposite effects are partly balanced by the decreased protonation in solution, which tends to increase the reactivity and hence the sorption of the anion.

Desorption in Estuaries and Coastal Waters: Model Calculations

Upon mixing of particle-bearing fresh water with seawater, all the individual effects discussed above occur simultaneously. To examine which of these effects is dominant where, we performed equilibrium computations, mixing in various proportions the fresh water and seawater endmembers shown in Table 1. In one series of computations, the pH of the fresh water endmember was assumed equal to that of seawater and thus maintained constant at 8.0; in the other, the pH was varied from 6.0 to 8.0. The calculations were performed in the presence and absence of a sorbing anion representing organic matter, whose sole function (in the model) was to impart a net negative charge to the hydrous ferric oxide (Hunter 1980; Tipping and Cooke 1982). The following discussion and figures are based chiefly on the calculations including this model organic acid. These calculations embody

TABLE 1. Fresh water and seawater endmember compositions employed in equilibrium calculations.[a,b]

Component	[−log (MOLAR TOTAL CONCENTRATION)] Fresh Water	Seawater
Na^+	3.5	0.33
K^+	4.2	1.99
Sr^{2+}	6.0	4.20
Ba^{2+}	6.7	7.30
Ca^{2+}	4.0	1.99
Mg^{2+}	3.8	1.27
Fe^{3+}	2.95[c]	4.95[c]
$\equiv Fe^sOH^o$	5.25[d]	7.25[d]
$\equiv Fe^wOH^o$	3.65[d]	5.65[d]
Ni^{2+}	7.0	8.64
Cu^{2+}	7.0	8.92
Zn^{2+}	7.0	10.22
Pb^{2+}	9.0	10.00
Cd^{2+}	8.7	11.70
Hg^{2+}	9.0	11.52
F^-	6.0	4.15
Cl^-	3.7	0.26
$B(OH)_4^-$	6.0	3.40
CO_3^{2-}	3.0	2.68
SO_4^{2-}	4.0	1.55
SiO_3^{2-}	3.7	5.50
PO_4^{3-}	5.7	6.70
HVO_4^{2-}	7.3	7.30
DOC	4.0[e]	4.6[e]
Ionic strength	2.5	0.22

[a]The calculations were performed according to the generalized two layer sorption models and corresponding intrinsic sorption constantly presented in Dzombak and Morel (1990). A modified version of the chemical equilibrium program MINEQL (Westall et al. 1976) was used for the calculations.

[b]Calculations were performed for the following fresh water:seawater dilutions—1:0, 10:1, 3:1, 2:1, 1:1, 1:2, 1:3, 1:10, 0:1. In a first series of calculations, pH was fixed at 8.0 for each case. In a second series of calculations, pH was fixed at the following values for the various cases (in the order cited above): 6.0, 6.2, 6.4, 6.6, 6.8, 7.0, 7.2, 7.4, 8.0.

[c]Fresh water iron concentration = 100 mg/l; seawater iron concentration = 10 mg/l. At the pH values of interest, essentially all the iron exists as hydrous ferric oxide.

[d]$\equiv Fe^sOH^o$ ($= 0.005 \times Fe_T$) represents high affinity surface sites on hydrous ferric oxide. $\equiv Fe^wOH^o$ ($= 0.2 \times Fe_T$) represents low affinity surface sites.

[e]Fresh water DOC concentration = 10 mg/l; seawater DOC concentration = 2 mg/l. A proton exchange capacity of 10 meq/g was assumed for the DOC, which was modeled as a diprotic acid with $pK_{a1} = 0.7$ and $pK_{a2} = 4.4$ (Tipping 1981). The sorption reactions involved the formation of $\equiv FeL^-$ and $\equiv FeOHL^{2-}$ surface species with intrinsic (log) formation constants of 14.1 and 7.1 respectively (with L^{2-} and $\equiv FeOH$ as reactants).

our present quantitative knowledge of sorption and they should provide insight into coastal processes, though they are clearly exemplary, not factual.

Under constant pH conditions, the variation in the calculated dissolved concentration of a trace cation such as Cu, Cd, Ni, or Zn upon dilution in seawater exhibits three different regions (see Fig. 2). At low dilution of fresh water into seawater the dissolved concentration increases sharply; at intermediate dilution it remains roughly constant before finally decreasing at high dilution.

According to the model calculations, initial desorption at low dilution results from three separate factors. First, the decrease in the activity coefficients of divalent ions with increasing ionic strength accounts for an increase by a factor of approximately three in the dissolved/particulate ratio (see Eq. 4). Another increase by a factor of four or so is due to a decrease in the coulombic

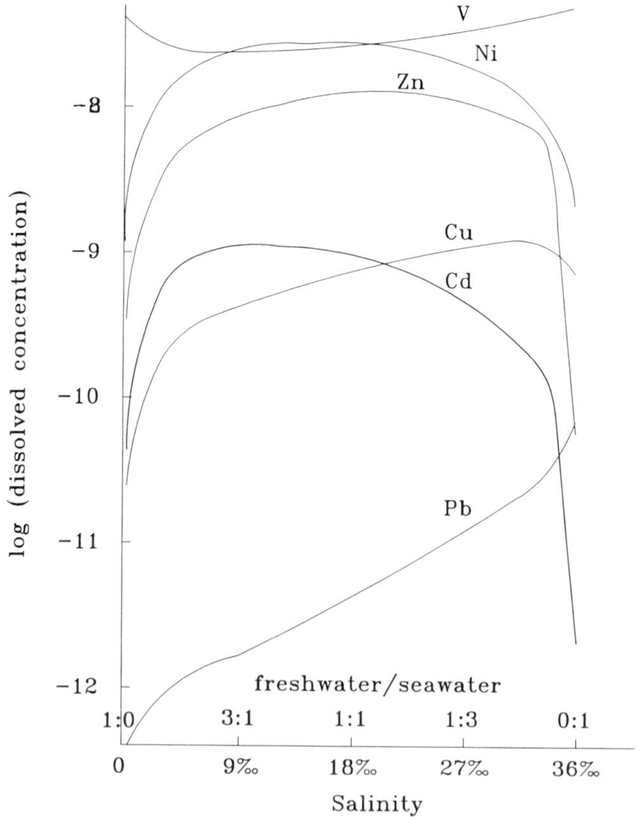

Fig. 2—Dissolved concentrations of trace cations as a function of dilution under conditions of constant pH.

attraction by the surface. Adsorption of Ca and Mg accounts for nearly 60% of the surface sites at a 3:1 dilution (fresh water/seawater), thus decreasing markedly the negative charge of the surface, consistent with what has been observed experimentally (Beckett and Le 1990). In addition, K_{coul} is also decreased by the shielding effect of higher ionic strength. The competition for high affinity binding sites by alkaline earth cations increases this desorption further such that the total effect is an increase in dissolved concentrations by a factor ca. 20. Finally, for Cd, desorption is enhanced initially by chloride complexation and most of the Cd becomes dissolved at 3:1 dilution. Such an effect has been observed in model laboratory studies (Comans and van Dijk 1988) and in field data in the Amazon (Boyle et al. 1982).

At intermediate dilutions, the dissolved concentrations calculated for Cu, Cd, Ni, or Zn remain nearly constant, increasing by only a factor of 2 or so between dilution ratios of fresh water over seawater of 3:1 and 1:10. This is due to the buffering effect by desorption upon dilution discussed above (Fig. 1b). When the dilution is high enough for the dissolved concentrations to account for most of the respective elements (dissolved/particulate > 1), conservative dilution occurs and results in the calculated decrease in dissolved metal concentration. In the model calculation, this happens at a dilution of fresh water by seawater of 3:1 for Cd, 1:1 for Ni and Zn, and only at very high dilution for Cu. As mentioned above, this last result would be affected by organic complexation of Cu in solution. The slight calculated increase in the dissolved concentration of Cu, Ni, and Zn at intermediate dilutions is due to competition for binding sites by Mg^{2+}, Ca^{2+}, and the organic acid.

Under variable pH conditions, neither the large initial desorption at low dilutions nor the further slight desorption at mid-dilutions is calculated to occur for Cu, Cd, Ni, or Zn (Fig. 3). Under the conditions of our calculations, the oxide has an increased tendency to sorb cations because of the decrease in H^+ competition and an increasing negative charge on the surface upon dilution. At the same time there is an increased tendency to desorb trace metals resulting from the decrease in dissolved ion activities and competition by sorbing alkaline earth cations. Except for Cu, which exhibits first a modest desorption and then a readsorption at low dilutions, these two effects balance each other almost perfectly. Thus the dissolved fractions of the metals increase only slightly at first and remain almost constant through most of the dilution range: 1–2% for Cu, 90–95% for Ni, and 70–80% for Zn. For Cd^{2+}, complexation by Cl^- increases the dissolved fraction from 76 to 100% at low dilution. Since the fraction of sorbed Cd is small in any case, the result for all four metals is quasi-conservative behavior, as if the dissolved fraction were simply diluted by seawater. At very high dilutions all the metals desorb completely except for Cu, which remains 40% sorbed.

Other elements, Ba, Pb, Hg, and V, follow patterns that are at variance with those of the four metals just discussed. Barium, which is calculated to be

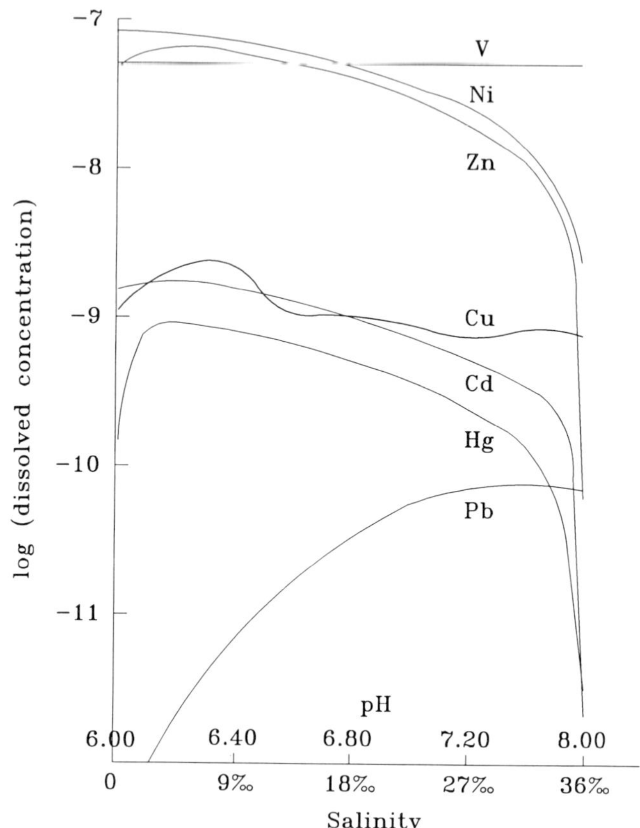

Fig. 3—Dissolved concentrations of trace cations as a function of dilution under conditions of variable pH.

only about 10% sorbed in fresh water under high pH conditions, desorbs totally (because of decreased activity and decreased electrostatic attraction) upon initial mixing with seawater (fresh water/seawater = 1:10) and behaves, of course, conservatively afterwards. Lead is very strongly sorbed initially but the formation of chlorocomplexes effects a gradual desorption over the whole dilution range, reaching 66% dissolution for the seawater endmember. Mercury, which is much more strongly sorbed than Ba, behaves like it, desorbing totally upon initial dilution in seawater. In this case, desorption is caused by the formation of chloro-mercuric complexes. Finally, the behavior of the vanadate anion, VO_4^{3-}, is dependent upon the assumed pH regime of the mixing region. At fixed pH, vanadate adsorbs upon mixing with seawater, due to the decrease in coulombic repulsion by the less negative surface, and then desorbs due to competition by major seawater ions. In the variable pH calculation, vanadate is initially only 3% sorbed despite the low pH because of competition by the organic acid and the resulting negative surface charge. Upon dilution vanadate dissolves primarily due to the increasing pH. This variable sorption behavior of vanadium is in accord with the available field data, which show that vanadium is mostly dissolved in estuaries and usually behaves conservatively, but sometimes exhibits a mid-salinity minimum depending on seasons and estuaries (Shiller and Boyle 1987). We note that the calculated sorption behavior of vanadate is very much affected by the presence of the model organic acid. Due to both coulombic repulsion and site competition, the organic acid causes the vanadate anion to desorb almost completely from the oxide. In contrast, for trace metals, the organic acid is calculated to cause only a modest and paradoxical desorption at high pH because of increased competition by major seawater cations.

Besides the detailed behavior of individual elements, three general lessons can be extracted from these model calculations.

1. The most dramatic changes in sorption are to be expected upon initial mixing of fresh water into seawater, at very low salinities; major changes in the characteristics of surfaces occur at these low dilutions. The available field data on Ba support this contention. Depending on the particular estuary, the Ba concentration exhibits a maximum between 2 and 10‰ salinity (Edmond et al. 1978).
2. The complexation of metals in solution can greatly enhance desorption. The calculations show this to be true for Pb, Hg, and Cd. Both field and laboratory data support the role of the chlorocomplexes of Cd in estuarine desorption (Comans and van Dijk 1988; Boyle et al. 1982).
3. Apparent conservative behavior of trace elements may be observed over a wide range of dilutions through a remarkable balance of increased and decreased tendency to sorb even when an element is largely in the particulate fraction. This is well exemplified by Zn, Cu, and Ni in the variable pH calculation. Thus one should be careful to interpret experimental data showing quasi-conservative behavior in coastal waters, particularly when such data encompass a limited salinity range. For example, it is conceivable that the apparent conservative behavior of Cu, Ni, and Cd in the salinity range 30 to 36% in the Gulf of Mexico (Boyle et al. 1984) and in the western North Atlantic (Bruland and Franks 1983) could reflect processes more complicated than simple mixing of fresh and marine water endmembers. Mixing is the likely explanation, though, since all three of these metals are chiefly desorbed at such high salinities. (Recall that the discrepancy in the calculation for Cu reflects the absence of dissolved organic complexing agents in the model.) Nonetheless, as noted above, extrapolation of the dissolved

concentrations to zero salinity may not yield the average "continental" endmember if desorption has occurred at lower salinities.

OTHER REACTIONS

Though important, sorption and desorption are not the only reactions involving particles in estuarine and coastal waters. Formation and dissolution of minerals, coagulation and sedimentation, and biological processes all contribute also to transferring chemicals between solutions and sediments, mostly from solution to sediments.

Dissolution–Precipitation

The conditions that lead to the dissolution of rocks from continental sources or precipitation of authigenic minerals have been well studied. In general the increased carbonate, sulfate, and hydroxide concentrations of marine waters lead to the saturation of a number of mineral phases. Some of the corresponding reactions such as precipitation of $CaCO_3$, SiO_2, and MnO_2 are catalyzed by organisms. In salt marshes and anoxic basins, precipitation of sulfide minerals is well documented and leads to the trapping of a number of trace elements.

Recent interest has been focused on the possible role of heterogeneous photochemical reactions in promoting the dissolution and transformation of minerals. In semiconductors, initial light absorption might be effected by the mineral itself, and in all solids it may be effected by surface species, sometimes involving adsorbed organic compounds (Sulzberger 1990). Through migration of separated holes and electrons or through surface reactions—typically a charge transfer process—a redox reaction is eventually promoted at the surface in which a metal ion is reduced, (e.g., $Fe[III] \rightarrow Fe[II]$), and some other chemical such as the sorbed organic ligand is oxidized. In some cases the reduced species may be sufficiently long lived to reach measurable dissolved steady-state concentrations. This is certainly the case for Mn(II) (Sunda et al. 1983), although the exact photoreduction mechanism for MnO_2 is not yet understood. Despite claims to the contrary, this is not likely the case for iron, however, for the released Fe(II) is rapidly reoxidized and reprecipitated at the surface. The net result is the formation of an amorphous hydrous oxide coating rather than a net solubilization. Nonetheless, the resulting increased lability of the iron particle may have great impact with respect to the biological availability of iron in coastal waters (Wells 1989; Rich and Morel 1990). Much should be learned over the next few years regarding the extent and mechanisms of heterogeneous photochemical processes and their geochemical and biological consequences.

Coagulation–Sedimentation

From the point of view of geochemical transport, it is not so much the transfer of a chemical to a solid phase that is important as its transfer to the sediments. From the early work of Boyle and Sholkovitz (Sholkovitz 1976; Boyle and Edmond 1977) we know that coagulation and sedimentation of iron and organic matter are the dominant trapping processes for many associated elements in estuaries. Coagulation of particles in the water column is easily explained in principle, if not in detail, by the concomitant effects of the decrease in negative surface charge by sorption of calcium and magnesium (vide supra) and of the shielding of particle repulsive charges by high ionic strength. However, the influence of chemistry on coagulation kinetics and particle size distributions is not well understood. O'Melia (1987) has presented an insightful summary of what is known. A good example of the limits of our current understanding is the lack of knowledge of the possible importance of peptization, i.e., the reverse of coagulation, in natural waters (Morel and Gschwend 1987). This may serve to maintain a large colloid concentration capable of keeping trace elements in the water column through an estuary. Obviously much still needs to be done, however, to understand quantitatively the kinetics of coagulation in natural waters, including coastal waters.

Another related and complicated process is the transfer of particles from the water column to the sediments. In its simplest form, the question is to understand which of simple gravity settling (before or after coagulation), of biologically induced settling (through feeding of zooplankton and benthos, and phytoplankton sinking), or of "capture" by the sediment bed is the dominant mechanism to bring suspended particles into the sediments. Farley (1990) has recently proposed a coagulation/settling model applicable to stratified systems. In well-mixed shallow coastal waters, field experiments appear to implicate the bed capture mechanism (Newman et al. 1990), and a hypothesis of filtration of the suspended water column particles by the benthic "fluff layer" has been proposed (Stolzenbach, in preparation). This question of the fate of coastal particles will receive more than academic attention over the next several years, as it is a prime determinant of the impact of anthropogenic waste in coastal waters.

Biological Processes

An important fraction of marine particles is alive, and we need to take stock of the heterogeneous reactions promoted by planktonic microorganisms in coastal waters. Such reactions include uptake of major and trace nutrients, release of metabolites, and the surface catalysis of chemical reactions.

The role of biological uptake in regulating the chemistry of seawater is best illustrated in the depletion of bioactive elements from surface open ocean waters. Thus, both major algal nutrients (N, P, Si) and many trace elements (e.g., Fe, Ni, Cd, V) exhibit characteristic nutrient-like profiles. The same processes are at work in coastal waters, however, even if higher concentrations or efficient vertical mixing often hide or erase the telltale vertical concentration gradients. From the point of view of trace elements, the most interesting new development is the realization that many elements have a biological role and that this role depends on the availability of other essential elements. For example, nickel is an essential nutrient when urea is a nitrogen source (Price and Morel 1991); more surprisingly, cadmium can replace zinc when the latter is limiting (Price and Morel 1990a).

In the same way, the activity of surface enzymes, which release inorganic nutrients from organic sources (phosphatases, deaminases), reduces certain metal species (transmembrane reductases) and releases redox intermediates such as H_2O_2, depending on the availability of particular types of major nutrient sources and trace elements (Price and Morel 1990b). For example, extracellular amino-acid oxidases which release NH_4^+, H_2O_2, and ketoacids are only active in the absence of ammonium (and presence of AA's), and some apparently require copper (Palenik and Morel 1990).

We know little about the release of extracellular metabolites by microorganisms in coastal waters, particularly concerning their effects on trace element chemistry. There is evidence from laboratory cultures that strong complexing and reducing agents can be released. Again, such processes are dependent on major and trace nutrient availability, as exemplified by the release of siderophores by some algae, bacteria, and fungi when the iron supply is insufficient (Neilands 1981; Trick et al. 1983). This topic of chemical control of their external milieu through release of specific compounds by marine microorganisms is clearly an intriguing research topic with much speculation and no firm data.

Regarding the known effects of the biota on the chemistry of coastal waters, what clearly emerges is that it can be both profound (e.g., control the geochemical cycle of nutrients) and subtle (e.g., changing the redox state of trace elements), and that it can result in unexpected and unstable chemical species such as reduced metal ions (in oxic water), hydrogen peroxide, or strong chelates. Even more importantly, the biochemical interdependency of uptake and assimilation of bioactive elements promotes a tight geochemical interdependency with no obvious chemical basis. The exquisite correlation between cadmium and phosphate concentrations in seawater is a case in point. How the biochemistry and geochemistry of elements from carbon to mercury are linked to each other is a conundrum whose simplest expression is an extended Redfield ratio and whose actual unraveling will fascinate this and the next generation of biologists and geochemists.

Acknowledgements. This work was supported in part by the National

Science Foundation (Grant # 8615545-OCE), the Office of Naval Research (Grant # N00014-86-K-0325), and SeaGrant (# NA86AA-10-SG089).

REFERENCES

Balistrieri, L.S., and J.W. Murray. 1987. The influence of major ions of seawater on the adsorption of simple organic acids by goethite. *Geochim. Cosmochim. Acta* **51**:1151–1160.

Beckett, R. and N.P. Le. 1990. The role of organic matter and ionic composition in determining the surface charge of suspended particles in natural waters. *Colloid & Surfaces,* **44**: 35–49.

Boyle, E.A., R. Collier, A.T. Dengler, J.M. Edmond, A.C. Ng, and R.F. Stallard. 1974. On the chemical mass balance in estuaries. *Geochim Cosmochim Acta* **38**: 1719–1778.

Boyle, E.A. and J.M. Edmond. 1977. The mechanism of iron removal in estuaries. *Geochim. Cosmochim. Acta* **41**:1313–1324.

Boyle, E.A., S.S. Huested, and B. Grant. 1982. The chemical mass balance of the Amazon Plume II. Copper, nickel and cadmium. *Deep Sea Res* **29**:1355–1364.

Boyle, E.A., D.F. Reid, S.S. Huested, and J. Hering. 1984. Trace metals and radium in the Gulf of Mexico: an evaluation of river and continental shelf sources. *Earth Planet. Sci. Lett.* **69**:69–87.

Bruland, K.W., and R.P. Franks. 1983. Mn, Ni, Cu, Zn, and Cd in the western North Atlantic. In: Trace Metals in Seawater, ed. C.S. Wong, E. Boyle, K.W. Bruland, J.D. Burton, and E.D. Goldberg, vol. 9, pp. 395–414. New York: Plenum.

Comans, R.N.J. and C.P.J. van Dijk. 1988. Role of complexation processes in cadmium mobilization during estuarine mixing. *Nature* **336**:151–154.

Dempsey, B.A., and P.C. Singer. 1980. The effects of calcium on the adsorption of zinc by MnO_x(s) and $Fe(OH)_3$ (am). In: Contaminants and Sediments, ed. R.A. Baker, vol. 2, pp. 333–352. Ann Arbor, MI: Ann Arbor Science.

Dzombak, D.A., and F.M.M. Morel. 1990. Surface Complexation Modeling: Hydrous Ferric Oxide. New York: Wiley.

Edmond, J.M., E.A. Boyle, D. Drummond, B. Grant, and T. Mislick. 1978. Desorption of barium in the plume of the Zaire (Congo) River. *Neth. J. Sea Res.* **12**:324–328.

Farley, K.J. 1990. Predicting organic accumulation in sediments near marine outfalls. *J. Env. Eng.* **116**:144–165.

Hingston, F.J., A.M. Posner, and J.P. Quirk. 1971. Competitive adsorption of negatively charged ligands on oxide surfaces. *Disc. Faraday Soc.* **52**:334–342.

Hunter, K.A. 1980. Microelectrophoretic properties of natural surface-active organic matter in coastal seawater. *Limnol. Ocean.* **25**:807–822.

Morel, F.M.M., and P.M. Gschwend. 1987. The role of colloids in the partitioning of solutes in natural waters. In: Aquatic Surface Chemistry, ed. W. Stumm, pp. 405–422. New York: Wiley.

Neilands, J.B. 1981. Microbial iron compounds. *Ann. Rev. Biochem.* **50:**715–731.

Newman, K.A., F.M.M. Morel, and K.D. Stolzenbach. 1990. Settling and coagulation characteristics of fluorescent particles determined by flow cytometry and fluorometry. *Envir. Sci. Tech.* **24**:506–513.

O'Melia, C.R. 1987. Particle-particle interactions. In: Aquatic Surface Chemistry, ed. W. Stumm, pp. 385–403. New York: Wiley.

Palenik, B., and F.M.M. Morel. 1990. Amino acid utilization by marine phytoplankton: a novel mechanism. *Limnol. Ocean.* **35**:260–269.

Price, N.M., and F.M.M. Morel. 1990a. Cadmium and cobalt substitution for zinc in a marine diatom. *Nature* **344**:658–660.

Price, N.M., and F.M.M. Morel. 1990b. The role of extracellular enzymatic reactions in natural waters. In: Aquatic Chemical Kinetics, ed. W. Stumm, pp. 235–257. New York: Wiley.

Price, N.M., and F.M.M. Morel. 1991. Colimitation of phytoplankton growth by nickel and nitrogen. *Limnol. Ocean.*, in press.

Rich, H.W., and F.M.M. Morel. 1990. Bioavailability of well-defined iron colloids to the marine diatom *Thalassiosira weissflogii*. *Limnol. Ocean.* **35**: 652-662.

Shiller, A.M., and E.A. Boyle. 1987. Dissolved vanadium in rivers and estuaries. *Earth Planet. Sci. Lett.* **86**:214–224.

Sholkovitz, E.R. 1976. Flocculation of dissolved organic and inorganic matter during mixing of river water and seawater. *Geochim. Cosmochim. Acta* **40**:831–845.

Sulzberger, B. 1990. Photoredox reactions at hydrous metal oxide surfaces: a surface coordination chemistry approach. In: Aquatic Chemical Kinetics, ed. W. Stumm, pp. 401–429. New York: Wiley.

Sunda, W.G., S.A. Huntsman, and G.R. Harvey. 1983. Photoreduction of manganese oxides in seawater and its geochemical and biological implications. *Nature* **301**:234–236.

Tipping, E. 1981. Adsorption by geothite (α-FeOOH) of humic substances from three different lakes. *Chem. Geol.* **33**:81–89.

Tipping, E., and D. Cooke. 1982. The effects of adsorbed humic substances on the surface charge of geothite (α-FeOOH) in fresh waters. *Geochim. Cosmochim. Acta* **46**:75–80.

Trick, C.G., R.J. Andersen, A. Gillam, and P.J. Harrison. 1983. Prorocentrin: an extracellular siderophore produced by the marine dinoflagellate *Prorocentrum minimum*. *Science* **219**:306–308.

Wells, M.L. 1989. The lability of iron in seawater and its relationship to phytoplankton. Ph.D. diss., University of Maine.

Westall, J.C., J.L. Zachary, and F.M.M. Morel. 1976. MINEQL: A Computer Program for the Calculation of Chemical Equilibrium Composition of Aqueous Systems. MIT Technical Note No. 18, Cambridge, MA.

Accumulation and Regeneration: Processes at the Benthic Boundary Layer

T.H. Blackburn

Department of Ecology and Genetics
University of Aarhus
8000 Aarhus C, Denmark

Abstract. There is a correlation between the rate of sediment accumulation and the proportion of the organic flux that is not mineralized in the sediment. It is probable, in many situations, that this decreased mineralization is due to SO_4^{2-} replacing O_2 as principal oxidant. It is not, however, certain that all mineralization processes involving SO_4^{2-} reduction are inherently slower or less complete than those involving O_2. Other factors, which determine the rate and extent of organic matter mineralization at the benthic boundary layer, are discussed: the nature of the organic matter, the inorganic material co-sedimenting with the organic matter, the presence of benthic fauna, and temperature.

INTRODUCTION

The aim of this chapter is to examine the factors that determine the rate and extent of organic detrital mineralization at the benthic boundary layer. The benthic boundary layer is not well defined. In a symposium devoted to a discussion of the benthic boundary layer, the contributors have considered it to extend meters below the sediment–water interface. An operational definition might be that it extends downward to a point where biological activity is absent, or nearly absent ($<$ 1% of the rate of mineralization that is measured at the sediment surface). Within the context of global change, it is important to determine whether increased primary production in the overlying water, followed by increased organic flux to the sediment, can lead to increased rates of organic burial and to increased rates of denitrification. Alternatively, will mineralized nutrients again be made available for further algal growth in the overlying water?

Ocean Margin Processes in Global Change
Edited by R.F.C. Mantoura, J.-M. Martin and R. Wollast

ANOXIC MINERALIZATION

Henrichs and Reeburgh (1987), in an excellent review article, examined data from a wide variety of sources; they concluded that there was no evidence for the proposition that more burial of organic material in marine sediments occurred under anoxic (SO_4^{2-}-reducing) conditions. This deduction was based on a plot of carbon burial efficiency against the rate of sedimentation (Fig. 1). The efficiency of carbon burial (E) was calculated from Eqs. 1 and 2:

$$F_b = (\%C\infty)(\omega)(\rho_s)(1-\phi)(100) \quad (1)$$

$$E = F_b/F_c \quad (2)$$

where F_b is the burial flux of organic carbon (g C m^{-2} yr^{-1}), $\%C\infty$ is the

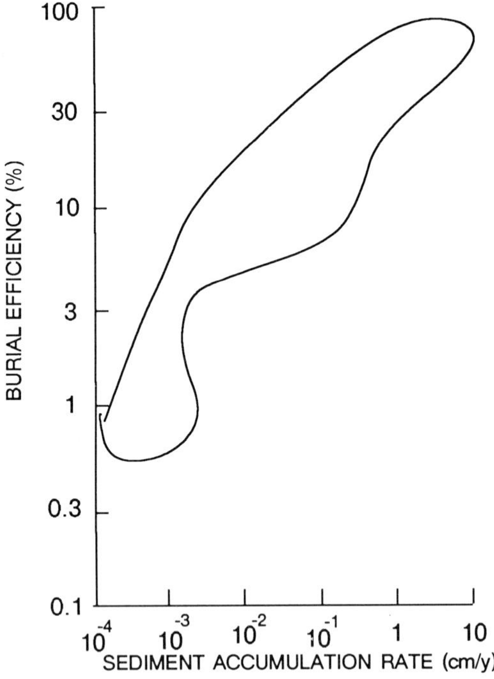

Fig. 1—Percent burial of organic carbon plotted against rate of sedimentation. Percent burial is the rate of burial of organic carbon divided by the rate of carbon input to the sediment, multiplied by 100. The 32 data points lie within the enclosed area. Redrawn from Henrichs and Reeburgh (1987).

carbon content (weight %) in the sediment where mineralization has ceased, ω is the sediment accumulation rate (cm yr^{-1}), ρ_s is the dry sediment density (g dry weight cm^{-3} dry sediment), ϕ is the porosity, and F_c is the flux of organic carbon to the sediment surface (g C m^{-2} yr^{-1}).

The data in Fig. 1 originated from a wide variety of sediment types, from oxic and suboxic pelagic sediments to more anoxic shelf and estuarine sediments. Burial efficiency varied considerably for the different sediment types: 0.6%–16% for pelagic, 8%–28% for slope, and 8%–79% for shelf and estuarine sediments. It was noted that sediments which were anoxic below 2 cm tended to have high E values. There was not, however, a consistent relationship between the degree of anoxia and burial efficiency. Henrichs and Reeburgh (1987) established the following empirical relationship between burial and sedimentation rate:

$$E = \omega^{0.4}/2.1. \tag{3}$$

Canfield (1989) examined a larger data set, which included euxinic marine sediments (sediments permanently overlaid with anoxic water) and semi-euxinic sediments (those intermittently covered by anoxic water). A plot of integrated rates of SO_4^{2-} reduction against sediment burial rates indicated that euxinic sediments had higher rates of SO_4^{2-} reduction than those of normal marine sediments (Fig. 2). This was not unexpected as organic matter was exclusively mineralized by SO_4^{2-} reduction in these euxinic sediments, whereas O_2 was responsible for much of the oxidation of readily degradable carbon in normal oxygenated sediments. The data clearly indicate that a more complete oxidation of organic matter via O_2 occurs at low sedimentation rates. This corroborates the observations by Jørgensen (1982), who showed a relationship between F_c and the proportion of F_c that was mineralized by SO_4^{2-} reduction. The most obvious explanation is that O_2 does not get depleted in sediments with low organic loading; because of the low sedimentation rate, the organic matter spends a longer time at the oxic interface, and O_2 can account for all carbon oxidation. This explanation must always be partially true for situations where O_2 is the principal oxidant and may sometimes be the complete explanation; however, other factors probably also play a role, e.g., the ratio of organic to inorganic input may not be constant. Moreover, the quality of both inputs is likely to be important in determining their effect. These additional factors could explain the wide range of SO_4^{2-} reduction rates observed for any rate of sediment burial (Fig. 2), indicating that ω cannot be the only regulator of carbon oxidation. The importance of these other factors will be discussed in later sections.

The integrated quantity of carbon oxidized by O_2 and SO_4^{2-} reduction in sediments, which had different rates of ω, is seen in Fig. 3. This presentation

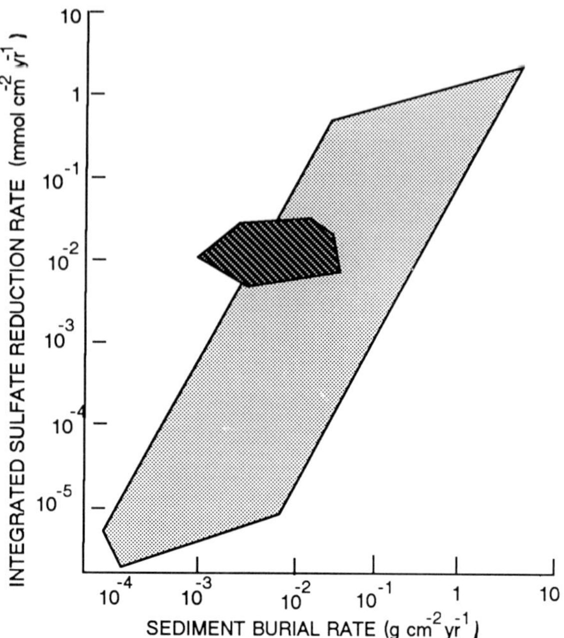

Fig. 2—Rates of sulfate reduction plotted against sediment burial rate. The points for normal sediments are enclosed by the large shaded area. The points for the euxinic sediments are inclosed by the small, cross-hatched area. Redrawn from Canfield (1989).

allows for a more direct comparison to be made between the two oxidative processes (the SO_4^{2-} one has twice the oxidizing capacity of O_2). The overlap between the two data sets (Fig. 3) indicates that approximately equal amounts of carbon are oxidized via O_2 and SO_4^{2-} at sedimentation rates > 0.1 g cm^{-2} y^{-1}. Sulfate reduction became progressively less important at sedimentation rates < 0.1 g cm^{-2} y^{-1} and was insignificant (< 1% carbon oxidized) at sedimentation rates < 0.001 g cm^{-2} y^{-1}. These deductions are consistent with the observations of Bender and Heggie (1984), who stated that SO_4^{2-} reduction was unimportant in the oxidation of carbon in sediments that underlie deep pelagic waters and have a low F_c. In contrast, Christensen (1989) has shown that much carbon oxidation in sediments, underlying water of 50 to 300 m depth, occurs via SO_4^{2-} reduction (20.4 g C m^{-2} y^{-1}). Approximately an equal quantity (22.7 g C m^{-2} y^{-1}) was oxidized by O_2, similar to the proportions noted by earlier investigations (Jørgensen 1982). Christensen (1989) noted that NO_3^- reduction accounted for little carbon oxidation (4 g C m^{-2} y^{-1}) and F_b was small (5 g C m^{-2} y^{-1}). Bioturbation complicates an assessment of the relative importance of O_2 versus SO_4^{2-} as

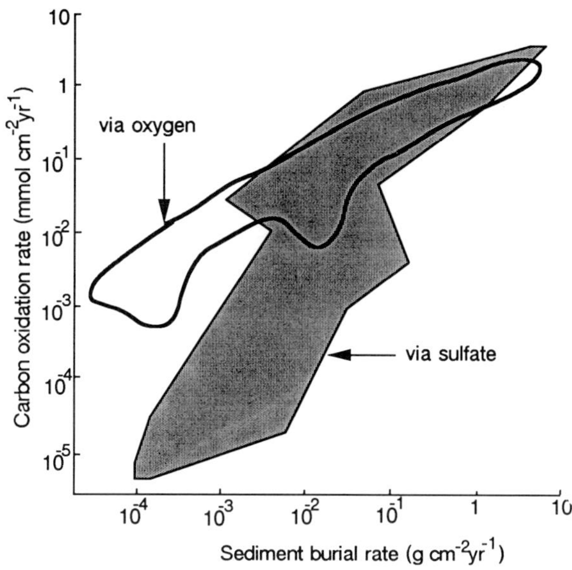

Fig. 3—Rate of organic carbon oxidation plotted against rate of sediment burial. The data points for the integrated rates of organic carbon oxidation via sulfate reduction are enclosed by the shaded area; those via oxygen are enclosed by the nonshaded area. Redrawn from Canfield (1989).

a terminal oxidizer of sedimentary organic carbon. Mixing of surface sediment downwards has the effect of transferring freshly sedimented material from the O_2-dominated zone to that dominated by SO_4^{2-} reduction. The net effect is to maintain a balance between O_2 and SO_4^{2-} reduction processes over a very wide range of sedimentation rates (0.1 to 10 g cm^{-2} y^{-1} [Fig. 3]).

Canfield (1989) made a comparison between the total rates of carbon oxidation and sedimentation rates (Fig. 4). The total rate of carbon oxidation is the sum of the carbon oxidized by O_2 and SO_4^{2-}. For the euxinic sediments there was no O_2 oxidation. It is assumed that other oxidants (nitrate, iron, and manganese) played minor roles in the mineralization of carbon. This is probably true for most of the sediment types in the data set of Henrichs and Reeburgh (1987). While the euxinic and semi-euxinic data points are not very different from those of the normal (oxic) sediment values, there is some indication that these euxinic points are lower than the mean; i.e., for a given rate of sediment burial they may have a lower rate of carbon oxidation. This lower rate of carbon oxidation would result in a higher rate of F_b, which is seen when E is plotted against ω (Fig. 5). E is greater in the euxinic sediments than in normal oxic and semi-euxinic sediments. The

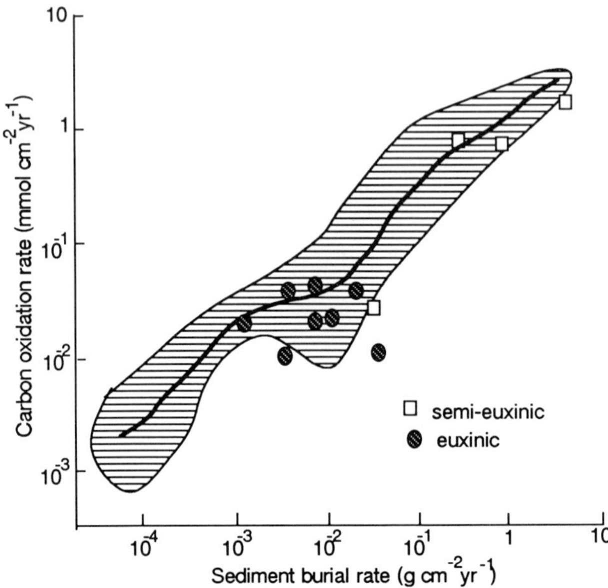

Fig. 4—Summed rates of organic carbon oxidation (O_2 and SO_4^{2-} reductions) plotted against rate of sediment burial. The data points for normal sediments are enclosed by the hatched area; the central line is a mean of these values. Redrawn from Canfield (1989).

data in Fig. 5 are evidence for a less efficient mineralization of carbon with SO_4^{2-}, as opposed to O_2, as the oxidant, assuming that the organic matter is the same in both situations. Canfield (1989) presents two interesting calculations to substantiate his thesis that lower rates of carbon oxidation under SO_4^{2-}-reducing conditions can result in an increased F_b. At MANOP site M (Bender and Heggie 1984) all mineralization was by O_2; E was 3% and the %C∞ was 1. If the site became euxinic and the rate of carbon oxidation by SO_4^{2-} reduction was 50% of the oxic rate, then 50% of the organic flux would be buried and %C∞ would be 15. If the rate of SO_4^{2-} reduction were 80% of the oxic rate, then %C∞ would be 7. Similarly, at MANOP site C, a 50% decrease in the rate of carbon oxidation would result in a change in %C∞ from 0.2 to 10.

All evidence that organic carbon oxidation is inherently slower via SO_4^{2-} than O_2 is ambiguous. There is, however, some geological evidence indicating that anoxia in North Atlantic basins during the Cenomanian-Turonian period has been associated with high rates of organic burial, even though sedimentation rates were low (Stein et al. 1986). On the other hand, rates of carbon mineralization by SO_4^{2-} reduction can be higher than by O_2 oxidation in a microcosm experiment (Kristensen and Blackburn 1987). It

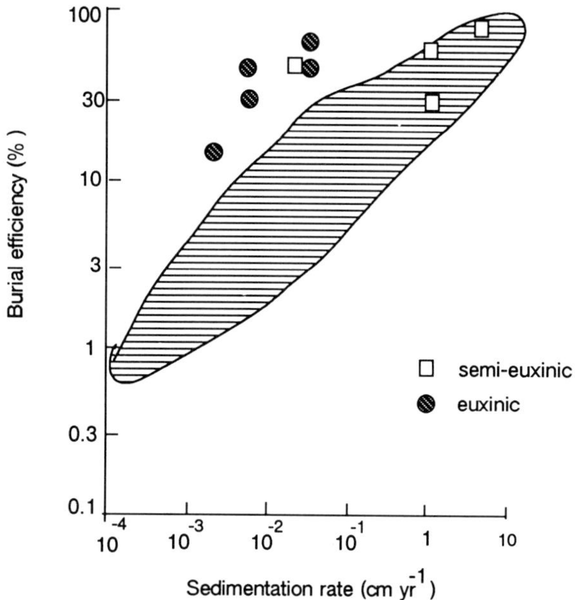

Fig. 5—Burial efficiency percent (E. 100) plotted against rate of sediment burial. The data points for normal sediments are enclosed by the hatched area. Redrawn from Canfield (1989).

is possible that the apparent equivalence of the two rates in these microcosms was due to the easily degradable nature of the substrate.

NATURE OF THE ORGANIC SUBSTRATE

In relation to the microcosm experiment, it was suggested that fresh, readily degradable organic material might be degraded at similar rates via O_2 or SO_4^{2-} reduction. The same conclusion can be drawn from the similar rates of degradation of *Scenedesmus* cells under aerobic and anaerobic (not SO_4^{2-} reduction) conditions in lake mud (Otsuki and Hanya 1972). There was, however, evidence for a greater accumulation of dissolved, nonoxidized organic products under anaerobic conditions (Otsuki and Hanya 1972). A wide variety of algal cell types was subjected to anaerobic (SO_4^{2-}-reducing) and aerobic conditions (Foree and McCarty 1970; Jewell and McCarty 1971). No difference in the rate of organic mineralization was observed between the two treatments, indicating that SO_4^{2-} was just as efficient an oxidant as O_2. There were considerable differences between the algal strains with regard to their degradability, but an average 40% of the carbon had not been oxidized after 200 day incubation. Jewell and McCarty (1971) calculated that the production of refractory organic matter from bacterial cells could not be more than 15% of the total refractory material present at the end of an aerobic incubation. The proportion would probably be less under anaerobic conditions, due to a lower efficiency of cell production. There is, however, the possibility that SO_4^{2-}-reducing cells may have components (e.g., lipids) that are relatively more refractory. Although the proportion of refractory material produced may seem to be small (ca. 10% of the degradable material), it could have very significant consequences in determining F_b and E. It should be remembered that most sediments have burial efficiencies of between 1% and 30% (Fig. 5).

More research has been directed to investigations of the kinetics of organic matter degradation rather than to a determination of the conditions leading to organic matter burial. Westrich and Berner (1984) have shown that organic matter in marine sediments was composed of three main fractions: Fraction 1 was readily decomposed, Fraction 2 was less easily decomposed, and Fraction 3 was resistant to decomposition. The decomposition of Fractions 1 and 2 was by first-order kinetics. First-order kinetics for organic matter degradation was also observed in many other sediment systems (Otsuki and Hanya 1972; Jewell and McCarty 1971; Foree and McCarty 1970; Kristensen and Blackburn 1987). The main problem remains: what is the nature of the organic matter which is resistant to decomposition? Does its character depend on the original source, and is it resistant to oxidation by SO_4^{2-}?

The actual composition of the resistant material has not been extensively investigated. It is likely that much of it can be classified as being fulvic acid and humic material (Nissenbaum and Kaplan 1972). The composition is probably independent of the origin of the sedimenting material, suggesting a common bacterial origin or considerable modification of the original organic matter. Organic material that is of terrestrial origin is characterized by its high content of lignin, a biopolymer that is absent in marine-derived detritus. The occurrence of lignin in sediments is limited to river mouths and is not of significant interest in most marine systems. It cannot be the origin of the humic material in most sediments.

Algal cells are extensively mineralized as they sediment through a deep water column (Suess 1980). This may have some effect in determining their subsequent rate of degradation in the sediment, because the partially mineralized cells should be more resistant to further bacterial attack; this, however, was not evident in the data examined by Henrichs and Reeburgh (1987) and Canfield (1989). Originally it was thought that SO_4^{2-}-reducing bacteria could only metabolize a limited range of small molecular compounds. These compounds were the products from other bacterial fermentations, e.g., lactate and hydrogen. Since 1979 it has been established that SO_4^{2-}-reducing bacteria are much more versatile and can oxidize a very wide range of complex molecules: aliphatics, aromatics, amino acids, long- and short-chain fatty acids (Widdel 1989). There may be compounds which SO_4^{2-}-reducing bacteria degrade so slowly that these compounds accumulate. Lignin fits into this category, as do humic substances (Widdel 1989). Bacterial lipids were mineralized more slowly under anoxic, SO_4^{2-}-reducing conditions than with O_2 as a terminal electron acceptor (Harvey et al. 1986). This study also showed that the presence of organic matter, in the form of humic material, reduced the rate of both aerobic and anaerobic oxidation. Harvey et al. suggested that adsorption of the lipid to the sediment's organic matrix prevented its breakdown by bacterial enzymes.

There may be other reasons for the slow breakdown of organic matter by SO_4^{2-}-reducing bacteria. The sulfide, resulting from the oxidation of organic matter, may itself be inhibitory to the SO_4^{2-}-reducing bacteria. The mechanism of inhibition might be by the precipitation of iron as insoluble sulfide. Sulfate-reducing bacteria have a major requirement for iron, but there is no evidence that an inhibition, due to iron deficiency, occurs in marine sediments.

In the context of the quality of the organic detritus, it is important to consider the effect of the size of the organic particle on its rate of breakdown or burial. Bacterial degradation of insoluble substrates begins by the attachment of extracellular enzymes to the substrate surface. Bacteria, with the exception of the Actinomycetes, have no capability for penetrating below the substrate surface by mycelial growth. This is unlike the terrestrial situation, where fungi can colonize solid substrates, which are then degraded by internal attack. This is much more efficient than surface attack by hydrolytic enzymes, particularly when the substrate has an unfavorably high C:N ratio. Because of this, most terrestrial decomposition of wood is caused by fungi, which can gain nitrogen externally and carbon (energy) from the center of the wood. Some fungi occur in marine sediments, but they are not common and are unimportant in particulate organic detrital degradation. Anaerobic fungi have not been extensively described and, again, the impression is that they are unimportant in sediments underlying both saltwater and fresh water. Wood, which has become buried in anoxic sediments, is not rapidly degraded, probably because of the absence of fungal attack. In addition, the lignin of wood is only slowly degraded in the absence of O_2. By analogy, it seems likely that other insoluble residues, if they have a low ratio of surface area to volume (i.e., are large), may resist bacterial decomposition in anoxic marine sediments.

INORGANIC SEDIMENTING PARTICLES

A large F_c, leading to anoxia and slow oxidation by SO_4^{2-}, is not the only explanation for high values of E. Associated with the organic flux is a flux of mineral particles to the sediment surface. It is probably the accumulation of these mineral particles, rather than the high flux of organics, which gives the good correlation between ω and E (Fig. 1). Is it possible that this inorganic flux is responsible for the inhibition of organic degradation, resulting in high burial efficiencies? The adsorption of low molecular weight organic acids (acetic, succinic, glutamic, and citric) and of sugars to hydroxyapatite greatly slowed their rate of degradation by an aerobic bacterium *Vibrio alginoliticus* (Gordon and Millero 1985). In addition to adsorption on minerals, organic–metal ion interactions are important in determining the rate of organic matter degradation (Degens and Mopper 1975). These authors observed that only 20–30% of organic matter was complexed in reducing sediments, compared with 50–80% in oxidized sediments. This would suggest that inhibition of degradation could be greater in oxic environments.

A number of studies have shown that simple molecules in pore water (e.g., acetate) are complexed with other molecules making them nondegradable (Christensen and Blackburn 1982). This effect is in addition to the adsorption of acetate to sediment particles, which also reduces its rate of mineralization. The importance of this type of inhibition has not been established for a large number of small molecules, nor has it been established whether it occurs predominantly in anoxic environments. Assays under aerobic and anaerobic conditions could be easily devised to answer general questions about biodegradability. For example, is there evidence for the burial of organic matter in SO_4^{2-}-dominated environments that might not have occurred had O_2 been present? This question could readily be answered by subjecting buried organic matter to oxic conditions suitably inoculated with aerobic bacteria. Westrich and Berner (1984) performed an experiment of this type, but the organic matter was not old, "nondegradable" material. The first-order rate constants for degradation by O_2 and by SO_4^{2-} were quite similar.

Henrichs and Reeburgh (1987) observed a close correlation between organic flux to the sediment and rate of sedimentation:

$$\log F_c = 0.69 \log \omega + 2.27 \quad (r = 0.95). \tag{4}$$

It might be deduced that a high F_c would always be associated with a high ω, and that one or both of these factors determines F_b and E. The most generally accepted argument is that a high F_c leads to a limited O_2 penetration below the sediment–water interface; the result is that most of the organic input is oxidized slowly by SO_4^{2-} (Canfield 1989). A high F_c is not, however, always linked to high ω. An example of this is the north Bering Sea or Chirikov Basin (Walsh et al. 1989). The center of the basin has a very high sand content, with virtually no ω, yet it receives ~150 g C over a three-month period: the %C∞ is only 0.4. There is thus no evidence for organic burial. Possibly the absence of co-sedimenting mineral particles, which may normally give high E values, is the reason for the low %C∞. There is, however, an additional factor in the form of a very active and large population of benthic animals, whose role will be discussed later.

The burial of carbon is associated with the concurrent burial of organic nitrogen. While burial of carbon is of interest in making a global budget and accounting for the removal of carbon from circulation, losses of nitrogen are even more important. This is because nitrogen is usually a limiting nutrient for marine phytoplankton growth, whose blooms can be a serious environmental problem in coastal waters. It seems probable, in view of the previous discussion, that increased algal growth would lead to an increase in F_c, ω, and E. The degree to which E is dependent on a high F_c is unknown, but inevitably a high F_c would result in O_2 depletion. A simulation was run showing how increasing inputs and mineralization rates of organic C and N in the sediment can affect denitrification, the second major sink for nitrogen (Fig. 6). In the simulation, NO_3^- was not allowed to increase in the water; therefore, denitrification was completely dependent on the NO_3^- produced in the sediment. Nitrification increased in parallel with increases in N mineralization rate (NH_4^+-availability), up to ~7 mmol m^{-2} day^{-1}. Rates of nitrification and denitrification were approximately 40% and 15% of this rate of N mineralization, respectively. At higher rates of N mineralization there was a decrease in nitrification rate followed by a

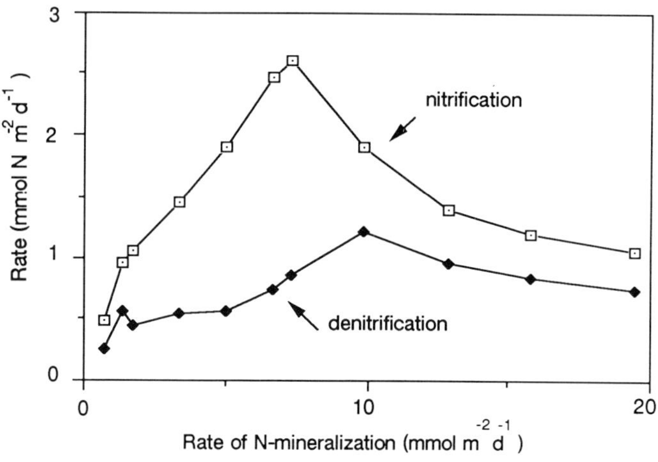

Fig. 6—Simulated rates of nitrification and denitrification plotted against rate of organic nitrogen mineralization. The C:N ratio of the organic substrate was 4:1, no depletion of substrate occurred, and there was no NO_3^- in the overlying water. Redrawn from Blackburn (1990).

decrease in the denitrification rate. Because of the dependence of denitrification on oxic (to generate NO_3^-) and anoxic conditions (to reduce NO_3^-), rates of denitrification did not vary greatly over large changes in the rates of N and C mineralization. Unfortunately for environmentalists, the model predicts that denitrification is a poor sink for nitrogen, especially at high fluxes of organic N to sediment. The situation would be very different if external sources of NO_3^- were present, as would occur in sewage treatment systems and in natural water systems receiving agricultural runoff.

EFFECT OF ANIMALS

The effect of benthic infauna in increasing the flux of O_2 into marine sediments, and the effect of this on nitrogen and carbon cycling has been well documented (e.g., in Blackburn and Sørensen 1987). Benthic animals can increase the rate of mineralization in excess of the effect that could be expected from the presence of increased O_2 concentrations (Kristensen and Blackburn 1987). These authors found that the polychaete *Nereis* greatly increased the rate of C and N mineralization in a sandy sediment microcosm. It was suggested that the increased rate of mineralization was caused by the exposure of a larger surface area to bacterial attack, due to substrate maceration. In this instance, there was preferential N mineralization presumably because N-containing compounds were more easily degraded.

The holes bored by Toredo worms in wood pilings and piers in marine environments are dramatic evidence of the importance of larger animals in attacking larger structures. The worms are obligate aerobes; no attack on wood is possible in anoxic sediments. The worms themselves do not contain cellulases. There is a mutualistic relationship between the host and its gut bacteria: the host provides macerated substrate to the bacteria, which otherwise would be unable to attack the intact wood. The bacterial cells provide a digestible N and C source for the host, which cannot digest wood. The situation is exactly analogous to the relationship between bacteria and other hosts, e.g., ruminants or termites.

A somewhat similar situation has been described for bacteria within the gut of the deposit-feeding lugworm, *Abarenicola vagabunda* (Plante et al. 1989). The bacteria grew in the hindgut, almost doubling every hour, a much more rapid growth rate than on the same substrate, in the sediment. Plante et al. suggest that there is increased contact between bacteria, their hydrolytic enzymes, and the substrate because of containment within the hindgut. The animals' role may be to act as a chamber in which bacterial growth is optimized, but there may be some additional exposure of substrate surface by animal maceration.

The presence of animals implies that there will be an increase in the animal biomass throughout the growing season. There is thus a partitioning of the primary resource food between the sediment infauna and the sediment bacteria. The infauna may also utilize the bacteria as a food source. In the Chirikov Basin, as much as 70% of the sediment community respiration may be due to the dense populations of amphipods (Blackburn 1987). This sediment is very sandy; it has a high F_c yet a very low organic content. Earlier it was suggested that the low E was due to the lack of mineral accumulation in the sediment. Possibly the large amphipod population might also have an influence by diverting the carbon flow through animal biomass instead of through bacteria. I suggested (Blackburn 1987) that soft animal biomass would be more degradable than bacterial biomass, whose mucopeptide cell walls are not readily hydrolyzed. It is difficult to distinguish between cause and effect in this system: the high sand content may alone be responsible for the low F_b or perhaps the amphipods, which require a sandy sediment, may be the cause of the very efficient mineralization. Downward mixing, by the infauna, of fresh organic material from the sediment surface to anoxic regions may result in a slower rate of mineralization by SO_4^{2-} reduction, for the reasons already discussed.

TEMPERATURE

At Cape Lookout Bight, rates of sediment SO_4^{2-}-reduction varied from 3.14 mM day^{-1} in summer at 27°C to 0.006 mM day^{-1} at 11°C in winter (Crill and Martens 1987). In the absence of other effects (e.g., input of fresh detritus), temperature would seem to play an important role in determining the rate of mineralization in this and in many other sediment systems. However, in sediments from the Chirikov Basin, where temperatures are always approximately at freezing point, there is no lag in the degradation of fresh detritus by either the bacteria or amphipods (Blackburn 1987). Low temperature does not appear to be an intrinsic barrier to the complete degradation of phytodetritus (zero F_b). In a situation where there are seasonal changes in temperature, there may be a lag while the resident population adjusts to the new temperature or while new strains grow.

Temperature and pressure effects are closely related. There is not much evidence that high pressure inhibits the rate of organic matter degradation, other than the mystery of the everlasting Alvin sandwich.

CONCLUSIONS

There is much circumstantial evidence which points to SO_4^{2-} being a less efficient and slower oxidant than O_2. This should now be investigated more directly by subjecting buried organic matter to decomposition in aerobic environments. The resistance of dissolved and particulate organic complexes to degradation needs further research. The mechanisms whereby sediment infauna increase mineralization rates requires explanation.

Acknowledgement. I wish to thank the Division of Polar Programs, National Science Foundation U.S.A. for financial support through the ISHTAR project grant DPP-B605659.

REFERENCES

Bender, M.L., and D.T. Heggie. 1984. Fate of organic carbon reaching the deep sea floor: a status report. *Geochim. Cosmochim. Acta* **48**:977–986.

Blackburn, T.H. 1987. Microbial food webs in sediments. In: Microbes in the Sea, ed. M.A. Sleigh, pp. 39–58. Chichester: Ellis Horwood.

Blackburn, T.H. 1990. Denitrification model for marine sediment. In: Denitrification in Soil and Sediment, ed. J. Sørensen and N.P. Revsbech. London: Plenum.

Blackburn, T.H., and J. Sørensen, eds. 1987. Nitrogen Cycling in Marine, Coastal Environments. Chichester: Wiley.

Canfield, D.E. 1989. Sulfate reduction and oxic respiration in marine sediments: implications for organic carbon preservation in euxinic environments. *Deep-Sea Res.* **36**:121–138.

Christensen, D., and T.H. Blackburn. 1982. Turnover of ^{14}C-labelled acetate in marine sediments. *Marine Biol.* **71**:113–119.

Christensen, J.P. 1989. Sulfate reduction and carbon oxidation rates in continental shelf sediments, an examination of offshelf carbon transport. *Cont. Shelf Res.* **9**:223–246.

Crill, P.M., and C.S. Martens. 1987. Biogeochemical cycling in an organic-rich marine basin. VI. Temporal and spatial variations in sulfate reduction rates. *Geochim. Cosmochim. Acta* **51**:1175–1186.

Degens, E.T., and K. Mopper. 1975. Early diagenesis of organic matter in marine soils. *Soil Sci.* **119**:65–72.

Foree, E.G., and P.L. McCarty. 1970. Anaerobic decomposition of algae. *Envir. Sci. Tech.* **4**:842–849.

Gordon, A.S., and F.J. Millero. 1985. Adsorption mediated decrease in the biodegradation rate of organic compounds. *Microbiol. Ecol.* **11**:289–298.

Harvey, H.R., R.D. Fallon, and J.S. Patton. 1986. The effect of organic matter and oxygen on the degradation of bacterial membrane lipids in marine sediments. *Geochim. Cosmochim. Acta* **50**:795–804.

Henrichs, S.M. and W.S. Reeburgh. 1987. Anaerobic mineralization of marine sediment organic matter: rates and the role of anaerobic processes in the oceanic carbon economy. *Geomicrobiol. J.* **5**:191–237.

Jewell, W.J., and P.L. McCarty. 1971. Aerobic decomposition of algae. *Envir. Sci. Tech.* **10**:1023–1031.

Jørgensen, B.B. 1982. Mineralization of organic matter in the sea bed—the role of sulphate reduction. *Nature* **296**:643–645.

Kristensen, E., and T.H. Blackburn. 1987. The fate of organic carbon and nitrogen in experimental marine sediment systems: influence of bioturbation and anoxia. *J. Marine Res.* **47**:231–257.

Nissenbaum, A., and I.R. Kaplan. 1972. Chemical and isotopic evidence for the *in situ* origin of marine humic substances. *Limnol. Ocean.* **17**:570–582.

Otsuki, A., and T. Hanya. 1972. Production of dissolved organic matter from dead green algal cells. II. Anaerobic microbial decomposition. *Limnol. Ocean.* **17**:258–264.

Plante, C.J., P.A. Jumars, and J.A. Baross. 1989. Rapid bacterial growth in the hindgut of a marine deposit feeder. *Microbiol. Ecol.* **18**:29–44.

Stein, R., J. Rullkötter, and D.H. Welte. 1986. Accumulation of organic-carbon-rich sediments in the late Jurassic and Cretaceous Atlantic Ocean—a synthesis. *Chem. Geol.* **56**:1–32.

Suess, E. 1980. Particulate organic carbon flux in the oceans—surface productivity and oxygen utilization. *Nature* **288**:260–263.

Walsh, J.J., C.P. McRoy, L.K. Coachman, J.J. Goering, J.J. Nihoul, T.E. Whitlidge, T.H. Blackburn, P.L. Parker, C.D. Wirick, P.G. Shuert, J.M. Grebmeyer, A.M. Springer, R.D. Tripp, D. Hansell, S. Djenidi, E. Deleersnijder, K. Henriksen, B.Aa. Lund, P. Andersen, F.E. Mueller-Karger, and K. Dean. 1989. Carbon and nitrogen cycling within the Bering/Chukchi Seas: source regions for organic matter effecting AOU demands of the Arctic Ocean. *Prog. Ocean* **22**: 277–359.

Westrich, J.T. and R.A. Berner. 1984. The role of sedimentary organic matter in bacterial sulfate reduction: the G model tested. *Limnol. Ocean.* **29**:236–249.

Widdel, F. 1989. Microbiology and ecology of sulfate-reducing bacteria. In: Anaerobic Bacteria, ed. A. Zender, pp. 469–585. Chichester: Wiley.

Dissolved Organic Carbon Enigma: Implications for Ocean Margins

Y. Suzuki and E. Tanoue

Geochemical Laboratory
Meteorological Research Institute
Nagamine 1-1, Tsukuba 305, Japan

Abstract. Dissolved organic carbon (DOC) concentrations in seawater and fresh water in the estuary, continental shelf, and open ocean have been determined by means of a high temperature catalytic oxidation (HTCO). DOC concentrations in the Delaware Estuary (U.S.A.) are high in the fresh and brackish waters (400–480 μmol C l^{-1} in river water), and appear to decrease nonconservatively with the increase in salinity (240 μmol C l^{-1}). This contrasts with a conservative distribution of DOC obtained by persulfate oxidation. DOC concentration in surface continental shelf waters of the East China Sea averaged 427 μmol C l^{-1} and decreased toward the open sea (255 μmol C l^{-1}). Both methods (HTCO and persulfate oxidation) yield quite different results between the continental margin and open sea. The concentration difference between both methods is small (ratio of HTCO to persulfate oxidation—1.2) in the continental margin and quite large (ratio of HTCO to persultate oxidation—3.2) in open sea. From the characteristic vertical distribution of HTCO–DOC in the open sea we presume that most of the DOC not oxidized by persulfate oxidation is biologically reactive and of relatively recent marine origin. However, most of the DOC oxidized by persulfate oxidation in the estuary and continental margin is of terrestrial origin and is more biologically resistant. DOC from terrestrial origins is also possibly removed within the continental margin in the East China Sea, in particular, at the frontal zone between the continental shelf and open sea. DOC from the marine biological origin in the continental margin is recycled by the heterotrophic organisms within these areas and is exchanged by upwelling and downwelling of water at the continental shelf. As a result, the contribution of lateral flux of DOC to the oceanic biogeochemical cycle of carbon, from the continental margins is possibly very small.

INTRODUCTION

The inventory of inorganic carbon in the ocean is more than 60 times larger

Ocean Margin Processes in Global Change
Edited by R.F.C. Mantoura, J.-M. Martin and R. Wollast

than the amount of carbon dioxide in the atmosphere. The cycle of carbon in the ocean is controlled by physical (advection, mixing, and air–sea exchange), biogeochemical (production and regeneration), and inorganic chemical (mineralization and dissolution of calcium carbonate) processes. To evaluate quantitatively these parameters in the oceanic carbon cycle is one goal of the Joint Global Ocean Flux Study (JGOFS). Of these processes controlling the oceanic carbon cycle, the study of the carbon cycle in relation to the biogeochemical processes (in particular, the conversion of inorganic carbon to organic carbon by photosynthesis and organic carbon to inorganic carbon by decomposition) appears most important. However, the flux and turnover time of carbon between both reservoirs are still not well known. The study of organic carbon (particulate and dissolved forms) is a key for understanding the biogeochemical cycle of carbon in the ocean.

Oceanic organic matter is classified mainly as dissolved and particulate. The major part of organic matter in the ocean is DOC with particulate organic carbon (POC) accounting for about 10% or less of the total. DOC includes the truly dissolved (particles size below 0.001 μm) and colloidal (0.001 to 0.1 μm) fractions. In the biogeochemical and trophodynamic aspects, DOC occupies an intermediate position between living organisms and inorganic nutrients; POC contains both living and nonliving organic matter. Organic matter produced by phytoplankton is incorporated into the marine food web. However, the quantitative explanation on the source and fate of dissolved organic matter (DOM) in seawater has not yet been given. Although there have been numerous studies and discussions on the methodology, concentration, distribution, source, transformation, role, and characterization of DOC over the past decades (reviewed by Mackinnon 1981; Romankevich 1984), it was believed that DOC might not play an important role in the oceanic carbon cycle. This was because most DOC in seawater was thought to consist of "refractory" compounds with an "old" apparent age of more than 3000 yr. (Williams and Druffel 1987).

The estuarine and coastal chemistry of terrestrial DOC is also poorly understood. Some fractions (e.g., humic acids) of river-transported DOC behave nonconservatively and are removed within estuaries by flocculation, precipitation, and adsorption onto particulate matter from mainly laboratory experiments (Sholokovitz et al. 1978). However, the distribution of DOC observed by photooxidation in the estuarine environment shows that DOC appears to behave conservatively (Mantoura and Woodward 1983).

The oxidation methods of organic matter could be classified broadly into either wet oxidation (potassium persulfate, photooxidation with ultraviolet irradiation, etc.) or high temperature combustion (Mackinnon 1981). Although some results of DOC obtained by a high temperature combustion are 15%–100% higher than those obtained with wet oxidation methods, it was previously concluded that the higher values might be due to the contamination or the system blank (Mackinnon 1981). There are still some serious problems yet to be resolved on the methods; however, much of the discussion on DOC behavior in the ocean was based on results obtained by wet oxidation methods.

Recent new measurements of DOC and DON in seawater using HTCO unveiled new aspects of DOM, both in terms of the concentration and correlations with the apparent oxygen utilization (AOU) (Suzuki et al. 1985; Sugimura and Suzuki 1988; Suzuki et al. 1990, submitted). Concentrations of DOC and DON in the western North Pacific indicate that (a) these compounds are three or four times more abundant in surface water, and twice as high in deep water than previously thought; (b) they are highly correlated with AOU; and (c) DOC obtained by wet oxidation methods is only a part of DOC in seawater.

In this chapter we report the results of DOC distributions as measured by a HTCO with those of wet oxidation in the estuary, continental shelf, and open ocean. We would also like to discuss the behavior of DOC from the terrestrial and marine biological origin within the ocean margin.

DISTRIBUTION AND FATE OF DOC IN THE CONTINENTAL MARGIN

Ocean margins are the sites of most intense primary production of organic matter in the ocean and are thus a most appropriate area for field studies of biogeochemical processes in the oceanic carbon cycle. Ocean margins are also ocean boundaries on the interior ocean; they are of special interest because their waters are enriched by the upwelling and intrusion of nutrient-rich waters associated with ocean boundary currents, as well as riverine input.

Recently, the importance of lateral flux of organic matter from the continental margins to the interior ocean has been discussed as well as the transport of trace metals and radionuclides (Bacon 1987). Mantoura and Woodward (1983) showed that riverine DOC, measured by photooxidation, was transported conservatively through the estuary to the open ocean; they calculated that up to 50% of oceanic DOC may be of riverine origin. Meyers-Schulte and Hedges (1986), however, have used lignin-derived phenols and C-13/C-12 isotope ratio analyses of organic matter in seawater to show the presence of a small terrestrial component of DOC, with less than 10% in the open ocean.

The relationship between DOC concentration measured by HTCO and salinity in the Delaware Estuary for May, 1988 is shown in Fig. 1; the relationship between carbon to nitrogen ratio in DOM and salinity is an example of DOC estuarine chemistry. DOC concentration in the Delaware River ranges from 439 μmol C l^{-1} to 529 μmol C l^{-1} and is 40–50% higher

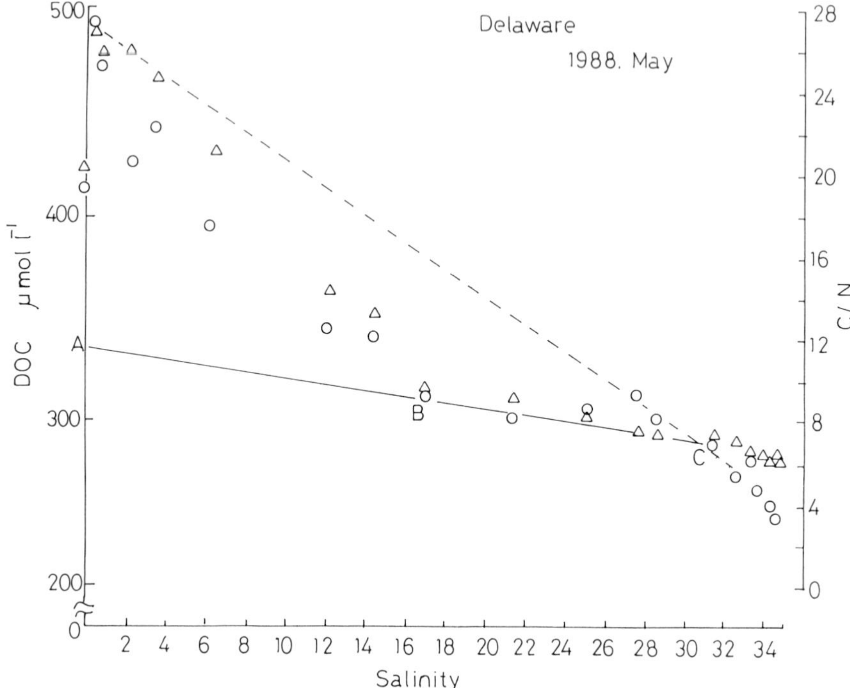

Fig. 1—DOC (salinity and C/N ratio in DOM)–salinity regressions for the Delaware Estuary on May, 1988. ○: concentration of DOC; △: C/N ratio in DOM); dotted line: conservative; solid line: estimation of DOC amount by removal; A: riverine DOC concentration estimated after removal; B: DOC concentration observed; C: DOC concentration in seawater.

than that measured by persulfate oxidation (Sharp et al. 1988). DOC concentration decreases with the increase of salinity, but its behavior is not conservative. In contrast, the behavior of DOC measured by persulfate oxidation (Sharp et al. 1988) and photooxidation (Mantoura and Woodward 1983) is conservative.

DOC concentration in the estuary is controlled by the dilution between river and seawaters and by precipitation, flocculation, and adsorption onto the particulate matter. If DOC is not removed, the concentration of DOC will fall along the dilution line between river water (490 μmol C l^{-1}) and seawater (290 μmol C l^{-1}). However, it is clear from the DOC–salinity curvature in Fig. 1 that DOC losses are occurring in the estuary. The low salinity extrapolation of the DOC–salinity line between 17–34‰ yields an apparent riverine DOC of 338 μmol C l^{-1} (Fig. 1, pt. A) rather than the observed value of 490 μmol C l^{-1}. It can thus be calculated that under steady-state conditions approximately 31% of the riverine DOC, measured by HTCO, is removed in the Delaware estuary. Therefore, ~69% of the riverine DOC could be transported to the coastal area or continental shelf. This result is supported by the C/N ratio in DOM, as shown in Fig. 1. The C/N ratio in river water is higher (26.6) and decreases with the increase of salinity to a value of 6.7 in seawater. The difference of C/N ratio between river water and seawater is due to the chemical nature difference of DOM in both waters.

The distribution of DOC in surface waters of the continental shelf from the East China Sea and the Kuroshio regions (from June to July, 1985) are shown in Fig. 2. The East China Sea is a marginal sea of the North Pacific Ocean and is characterized by inputs from Chinese rivers (Yangtze River and Yellow River), the continental shelf water, and the Kuroshio. The DOC concentrations in surface water ranges from 184–483 μmol C l^{-1}, with an average of 255 μmol C l^{-1} in the Kuroshio waters and 427 μmol C l^{-1} in the continental margin. DOC concentrations in the continental margin are higher than those in the Kuroshio region and decrease toward the open sea. The relationship between DOC and salinity in the East China Sea is complicated by shelf production, as shown in Fig. 3. Between 30.4 to 32.4‰ of salinity, DOC concentration is nearly constant, decreasing linearly at higher salinities. It seems that most DOC is not removed by the physicochemical processes in the continental shelf.

The vertical oceanic distribution of DOC obtained by HTCO (Sugimura and Suzuki 1988; Suzuki et al. 1990) and persulfate oxidation (Strickland and Persons 1972) are shown in Fig. 4. The main features of DOC distribution between both methods are as follows.

1. The major portion (about 65–70%) of DOC in the surface water is not oxidized by persulfate oxidation.
2. The difference of DOC concentrations between both methods reduces with depth toward the oxygen minimum layer where the difference between methods is only 1% to 15%.
3. About 40% of DOC in deep waters is not oxidized by persulfate oxidation, although DOC concentration is lower than that in surface waters.

It should be noted that the DOC fraction not oxidized by persulfate oxidation within the thermocline shows a large gradient of DOC concentration. This indicates that most of DOC is produced and is broken down within a short time comparable to or less than the vertical mixing time of water in the upper ocean layer. This suggests that it has a recent biological origin and is utilized relatively rapidly by the marine microbial community. Therefore, we think that the DOC fraction not oxidized by the persulfate

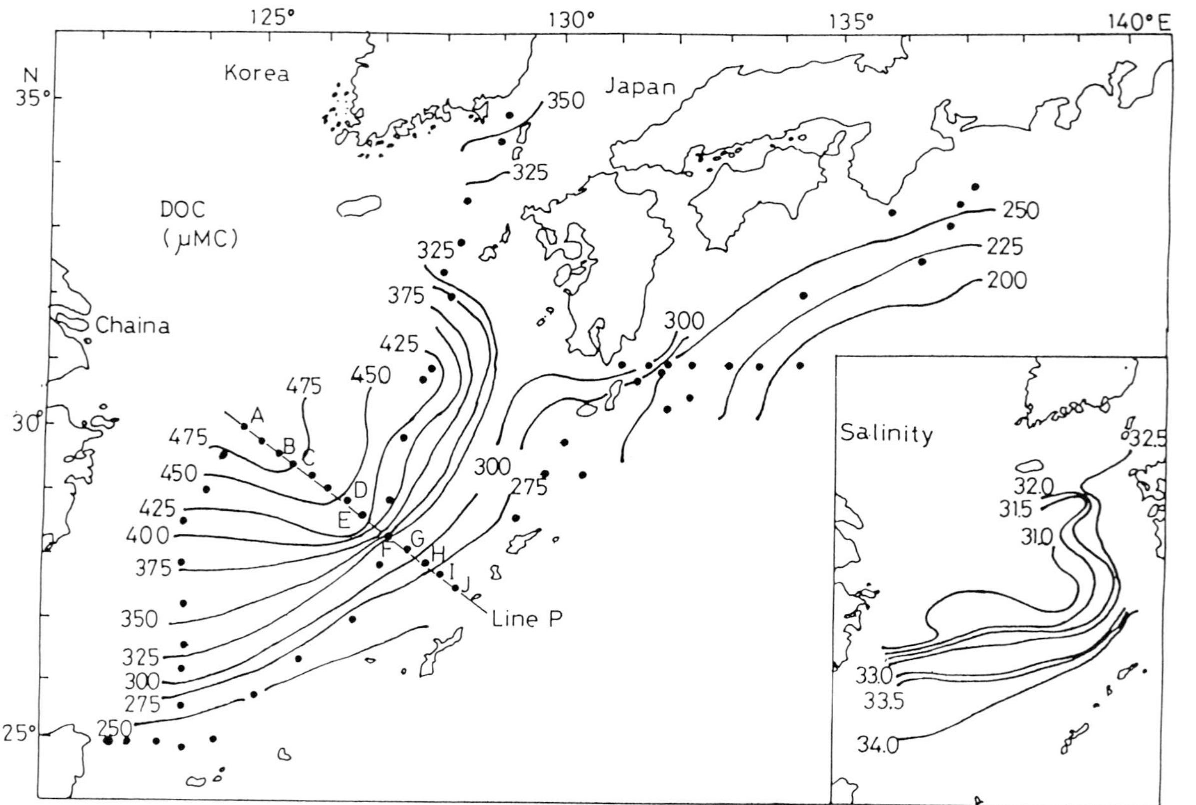

Fig. 2—The surface distribution of DOC and salinity in the East China Sea (July, 1984 and 1985). After Sugimura and Suzuki (1988).

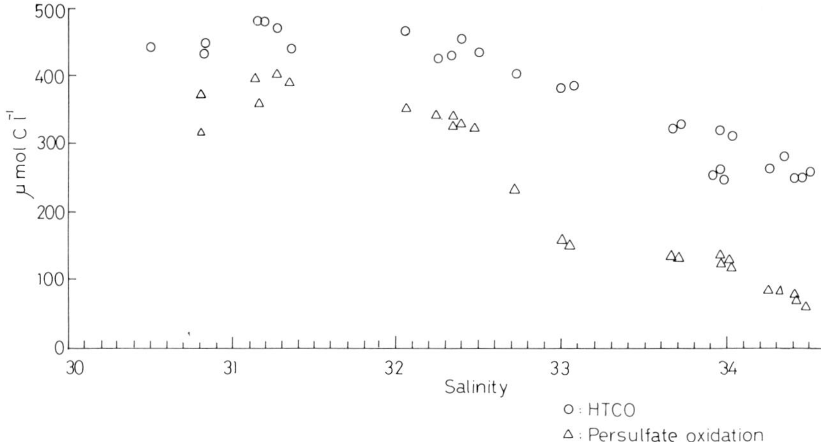

Fig. 3—DOC–salinity regressions for the East China Sea on July, 1984 and 1985. ○: HTCO, △: persulfate oxidation.

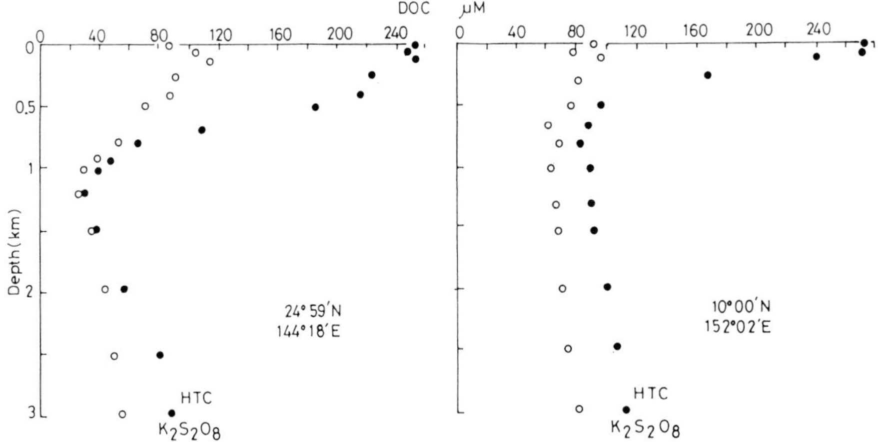

Fig. 4—The vertical distributions of DOC in the western North Pacific Ocean. ○: HTCO, ●: persulfate oxidation.

oxidation method is reactive biologically but paradoxically is inert chemically. We tentatively call this fraction bioreactive DOC. Most bioreactive DOC is supplied from the marine biological processes. In contrast, it is generally accepted that DOC extracted from surface to deep waters by persulfate oxidation or photooxidations is relatively "old" and thus inert to rapid biochemical utilization, but is chemically reactive. This photooxidizable DOC has an "apparent" age of 1300 yr. at the surface and 6000 yr. from

900 to 5720 m depth (Williams and Druffel 1987). We tentatively define this fraction oxidized by persulfate or UV oxidation as bioresistant DOC. The origin of bioresistant DOC assumes that 90% is also of marine biological origin in the open sea and 10% is of terrestrial origin (Williams and Druffel 1988). Our observations that most of the DOC at the oxygen minimum layer comprises the bioresistant DOC implies that the oxygen minimum layer is relatively much older than either the underlying or overlying water layers.

DOC distributions determined by means of HTCO and persulfate oxidation in the continental margin and the Kuroshio regions as a function of distance from the Yangtze River mouth (along line P in Fig. 2) are shown with that of salinity in Fig. 5. DOC concentrations (bioresistant DOC) measured by persulfate oxidation are abundant in the continental margin relative to the open sea, ranging between 345–400 μmol C l^{-1} in the continental margin to 93–108 μmol C l^{-1} in the open sea. The concentrations of bioresistant DOC decrease rapidly at the frontal zone (Fig. 5, stations E–G).

Summarizing some results of DOC distributions in the continental margin, we find: (a) about 90% of DOC in the surface waters of the continental margin is bioresistant while in the open sea about 65% of DOC is bioreactive; (b) the concentrations of bioresistant DOC decrease rapidly at the frontal zone of the continental margin whereas the concentration of bioreactive DOC increases toward the open sea. It should be noted that the bioreactive DOC concentration is less (HTCO/persulfate oxidation ratio = 1.2) in the continental margin waters but greater (HTCO/persulfate oxidation ratio = 3.2) in open sea waters. If we assume the concentration of bioresistant DOC from marine biological origin to be about 90 μmol C l^{-1}, i.e., that 90% of bioresistant DOC in the open sea is assumed to be of marine origin (Williams and Druffel 1988), then we calculate that 75% of the bioresistant DOC in the continental margin is of terrestrial origin (from river water and coastal sediment). As shown in Fig. 3, the concentration of DOC from terrestrial origins is apparently balanced between supplies from sediment or river water and bacterial and chemical removal from water in the continental margin, particularly at the frontal zone between the continental shelf and open sea.

It is commonly accepted that the ocean margin is the site of most intense primary production of organic matter in the ocean, as shown in Table 1, and that higher productivity would produce a higher DOC concentration with POC (Romankevich 1984). If the bioreactive DOC is produced by biological production, we expect DOC concentration to be much higher in the continental margin than in the open sea. However, the results in Fig. 5 show that the steady-state concentration of bioreactive DOC is lower in the continental margin than that in the open sea. We suppose that the causes of disappearance of bioreactive DOC might be due to rapid

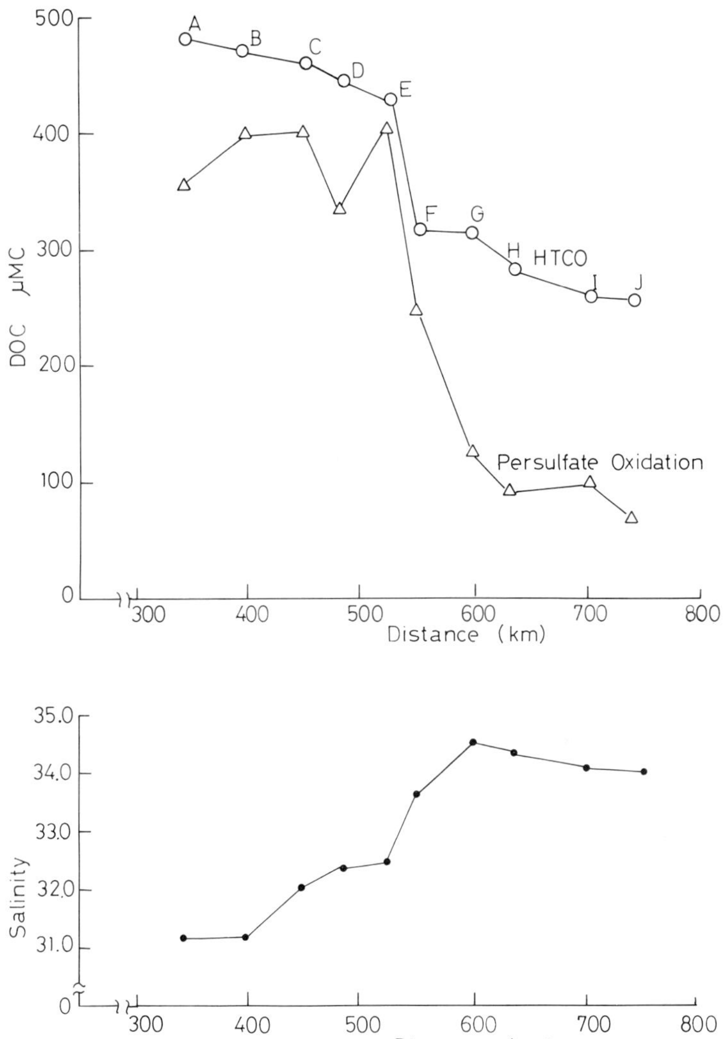

Fig. 5—The distribution of DOC and salinity as a function of distance from Yangtze River mouth (along line P in Fig. 2). ○: HTCO, △: persulfate oxidation.

consumption or decomposition of DOM by marine heterotorophic activity rather than by physicochemical removal processes. It seems as if the bioreactive DOC in shelf water is recycled with much shorter turnover times than in the ocean. We think that the difference of concentration of bioreactive DOC might be reflected by the difference of rate of DOM turnover time.

TABLE 1. Global net primary production and standing crop (Romankevich 1984).

Ecosystem type	Area 10^6 km^2	Total net primary production 10^9t C_{org}yr^{-1}	$gC_{org}m^{-2}$ yr^{-1}	Total plant mass of carbon 10^9tC_{org}
Continental ecosystems	149.0	52.8	354.4	826.5
Marine ecosystems including:	361.0	24.7 38.1*	68.7	1.7
Algal bed and reef	0.6	0.7	1166.7	0.54
Estuaries	1.4	1.0	714.3	0.63
Upwelling zones	0.4	0.1	250.0	0.004
Continental shelf	26.6	4.3	161.6	0.12
Open sea	332.0	18.7 32.0*	56.3 96.0*	0.45

*The values calculated using HTCO.

OCEANIC CARBON CYCLE AND IMPLICATIONS FOR OCEAN MARGINS

The carbon cycle in the ocean mainly takes place through two processes: through the motion of seawater and biological activity. In addition, there are processes of carbon supply from outside of the ocean as well as removal from seawater by sedimentation to the bottom.

Oceanic HTOC–DOC profiles mainly show four specific features: (a) the large DOC gradient in the oceanic thermocline, (b) the minimum value of DOC at the oxygen minimum layer, (c) a high correlation with AOU, and (d) the geographically significant N–S distribution in the western North Pacific Ocean (Suzuki et al. 1990). Such remarkable features were never found in the DOC distributions made by persulfate oxidation or photooxidation. The average value of DOC is 240 μmol C l^{-1} in the surface water, of which 65–70% is bioreactive DOC averaging 100 μmol C l^{-1} in deeper water, of which 50% is bioreactive DOC.

The inventory of bioreactive and bioresistant DOC to a depth of 200 m from the surface is 3.4×10^7 μmol C m^{-2} and 1.4×10^7 μmol C m^{-2}, respectively. The corresponding inventories between depths of 200 m to 4200 m are 20×10^7 μmol C m^{-2} and 20×10^7 μmol C m^{-2}, respectively.

Assuming that the rate of primary organic production is 8.0×10^6 μmol C m^{-2} yr^{-1} and 80% of oceanic primary production is recycled within the upper ocean layer (0 to 200 m), the recycling flux of organic matter is about 6.4×10^6 μmol C m^{-2} yr^{-1}. Assuming that most organic matter produced is formed into DOM, the turnover time of bioreactive DOC is about five years.

The annual flux of riverine DOC is calculated using the average concentration of DOC (490 μmol C l^{-1} is used) in river water and the annual discharge of river water (80 l m^{-2} yr^{-1} [Livingstone 1963]). The annual flux of riverine DOC is about 3.9×10^4 μmol C m^{-2}. Assuming that about 80–90% of riverine DOC is removed in the continental margin, the influx of riverine DOC to the open ocean is only 4×10^3 μmol C m^{-2}.

The flux of riverine DOC to the oceanic DOC pool is very small in comparison with that of oceanic biological flux of surface DOC as measured by HTCO. The oceanic carbon cycle is mainly controlled by the biological processes and the motion of water. The challenge is to evaluate the transport of bioreactive DOC produced in the ocean margin to the open sea. The primary production of carbon in the ocean margin (estuaries and continental shelf) is about 80×10^6 μmol C m^{-2} yr^{-1} and is ten times higher than that in the open sea. However, we suppose that since the concentrations of bioreactive DOC is less compared to the open sea, it is being more efficiently recycled within the ocean margins.

This suggests that the lateral flux of organic matter from the ocean margin to the interior ocean is not as important as previously believed. Other studies on continental fluxes of inorganic species (heavy metals) and radionuclides (Th-234, Th-228, Pb-210, etc.) emphasize the importance of lateral fluxes to the boundaries rather than vertical fluxes of particles in the interior ocean (Bacon 1987). In previous estuarine studies, DOC was found to be conservative (Mantoura and Woodward 1983) and terrestrial organic biopolymers (e.g., lignins) have been found in oceanic DOC (Meyers-Schulte and Hedges 1986). However, the results measured by HTCO indicate that most of riverine DOC is not conserved in the shelf; most of the marine biological DOC produced in the ocean margins is also recycled within the ocean margins. We recognize that radionuclides are useful as a tracer for understanding chemical cycling and evaluation of lateral flux of inorganic materials from the ocean margin to the interior ocean; however, we think that the question of lateral flux of organic matter should be considered separately, because the cycling time between organic and inorganic matters in the ocean might be different. Although we have no quantitative explanation as to whether the lateral flux of organic matter to the interior ocean from the ocean margins is important or not, we suppose that the process of transportation of the organic matter over the boundaries may not be important.

CONCLUSION

Ocean margins are divided from three areas: estuary, continental shelf, and continental slope. The concentration of DOC in river water as well as that in the open sea is 40–50% higher than previously measured. Significant proportions (~30%) of riverine DOC are removed in the estuary, and its behavior is not conservative. A small portion of riverine DOC is transported to the continental shelf and open sea. In the continental shelf and open sea, we distinguish two fractions of DOC on the basis of the difference of DOC distributions measured by HTCO and persulfate oxidation. Although the difference of DOC concentrations measured by both methods is due to the difference of oxidative capacity between both methods for DOC, these two fractions of DOC extracted by both methods appear to have biogeochemically and oceanographically significant distribution. We tentatively defined two DOC fractions as bioreactive (not oxidized by persulfate oxidation) and bioresistant (oxidized by persulfate oxidation) from the specific features of these distributions. In the continental shelf, bioresistant DOC (about 75% is of terrestrial origin) is in quasi steady state between supply from rivers and sediments and removal by physicochemical processes, whereas bioreactive DOC is recycled with a short turnover time relative to the biological processes within the continental shelf. At the continental slope, most bioresistant DOC of terrestrial origin is removed, and a small portion (less than 10%?) of the terrestrial DOC is transported to the open ocean. Bioresistant DOC of marine biological origin is supplied to the continental shelf by water exchange processes, including buoyancy flows, upwelling, and boundary currents.

We hypothesize that the lateral advective flux of DOC from the continental margin to the interior ocean is not large and so the biogeochemical cycle of organic carbon in the interior ocean is determined principally by the internal cycling of oceanic carbon in the ocean. Future studies should focus on the isotopically constrained dynamics of the oceanic DOC in relation to coupled biogeochemical processes, particularly the flux and the rate constant for the interconversion of inorganic to organic carbon in the upper ocean. From this we finally propose a new concept of “biodynamic converter,” relating the flux and rate constant of production and decomposition of DOC in the upper ocean layer. We think that a quantitative explanation of this “biodynamic converter” is required as well as that of the “biological pump” used to explain the vertical transport of particle organic matter in the oceanic carbon cycle.

Acknowledgements We thank Dr. Yukio Sugimura for his grateful help and also Drs. Toshio Hirano (KH8402), Takahisa Nemoto (KH8503), and Ushio Shimizu (KH8801) of the Ocean Research Institute, University of Tokyo and colleagues for their help on board during the Hakuho-maru cruise. We thank Dr. J.H. Sharp of Delaware University for offering their valuable samples with good comments.

REFERENCES

Bacon, M.P. 1987. Scavenging of radionuclides at ocean margin and its effect on fluxes measured in the ocean interior. In: Ocean Margins in JGOFS, ed. G.A. Knauer, pp. 69–79. Washington, D.C.: US GOFS Rep. 6.

Livingstone, D. 1963. Chemical Composition of Rivers and Lakes. Washington, D.C.: U.S. Geolo. Surv. Paper: 440-G.

Mackinnon, M.D. 1981. The measurement of organic carbon in seawater. In: Marine Organic Chemistry, ed. E.K. Duursma and R. Dawason, pp. 415–443. Amsterdam: Elsevier.

Mantoura, R.F.C., and E.M.S. Woodward. 1983. Conservative behavior of riverine dissolved organic carbon in the Severn Estuary: chemical and geochemical implications. *Geochim. Cosmochim. Acta* **47**:1293–1309.

Meyers-Schulte, K.J. and J.H. Hedges. 1986. Molecular evidence for a terrestrial component of organic matter dissolved in ocean water. *Nature* **321**:61–63.

Romankevich, E.A. 1984. Geochemistry of Organic Matter in the Ocean, pp. 334. Berlin, New York: Springer-Verlag.

Sharp, J.H., Y. Suzuki, and W.L. Mundy. 1988. Intercomparison of dissolved organic carbon analyses in estuarine and coastal waters of the North Atlantic ocean. *EOS* **69**:1134.

Sholkovitz, E.R., E.A. Boyle, and N.B. Price. 1978. The removal of dissolved humic acids and iron during estuarine mixing. *Earth Planet. Sci. Lett.* **40**:130–136.

Strickland, J.D.H., and T.R. Persons. 1972. A practical handbook of seawater analysis, 2nd ed. *Bull. Fish. Res. Board Can.* **167**:311.

Sugimura, Y., and Y. Suzuki. 1988. A high temperature catalytic oxidation method of non-volatile dissolved organic carbon in seawater by direct injection of liquid samples. *Mar. Chem.* **24**:105–131.

Suzuki, Y., Y. Sugimura, and T. Ito. 1985. A catalytic oxidation method for the determination of total nitrogen dissolved in seawater. *Mar. Chem.* **16:**83–97.

Suzuki, Y., E. Tanoue, and Y. Sugimura. 1990. Distribution of dissolved organic carbon in subtropical and tropical Pacific Ocean waters. *Deep Sea Res.*, in press.

Williams, P.M., and E.R.M. Druffel. 1987. Radiocarbon in dissolved organic matter in the central North Pacific Ocean. *Nature* **330**:246–248.

Williams, P.M., and E.R.M. Druffel. 1988. Dissolved organic matter in the ocean: comments on a controversy. *Oceanogr. Mag.* **1**:14–17.

Standing, left to right:
Henry Blackburn, Fauzi Mantoura, Doeke Eisma, Chris Martens, Tim Jickells, Daniel Vaulot, Andreas Moll
Seated, left to right:
Scott Fowler, Jack Blanton, Renate Scharek, Yoshimi Suzuki

Group Report: What Determines the Fate of Materials Within Ocean Margins?

T.D. Jickells, Rapporteur
T.H. Blackburn
J.O. Blanton
D. Eisma
S.W. Fowler
R.F.C. Mantoura
C.S. Martens
A. Moll
R. Scharek
Y. Suzuki
D. Vaulot

INTRODUCTION

When this diverse group of physicists, chemists, biologists, and geologists first met, a wide range of issues were prepared for discussion and these were subsequently crystallized to six questions that incorporated most of our concerns over what we felt to be important but poorly understood factors controlling the fate of materials in margins. These questions were discussed and are presented approximately in a flow pattern, from the physics of energy exchange through water column processes to the sediments.

From the outset of our deliberations, it was evident that physical forcing by rivers, the atmosphere, and the oceans is a major determinant of all processes on continental margins. Furthermore, the type of forcing varies greatly from place to place, depending on the balance of the forcing functions and the geomorphology of the shelf. The group therefore began by first considering the physical forcings and their interactions with geomorphology before moving on to discuss biogeochemical cycles on and around ocean margins within their physical and geomorphological context. A theme of our deliberation was often to find common factors in shelves in order to simplify these complex systems. There is a danger that such attempts at rationalization will oversimplify the system, but it is also evident that some sort of simplification is necessary in order to begin to comprehend these complex and diverse systems.

Ocean Margin Processes in Global Change
Edited by R.F.C. Mantoura, J.-M. Martin and R. Wollast

HOW DO BATHYMETRY, HYDRODYNAMICS, AND DISCHARGE CONTROL BIOGEOCHEMICAL CYCLES IN OCEAN MARGINS?

In order to consider this question, the group first considered external physical forcing before discussing the mechanisms by which this forcing interacted physically with water on the shelf. Biogeochemical cycles in marine systems are controlled mainly by interactions between dissolved and particulate phases. Therefore, the group next discussed the interactions of physical forcing on the two sources of particulates: the biota and the sediments.

External Forcing

The major external physical forcing functions on coastal margins are river discharge, wind-induced upwelling/downwelling, migrating atmospheric low pressure systems, the interactions of internal oceanic waves and boundary currents with shelf edges (Blanton, this volume). The effect of these forcings varies greatly depending on the character of the shelf itself, notably factors such as the size and topography of the shelf and the degree of stratification of the water on the shelf.

Recognizing this, an attempt was made to classify shelves in terms of their size in comparison to the forcing functions; in doing this we built on an approach adopted by the U.S. GOFS community (USGOFS 1987b). We considered a simplified shelf system only in a two-dimensional fashion, with inputs of fresh water and exchange of water from offshore.

In a general sense, the ratio of the river input to the ocean water exchange flux provides a measure of the extent to which a shelf is dominated by river inputs or by exchange with offshore waters. Control is also exerted by the size of the shelf over which these forcings are dispersed and by the kinetic energy of the exchange forcings themselves. Since these forcings occur at the margins, in a large margin with a long water residence time, internal biogeochemical cycles will be of great importance, while on small margins with short water residence times, exchange may be so fast that unique biogeochemical systems are not developed. Using this classification idea, it is clear that various extremes can be identified. The Amazon is a system with river exchange dominating a relatively small margin. At the other extreme, there is the large shallow Indonesian margin in which internal cycling likely dominates across a large part of the margin since the distance between the inner and outer boundaries is large. On some margins it is possible to consider an inner and outer region: the outer dominated by margin/ocean exchange and the inner by land/ocean interactions. This distinction is enhanced in certain areas where water flows are along the margin. In view of this distinction, mechanism of cross-margin water exchange are of particular importance and some of these are illustrated in Fig. 1. Important mechanisms of exchange are canyons, which focus water and sediment flows, and embayments or headlands, which deflect along shelf currents to encourage across-margin exchange. In short shelves the distinction breaks down and the case of large rivers driving cross-margin transport has been noted.

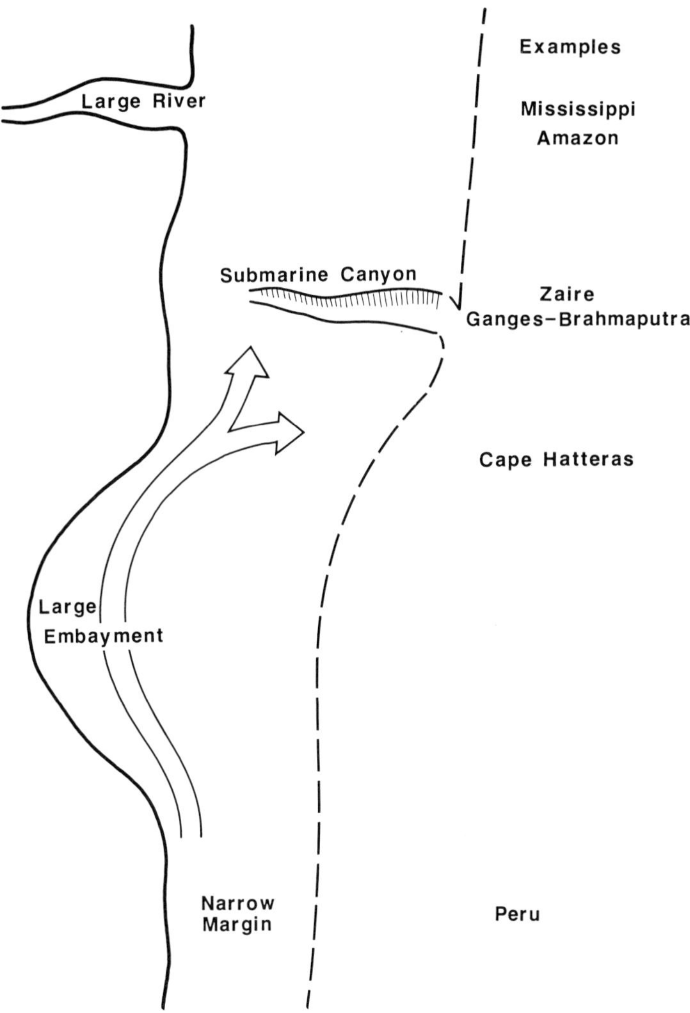

Fig. 1—Margin configuration/situations in which inner margin material exchange with the outer margin is enhanced together with examples of such areas.

In attempting such a classification, several problems become evident. First, time and space scales over which such classifications should be done are uncertain. An annual scale may be appropriate, but rare extreme events may have major effects on a margin and the effects may persist for long periods. If seasonal water column stratification occurs, a seasonal time scale may be more appropriate. In terms of spatial scales, it is also clear that river plumes will effect only certain areas of a shelf, so that the geographic shelf boundaries may not be the relevant one. The group therefore concluded that this classification is potentially useful but requires refinement, which was not possible during the meeting. The group also noted that a compilation of the shelf areas in terms of the physical forcing functions and geomorphological types was not available in the scientific literature, but that this could be produced and then used to test the applicability of the approach.

Stratification

The effect of the forcing functions on a shelf will depend in part on the water structure on the shelf.

Horizontal density gradients (Fig. 2a) suggest efficient vertical mixing from surface to bottom; horizontal exchange across isopycnals is inhibited to some extent. On the other extreme, vertical density gradients (Fig. 2b) support efficient horizontal transport, and vertical exchange across isopycnals is inhibited. This is a particularly important state of stratification because it may allow efficient cross-shelf exchange of material between inner margin sites with those on outer margins. Spatial gradients in vertical mixing and inputs of buoyancy plus the influence of eddies and advection (e.g., wind-generated upwelling or downwelling) may produce a combination of horizontal and vertical gradients in which a secondary circulation regime has potentially important horizontal and vertical components (Fig. 2c, d;

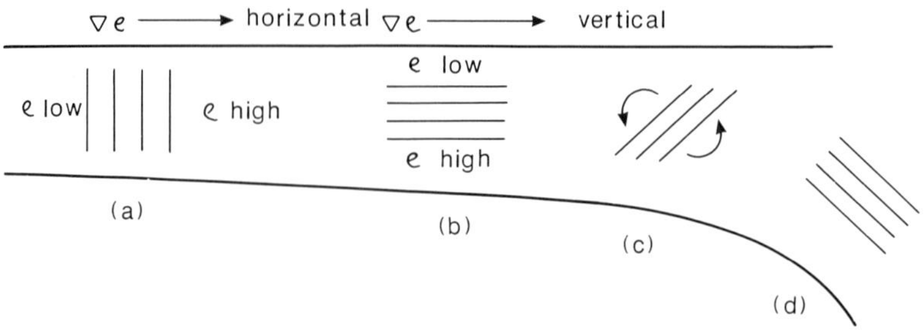

Fig. 2—Schematic view of vertical and horizontal density gradients (ρ is density).

Blanton, this volume). Mixing along and across isopycnals occurs to different extents in these different density regimes.

Inputs of buoyancy tend to stratify the system horizontally. The more obvious processes that increase buoyancy include solar radiation, freshwater runoff, rainfall, and melting ice. The time scales of variation are on the order of one day to annual.

Work by frictional and convective processes at the surface and sea bottom redistribute buoyancy (Fig. 3) and, if the work is sufficiently strong, horizontal stratification is destroyed. Simpson and Hunter (1974) have derived a stratification model which synthesizes the opposing work done by buoyancy fluxes versus work done by mechanical energy. In the absence of advection of buoyancy and neglecting buoyancy fluxes from lateral boundaries, such as ice melt at the shore or river runoff, a *minimum* set of variables for a given margin can be incorporated in a dimensionless stratification number (SN), defined as

$$\mathrm{SN} = \frac{\mathrm{BH}}{\mathrm{KU}^3}$$

where B = buoyancy flux (L^2T^{-3}), H = water depth (L), U is a velocity scale (LT^{-1}), and K is a dimensionless constant which includes an efficiency factor for the proportion of mechanical energy that is used to mix buoyancy. With constant buoyancy flux and technical efficiency, the ratio H/U^3 can be used to define the stratification regime. In shallow water (H small), and using reasonable values of U, SN is small and stratification could be weak or even destroyed at suitable levels of U^3. For large H, constant B, and U^3, SN is large and would represent a stronger degree of stratification.

Sources of mechanical work include friction by bottom currents generated by tides and winds and surface friction resulting from wind stress and surface waves. The time scales here range from seconds to days, considerably shorter than time scales of buoyancy generation. Convection is also an

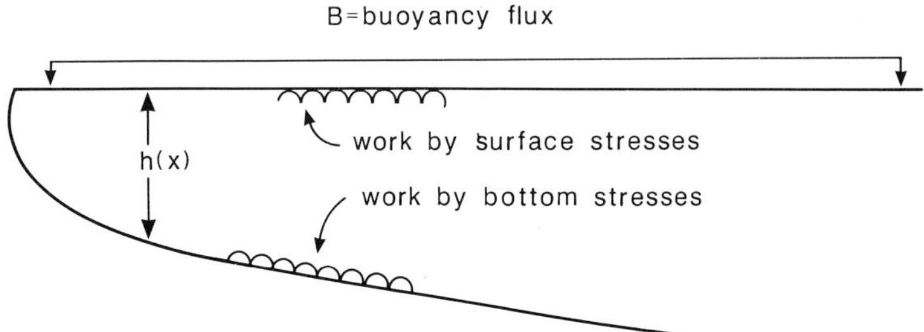

Fig. 3—Hypothetical margin where water depth is a function of offshore width.

efficient source of mechanical mixing ($B < 0$) and is represented by processes such as cooling of surface water by air and the production of brine by freezing seawater. These time scales range from day to annual, comparable to those of buoyancy generation.

An example of information to be derived from SN is given by Fig. 4. Times during which high values of SN are created (e.g., high net solar radiation and low wind stress) are favorable for enhanced exchange between inner and outer portions of continental margins.

Sediments

Sediments are an important component of the shelf system, but they behave in fundamentally different ways compared to the water, although they are under the influence of similar physical forcing factors within the margin system (Eisma 1988).

Only relatively small amounts of suspended matter coming from the continents reach the outer shelf, slope, or continental rise. Turbid waters and muddy deposits are mainly found on the inner shelf. Most suspended material that reaches the sea from land in most areas comes from rivers, directly or indirectly through resuspension. There are mainly five processes that retain suspended particles on the shelf (Fig. 5, left side). The different processes may act simultaneously in the same area and, as far as this has been studied, often do so, thus making the estimation of their relative importance in field experiments difficult if not impossible. Carefully constructed mathematical models are often the only way to separate the different processes that move sediments.

A minimum set of variables (not only physiographic ones) that determine dispersal and retention of suspended material on the shelf are indicated on the right side of the scheme. The complexity of the circulation is reflected in the relatively large number of variables that are involved (river flow, tides, development of density gradients and their intensity, wind/waves

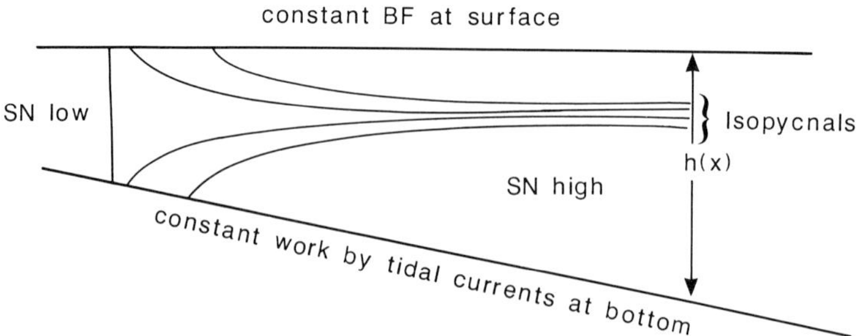

Fig. 4—Hypothetical margin showing possible relation between SN and water depth.

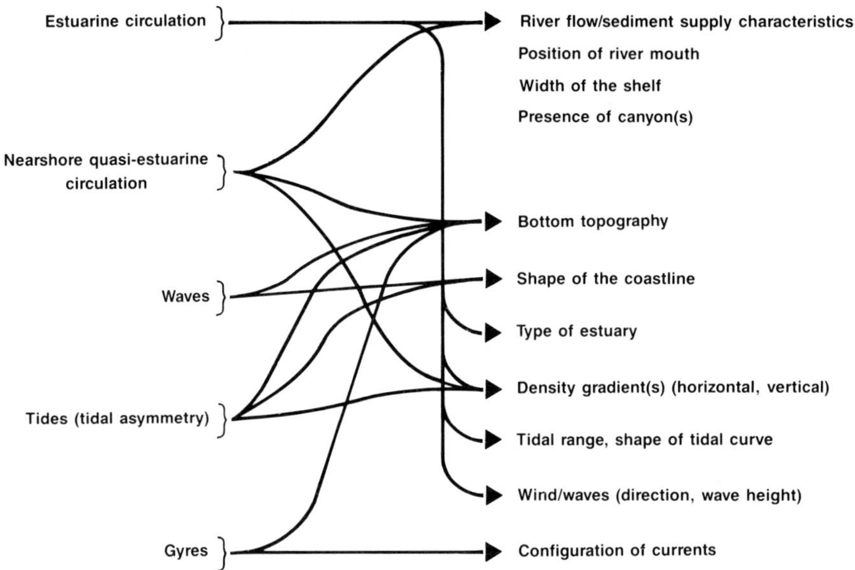

Fig. 5—Processes operating in a margin to retain suspended matter.

particularly through wind-forcing, and type of estuary), which also includes variables of minor or local importance. The nearshore quasi-estuarine circulation is largely determined by freshwater supply, the density gradients, and water depth. When the water becomes deeper, the strength of bottom transport diminishes. In general, stratification produces areas of low velocity where sediment accumulates. Wave effects are largely dependent on bottom topography and the shape of the coastline; waves act mainly through resuspension of bottom sediment and by keeping suspended material in suspension. Tidal asymmetry develops under the influence of the bottom topography, the shape of the coastline, and density gradients; in general, it serves to retain sediments by moving them inshore. Gyres also develop under the influence of the bottom topography and the shape of the coastline, as well as through configuration of the currents; the center of gyres act as areas of low velocity where sediments accumulate.

Estuarine circulation occurs in all river mouths in one form or another; in nontidal estuaries it may result in the formation of turbid bottom layer flow over the shelf. The quasi-estuarine circulation develops on all coasts with freshwater runoff. Wave effects also occur worldwide. Asymmetrical tides occur in embayments and tidal flat areas. Gyres occur anywhere on the shelf where residual and persistent currents are present. They can reach dimensions of 50–100 km. Thus, physical forcing acts to retain sediments within ocean margins, but this retention can be destroyed by extreme events such as hurricanes. A further factor is that physical forcing often acts to resuspend sediment continually, which introduces sites for the adsorption of materials to and remobilization from suspended solids and reduces light penetration.

Biota

Physically driven water exchange with the open ocean imposes constraints on phytoplankton systems. Ocean margins offer environments rich in nutrients that can be highly productive, but this requires that the phytoplankton be retained within the margin and within the euphotic zone in that margin. Since light penetration is a function of depth and suspended solid concentration, the persistence of phytoplankton on a margin interacts with both sediment transport and physical forcing. While many plankton are passive to transport processes, there is evidence that some are active in attempts to remain within a system. Examples are dinoflagellate and zooplankton migrations. The question of sustaining a plankton boom is considered in the next section, but here the group noted the strategies of retaining a relic overwintering population in temperate shelves to provide a seed population for the spring bloom. This usually involves the deposition of resting cells (diatoms), resting cysts (dinoflagellates), or eggs (copepods) in sediment or ice, thus preventing cells being flushed out with the water. These strategies have risks, since it is possible that the resting phases could become deeply buried and hence unable to escape from the sediments or buried in anoxic sediments (and hence killed). However, the success of these strategies is evident from the regular seasonal succession of algal populations seen in many areas. The risk is undoubtably mitigated by the strategy of releasing very large numbers of the resting stages. In tropical latitudes where overwintering is unnecessary, there are suggestions that plankton operate strategies to retain themselves in the coastal zone by sinking from outflowing surface water into onflowing deep waters; however, these communities are much less well studied. A similar strategy employed by *Euphausids* in the Southern Ocean of Antarctica is relatively well understood (Steidinger and Walker 1984).

In some areas where light penetration is great enough, benthic primary production is also important. In addition, there are highly productive margin edge communities which actively modify the physical environment, as in the case of mangroves, marshes, sea grasses, and coral reefs. They occupy physical locations which optimize their productivity by allowing accumulation of sediments (mangroves/marshes) or a location where light and nutrient supply is optimal (coral reefs, sea grasses). The controls on these communities are physical, i.e., light and habitat. All are under threat from coastal development, eutrophication, or disturbance. The margin edge communities may not always be of global importance as material reservoirs, but they are of great socioeconomic value as nurseries for fisheries and coastal protection. Such communities are likely to be affected to a considerable extent by sea level change.

HOW DO THE FACTORS WHICH LIMIT PRIMARY PRODUCTIVITY VARY WITHIN AND BETWEEN OCEAN MARGINS?

The factors controlling primary productivity have been reviewed in an earlier Dahlem Workshop Report (Berger et al. 1989) and in a background paper for this conference (Smetacek et al., this volume). Thus, the group simply listed the major factors of importance such as light, temperature, nutrients (N, P, Si, Fe, and possibly other trace metals), mixing, and grazing without debate. It was noted that the limiting factor on autotrophic biomass must ultimately be chemical when considering large enough spatial and temporal scales for which quasi-steady state is a reasonable assumption, e.g., in the context of global biogeochemical budgets. The term "regulating factor" was adopted as more useful in the context of shelf water, where physical as well as chemical factors are important on shorter time scales.

Differences between Ocean and Ocean Margin Primary Productivity

It is evident from global compilations of phytoplankton primary production rates that margin areas are generally productive in comparison to open ocean areas (e.g., Berger 1989). It was accepted that this characteristic of margin phytoplankton systems derives from the physical setting in which they operate. In comparison to ocean systems in general, much greater nutrient supplies (derived from riverine, atmospheric, and upwelling sources) produce a relatively large nutrient reservoir, and the presence of a shallow sediment boundary to the system allows rapid recycling to the euphotic zone. In many systems this benthic recycling is sufficiently rapid to allow inclusion of the upper few cm of sediment as essentially part of the water column.

There are other distinct characteristics of shelf systems. In addition to high average primary productivity, the formation of intense blooms terminated by nutrient exhaustion appears to be characteristic of shelf systems, though it was recognized that observations in the open ocean could be sufficiently infrequent as to miss such blooms there. Another characteristic appears to be different plankton community structures. These, in part, reflect the increased nutrient supply favoring certain classes of phytoplankton; in addition, however, the availability of a sediment boundary allows zooplankton to use different strategies for resting phases (e.g., eggs) compared to open ocean species. The high productivity and rather simple food webs in shelf systems yield large carnivore populations relative to those of herbivores, as compared to open ocean systems. The balance of the activity of these herbivores and the timing of egg hatching is a controlling factor on bloom formation.

Offshore, the concept of "new" production has recently been widely used (USGOFS 1987a; Berger et al. 1989) to quantify the export of carbon from the euphotic zone to deep water. This is often defined in terms of the proportion of primary production fuelled by nitrate (new production) and ammonium (recycled production). In reality, the new production in margins can be simply expressed as summed sources of nitrogen to the margin from the open oceans, atmosphere, and rivers. However, in general, the group felt that "new" production requires redefinition to become a useful concept in margin systems.

Differences in Primary Productivity between Margins

Margins in different physical settings will have markedly different primary productivity with the key factors being nutrient supply, light penetration, and the flushing of cells offshore. Extreme examples are:

1. Upwelling margins on short shelves, e.g., Peruvian shelf, where very high rates of primary production are sustained at the shelf break or just offshore with major export to the underlying sediments;
2. River dominated margins where productivity is high in the river plume, assuming light is not the controlling factor;
3. A large shelf with limited upwelling or river inputs where productivity is fuelled predominantly by internal recycling, and where benthic primary production rates may be high if suspended solid concentrations are low enough to allow light penetration;
4. Ice covered shelves where productivity is very low unless there is a short ice-free season.

However, many shelves incorporate several of these features and a more complex classification may be required to describe any single system. Nevertheless, this classification emphasizes that the differences between shelves are again the result of physical forcing functions. The crude classification is also a basis upon which to judge the types of shelves that have been intensively studied. Thus, for example, some medium-sized, river-dominated margin systems (e.g., Rhine, St. Lawrence) are reasonably well studied while some very large systems, such as the Amazon or Zaire (Congo), are not. The Peru and Californian upwelling systems are well studied but related Indian Ocean systems are not. Similarly, the large shelf area of the Indonesian region is little studied.

Differences in Primary Productivity within Margins

In general, primary productivity in freshwater systems is considered to be regulated by phosphorus supply (Billen, this volume). However, in oceanic systems, fixed nitrogen is normally considered the regulating nutrient (Codispoti 1989). Thus, margins may represent a transition zone between these two systems, although nitrogen is generally assumed to be limiting outside of estuarine or inner margin systems. Lowered phosphorus recycling due to sediment adsorption during winter months may produce seasonal phosphorus limitation during that season in some areas or phosphorus limitation may occur if nearshore nitrogen inputs have been enhanced (e.g., Bermuda, Jickells et al. 1986).

However, as primary productivity and biomass vary within margins on short time (day-week) and spatial (10–100 km) scales in response to physical forcing, it is generally difficult to assess which specific factor limits at a given place and time. An example is provided by large river plumes, such as the Changjiang (Fig. 6), where biomass is regulated by opposite gradients of light and nutrient availability resulting from physical processes of mixing and sedimentation (Ning et al. 1988).

The transition may not be gradual, but rather at some point a change in regulating factor may drive a switch from one quasi-stable state to another (a bifurcation in the terminology of Legendre and Le Fèvre [1989]). This is evident in the temporal variability in some systems, e.g., in temperate

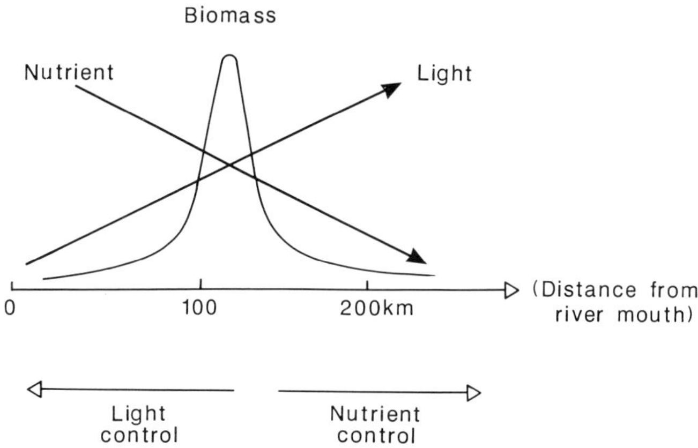

Fig. 6—Relationship between primary productivity and distance offshore in the plume of the Changjiang.

margins where seasonal water column stratification occurs. The effect of stratification on the formation and sequence of plankton blooms depends on the meteorological forcing in a nonlinear manner (Radoch and Moll 1990). The transition from one system to another occurs during the spring, from a diatom-dominated system driven by the nutrients regenerated over winter to a flagellate community that recycles nutrients very efficiently. In autumn there may be another transition to return to a system dominated by diatoms. The transitions are forced by the availability of light, nutrients, and mixing. In this case, Si is a potential controlling nutrient because of its requirement by diatoms. The timing of the shift from one system to another and whether there is an autumn bloom or not depends on the latitude in which the area is situated and on the topography of the shelf. For example, during a stratification period in summer, when the flagellate system has established itself, erosion of the nutracline (pycnocline) due to storms or increased bottom friction can occur. This introduces nutrients regenerated below the seasonal thermocline into the upper mixed layer, and populations of diatoms can build up again. The two systems are clearly distinguishable by their dominant organisms. Understanding how continental margin primary production communities switch from one state to another is of critical importance in the case of potentially harmful blooms (e.g., toxic dinoflagellates, *Phaeocystis pouchetii* [Smetacek, this volume; Smetacek et al. 1984]).

Although we have been able to identify several factors that influence primary productivity in margin areas, the data base for areas apart from those close to U.S. and European coasts is very limited. Productivity studies of margin areas of various types over at least annual cycles, and preferably long time series, are an urgent research priority. In addition, the importance of benthic vs. phytoplankton primary productivity requires evaluation, as does the relative importance of $CaCO_3$ and organic carbon production.

Macro and micro benthic primary production and coastal marsh or mangrove communities are unique, diverse, and complex communities within margins; they are of major importance for the functioning of the whole ecosystem. Their quantitative role in margin biogeochemistry is poorly understood and requires evaluation, particularly because of the sensitivity of these environments to global change over the near future.

WHERE DOES REGENERATION OCCUR?

High rates of particulate organic matter (POM) and carbon (POC) production are found in ocean margins. Since not all of this POM is buried (Wollast, this volume), it follows that rates of regenerative processes will also be high in these zones. Upon production (and death for biological particles), sinking will begin to transport the produced material to the sediments (Fowler, this volume). An important control on the rates and type of particle formed and its subsequent fate is the biological community in which it is formed since, as noted earlier, certain biological communities favor the production and settlement of particles while others favor intense water column recycling. Following production, biogenic particles whether primary or secondary aggregates (Fowler, this volume) will undergo regeneration of their component elements and compounds by such processes as microbial degradation, physical destruction through grazing or particle solubilization, element desorption, and zooplankton and bacterial metabolic excretion. The rates associated with these processes will be controlled to a large extent by parameters such as temperature, salinity, energy, and the redox chemistry of the surrounding waters. These parameters vary widely in coastal margins. In addition, regeneration rates, and hence where regeneration occurs in the water column, will be a function of POM composition and the degree to which it is labile since breakdown rates of different organic components vary, as do rates of dissolution of mineral phases such as different carbonates and silica.

From open ocean sediment trap flux profiles it has been demonstrated that the majority of POC and PON (particulate organic nitrogen) regeneration in sinking detritus occurs in the upper few hundred meters, i.e., at depths which often bracket the entire water column in a margin zone. However, recent sediment trap work has suggested that the regeneration rates derived from such particle flux experiments can be grossly overestimated if all living POM which actively entered the collector (i.e., zooplankton "swimmers") is not removed from traps in the upper water column (Fowler, this volume). Furthermore, in high energy margin systems, mass and POM flux do not always follow the classical pattern of exponential decrease with depth seen in the open ocean, because the vertical flux of sinking particles is often confounded by an input of laterally advected particles originating from the margin. Thus, it may be far more difficult to derive element regeneration rates from particle flux profiles in margin waters. The development of new techniques to determine net fluxes to the sediments in margin areas uncorrupted by lateral advection and resuspension is a priority.

Because ocean margins are relatively shallow compared to the open ocean, biogenic particles sinking at speeds of tens to hundreds of meters per day have a greater probability of reaching the sediments with much of their organic matter intact. The alternative fate of a particle being transported off the shelf is dependent on the flushing rate of the shelf in comparison to this particle lifetime of a few days within the water column. Thus, regenerative processes, such as aerobic oxidation and sulfate reduction in or on sediments, will play a major role in resupplying nutrients and trace elements to the overlying waters as well as carbon (Wollast and Blackburn, both this volume). That the benthic boundary layer in margins is an active site for the remineralization of more refractory materials is inferred from studies which show much higher concentrations of certain fallout radionuclides and chlorinated hydrocarbons in sinking particles collected near the sea bottom and in benthic flocs than in the surface layer of the underlying sediments.

In addition to regenerative processes in the water column, active removal of some elements and compounds from seawater by scavenging processes onto particles occurs (Fowler, this volume). The rate of this process depends on the rate of scavenging and on the size and possibly the character of particles in the water column. The number of particles in the water column depends on rates of supply from river and coastal erosion, the rate of production by biological processes, and the rate of resuspension—again all factors controlled by physical forcing. For certain dissolved substances not removed by biological cycling, the margins can act as a sink, if the rates of removal by scavenging are fast compared to the rates of flushing of water off the margin. Such removal is evident for many particle reactive elements delivered to the margins by rivers or the atmosphere (GESAMP 1987, 1989). In addition, the margins are a sink for some elements that are delivered to the open ocean by atmospheric inputs but where removal in the open ocean is slow due to the low particle concentration. For these substances (which include Pu, Th, and Pb [US GOFS 1987b]), transport to the margin with subsequent removal by scavenging in the higher particle concentrations found there is a globally significant sink.

WHAT ARE THE MOLECULAR NATURE, SOURCES, AND FATE OF DOM AND POM IN OCEAN MARGINS?

Although the boundary fluxes of DOC and POC in ocean margins have been estimated (Spitzy et al. and Wollast, both this volume), these are very uncertain. Fundamentally, it is recognized that the molecular nature of DOM and POM ultimately regulates its environmental reactivity (oxidation, coagulation, hydrolysis, adsorption) and its bioavailability to degrading organisms. However, despite tremendous improvements in marine organic analytical chemistry, we have still chemically identified less than 10% of DOM and POM. We suspect that subcolloidal (0.02–0.2 μm) high molecular weight terrestrial and marine biopolymers (e.g., polysaccharides, lignin, protein) and altered heteropolysaccharides (humics, fulvics, gelbstoff) account for the uncharacterized fraction of the organic matter. The different organic compounds degrade at different rates. This makes it very difficult to verify carbon mass balances in the absence of information on the molecular nature of the organic matter.

In order to integrate both the molecular diversity and the very different mechanisms and reaction time scales for low molecular weight (LMW) and high molecular weight (HMW) organics dissolved in the sea, Gough and Mantoura (1990) proposed a Molecular Weight Distribution Model depicted in Fig. 7a. The model arises by considering the biological and physicochemical selectivity which occurs during the production and removal process of organic molecules in the sea.

The starting point of the model (Fig. 7a) shows a hypothetical molecular weight distribution for dissolved and colloidal organic production or input from phytoplankton and river sources to the sea. The log MW vs. C weight distribution curve is meant to depict a relative rather than an absolute distribution curve, but it does attempt to integrate the marine organic chemistry information available. The production processes for LMW organics (e.g., dissolved free amino acids, methane, monosaccharides) include biological exudation and excretion of microbial metabolites, bacterial hydrolyses of complex organics, etc. and have been studied extensively by marine organic chemists.

For complex HMW organics production there are several potential sources and transformation processes in the sea. These include: extracellular polysaccharides and proteins being produced by phytoplankton, bacterial enzymatic hydrolysis of POM into soluble oligopolymers and condensation reactions (e.g., Melanoidins [Kalle 1966]), and photooxidative cross-linking (Harvey et al. 1983).

The next stage (Fig. 7b) considers the molecular weight selectivity trends for the principal removal process. If we consider a homologous series of LMW biogenic organics, then physicochemically controlled losses (such as volatilization, hydrolysis, defunctionalization) would be expected to increase with a decrease in molecular weight (Lyman et al. 1982). Similarly, the evidence in the marine microbiological literature (Hobbie and Williams 1984) suggests that biomonomers are more readily absorbed and assimilated than their oligomeric and polymeric analogues. For example, dissolved free amino acids are rapidly taken up by microheterotrophs compared to polypeptides or proteins, since the latter require extracellular hydrolases to slice off assimilable biomonomers.

With increasing molecular weights, interfacial removal processes become more important. These include adsorption on mineral surfaces, on air bubbles and at the air–sea interface, micellization, and coagulation through complexation with major cations in seawater. All these processes yield "POC" with the potential for bacterial colonization. More recently it has been shown that chromophoric properties of deep ocean DOC renders it easily photooxidizable under surface euphotic conditions to yield a variety of LMW carbonyl compounds that are then easily assimilated by marine bacteria (Mopper and Stahovec 1986). The cycle for the extremes of the

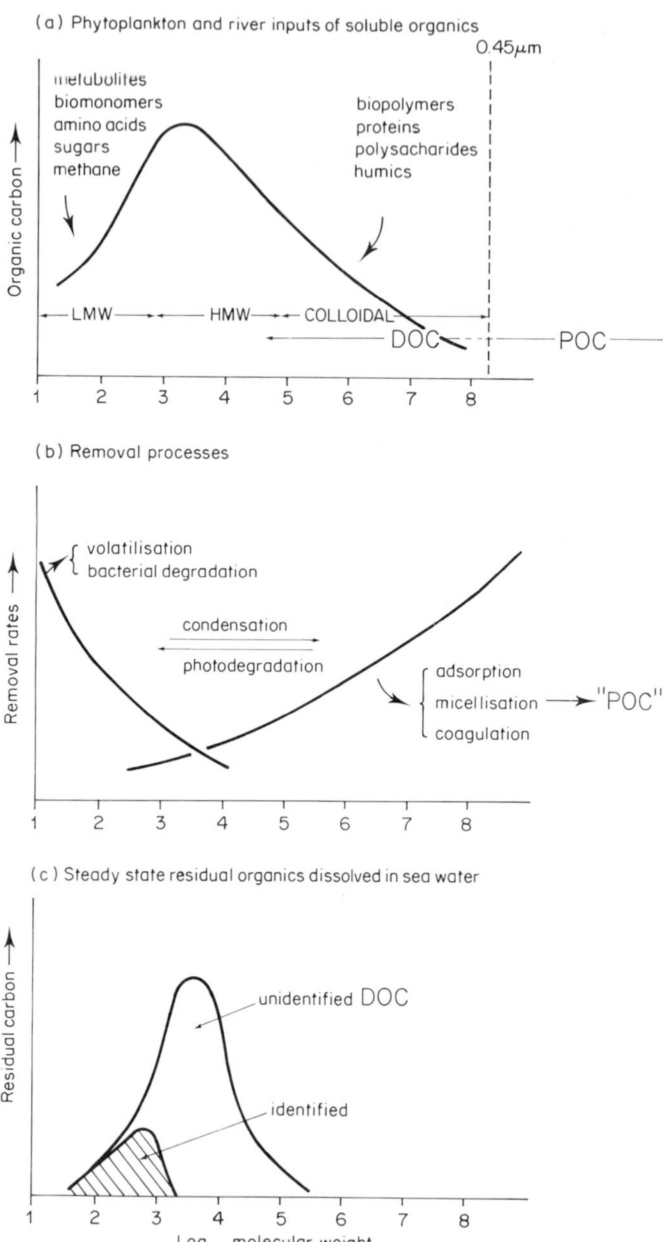

Fig. 7—(A) Hypothetical molecular weight distribution of DOC. (B) Molecular weight dependence of removal processes acting on DOC. (C) Hypothetical resultant residual molecular weight distribution of DOC.

LMW and HMW compounds is potentially complete, except for a residual ensemble (Fig. 7c) of intermediate molecular weight organics which we hypothesize to escape the LMW and HMW removal processes.

As the new high temperature catalytic oxidation (HTCO) technique (Suzuki and Tanoue, this volume) becomes widely available, priority research should include molecular size and biodegradability studies of DOM in contrasting ocean margins. By linking such studies to isotope studies (^{13}C, ^{15}N, and ^{14}C dating), it should be possible to constrain the sources, cycling, and fate of DOM and POM in ocean margins. However, much innovative chemical research focusing on biopolymer chemistry of DOM is needed, and this should be possible following recent advances in chromatographic and spectrophotometric techniques.

Unpublished riverine DOC results (Suzuki et al.) using the HTCO technique suggest previous measurements underestimate DOC by 25–30%. However, this additional (25–30%) DOC fraction appears to be removed nonconservatively in estuaries, unlike the DOC determined by previous techniques. Thus the net effect on the carbon budget of ocean margins may be modest, though the effect on CO_2 budgets may be significant.

At ocean margins there is a steep gradient in salinity, light, ventilation, and bacterial activity; these regions are therefore expected to be areas of degradation of DOM from terrestrial and marine sources, as suggested by data presented in Suzuki and Tanoue (this volume). If further studies confirm these patterns, ocean margins may be globally significant sinks for DOM from both sources. In addition, HTCO–DOC and DON studies (Suzuki, unpublished) indicate that the C/N ratios of surface marine and riverine DOM is 7.6 and 25, respectively. Thus the degradation of this material should result in the release of nitrogen, which will serve to increase the primary productivity on margins. Photolysis of DOM may be efficient in transforming refractory molecules into bioavailable products, though the mechanism and magnitude of these processes is poorly known.

Geological evidence suggests fluvial sediments are efficiently trapped in nearshore sediments and deltas (Milliman, this volume); however, the extent to which the more buoyant finer POM is trapped is uncertain. The large Asian river systems, which discharge most particulate material (Milliman, this volume), require further study. The biodegradability of fluvial POC in ocean margins is unknown. Mass balances suggest that organic carbon deposition on margins is of a similar order of magnitude to fluvial inputs. However, carbon isotope data suggests that sediment carbon on the shelves is predominantly of marine origin. This suggests that either the fluvial POC escapes offshore or is degraded over relatively rapid time scales compared to sediment processes, i.e., tens and hundreds of years.

WHY DOES ORGANIC MATTER ACCUMULATE IN MARGIN SEDIMENTS? DOES ORGANIC MATTER SUSCEPTIBLE TO MICROBIAL DEGRADATION BECOME BURIED?

Current estimates of the sources of particulate carbon accumulating in margin sediments suggest that the major portion is derived from algal cells that have been produced in the overlying water. Little of the sediment organic carbon in margin sediments appears to be of terrestrial origin.

The sediment water interface is a very active area of organic matter degradation; however, a significant fraction of the carbon is still broken down within the sediments themselves. This organic matter consumes oxygen until the supply is exhausted and then a series of oxidants dominated by sulfate are utilized in a microbially mediated sequence (Blackburn, this volume). Geological evidence suggests that preservation of organic matter is more complete under sulfate reducing conditions, but biological evidence suggests that sulfate reduction is both a rapid and efficient route for organic matter degradation, except possibly under euxinic conditions (Martens and Klump 1984).

The oxidation of the organic carbon within the sediments drives a range of important reactions, the products of which affect the sediment and overlying water quality. The oxidation of the carbon produces CO_2, which ultimately can cause the dissolution of inorganic carbonate phases in the sediments. The intimate linkage of organic carbon and calcium carbonate cycles in sediments has been little studied and since both represent important carbon reservoirs, this cycle requires further study. The breakdown of the organic carbon consumes oxygen if it is available. This is widely known to affect oxygen levels in the overlying water, and hence the ocean waters in the case of slope sediments. The reaeration of overlying waters in margins by the atmosphere depends on the stratification of the water, since the oxygen is supplied to surface waters by gas exchange and primary production, and this oxygen-rich water must subsequently be mixed down to the sediment surface. Thus the nature of organic decomposition in sediments is, in part, linked to the physical forcing functions on the margin and the stratification of the waters. In the absence of oxygen, the organic carbon is oxidized by sulfate reduction with the production of sulfides. This is predominantly an ocean margin process and presently represents the dominant geological sink for sulfur in the oceans. The breakdown of organic matter releases nutrients, though the extent to which these are returned to the water column depends on the geochemistry of the sediments. Under oxic conditions, phosphorus is trapped in the sediment through adsorption while nitrogen can escape under reducing conditions. Denitrification (Billen et al., this volume) is a minor process in carbon oxidation but is a significant component of the nitrogen cycle on a global basis. Also under reducing conditions, iron phosphates dissolve, returning phosphate to the pore water from which it can escape to the overlying water if there is no oxic layer of sediment overlying the reducing zone, or if sediment disturbance by physical or biological processes pump the pore water into the overlying water (Billen et al., this volume). Benthic autotrophic and heterotrophic communities can greatly modify nutrient fluxes out of sediments to the overlying water (Henriksen and Kemp 1988), both by modifying material fluxes across the sediments and by actively pumping water and sediment across the boundary (Rhoods 1974; Aller 1982; Hicks 1988).

The microbial degradation of this organic matter drives these biogeochemical transformations but also sustains the rich benthic communities that characterize some areas of the margin. These communities, in turn, participate in the biogeochemical cycle by bioturbating the sediments.

It is observed that despite the range of microbial processes operating, a significant fraction of sedimented organic material can be buried for geologically significant time scales, undergoing no further biological degradation. In order to explain the burial of organic carbon in shelf sediments, one cannot postulate a strictly refractory composition for all the input material. The magnitude of the labile and refractory organic matter fraction preserved varies from place to place, though the reasons for this are not clear (Blackburn, this volume); factors believed to be important include the nature of the organic matter, bioturbation, and the rate of supply of inorganic particles.

A notable feature of the U.S. east coast margin is the high concentration of organic carbon at intermediate depths on the margin slope (e.g., Walsh et al. 1988). The origin of this maximum is uncertain, probably representing a complex balance between input and degradation processes. The source of the carbon may be from warm-core rings or export from the shelf (Milliman, in preparation).

The role of ocean margin sediment as sources or sinks for other substances was discussed briefly. Many other components transported in dissolved phases can be transferred to particulate phases in ocean margins by incorporation with biogenic particles or by adsorbing onto particles that are usually coated with organic matter. The breakdown of organic matter will result in the release of these components, though they may subsequently become associated with other sedimentary phases if they are highly surface reactive. If not, they will be returned to the water phase. If the sediments become reducing at depth, iron and manganese oxyhydroxide phases will dissolve, releasing associated components into the pore water. The release of the iron manganese and associated components to the overlying water depends on the rates of transport versus oxidation, as discussed above for phosphorus.

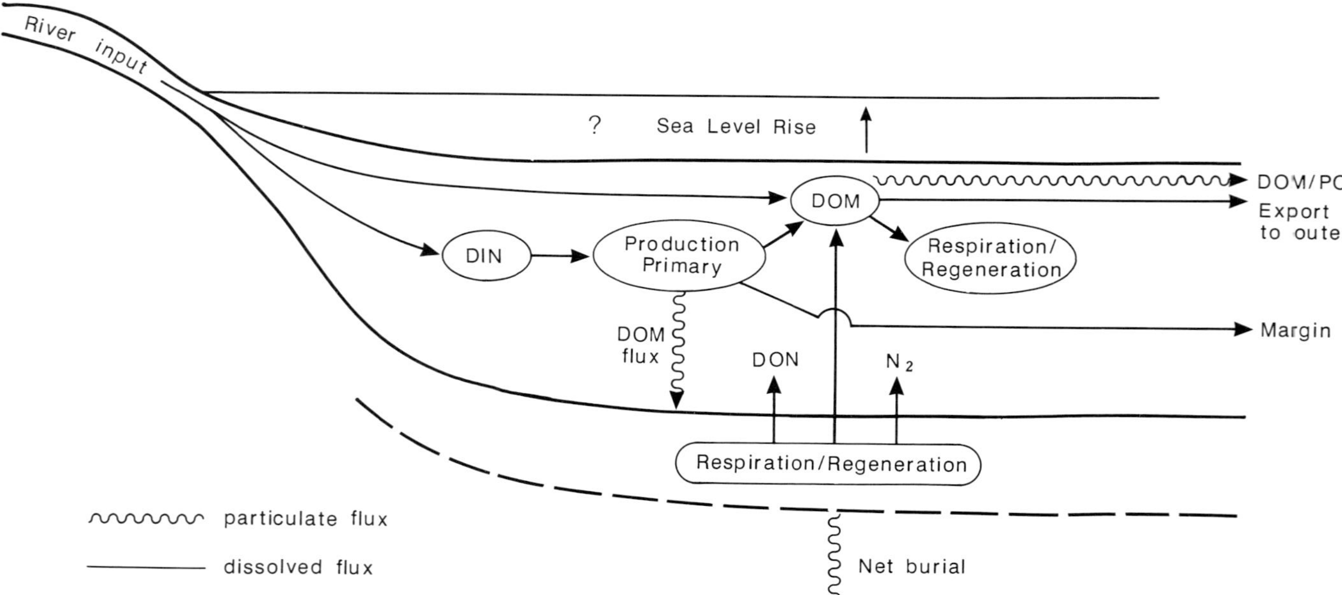

Fig. 8—Key processes of the inner margin requiring further study.

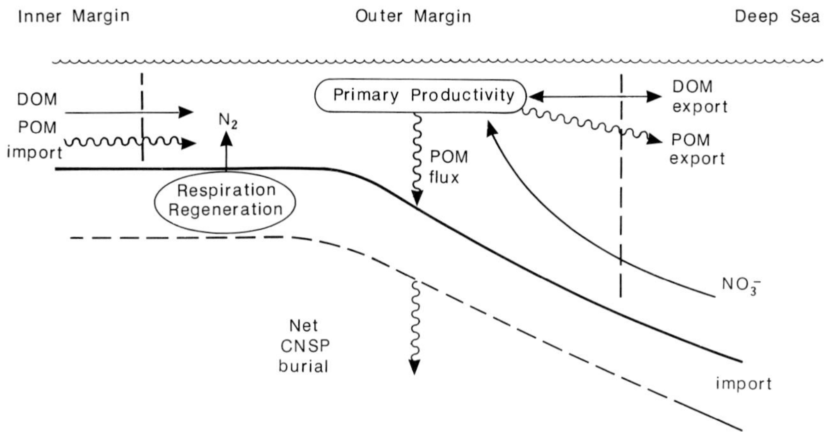

Fig. 9—Key biogeochemical fluxes at the outer margin.

DISTINGUISHING NET FROM GROSS FLUXES: HOW CAN WE LINK INSTANTANEOUS FLUX MEASUREMENTS WITH MASS BALANCES?

The problem of distinguishing net from gross fluxes arises at all boundaries. Examples include the recycling of components between marine water and the atmosphere, the recycling of carbon and nitrogen between the water column and the sediments, "new" and gross primary productivity, and net and gross water and material fluxes across the boundary between the margin and ocean. The problem of distinguishing the fluxes depends on the boundaries considered. If there is a clear physical boundary, such as the sediment–water or water–atmosphere interface, the determinations are relatively straightforward. Where there is no simple boundary, as in the case of margin–ocean exchange in many cases, it is difficult to distinguish the net fluxes from the gross. For example, the flow of water on and off a margin may occur in different places and over different time scales even at a steady state; the importance of rare extreme events complicates the matter further. This example is of interest because the group identified the ocean–margin exchange process as a critical determinant of shelf processes. It is possible to define boundaries in a computer model and to predict water and material exchanges; however, the physical verification of these estimates is an extremely difficult task. Measurements must be made at specific sites and times, and it is rarely obvious if these are the relevant space and time scales in which measurements are needed. The development of strategies to make these types of exchange measurements, particularly at the outer margin boundary, is clearly a high priority but will require the development of novel techniques. Quantitative assessments of chemical fluxes at the outer margin will require particularly difficult combined water, dissolved and particulate chemical flux measurements.

CONCLUSIONS

Throughout our discussions the overwhelming importance of physical forcing on all processes in the ocean margin was a common theme. It controls the retention of sediments, the rates and form of biological activity, and also exerts a strong effect on sediment recycling processes. While we can often quantify the physical forcings on some margins, we cannot yet quantitatively predict how they drive the biogeochemical cycles on the margins. In particular, we do not know how the rates of primary productivity are controlled in many shelf areas and cannot predict them from physical variables. Our understanding of the cycle of organic carbon in margin systems is very limited, and, while it is clear that organic carbon is being deposited in large amounts in margins, we do not understand what controls the amount buried.

It is possible to distinguish two general areas of margins on the basis of the dominance of certain physical forcing. Thus at the outer oceanic boundary of the margin, exchange with the open ocean is the dominant process and this has implications throughout the cycling and preservation of material. At the other extreme of the inshore boundary, the interaction with fluvial systems dominate, producing another unique biogeochemical system. These interactions are illustrated in Figs. 8 and 9. In some locations these two systems will be physically separated by, e.g., a large margin shelf and hence do not interact. At the other extreme they may overlap totally and linkages in intermediate cases were discussed earlier. Our aim is not to force this classification on all systems but rather to use this classification where appropriate as a means to emphasize key processes and fluxes upon which our discussion focused.

It is clear from our report that much more research is needed on ocean margins and, in particular, the large numbers of margins around the world which have to date been little studied. The group produced a long list of future research suggestions; these have been distilled to a few major themes below.

Research Recommendations

The few points below attempt to synthesize the major unknowns identified by our discussion.

1. The physical transport of fresh and saltwater on, off, and over shelves

and the interactions of this transport with sediments is clearly a major control on margin process and yet is poorly understood.

2. The role of recycling of nutrients within the water column and on the sediments, and the way this process interacts with water mixing processes to control primary productivity is central to the understanding of the cycling of organic matter on margins.
3. DOC and POC cycling in margins is still very poorly known and requires study.
4. The factors regulating the preservation of organic and inorganic carbon in margins need to be understood in order to predict the future role of margins as carbon storage sites.

REFERENCES

Aller, R.C. 1982. The effects of macrobenthos on chemical properties of marine sediments and overlying waters. In: Animal-Sediment Relations, ed. P.L. McCall and M.J. Tevesz, pp. 53–102. New York: Plenum Press.

Berger, W.H. 1989. Global maps of ocean productivity. In: Productivity of the Ocean: Past and Present, ed. W.H. Berger, V. Smetacek and D. Wefer, pp. 429–457. Dahlem Workshop Report LS 44. Chichester: Wiley.

Berger, W.H., V. Smetacek, and D. Wefer, eds. 1989. Productivity of the Ocean: Past and Present. Dahlem Workshop Report LS 44. Chichester: Wiley.

Codispoti, L.A. 1989. Phosphorus vs. nitrogen limitation of new and export production. In: Productivity of the Ocean: Past and Present, ed. W.H. Berger, V. Smetacek, and D. Wefer, pp. 377–394. Dahlem Workshop Report LS44. Chichester: Wiley.

Eisma, D. 1988. Transport and deposition of suspended matter in estuaries and the nearshore sea. In: Physical and Chemical Cycles, ed. A. Lerman and M. Meybeck, NATO ASI Series C, vol. 251, pp. 273–298. Dordrecht: Kluwer Academic.

GESAMP. 1987. IMO/FAO/UNESCO/WMO/WHO/IAEA/UN/UNEP Joint group of experts on the scientific aspects of marine pollution. Land/sea boundary flux of contaminants: contributions from rivers. Reports and Studies GESAMP No. 32.

GESAMP. 1989. IMO/FAO/UNESCO/WMO/WHO/IAEA/UN/UNEP. Joint group of experts on the scientific aspects of marine pollution. The atmospheric input of trace species to the world ocean. Reports and Studies GESAMP No. 38.

Gough, M.A., and R.F.C. Mantoura. 1990. Advanced analytical methods for the characterization of macromolecular marine organic matter. Proc. of Particle Desorption Mass Spectrometry for Marine Organic Chemistry. Third International Workshop on the Physics of Small Systems, ed. E.R. Hilfe and W. Tuszguski, pp. 114–130. F.R. Germany.

Harvey, G.R., D.A. Boran, L.A. Chesal, and J.M. Toker. 1983. The structure of marine fulvic and humic acids. Marine Chem. **12**:119–132.

Henricksen, K., and W.M. Kemp. 1988. Nitrification in estuarine and coastal marine sediments. In: Nitrogen Cycling in Coastal Marine Environments, ed. T.H. Blackburn and J. Sorensen, Scope 33. Chichester: Wiley.

Hicks, G.R.F. 1988. Sediment-rafting: a novel mechanism for the small-scale dispersal of intertidal estuarine meiofauna. Mar. Ecol. Prog. Ser. **48**:68–80.

Hobbie, J.E., and P.J.LeB. Williams. 1984. Heterographic Activity in the Sea. New York: Plenum.

Jickells, T.D., A.H. Knap, and S.R. Smith. 1986. Trace metal and nutrient fluxes through the Bermuda inshore waters. *Rapp. P.-v. Cons. Int. Explor. Mer.* **186**:251–262.

Legendre, L., and J. Le Fèvre. 1989. Hydrodynamical singularities as controls of recycled versus export production in oceans. In: Productivity of the Ocean: Present and Past, ed. W.H. Berger, V. Smetacek, and D. Wefer, pp. 49–64. Dahlem Workshop Report LS 44. Chichester: Wiley.

Lyman, W.J., W.F. Rheel, and D.H. Rosenblatt. 1982. Chemical Properties Estimation Methods—Environmental Behaviour of Organic Compounds. New York: McGraw-Hill.

Martens, C.S., and J.V. Klump. 1984. Biogeochemical cycling in an organic-rich coastal marine basin 4. An organic carbon budget for sediments dominated by sulphate reduction and methanogenesis. *Geochim. Cosdmochim. Acta* **48**:1987–2004.

Mopper, K., and W.L. Stahovec. 1986. Sources and sinks of low molecular weight organic carbonyl compounds in seawater. *Mar. Chem.* **19**:305–321.

Ning, X., D. Vaulot, Z. Liu, and Z. Liu. 1988. Standing stock and production of phytoplankton in the estuary of the Changjiang (Yangtze River) and the adjacent China Sea. *Mar. Ecol. Prog. Ser.* **49**:141–150.

Radach, G., and A. Moll. 1990. The importance of stratification for the development of phytoplankton blooms—a simulation study. In: Estuarine Water Quality Management—Monitoring, Modelling and Research, ed. W. Michaels, Coastal and Estuarine Studies, vol. 36. Berlin: Springer-Verlag.

Rhoods, D.C. 1974. Organism-sediment relations on the muddy sea floor. *Oceanogr. Mar. Biol. Ann. Rev.* **12**:263–300.

Simpson, J.H., and H.R. Hunter. 1974. Fronts in the Irish Sea. *Nature* **250**:404–406.

Smetacek, V., B.v. Bodungen, B. Knoppers, R. Peinert, F. Pollehne, P. Stegmann, and B. Zeitzschel. 1984. Seasonal stages characterizing the annual cycle of an inshore pelagic system. *Rapp. P. V. Reun. Cons. Int. Explor. Mer* **183**:126–135.

Steidinger, K.A., and L.M. Walker. 1984. Marine Plankton Life Cycle Strategies. Boca Raton, FL: CRC Press.

US GOFS. 1987a. Workshop on Modelling in GOFS, Rep. No. 4, WHOI. Woods Hole, MA: US GOFS Planning and Coordination Office.

US GOFS. 1987b. Workshop on the Impact of Ocean Boundaries on the Interior Ocean, Rep. No. 6, WHOI. Woods Hole, MA: US GOFS Planning and Coordination Office.

Walsh, J.J., D.A. Dieterle, and M.B. Meyers. 1988. A simulation of the fate of phytoplankton within the mid-Atlantic Bight. *Cont. Shelf Res.* **8**:757–787.

Fractal Theory and Time Dependency in Ocean Margin Processes

K.J. Hsü

Geological Institute, ETH-Zentrum
8092 Zürich, Switzerland

Abstract. Long-run records of temporal variation of magnitude of natural events show a fractal geometry: there is a common tendency of clustering the extremes (Joseph's effect of Mandelbrot), and there is inevitability of the improbable (actualistic catastrophism of Hsü, or Noah's effect of Mandelbrot). This time-dependent probability renders the common practice of assuming Gaussian probability in evaluating magnitude/frequency relationship obsolete. Fractal geometry is a new statistical method, devised by Benoit Mandelbrot on the basis of evaluating long-run records, to predict non-Gaussian probability of natural perturbations. Assessment of responses of ocean–margin processes to natural or anthropogenic perturbations should take into consideration such fractal geometry as indicated by long-run paleoceanographical and geological records.

INTRODUCTION

Several years ago I was on a panel of interdisciplinary experts to discuss the safety (or lack of it) of disposing atomic wastes in seabed. I was astonished to learn, from a colleague who claimed, that ocean-bottom water could never be carried back to the surface by a physical process; only living organisms could overcome gravity and transport radioactive contaminants from deep to shallow waters. The expert is a mathematician, and his far-reaching conclusion is based upon sophisticated and expensive mathematical modeling. In blind admiration for the so-called "exact science," the conclusions were greeted with enthusiasm by delegates to the London Dumping Convention of the International Marine Organization. It seemed that dumping of atomic wastes should be harmless, except for the rare deep-swimming animals.

"There was a time, not so long ago, when the deep ocean was thought

Ocean Margin Processes in Global Change
Edited by R.F.C. Mantoura, J.-M. Martin and R. Wollast

to be a quiet, calm environment" (Seibold and Berger 1982, p.100). Oceanographers now know better. Surface waters go down to the depth, the ^{14}C content of deep waters proves their surficial origin. Material balance requires that deep waters have to come up. In fact, using ^{14}C as a tracer, it was found that the ocean turnover time is about 1600 years (Broecker and Peng 1982).

If it takes 1600 years for oceans to turn over, one might still conclude that the waste dumped in today will not come back to haunt us until all of us are long dead. Yet most of us do think of our responsibility to posterity. Plutonium 244 is not only poisonous but also radioactive, with a long half-life of 80.5 million years. The disposal problem is thus not only to predict what might happen today or tomorrow, but what may happen in millions of years. How do we model the possible dynamics of the oceans during the next hundred, thousand, or million years? What kind of data must be fed in? Is it at all possible?

Variables influencing ocean dynamics are time dependent. Take, for instance, a well known example of vertically upward motion in the ocean, namely, upwelling: Off coasts with eastern boundary currents, surface water has a tendency to move seaward due to the deflection of Coriolis force, and this seaward motion is reinforced when the winds blow offshore so that outward-bound surface waters are replaced by denser bottom waters. The flux rate of cold water upwelling is a function of wind strength. If the measurements on wind directions fed into the computer are taken during seasons of winds blowing onshore, or during years of El Niño, the possibility of upwelling would not come out of the mathematic modeling. Even if we feed in data on winds blowing offshore, we still face a problem of temporal variation of the wind strength.

Strong winds are rare and mild winds are common; an inverse correlation of wind strength and frequency occurrence of a wind of a given class is well known. Various parameters have been used to define the wind magnitude, such as wind speed, pressure difference, wind-induced wave heights, etc. Assuming Gaussian probability to relate frequency and magnitude, long-term behavior could be predicted on the basis of short-run record (Simpson and Rieh 1981). The assumption of time-independent temporal variation of magnitude is the basis of the uniformitarianism: the present is the key to interpret the past and to predict the future. Studying geological records I came to the conclusion, however, that catastrophes of extreme magnitude have taken place during the long history of the Earth. I proposed, therefore, the principle of actualistic catastrophism to predict rare and unusual events. (Hsü 1983). The mathematical basis of this scientific philosophy is the fractal geometry of Benoit Mandelbrot (Mandelbrot 1977). This paper is a brief introduction to the fractal geometry and its relevance to predict the time dependency of ocean margin processes.

SUBSTANTIVE UNIFORMITARIANISM AND ACTUALISTIC CATASTROPHISM

The debate of catastrophists and uniformitarianists of the 19th century could not be resolved when there was no method to measure geological time. Lyell's substantive uniformitarianism is an Ockham's razor: we had to assume uniform rate, if it could not be proven otherwise. Since the discovery of radioactivity, dating methods have been developed, and rates of natural processes could now be evaluated. Unfortunately, science like religion has a tendency to grow dogmatic. Lyell has been honored the Father of Geology, "the first pope," and what he assumed became an infallible axiom long after his assumption of uniform rates had been falsified.

I ran into this dogmatic mentality when I proposed that the Mediterranean Sea was desiccated 5 million years ago (Hsü 1971). Is it physically possible? One could calculate the influx of river and Atlantic waters, the evaporative loss, and the outflow of Mediterranean water. This inland sea cannot be desiccated if the hydrologic budget has always been, or will always be, what it is. This conclusion is, of course, based upon the indefensible assumption that the fluxes are not time dependent. Five million years ago, when the influx of the Atlantic water was much reduced, the Mediterranean basins were turned into salt pans.

When I postulated that a large bolide impact occurred to interpret the terminal Cretaceous mass extinction, I again ran into opposition from substantive uniformitarianists (Hsü 1980). The postulate was dismissed by some as being unscientific; a collision with a Halley-sized comet is, in the opinion of nearsighted astronomers, physically impossible.

Physical impossibility could be defined as a probability of less than 1 in 10^{10} years. Is a large bolide (10^{18}g) impact so improbable that it is not likely to have taken place during the 4.5 billion years of the Earth's history? What are the probabilities of bolide impacts? A large meteorite hit Arizona several thousand years ago and dug the Meteorite Crater 1 km in diameter. A very large meteorite hit Canada 200 million years ago and made a crater some 100 km across. Astronomers studying the frequency of cratering on Earth and on Moon came up with an empirical relation relating frequency of impact to crater size (Hughes 1977):

$$f = 1/1400\ D^2 \tag{1}$$

where f is the rate (y^{-1}) of a meteorite impact which produces a crater with a diameter D (km). Now a Halley-sized comet could have enough impact energy to create a crater up to 500 km across. Using Eq. 1 I could conclude that there should have been at least one impact with such a comet during the 600 million years of the Phanerozoic history (Hsü 1980).

The inversely linear log–log relationship relating the frequency and magnitude of impacts has in fact been noted by seismologists to predict occurrences of earthquakes (Fig. 1). Furthermore, this relation seems to be generally applicable to describe patterns of frequency/magnitude relationship of natural perturbations. I proposed, therefore, that the Lyellian substantive uniformitarianism should be replaced by the concept of actualistic catastrophism (Hsü 1983). The essence of the new philosophy is to emphasize the inevitability of the improbable in geology.

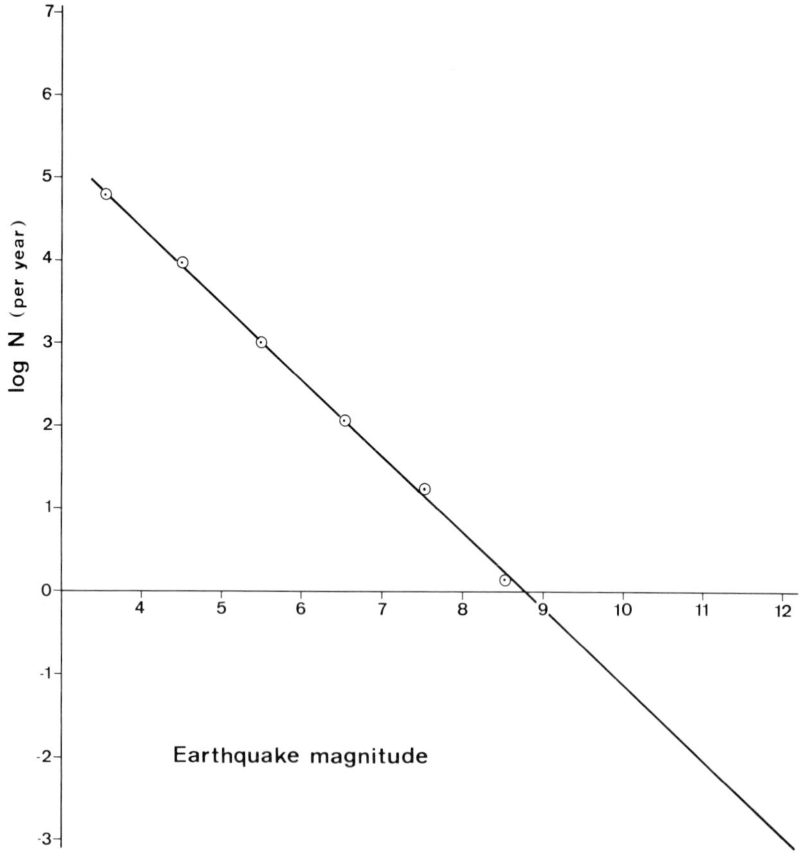

Fig. 1—Inversely linear log–log relation between frequency occurrences and magnitude of earthquakes. Data from Swiss Earthquake Service (Mayer-Rosa 1985).

NONLINEARITY OF GEOLOGICAL TIME

The substantive uniformitarianism assumes that

$$W = \int_{o}^{T} W_o \cdot dt = W_o T \tag{2}$$

where W is the total energy output or work accomplished during an interval of geologic time T and W_o is the average rate during this interval. Using the present annual rate of salt supply to the ocean and this linear relationship to estimate the age of the Earth, geologists of the last century obtained a figure more than one order of magnitude too small.

Geologic time is not linear because rates of processes are time dependent. Geological age is overestimated because of an extrapolation of an unusually high rate of the present: the rate of salt storage in ocean water was less than usual or even negative during times of catastrophic desiccation, such as when the Mediterranean dried up. Eq. 2 is obviously not applicable. Geological age has also been overestimated: an assumption of linear rate of cliff retreat in southern England led Darwin to an estimate of a 300 million-year-old Middle Cretaceous. We now know that the retreat took place in a much shorter time of some 30 million years; the cliff fell back at much faster rates (by landsliding, for example) on rare occasions. The work of nature is thus not only an integration of continuous function of linear increments, but also the sum of discrete additions resulting from events.

Steady-state accumulative effect could, in fact, be overshadowed by that of one catastrophic event: For example, the total energy output by all earthquakes during the first 50 years of the 20th century, is 50×10^9 kWh, at an annual average of about 1×10^9 kWh. Yet the largest event of the century, the 1960 Chile Earthquake, has an energy output of 300×10^9 kWh, or six times as much as the total of the first half century (Mayer-Rosa 1985). In this case, we see that the use of short-run averages could be seriously misleading.

This is, of course, common sense. Even uniformitarians admit that there are events of larger or smaller magnitude; we see that in our life time. When we speak of the flood of the decade or the hurricane of the century, we are expressing temporal variations of the magnitude of natural processes. There is a difference, however, between uniformitarianism and catastrophism, and this is expressed in the language of mathematics in terms of Gaussian or non-Gaussian probability. In the Gaussian world, the maximum magnitude is limited, and the recurrence of this maximum can be predicted on the basis of short-run records (Mandelbrot 1977; Simpson and Rieh 1981). In the world of fractal geometry, or of non-Gaussian probability, the maximum has an open end and is time dependent (Mandelbrot 1977). Actualistic catastrophism assumes a non-Gaussian probability of rare events, occurring over a period of time much longer than the duration of a life time.

THEORY OF FRACTAL GEOMETRY

As Freeman Dyson told us (1978), "*fractal* is a word invented by Mandelbrot to bring together under one heading a large class of objects that have played an historical role in the development of pure mathematics. A great revolution of ideas separated the classical mathematics of the 19th century from the modern mathematics of the 20th. Classical mathematics had its root in the regular geometric structures of Euclid and the continuously evolving dynamics of Newton. . . . A revolution was forced by the discovery of mathematical structures that did not fit the patterns of Euclid and Newton. These new structures were regarded . . . 'pathological', showing that the world of pure mathematics contains a richness of possibilities going far beyond the simple structures which they saw in Nature. Now, . . . Nature has played a joke on the mathematicians . . . The same pathological structures invented to break away loose from 19th century naturalism turn out to be inherent in familiar objects all around us."

Fractal geometry could be defined by the power function

$$N = c/r^D \tag{3}$$

where N and r are two variables related by a fractal relation, c is a constant of proportionality, and D is called fractal dimension (Mandelbrot 1977). Mandelbrot coined *fractal* from the Latin adjective *fractus*, from the consideration that the fractal dimension is almost always a fraction. The fractal dimension cannot be a whole number because the relation in that case would be represented by Euclid geometry.

Eq. 3 can be restated as

$$\log N = c - D \log r. \tag{4}$$

This is, of course, the inversely linear log–log relationship discovered by geophysicists to relate the magnitude and frequency of earthquakes (Hsü 1983). Mandelbrot cited one example after another to advance his postulate that fractal geometry is a manifestation of the harmony of nature. By that he meant that natural processes are not chaotic. They are not deterministic nor Gaussian, rather they are statistically predictable according to the law of fractal geometry.

That the fluxes of material and energy across boundaries of ocean margin vary with time is well known. In discussing the magnitude and frequency of forces in geomorphic processes, Wolman and Miller (1960) posed the question of whether annual or rare events have performed more work in sediment transport of rivers. They assumed a log–normal relation of frequency/magnitude and concluded that the greatest bulk of sediment is, as a rule, transported by more moderate and more frequent events. "Pulses" of unusual river discharges have been recognized on a century-long basis. In such instances, more accumulative transport seems to have been accomplished by the few rare events (Graf 1988).

An assumption of a Gaussian distribution of frequency vs. magnitude is consistent with short-run records (Carling and Hurley 1987; Wolman and Miller 1960). Where a sufficiently long-run record is kept, a fractal relation is apparent. Figure 2 is a log–log plot of the data on peak flood recurrences of River Gail at Rattendorf, Austria, during the period 1890–1952. The inversely linear relationship is an expression of the fractal geometry.

JOSEPH'S AND NOAH'S EFFECT

The fractal geometry of Nile discharges has been established by one of the longest kept historical records. Mandelbrot (1977) noted an erratic pattern of annual variations (Hurst 1955; Mandelbrot and Wallis 1969) and claimed that this pattern has been known since the biblical times. He cited the Bible: "There came seven years of great plenty throughout the land of Egypt. And there shall arise after them seven years of famine" (Genesis 41: 29–30). The biblical "seven and seven" is, in his opinion, a poetic oversimplification of reality. Those are pulses of years of floods and of droughts. Mandelbrot used the term "Joseph's effect" to designate this tendency of clustering wet and dry years in the Middle East.

Another aspect of fractal geometry is the postulate of actualistic catastrophism: given enough time, anything that can happen will happen. Again Mandelbrot cited the Bible: ". . . were all the foundations of the great deep broken up, and the windows of heaven were opened. And the rain was upon the earth forty days and forty nights" (Genesis 6: 11–12). The rare catastrophic maximum was called "Noah's effect."

To design a reservoir dam, one needs to know the temporal variation of river discharges (Hurst 1955). If we underestimate the magnitude and frequency of droughts, the reservoir might become dry, leading to famine when water for irrigation is insufficient. If we underestimate the magnitude and frequency of floods, excess water might spill over causing erosion and eventual collapse of the dam. A reservoir performs ideally if it ends as full as it begins, if it is never empty and never overruns, and if it produces a uniform flow. That would have been no problem if the temporal variation is not time dependent; the maximum deviation from sample variance could then be calculated on the basis of short-run records, assuming Gaussian

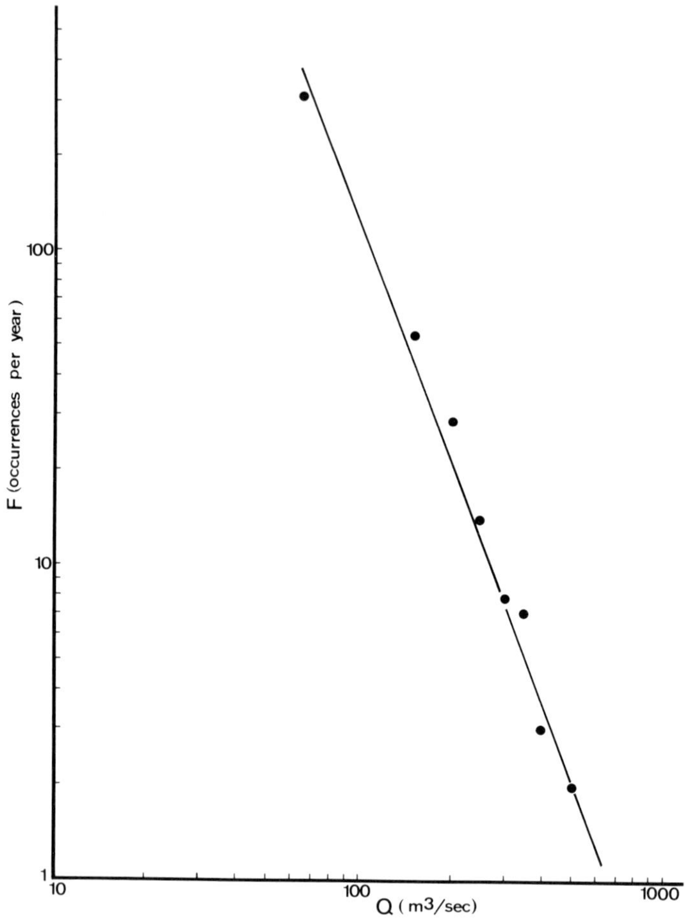

Fig. 2—Inversely linear log–log relation between frequency occurrences per year and the peak flood discharges exceeding Q of River Gail at Rattendorf, Austria. Data from Scheiddeger (1975).

probability. If, however, there are the Joseph's and Noah's effects, we are faced with uncertainties: How do we know, from a short-run record, if our record is taken in a period of wet or of dry years? How do we know when and if a Noah's flood would come again after a rain of forty days and nights?

Harold Edwin Hurst devoted the greatest part of his life to studying the long-run record of the Nile river discharges so that an optimum design for the Aswan dam would be possible. Mandelbrot and Wallis developed a theoretical model of fractal geometry on the basis of Hurst's (1955) analysis in order to predict the long-run behavior of river discharges (Mandelbrot and Wallis 1969). They denoted X (t) as a record containing T readings uniformly spaced in time from t = 1 to t = T, and X*(t) Niles cumulative discharge between the beginning of year 1 and the end of year t, or Σ_{u+1}^{u+t} = X(u). They further adjusted X*(t) by subtracting the sample average discharge between these years. If the latter is the volume discharged from the reservoir, the adjusted X*(t) is the excess or deficit compared to the initial storage. Let R(t,s) be the difference between the maximum and minimum of the adjusted X*(t) as t ranges between t + 1 to t + s; s being the years of lag. R(t,s) is thus the capacity one should have attributed to a reservoir to insure ideal performance over this interval s, whereas the term $S^2(t,s)$ is the sample variance during these years. The ratio R(t,s)/S(t,s) has been called a rescaled range and its temporal variation can be evaluated by pox diagrams.

To construct a pox diagram, a sequence of values of lag s is selected and marked on the axis of abscissas on doubly logarithmic paper (Mandelbrot and Wallis 1969). For each s, one selects a number of starting points, t, and the values of R(t,s)/S(t,s) so obtained are plotted on the ordinates. Thus, above every marked value of s, several points (marked by + signs) are aligned. For each s, the sample average of the quantities R(t,s)/S(t,s) is marked by a little square box. The line connecting such boxes is found by Mandelbrot to be a straight line on a pox diagram.

The rescaled range R(t,s)/S(t,s) is, as one might expect, a function of time: the longer the time interval to be considered, the greater the range should be. The theoretical relation of R(t,s)/S(t,s) to time d, from a consideration of the Gaussian probability, is according to Feller (1951):

$$R(t,s)/S(t,s) = (d)^{\frac{1}{2}}. \tag{5}$$

The Nile data did, however, deviate significantly from the prediction of the Gaussian probability. Hurst (1955) found instead that the relation should be

$$R(t,s)/S(t,s) = d^{H} \tag{6}$$

and that the value of H is almost always greater than 1/2; the value varies between 0.9 and 0.5 for various rivers (Hurst 1955; Mandelbrot and Wallis 1969).

This discovery indicates that the temporal variation of natural fluxes cannot be evaluated on the basis of short-term record. Figure 3 shows, for example, the R/S of the Nile evaluated on the basis of a 20-year record through an assumption that Gaussian probability is about 20; however, the long-run record indicates that it should be almost 100, or the rescaled range

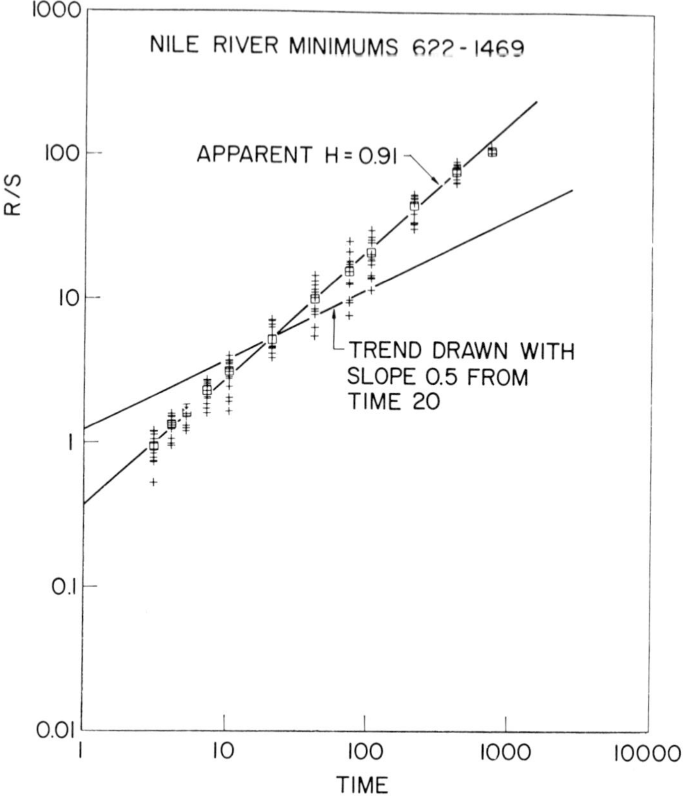

Fig. 3—Pox diagrams of log R/S vs. log s for the Nile-discharge minimums. Crosses are the R(t,s)/S(t,s) values for a number of starting points t; the little square is the average, calculated for lag-time s = 3, 4, 5, 7, 10, 20, 70, 100, 200, 400, 700 years. After Mandelbrot and Wallis (1969).

will be 100 times (not 20) the sample variance. Disregarding the fractal geometry of Nature could result in a grossly underestimated temporal variation of the Nile discharges and thus lead to famine or flood catastrophe.

FRACTAL GEOMETRY OF OCEAN MARGIN PROCESS

At this Dahlem conference, the group discussions centered around four themes: What regulates boundary fluxes at ocean margins? What determines the fate of materials within ocean margins? What is the response of ocean margin systems to perturbations? And what is the importance of ocean margin processes in global changes. In all those questions, the variable time must be considered. The discussions on Nile discharges illustrate the non-

Gaussian distribution of river (sediment and dissolved particulate) fluxes into coastal seas; one has to take into consideration the bias of short-term records caused by the Joseph's effect. The principle of actual catastrophism illustrates further the Noah's effect, or the potential magnitude of perturbations in the long run: Analyzing the consequence of a large oceanic meteorite impact, for example, it was found that the amount of moisture thrown into the air is more than enough to explain the rain of 40 days and nights described by the Bible (Hsü 1986). The conclusion is inescapable that we need data from long-run records to evaluate the time dependency in ocean margin processes.

Unfortunately, the history of *Homo sapiens* is very short, even the record of the Nile discharge is of the order of 10^3 years, some six orders of magnitude less than the history of life on Earth. We depend upon geology and paleoceanography for supplementary records of the Noah's and Joseph's effects.

Varves in proglacial lakes, for example, give a record of river discharges in prehistorical time. Figure 4 is a pox diagram constructed by Mandelbrot and Wallis (1969) for the thickness of varves in Timiskaming, Canada. They show practically the same pattern as that for the Nile discharge, with the value of H being 0.96. Other records, such as tree-ring thickness (Fig. 5), or paleo-temperature measurements could also yield information on the long-run behavior of natural processes. In order to evaluate the perturbations in ocean margin processes during the next decades or centuries, the records from studies of Quaternary geology are very important.

The Ocean "Landscape"

Fractal geometry is scale independent, or there is a self-similarity. Mandelbrot (1977) explained this phenomenon with the paradox of indeterminate length of state boundaries: L.F. Richardson, a British physicist, discovered to his astonishment that the lengths of the common frontier between Spain and Portugal, as reported in these neighbour's encyclopedias, differ by 20%. This so-called Richardson's effect is now expressed in terms of fractal geometry, relating the yardstick length e and the total measured length L(e), which could be state boundary, coastline, or any other linear dimension. This relation is shown by Figure 6.

In a paper on the ocean "landscape," Steele (1989) pointed out the complex physical structure of the ocean at all scales in space and time. Could we find fractal geometry such as the Noah's effect or the Richardson's effect in the ocean "landscape"?

The answer is provisionally positive. Figure 7 illustrates the power spectrum of temperature variance at 2000 m off Bermuda. The inversely proportional log–log linear relation between magnitude and frequency is an

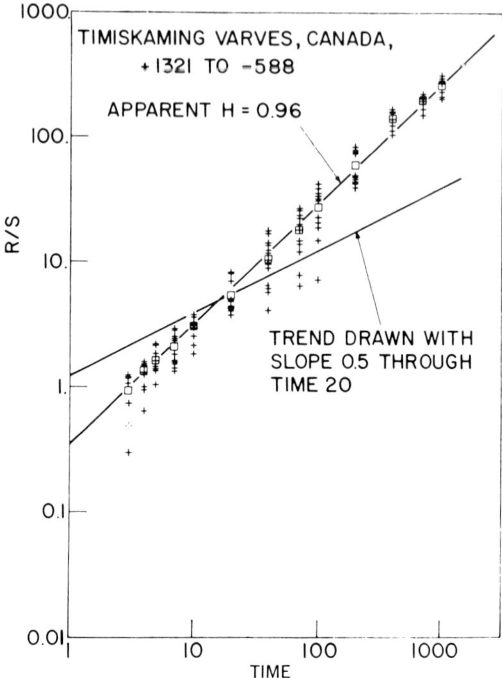

Fig. 4—Pox diagram of log R/S vs. log s for the varve thickness at Timiskaming, Canada. Crosses are the R(t,s)/S(t,s) values for a number of starting points t; the little square is the average calculated for lag-time s = 3, 4, 5, 7, 10, 20, 70, 100, 200, 400, 700 years. After Mandelbrot and Wallis (1969).

expression of the Noah's effect; greater variation happens less frequently. The value D, or that of the fractal dimension, varies however from 0.28 to 4.27 for different frequency ranges.

Figure 8 can be considered an illustration of the self-similarity of the ocean landscape. If the particle concentration is made equivalent to L(e) and the equivalent diameter of the marine organisms is considered a parametric expression of the yardstick e to measure the concentration, the inversely linear relation of the log–log plot can be considered an expression of the Richardson's effect: the irregularities even out when a big yardstick is used for surveying. Considering the radius of movement of whales and tunas, the particle concentration of seawater is close to that of the global average. Local "irregularity" or the unusually high local concentration such as needed to sustain plankton bloom would be overlooked.

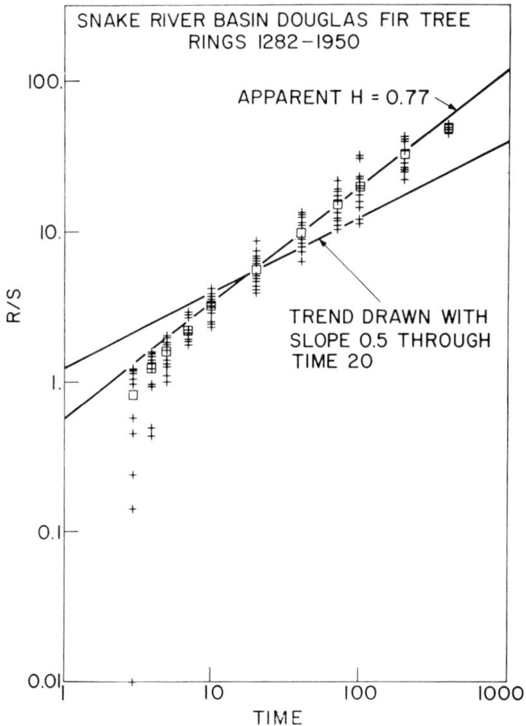

Fig. 5—Pox diagram of log R/S vs. log s for the tree-ring thickness, Douglas Fir, Snake River. Crosses are the R(t,s)/S(t,s) values for a number of starting points t; the little square is the average calculated for lag-time s = 3, 4, 5, 7, 10, 20, 70, 100, 200, 400 years. After Mandelbrot and Wallis (1969).

Relevance of Geology to Global Change

Global changes of concern to modern society are processes which are to take place during the next decades or centuries. Is there any relevance from the billion-year old record of geology?

I have given an emphatically positive answer to this question (Hsü 1990): The extremely long-run record provides means to evaluate extreme perturbations in Nature. Extremely unusual natural catastrophes are also extremely rare. A trillion-ton bolide triggering an environmental crisis and causing mass extinction is not likely to hit us in the next million years. We could be complacent that our historical records give us adequate estimates of the potential responses of ocean margin systems to natural perturbations during the near future: the behavior of Nile discharges at Aswan during the next millenium is predictable on the basis of the record of the last millenium. Such complacency may, however, be inappropriate if we want to evaluate

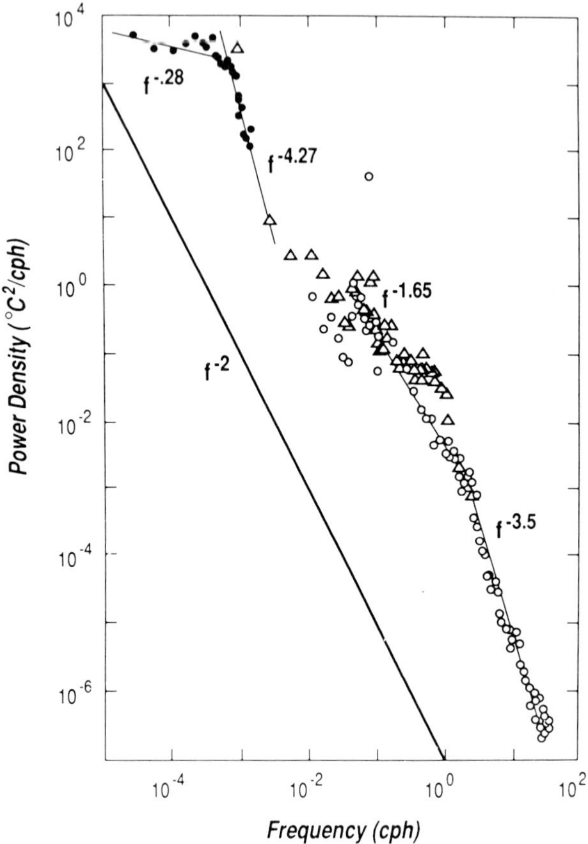

Fig. 6—Power spectrum of temperature variance at 2000 m off Bermuda. The variance, as a function of frequency (cycles per hour), was determined by moored instruments and ship observations. After Steele (1989).

the consequences of anthropogenic perturbations. In a discussion at the Interlaken Workshop on "Earth Processes and Global Changes," Digby McLaren used the expression "human bolide" to express the devastating influence of *Homo sapiens*. The present rate of species extinction may have already exceeded that at the end of the Cretaceous (after the bolide hit). The fluxes of material and energy across ocean margin boundaries are likely to reach catastrophic rates, such as those prevailing during that critical time in Earth's history (Hsü 1990). In order to properly assess the response of ocean margin systems to anthropogenic perturbations, one needs to consider environmental changes in times of biotic crisis (Hsü 1986).

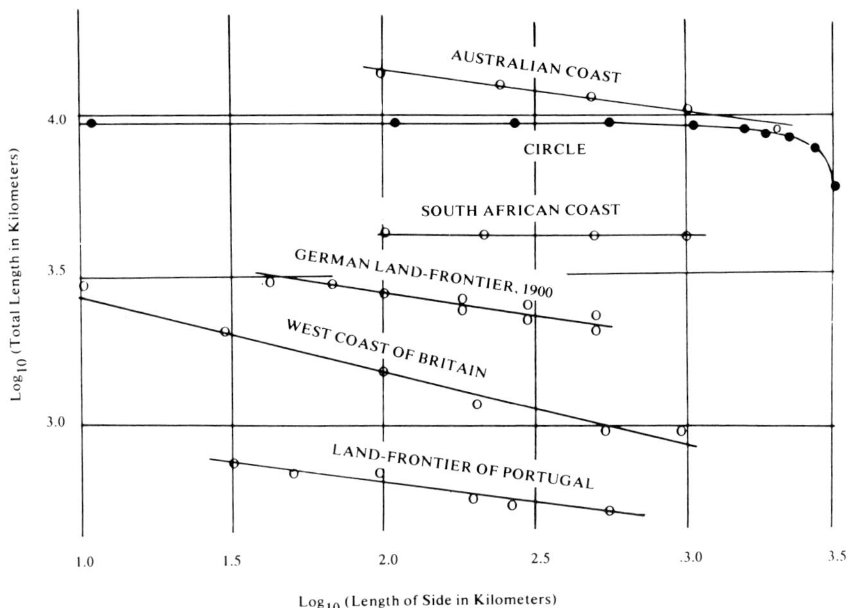

Fig. 7—Richardson's empirical data concerning the rate of increase of coastlines' length. After Mandelbrot (1977).

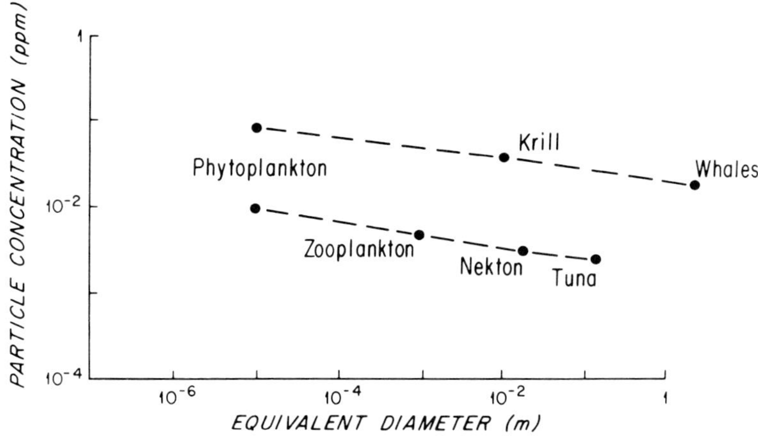

Fig. 8—The relation between trophic groups, size, and density for subtropical and antarctic communities. After Steele (1989).

REFERENCES

Broecker, W., and T.H. Peng. 1982. Tracers in the Sea. New York: Eldigo.

Carling, P.A., and M.A. Hurley. 1987. A time-varying stochastic model of the frequency and magnitude of bed load transport events in two small trout streams. In: Sediment Transport in Gravel-bed Rivers, ed. C.R. Thorne, J.C. Bathurst, and R.D. Hey, pp. 897–920. New York: Wiley.

Dyson, F. 1978. Characterizing Irregularity. *Science* **200**:677–678.

Feller, W. 1951. The asymptotic distribution of the range of sums of independent random variables. *Ann. Math. Stat.* **22**:427.

Graf, W.L. 1988. Fluvial Processes in Dryland Rivers. Berlin: Springer.

Hsü, K.J. 1971. When the Mediterranean dried up. *Sci. Am.* **277/6**:26–36.

Hsü, K.J. 1980. Terrestrial catastrophe caused by cometary impact at the end of Cretaceous. *Nature* **285**:201–203.

Hsü, K.J. 1983. Actualistic catastrophism: address of the retiring President of the International Association of Sedimentologists. *Sedimentol.* **30**:3–9.

Hsü, K.J. 1986. Environmental changes in times of biotic crisis. In: Patterns and Processes in the History of Life, ed. D.M. Raup, and D. Jablonski, pp. 63–76. Dahlem Workshop Report LS 36. Berlin: Springer.

Hsü, K.J. 1986. The Great Dying. San Diego: Harcourt, Brace & Jovanovich.

Hsü, K.J. 1990. Actualistic catastrophism and global change. Global & Planetary Change (in press).

Hughes, D.W. 1977. *Nature* **270**:558–560.

Hurst, H.E. 1955. Methods of using long-term storage in reservoirs. *Proc. Inst. Civil Eng.* Part 1: 519–577.

Mandelbrot, B.B. 1977. The Fractal Geometry of Nature. New York: Freeman.

Mandelbrot, B.B., and J.R. Wallis. 1969. Some long-run properties of geophysical records. *Wat. Res. Resear.* **5**:321–340.

Mayer-Rosa, D. 1985. Erdbeben, Entstehung, Risiko, und Hilfe. Zurich: Schw. Geoph. Komm.

Scheiddeger, A.E. 1975. Physical Aspects of Natural Catastrophes. Amsterdam: Elsevier.

Seibold, E., and W.H. Berger. 1982. The Sea Floor. Heidelberg: Springer.

Simpson, R.H., and H. Rieh. 1981. The Hurricane and Its Impact. Baton Rouge:Louisiana State Univ. Press.

Steele, J.H. 1989. The ocean 'landscape'. *Landsc. Ecol.* **3**:185–192.

Wolman, M.G., and J.P. Miller. 1960. Magnitude and frequency of forces in geomorphic forces. *J. Geol.* **68**:54–74.

Coastal Eutrophication: Causes and Consequences

V. Smetacek, U. Bathmann, E.-M. Nöthig, and R. Scharek

Alfred Wegener Institute for Polar and Marine Research
Postfach 120161, Am Handelshafen 12
2850 Bremerhaven, F.R. Germany

Abstract. Whereas estuaries and many smaller, enclosed water bodies along the coasts of industrialized countries are severely eutrophic, the degree of eutrophication of larger water bodies such as the North and Baltic Seas is still under discussion. As in the case of all natural processes influenced in a chronic manner by anthropogenic input, eutrophication is first evidenced as a shift in the range of variability inherent to the receiving ecosystem. Situations judged as extreme in the past tend to become more common, eventually leading to a more drastic or qualitative change in the structure and behavior of the perturbed system. Depending on the properties, buffering capacity, and resilience of the system, its response can be in the form of a gradual trend; however, it can also react with seemingly unpredictable oscillations. At present, it is difficult to diagnose to which degree the changes recorded in recent years in shelf systems are a result of excessive nutrient input, changing climatic cycles, or simply an effect of improvement in observational coverage. The feedback loops that can buffer or amplify anthropogenic impact are discussed here in the context of seasonal cycles. Based on this framework, we critically examine some recent trends and reports of unusual phenomena from the Baltic and North Seas that have been attributed to eutrophication. Finally, the role of "environmental consciousness" in the populations of industrial countries and its potential effect on future trends is discussed. We consider the detrimental effects of eutrophication to far outweigh whatever beneficial effects it might have, such as increasing productivity or removal of atmospheric CO_2 via burial of organic matter.

INTRODUCTION

In its widest sense, eutrophication of an aquatic system is caused by input of nutrients from terrestrial sources. Its primary consequence is an enhancement of algal productivity and accumulation of biomass. Depending

Ocean Margin Processes in Global Change
Edited by R.F.C. Mantoura, J.-M. Martin and R. Wollast

on the nutrient input in relation to its dilution in the receiving water and the subsequent fate of the "extra" algal biomass, the secondary effects of eutrophication will influence the structure of the planktonic community and eventually that of the benthos. In shallower lakes, increased algal biomass induces anoxia in the sediments resulting in release of nutrients, particularly phosphate which again has a positive feedback effect on algal biomass in surface layers. Noxious blooms of nitrogen-fixing cyanobacteria are the product of this vicious cycle. Eutrophication caused by anthropogenic input of nutrients has been regarded as a serious problem in fresh waters for several decades, because their restricted volumes renders the effects—discolored water and the odor of decaying algae—obvious to even a casual observer.

Comparable symptoms demonstrating ongoing eutrophication are less obvious in marine waters because of the dilution effect. Well-flushed coastal areas with a high nutrient load, but receiving substantial nutrient inputs, might well be less eutrophic than topographically restricted areas with lesser input. Therefore, the relationship between nutrient input rates and local hydrography and natural nutrient status of the receiving coastal area will determine the degree of eutrophication and eventual fate of the nutrient additions. In this short review we will consider the causes and consequences of eutrophication only in open coastal and shelf environments, with special emphasis on two widely differing water bodies: the southern North Sea and the Baltic Sea. The role of river discharge and processes occurring in the estuaries has been addressed by Billen et al., Halim, and Wollast (all this volume). The influence of eutrophication on mass balances of material cycling and sinks on the shelves has been reviewed by Walsh (1988). Here we discuss biological feedback loops that can buffer or amplify the impact of eutrophication in the context of seasonal cycles.

Public concern over the deterioration of coastal waters has been increasing significantly since the seventies, although the issue is still under dispute among marine scientists. The problem is one of discriminating between natural variability and anthropogenic effects. In recent years the debate has become emotionally charged as a result of increasing public sensitivity engendered by highly publicized events, such as unusual algal blooms and large-scale animal mortality (Krom, this volume). The questions being put to scientists nowadays are difficult to answer because they tend to be overly simplistic. However, also among scientists, the temptation to become champions of a popular environmental cause is now overcoming the discretion practiced only a few years ago. Because future trends will be significantly influenced by the growing environmental movement, we shall include the concern over eutrophication as an important part of the consequences of this process.

CAUSES OF EUTROPHICATION

The majority of marine ecologists believe that nitrogen rather than phosphorus is the limiting plant nutrient in the sea. On longer time scales it has been argued that P will be the crucial element in the sea, as is generally the case in lakes, because N_2-fixing organisms can couple combined nitrogen to ambient phosphate concentrations at the Redfield ratio of 16:1 (Smith 1984). Indeed, the surprising ubiquity of Redfield ratios in the sea at large suggests that such a mechanism must be in operation. However, as N_2-fixing organisms are not prominent in the marine realm, aerobic cycling must maintain this ratio in the long term. Ratios higher than 16:1 are rarely found whereas lower ratios, common in coastal environments, invariably signal the presence of anoxic conditions. Lower ratios can be due to denitrification (selective removal of combined N) or excess P input from anaerobic sediments. In either case, N shortage is much more common than N excess. Short-term experiments also indicate that addition of only N to natural marine plankton always stimulates growth whereas addition of only P does not have a comparable effect. Needless to say, addition of both nutrients results in the highest biomass yields. We can conclude that input of combined nitrogen to the sea is the single most important factor contributing to eutrophication. This input can occur either via a point source, such as a river, or diffusely through the atmosphere. The implications for the fate of these nutrients will differ in either case. In the following we list various causes that lead to or enhance nitrogen input to coastal systems.

Agriculture

Global manufacture of nitrogen fertilizer is estimated to contribute ca. 28% to annual nitrogen fixation by the entire biosphere (Walsh 1988). This is a staggering amount and one would intuitively expect it to have far-reaching consequences, apart from raising crop yields. In the U.S.A. and western Europe, fertilizer production increased threefold from the fifties to the seventies, although the area under cultivation has actually declined slightly (Barney 1980). For the Scheldt watershed Billen et al. (1985) estimated that from the nitrogen applied to fields as fertilizer, about 5% is leached out into the rivers and 43% is either denitrified or volatilized as NH_3. Another source of reactive nitrogen from agriculture is animal wastes, of which about 10% and 90% of the nitrogen is discharged directly into the rivers or applied to the fields, respectively. Modern practices of animal husbandry produce concentrated wastes which result in proportionately greater loss of ammonia than would be the case if animal excreta were distributed diffusely on the pastures directly.

Development of River Beds and Soil Erosion in Catchment Basins

Large-scale manipulation of the lower reaches of rivers in industrialized countries has resulted in drainage of wetlands and increasing speed of river flow. Deforestation and establishment of agriculture increases soil erosion and nutrient leaching from the soil. Uptake of dissolved nutrients by plankton and terrestrial vegetation is curtailed, also by increased turbidity of river water, and a correspondingly larger part of the nutrient load reaches the estuary directly. Suspended particles, whether mineral or organic, also stand a smaller chance of being deposited underway but rather tend to accumulate in the estuarine region. Because this poses a problem to shipping, the accumulating sludge is dredged and deposited in deeper water where nutrients can be leached out in the course of its sedimentation.

Inadvertent Nitrogen Fixation

Molecular nitrogen is also fixed, albeit inadvertently, by a variety of energy conversion processes involving heat and pressure, such as in internal combustion engines and power plants. The reactive nitrogen exits with the exhaust and is transported via the atmosphere until its subsequent deposition in precipitation. The amount of combined nitrogen fixed this way is difficult to estimate. Its contribution to N input in the vicinity of river mouths will be relatively insignificant ($< 10\%$), but in oligotrophic regions this value might well be 50% or more (Nelissen and Stefels 1988).

Pollution by Toxins

Input of industrial toxins (heavy metals, organic compounds such as PCBs or phenols, etc.) to rivers can suppress growth of freshwater plants and riverbank vegetation. The nutrient load of polluted river water then becomes available to marine phytoplankton after dilution has sufficiently reduced the adverse effects of the noxious chemicals on organisms. It is thus difficult to separate the effects of harmful wastes from those of nutrients. Some authors (Greve and Parsons 1977) suggest that pollutants can change the structure of planktonic systems although there is as yet no firm evidence that this is indeed happening in the coastal seas. We shall not discuss this effect here.

Tourism and Development of Sandy Beach Areas

The role of tourism in enhancing coastal eutrophication might not, in bulk terms, compare with the various causes listed above. However, the coastlines attractive to tourists tend to be located along the sandy beaches of oligotrophic regions such as the Mediterranean and the subtropical and tropical eastern coasts. Coral barrier reefs and islands, in particular, are becoming increasingly popular. These ecosystems in oligotrophic waters are sensitive to much lower levels of nutrient or seston input than are the temperate coastal systems bordering the industrially developed countries. Coral reefs can sustain damage or even be exterminated when subjected to the increased turbidity caused by the type of coastal development required by modern tourism. Apart from the event-scale perturbation caused by erection of the necessary infrastructure, long-term, low-level discharge of untreated wastes from a "tourist town" can have a seriously detrimental effect on the health of the coastal ecosystem.

Marine Aquaculture

Aquacultural techniques in use at present tend to favor restricted areas where the forces of winds and tides are constrained by topography. Such protected bays or fjords consequently undergo only little water exchange with the open sea. Mollusc farming has a long tradition and is generally carried out in shallow water. The animals concentrate natural food from the environment but their feces and pseudofeces (agglutinated food that is discarded instead of being ingested) tend to collect on the sea bottom and induce anoxia there. Thus, mussel cultures funnel plankton from a large volume to a small area of the benthos and promote eutrophication of their immediate environment.

Recent years have witnessed an explosive growth in aquaculture with the introduction of salmon rearing. This is actually not aquaculture in the strict sense of the term, since the fish are maintained in large, suspended cages and fed with artificial food. Uneaten food and fish excrement collects on the sea bottom below the cages and creates anoxic conditions there. From time to time the cages are moved to another site before the detrimental effects of anoxic water can endanger the fish. Fjords are best suited for this type of enterprise as they provide the deep water necessary to separate the fish from their wastes. Needless to say, the addition of nitrogen-rich fish food contributes significantly to eutrophication of the fjords, although their effect on adjacent seas is reported to be negligible (Ackefors and Enell 1990).

CONSEQUENCES

There are many consequences of coastal eutrophication. Nevertheless, as mentioned in the INTRODUCTION, demonstrating the effects of eutrophication outside of river mouths and enclosed water bodies is a difficult task and the object of much controversy. As eutrophication is a process that can only be judged relative to an earlier, pristine state of the receiving water body, we will not attempt a definition in absolute terms and apply this to the real world as has been done for lakes (Vollenweider 1982). Rather, we will assess the potential effect and eventual fate of the nutrients known to be entering the sea against the background of the present state of our knowledge of the seasonal cycles of coastal systems.

Seasonal Cycles of Coastal Systems

All coastal systems are characterized by a seasonal cycle in biomass and species succession of the various interacting compartments comprising the pelagic and benthic systems. On a very general level, two major phases can be distinguished in the pelagial: a resource-limited phase during which planktonic biomass remains stable or declines and a resource-replete phase characterized by net increase of biomass. The former can again be differentiated according to whether the limiting resource is light or a dissolved nutrient. One can accordingly differentiate three basic "types" of pelagic system that have been schematically depicted in Fig. 1:

Pelagic systems.

1. Resource-limited systems.
   a) A winter or light-limited system in which the rate of nutrient conversion to biomass by algae is lower than the breakdown rate of organic matter by heterotrophic bacteria and protists as well as metazoans. Biomass and accumulated organic matter reserves decline in this period accompanied by a resultant buildup of dissolved nutrient concentrations.
   b) A summer or nutrient-limited system arises in stratified water when available nutrients are tied up in organic matter. Phytoplankton growth is geared to nutrients regenerated by heterotrophs. Ideally, this is a balanced, recycling community in which the phytoplankton is dominated by small organisms, particularly flagellates that are grazed by a suite of protozoans and metazoans, the latter dominated by copepods. Although vertical flux of organic particles in this phase is low, some loss always occurs; hence the total amount of essential elements recycling in a stable, stratified layer decreases with time. It is obvious that these communities are the ones most likely to be immediately affected by eutrophication.
2. Resource-replete systems.
   The transition between light- and nutrient-limited systems occurs in the spring and is initiated by the seasonal increase in illumination of a nutrient-rich, winter water column; conversion of these nutrients into biomass invariably occurs as a rapid buildup of plant biomass (a phytoplankton bloom) followed subsequently by increasing heterotrophic biomass. In all

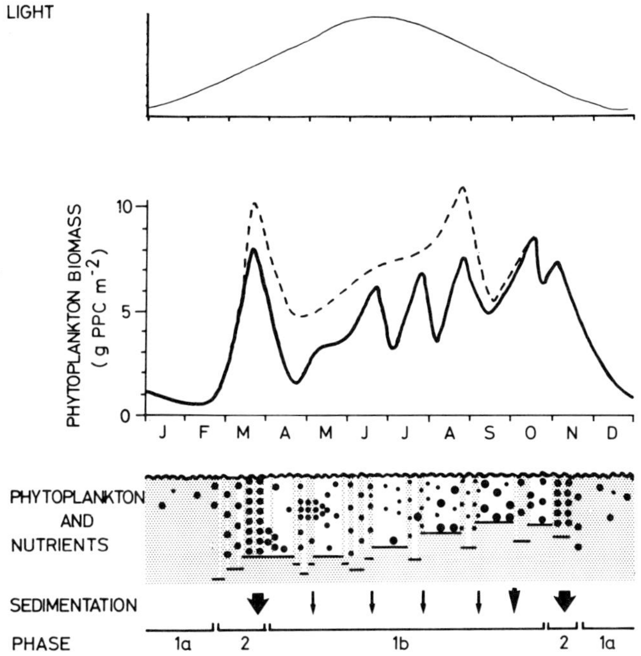

Fig. 1—Schematic representation of the seasonal cycle of light (top panel), integrated phytoplankton biomass in the water column (middle panel), and vertical distribution of phytoplankton (dots) and nutrients (gray areas) (bottom panel) for a temperate coastal system. The dashed line in the middle panel represents critical areas where biomass increase would occur as a result of eutrophication. The arrows indicate sedimentation events. The phases 1a, b, and 2 are discussed in the text.

coastal and shelf systems, biomass growth is eventually limited by exhaustion of an essential element, as a rule, nitrate. As medium-sized diatoms are almost always the initiators of the spring bloom, their growth is halted by exhaustion of dissolved silica. Under "normal" conditions this occurs concomitantly with the exhaustion of nitrate. Nutrient-exhausted diatom blooms tend to increase sinking rates dramatically. The process is accompanied by flocculation of diatom chains into aggregates which "sweep" suspended particles from the water column in the course of their descent. This "spring cleaning" of the water column can remove up to half the organic matter produced by the spring bloom.

Blooms can also be triggered by nutrient input to recycling communities, for instance through admixture of deeper water by storms. The species composition and hence the fate of such blooms can differ considerably from that of typical, diatom-dominated spring blooms. Because growth of these populations occurs in the presence of a large grazer community and under conditions of low silica concentrations, flagellates that are unpalatable to pelagic grazers tend to dominate. The dinoflagellate genus *Ceratium* is one of the characteristic groups that achieves dominance in the course of the summer and autumn in coastal areas. Their growth is eventually restricted by decreasing light. Biomass produced by late summer and autumn blooms tends to be diverted to the benthos rather than utilized in the water column. The above account is based on detailed studies of the pelagic system in Kiel Bight conducted in the seventies (Smetacek et al. 1984).

Benthic systems. The relationship between the supplies of food and oxygen determine structure, biomass, and the seasonal cycle of metabolism of the sublitoral benthos. Studies conducted in Kiel Bight (Graf 1987) indicate that event-scale deposition of organic matter emanating from phytoplankton blooms is rapidly utilized by the benthos (Fig. 2). Deposition of the spring diatom bloom, although the largest input event of the year, does not cause anoxia. During the course of summer stratification, oxygen levels in

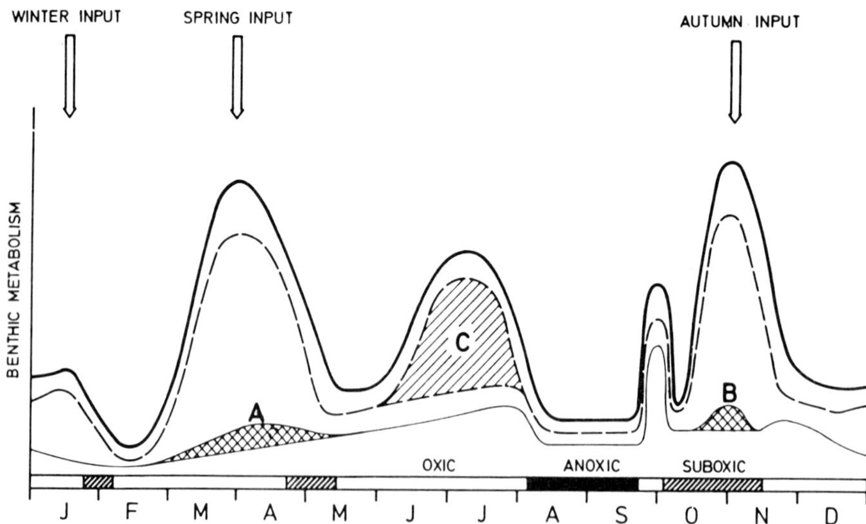

Fig. 2—Conceptual model of the seasonal cycle in food and oxygen supply and their effect on benthic metabolism for a sublitoral sediment station in Kiel Bight without benthic primary production. Total benthic metabolism is indicated by the solid line. The dashed line shows metabolism related to actual sedimentation rate. The thin line indicates metabolism based on stored organic matter in the sediment which depends on temperature and on the availability of oxygen. Areas A and B depict co-oxidation induced during bloom input pulses; area C depicts the influence of biodeposition by suspension feeding organisms in summer. From Graf (1987).

stagnating deeper water decline until the annual minimum is attained in late summer. Autumn storms generally break down bottom water stagnation, but the input of fresh organic matter from the autumn blooms of ceratia and diatoms again reduces oxygen levels significantly. Winter mixing aerates the water column and sediments, but resuspension and redeposition of organic matter and macroalgal debris provides additional input of food to the sublitoral benthos. Oxygen consumption increases as a result. Thorough ventilation is achieved by late winter and the redox discontinuity layer (RDL) in the sediment moves downward. Partly because of this winter ventilation and the high oxygen concentrations in cold winter water, sedimentation of the spring bloom has only a moderate effect on oxygen levels.

Influence of Eutrophication on Coastal Systems

The following is a hypothetical scenario of how nutrient input via eutrophication might affect coastal seasonal cycles. Nutrient input is a function of fresh water runoff and, in northern Europe, is fairly independent of the season. During the winter months, when uptake by plants is minimal, one would expect a gradual buildup of nutrient concentrations in coastal waters. River input from populated areas under intensive agriculture increases the N and P load but Si is not changed (Billen et al., this volume). If the N:Si ratio established prior to spring bloom development is high, one would expect the diatom bloom to culminate as a result of Si exhaustion, leaving behind a significant amount of unused N and P nutrients. The total amount of organic matter channeled to the benthos by the spring bloom would increase even with constant winter Si levels, because Si requirements among diatom species vary considerably and some species with lower Si demand would continue growth until their threshold values for Si are reached. Further, the higher the bloom biomass, the greater the proportion that sediments out (Smetacek 1985).

The excess N and P nutrients "left behind" in the water column would increase the biomass of flagellates that develop after the diatom bloom (Officer and Ryther 1980). The fate of this "excess" organic matter would depend on the behavior of the organisms concerned. Smetacek (1985) has pointed out that diatoms, because of their heavy silica frustule and their tendency to accelerate sinking rates under unfavorable conditions, are one of the major contributors to vertical particle flux in the ocean. The behavior of flagellates in this regard is not as well known and is likely to vary considerably from group to group. For instance, the coccolithophores (a flagellate group bearing heavily calcified plates and also known to form extensive blooms) appear to exhibit behavior similar to diatoms (Honjo 1982). Organic matter produced by flagellates lacking mineral skeletons is likely to be recycled within the pelagial and results in greater biomass of the recycling community that establishes itself following spring bloom sedimentation, indicated by the dashed line in Fig. 1. Alternatively, there could be more pulsed sedimentation events during the summer. Continued nutrient input to the recycling community would further contribute to maintenance of a large plankton biomass and eventually increase food supply to the benthos in late summer. The qualitative effects of eutrophication on species succession are discussed further below. We conclude here that eutrophication has its greatest effect on the recycling summer system; the autumn blooms, being light limited, would be less affected.

Eutrophication increases benthic food and also oxygen supply within the euphotic zone; however, increased water column turbidity reduces light penetration and constrains growth of deeper-living brown and red benthic algae. The higher nutrient levels promote growth of green algae that thrive closer to the surface. They tend to end up on the beaches where the decaying masses offend eyes and noses of holiday makers. Below the mixed layer, oxygen can only be supplied by advection which depends on local hydrography. When oxygen supply is adequate, macrobenthic biomass can achieve impressive proportions as exemplified by coral reefs, rocky coasts, and mud flats. As a rule, the biomass of oxygenated benthos appears to increase with its food supply. Thus, increasing spring sedimentation rates would increase benthic biomass because of favorable oxygen conditions prevailing at the sediment interface. Initially, the food supply of demersal fish increases.

Increasing benthic biomass results in an increasing oxygen demand. In poorly flushed areas, demand can eventually exceed supply. Many macrobenthic animals can survive prolonged periods of low oxygen levels by decreasing metabolic rates and switching to anaerobic respiration (Oeschger 1990). However, input of fresh organic matter stimulates the microbial community that can use nitrate and sulfate for their metabolism. Their byproducts, particularly hydrogen sulfide, eventually diffuse to the overlying water resulting in consumption of the remaining oxygen. Severe anoxia is the result and, when such a situation arises, mass mortality of the macrobenthos occurs. This provides an enormous quantity of high quality bacterial substrate, which greatly increases respiration of anoxic bacteria. Anoxia moves up the water column with upward molecular and turbulent diffusion of reducing agents, which affect the benthos at shallower depths. Horizontal movement of the anoxic water can also affect new areas and kill off the resident macrobenthos. The effect of first-time anoxia is particularly dramatic as a large standing stock of macrobenthos, built up over the years of incipient eutrophication, is killed off, thus radically amplifying the effect of eutrophication. It should be remembered that prolonged stagnation of bottom water, even at comparatively low input rates of organic matter, can have the same effect as a drastic increase in organic input. In temperate lakes, the latter factor is invariably the cause of first-time bottom anoxia whereas in the sea, a shift in the normal seasonal hydrography leading to longer residence time of bottom water can elicit the same result.

The two most conspicuous features of eutrophication are hence an increased incidence of phytoplankton blooms followed by increasing macrobenthic biomass, a trend which can culminate in mass mortality of the benthos due to anoxia. In recent years there appears to be an increase in these features in coastal areas adjoining industrialized countries. However, as long-term data sets are rare, it is difficult to categorize such unusual phenomena in terms of climatic variability or anthropogenic perturbation. The influence of increasing fishing pressure and the decimation of marine mammal populations are further factors to be considered (Elmgren 1989). An evaluation of the effects of eutrophication on the German Bight and the Baltic Seas follows next.

The southern North Sea. A comparison of the data presented in Figs. 3 and 4, collected at the island of Heligoland in the southern German Bight, with Fig. 1 clearly indicates how a "natural" seasonal cycle (early sixties) can be modified by continuous long-term nutrient input. Winter/spring nitrate values have increased while silica values decreased drastically in this period. An explanation for the latter puzzling trend is provided by Billen et al. (this volume), who report that silica uptake and deposition by diatom blooms in the river mouths has increased as a result of increased nitrate loading of river water. The resultant decline in diatom biomass in the southern German Bight and the increase in that of flagellates can be ascertained. One of the results not yet studied would be a change in sedimentation patterns, perhaps reflected in the shifting of the annual biomass maximum from spring (April/May) to late summer (July/August). Of particular interest is the extraordinarily wide range in N/P ratios, with highest values occurring in late spring—a consistent feature throughout the study period. N/P ratios in excess of 30 are common in this season at nitrate concentrations above 10 mmol m^{-3}. Phosphate limitation of algal production has recently been recorded in the North Sea (Veldhuis et al. 1986). This can have important implications for algal species composition, discussed below. The Heligoland data were collected in an area influenced by the plume of the Elbe River and hence were subject to great horizontal variability induced by discharge rate and wind field prevailing at the sampling date. However, as the data were collected at weekly intervals over 23 years, this effect should be averaged out to some extent. The role of eutrophication in inducing exceptional blooms is discussed further below.

In the summer of 1981 an exceptionally dense and extensive bloom of *Ceratium furca* was recorded in the German Bight. Oxygen concentrations

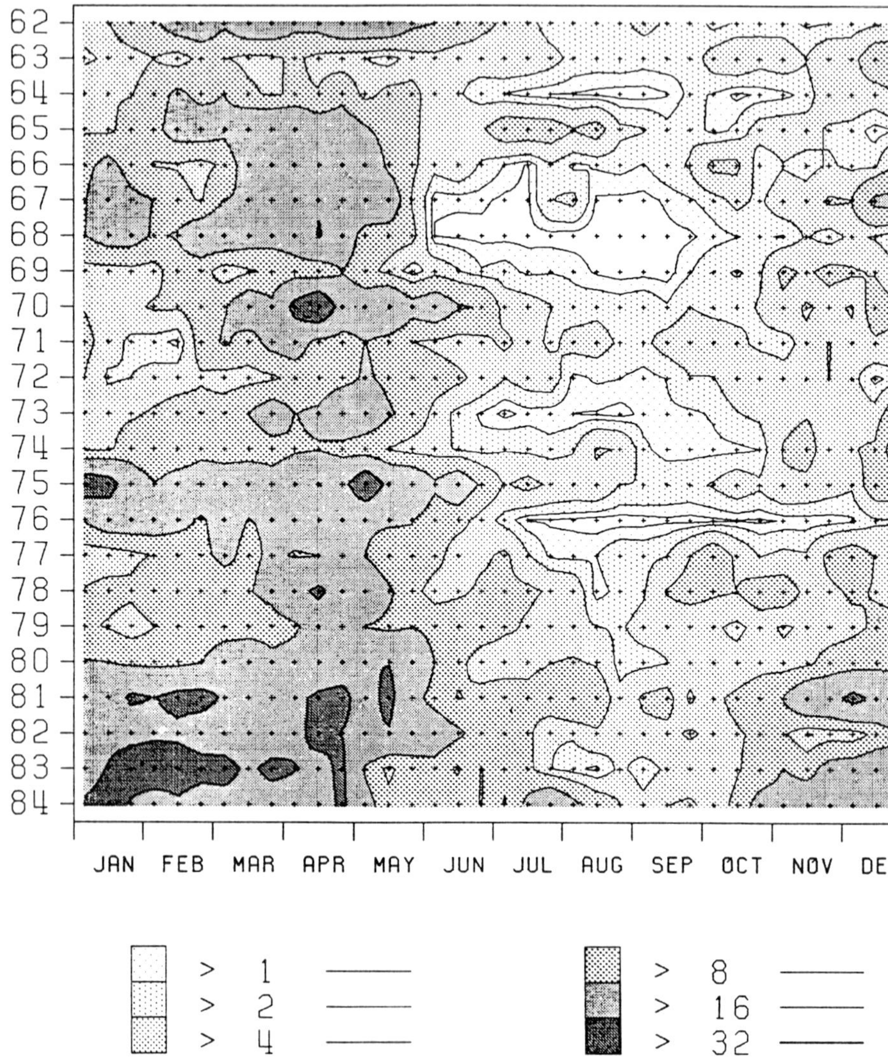

Fig. 3—Isopleth diagrams (in logarithmic intervals) of nutrient concentrations (10-day means) measured at weekly intervals from 1962–1984 at the station Heligoland Reede in the southern German Bight. Concentrations of nitrate (Fig. 3a, above), phosphate (Fig. 3b, opposite), and silicate (Fig. 3c, p. 264) are all given in mmol m^{-3}. From Radach et al. (1990).

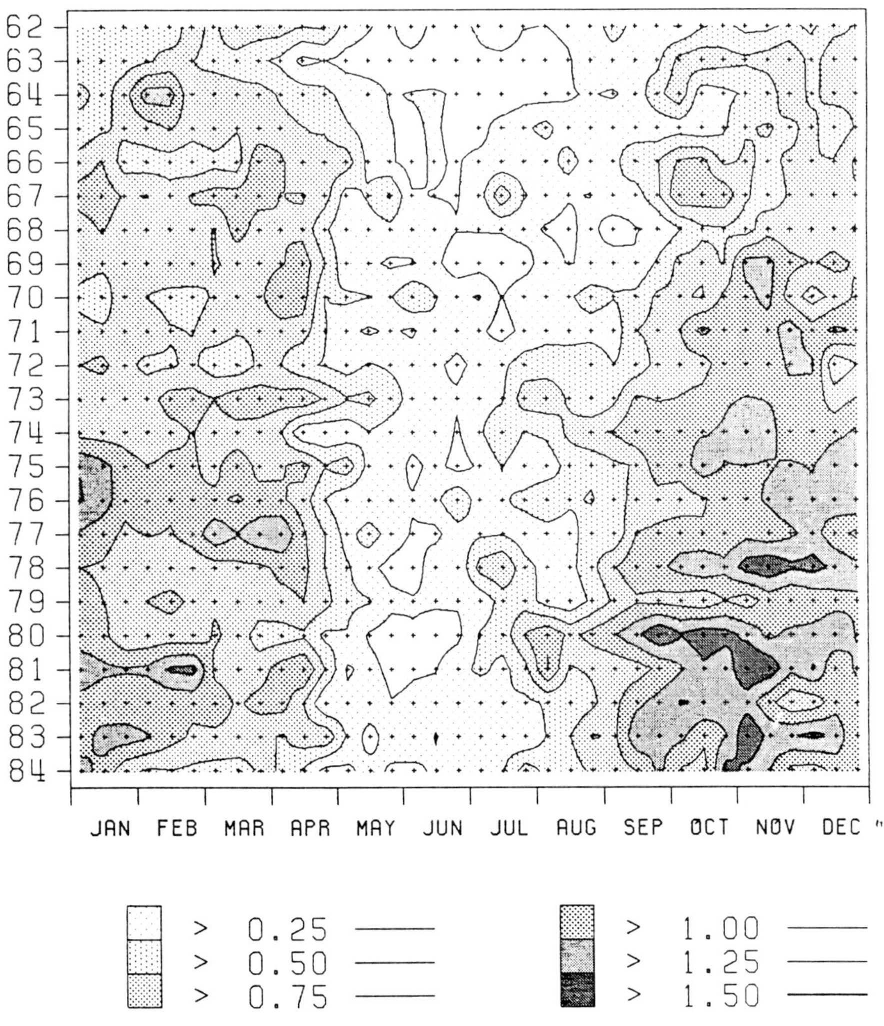

62
63
64
65
66
67
68
69
70
71
72
73
74
75
76
77
78
79
80
81
82
83
84
JAN FEB MAR APR MAY JUN JUL AUG SEP OCT NOV DEC
> 0.25
> 0.50
> 0.75
> 1.00
> 1.25
> 1.50

Fig. 3b

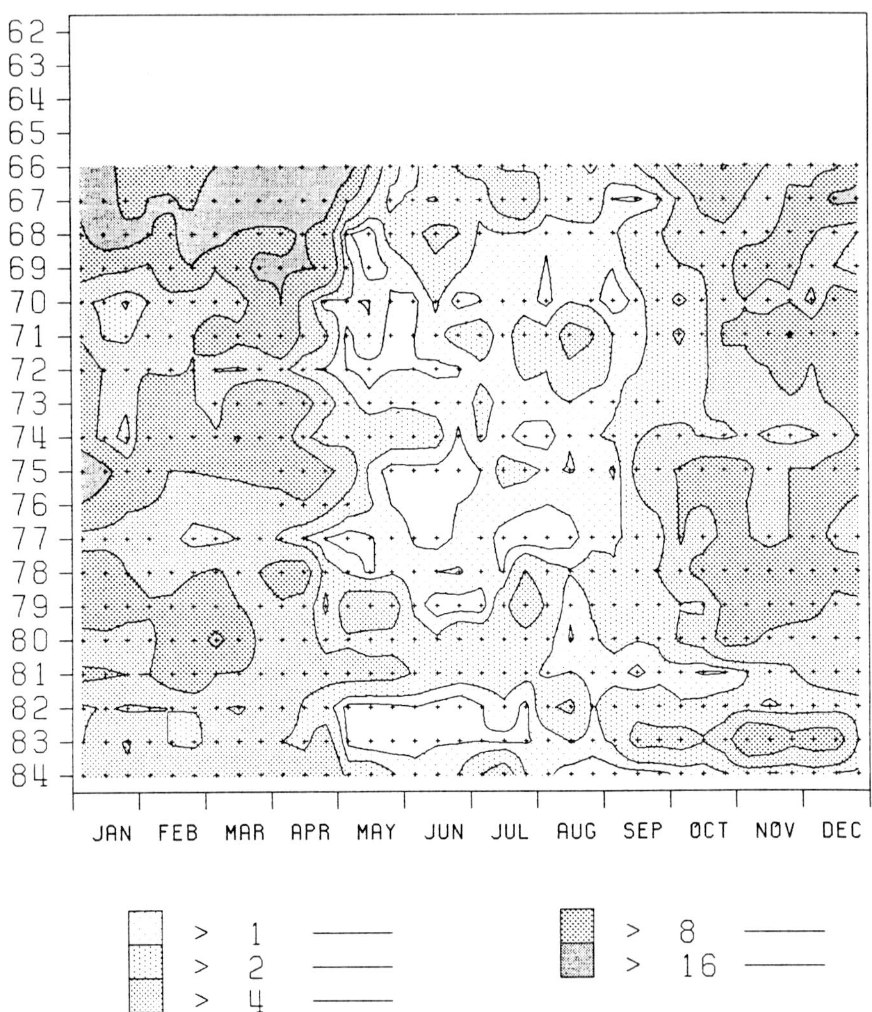

62
63
64
65
66
67
68
69
70
71
72
73
74
75
76
77
78
79
80
81
82
83
84
JAN FEB MAR APR MAY JUN JUL AUG SEP OCT NOV DEC
> 1
> 2
> 4
> 8
> 16

Fig. 3c

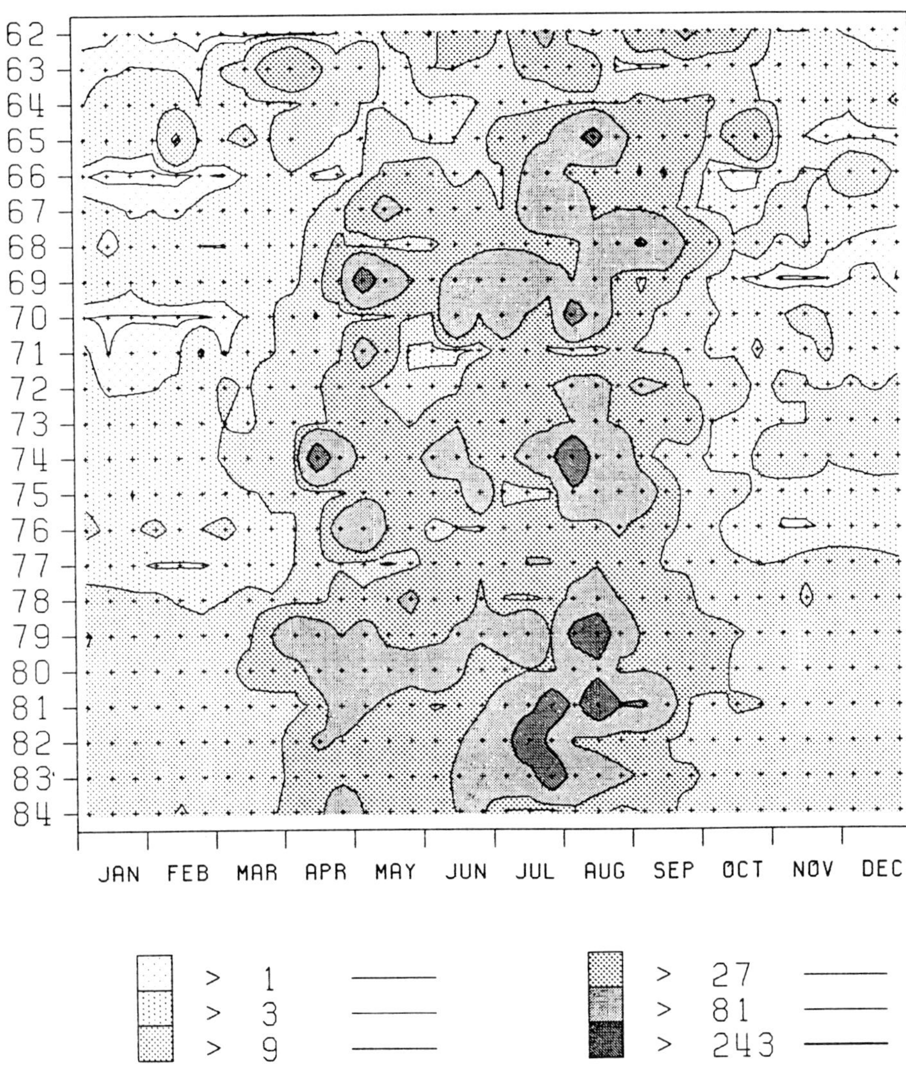

Fig. 4—Isopleth diagrams (in logarithmic intervals) of total phytoplankton (Fig. 4a, above), diatom (Fig. 4b, p. 266) and flagellate (Fig. 4c, p. 267), biomasses (10-day means), expressed in carbon units of mg m^{-3}. The values are derived from microscope counting of samples collected at weekly intervals from the Heligoland Reede from 1962–1984 with subsequent conversion to biomass units. From Radach et al. (1990).

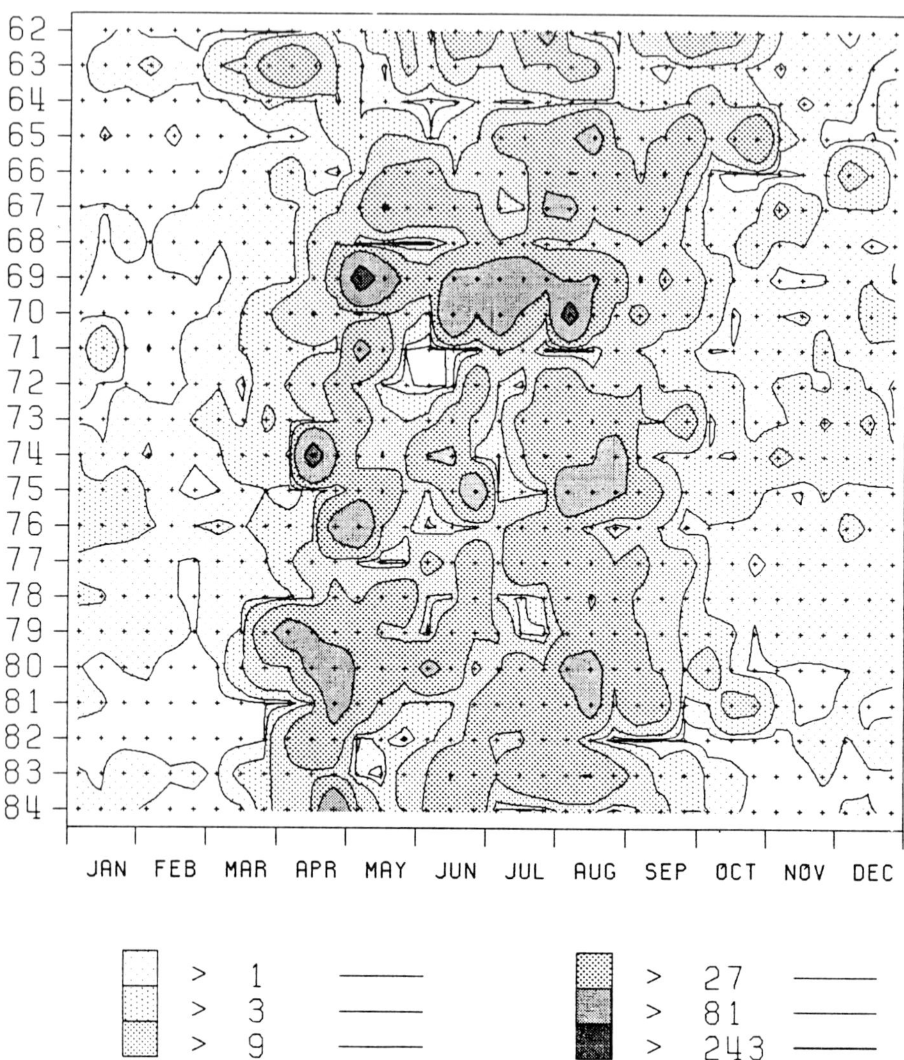

62
63
64
65
66
67
68
69
70
71
72
73
74
75
76
77
78
79
80
81
82
83
84
JAN FEB MAR APR MAY JUN JUL AUG SEP OCT NOV DEC
> 1
> 3
> 9
> 27
> 81
> 243

Fig. 4b

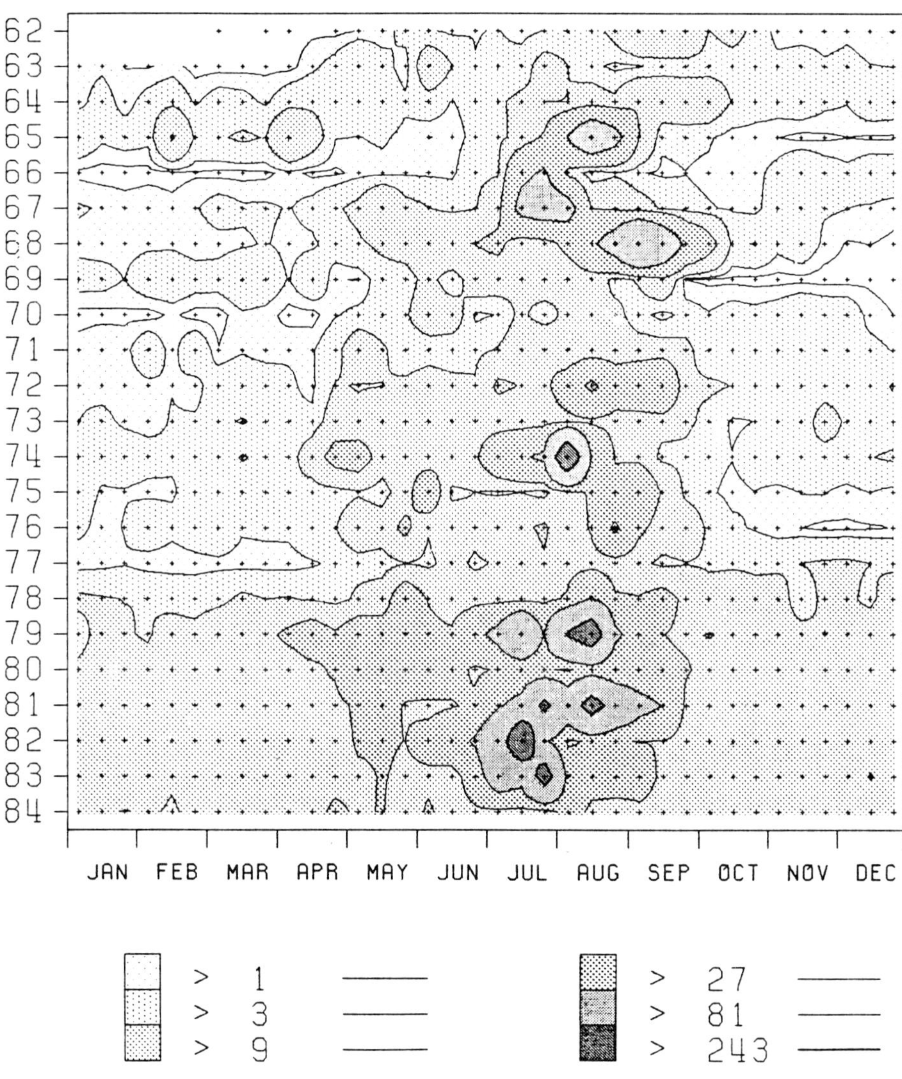

62
63
64
65
66
67
68
69
70
71
72
73
74
75
76
77
78
79
80
81
82
83
84
JAN FEB MAR APR MAY JUN JUL AUG SEP OCT NOV DEC
> 1
> 3
> 9
> 27
> 81
> 243

Fig. 4c

in late summer in the bottom water of deeper areas were found to be surprisingly low over an extensive area, a fact that received much publicity at the time. However, comparable measurements had not been carried out before 1981. Since then, intensive monitoring has revealed the presence of low-oxygen bottom water every year in late summer, albeit with considerable interannual variation in position and extent. No comparable ceratia blooms were recorded subsequently and no incident of benthic mass mortality has been recorded yet, although significant shifts in species composition and distribution have been observed (Nelissen and Stefels 1988). As Brockmann et al. (1988) point out, it is quite likely that late summer suboxia in deep water in the Bight is part of the "normal" seasonal cycle that has been overlooked before. However, it indicates that the area is sensitive to increased organic loading and that anoxia and mass benthic mortality could well eventually occur if the trend evident from the Heligoland data continues. Brockmann et al. (1988) estimate that the oxygen deficit (difference between 100% saturation concentration and measured values) in the affected water volume could be accounted for by respiration of 154,000 t of biomass, i.e., 30,000 t of nitrogen which is equivalent to only 12% of the annual mean Weser/Elbe River discharges. That anoxia has not occurred yet is surprising.

The Baltic Sea. In contrast to the German Bight data, the effect of eutrophication in the open Baltic Sea has not been reflected in any clear changes in the plankton community (Wulff et al. 1986). As many bays and fjords, however, show severe signs of eutrophication (Edler 1984; Stienen 1986), one can conclude that the excess nutrients are retained in the vicinity of the input site. In a concise review of the large and conflicting data, Elmgren (1989) points out that although N and P nutrient concentrations in late winter water of the open Baltic have increased two- to threefold since the late sixties, a concomitant increase in phytoplankton biomass has not been easy to demonstrate. Indeed, higher productivity measured in the open Baltic in the early eighties was followed by lower values in the following years. Possibly the size of the spring bloom and vertical flux has increased, as well as summer phytoplankton stocks. There is no clear trend in increase of zooplankton (Wulff et al. 1986). An as yet unexplained phenomenon is the decreasing trend in winter Si concentrations reported from the Kiel Bight by von Bodungen (1986).

Various studies conducted in the Baltic indicate that biomass of the macrobenthos in the oxygenated layer above the halocline has increased over the past decades. Elmgren (1989) argues that this is due to increased food supply rather than decline in flatfish predators due to heavy fishing. Anoxia below the halocline is linked to topography and the sporadic supply of new bottom water through the Belt Sea. As the Baltic is witnessing the longest stagnation period of bottom water ever recorded (since 1976), it is not possible to differentiate this effect from that of increasing eutrophication (Wulff et al. 1986).

Fate of Anthropogenic Nutrients: Deposition or Denitrification?

The fate of anthropogenic nutrients, particularly nitrogen, entering the sea is the subject of much debate. Some authors suggest that the bulk is denitrified within the sediments (Rönner 1985; Christensen et al. 1987) whereas others believe that a significant portion is deposited as organic sediment on the shelf slopes (Walsh et al. 1985). Either situation will have important but differing implications for NO_x and CO_2 content of the atmosphere, respectively. Another fate of seemingly little consequence would simply be dilution into the huge pool of oceanic nutrients. Obviously all three processes are occurring and large-scale, systematic measurements will be necessary to resolve the issue. Here we discuss some aspects of the processes induced by eutrophication that have a bearing on the fate of the nutrients.

In coastal waters, both nitrification (bacterial oxidation of ammonia to nitrate) and denitrification (reduction of nitrate to molecular nitrogen) occur primarily in the sediments (see Billen et al., this volume); however, the latter process is only supposed to attain importance when nitrate is advected to the denitrification site rather than produced in its vicinity (see Blackburn, this volume). In the deep waters of the open Baltic, the nitrate-rich water in intermediate layers between the seasonal thermocline and the permanent halocline comes extensively in contact with suboxic sediments. Denitrification is accordingly estimated to account for more than 60% of the total anthropogenic input of nitrogen to this sea (Wulff et al. 1986). In the shallow Kiel Bight, denitrification is of much lesser magnitude (Balzer and Kähler 1988) because substantial nitrate concentrations are only built up over the oxygen-rich winter months. During the rest of the year, ammonia is the main form of dissolved inorganic nitrogen. The North Sea is likely to occupy an intermediate position and approximately 20% of new nitrogen is suspected to be lost to denitrification (Brockmann, pers. comm). Balzer and Kähler (1988) have shown that sandy sediments are much more active sites of denitrification than muddy sand (by a factor of 4) and mud (factor of 10). In the Kiel Bight, increasing incidence and extent of anoxia in summer bottom water has resulted in encroachment of mud-covered areas (Eversberg 1990), an effect which will presumably reduce the rate of denitrification as the extent of muddy sediments increases. We hypothesize that whereas chronic anoxia promotes burial of carbon and nitrogen, intermittent anoxia promotes denitrification. In either case, combined nitrogen is lost from the system.

Walsh et al. (1985) suggest that deposition of organic matter on shelf slopes is enhanced significantly by eutrophication; they speculate that this sink might be partly responsible for removal of "missing" CO_2 that is not accounted for by mass budgets based on known sources and sinks of this gas. Conversion of N_2 in fertilizer factories burning fossil fuels to combined nitrogen occurs at a ratio of C:N = 1 (Finck 1979), whereas burial occurs at C:N ratios ranging between 6 and 12. Based on these considerations, one might speculate that eutrophication has a beneficial aspect as it binds CO_2 in the sediments. However, nitrogen geometabolism (nitrification and denitrification) leaks N_2O, which is a much more potent greenhouse gas than CO_2. Yoshida et al. (1989) claim that denitrification can well be a significant source of this gas. If this is true then sandy, near-shore sediments along continental margins would be a major source of N_2O. It is too early to speculate further on this topic; however, considering its global importance, research in this field should be encouraged.

Exceptional Phytoplankton Blooms

Monospecific blooms of very high cell density that can discolor large areas of surface water are a frequent feature of the annual cycle in some areas and rare occurrences in others. They often occupy only a thin upper layer, and as they are apparently the result of concentration by selective movement of the cells relative to the water they grew in, the total areal biomass is generally modest when compared to that of diatom spring blooms. These blooms—known as red tides in some cases—have always received special attention because they are sometimes composed of noxious or even highly toxic species and are hence of importance to aquaculture and fisheries. The number of reports of "first-time observations" of such phenomena have increased in recent years and there is a growing tendency to attribute them to insidious effects of eutrophication (Smayda 1989). The blooms occur invariably after the spring diatom bloom or in summer and are hence generally dominated by flagellates, sometimes by a single species. Some of the recent cases, particularly the "brown tides" of the northeastern U.S.A. coast, have been presented together with their impacts on aquaculture by Cosper et al. (1989). In the following, we discuss some of the spectacular cases from the European coasts.

Prymnesiophyte (formerly Haptophyte) blooms. The widely distributed genus *Phaeocystis* blooms shortly after but sometimes in conjunction with the spring diatom bloom (Lancelot et al. 1987). It is a regular feature of phytoplankton succession along the continental margin of the North Atlantic but also occurs abundantly at the ice edge in the Arctic and Antarctic. In its mass growth stage, *Phaeocystis* forms large gelatinous colonies which are not relished by zooplankton; organic matter emanating from this bloom is pounded by waves into a stiff, meter-high malodorous surf on the beaches of the southern North Sea and bears an alarming resemblance to detergent foam. The first "foam event" was reported in 1978 along the coasts of the German Bight and has occurred every year since (Bätje and Michaelis 1986). The appearance of foam on the beaches is now regularly reported by the newspapers. Dense blooms have been recorded as far back as the thirties, but there is no mention of foam on the beaches in earlier chronicles (Bätje and Michaelis 1986). However, long-term data from the Wadden Sea clearly indicate an increase in the abundance of *Phaeocystis* from 1973–1985 (Cadée and Hegeman 1986).

The occurrence of toxic algal blooms in the sea was brought to the attention of the European public by the notorious bloom of a little known, small prymnesiophyte flagellate, *Chrysochromulina polylepis*, in May 1988. Attention was first drawn to its presence by fish kills in aquaculture farms along the west coast of Sweden. Northward advection of the bloom in the coastal current threatened the many fish farms along the Norwegian coast, and the farmers in the path of the bloom started towing their cages deeper into the fjords because lower salinity reduced virulence of the alga. It has been suggested that toxicity of the alga increased with increasing N:P ratios (assuming that, at the time, N:P ratios increased with salinity) as has been demonstrated for the related *Prymnesium parvum*, also a notorious fish killer (Skjoldal and Dundas 1989). The maritime operation launched to save the caged fish was the largest in the peacetime history of the country. The progress of the bloom was tailored to the needs of the press: mass mortality of fish, macrobenthos, and even some macroalgae that came into contact with the bloom were reported in catastrophic terms. The bloom developed in the pycnocline between outflowing low-salinity Baltic and higher-salinity Kattegat water. The pycnocline surfaced in patches along the Swedish west coast, exposing the bloom as an ugly yellowish brown slick. Film sequences shown on primetime television of the active flagellate under the microscope followed by the algal carpets filmed from a helicopter conjured up visions of aggressively swarming algae. The term "killer algae" coined by the press established a connection with the spread of the notorious "killer bees" in South and Central America.

Mass mortality of seals on an island adjacent to the bloom was reported at the same time. These two events were at first linked as cause and effect, but as evaluation of the evidence showed no direct connection, anthropogenic perturbation of the environment was held responsible. The simultaneous spread of the seal pandemic and that of the "killer algal" bloom, albeit in different directions, begged a common explanation and most scientists interviewed at the time blamed the twin problems of pollution and eutrophication as ultimate if not proximate causes (Rosenberg et al. 1988). Since then the seal mortality has been shown to be due to a viral infection whose pattern of spread did not match the distribution pattern of pollutants. The absence of a bloom in May, 1989, in spite of similar weather conditions to those of 1988, also cast doubt on the explanation of a simple cause-and-effect relationship between increasing eutrophication and the "killer algal" bloom. However, if its toxicity was indeed related to high N:P ratios, then a connection between eutrophication and damage caused by the bloom would be established. Recall the high N:P ratios characteristic of late spring waters around Heligoland; this water is advected into the Kattegat with the coastal current.

Another exceptional prymnesiophyte bloom—the gelatinous colony-forming species, *Corymbellus aureus*—was recorded from the Fladen Ground in the northern North Sea in May, 1983. The species had only been recently described and this was the first time it was observed in any abundance (Gieskes and Kraay 1986). During a similar study in the same area conducted in 1976, a classic spring diatom bloom was observed when losses of primary produced material due to sinking amounted to 20–35% per day (Davies and Payne 1984). The *Corymbellus* bloom of 1983 was preceded by a smaller population of mixed diatoms. Sedimentation rates were much lower than in 1976 and amounted to only ca. 1% of the production and most of the material ended up in the DOC pool; grazing of *Corymbellus* cells was also low (Cadée 1986). This example clearly indicates the role of species composition in determining the fate of available nutrients; whether the difference between years can be attributed to eutrophication is not clear, rather it is unlikely as the Fladen Ground is far from coastal nutrient sources.

Extensive blooms of coccolithophorids (*Emiliana huxleyi*) occur during early summer along the outer margin of the European continental shelf and also in the North Sea and in fjords (Groom and Holligan 1986). In contrast to the other members of this class, their possible role in enhancing vertical flux of carbon via their carbonate coccoliths has been pointed out by authors. One might conclude from these observations that prymnesiophytes are particularly well adapted to the late spring situation and that we should expect increased abundance of these algae with advancing eutrophication.

Another common late spring species is the very widely distributed chrysophyte silicoflagellate, *Dictyocha speculum* (formerly *Distephanus*). A new form of this species without a skeleton attained bloom proportions in Kiel Bight in May, 1983 and has appeared repeatedly in May since then. Jochem and Babenerd (1989) interpret the high biomass of these blooms in a season previously characterized by low nutrients as the result of increasing eutrophication over the past years. The lack of a silica skeleton indicates a peculiar adaptation to increasing N/Si ratios. The species was previously considered innocuous as it is eaten by zooplankton. However, the skeletonless form has caused fish kills.

Dinoflagellate blooms. The vast majority of red tide-forming organisms belong to this class, and there is little doubt that we are witnessing a worldwide expansion in the geographic distribution of neritic members of this algal class (Smayda 1989). Many of the species contain potent toxins and are responsible for fish kills as well as human poisoning via shellfish; not surprisingly, their spread is looked upon with alarm and carefully monitored. The species in question tend to have life cycles involving benthic resting spores. It has been suggested that such resistant spores have been transported from harbor to harbor in the ballast tanks of ships and that new populations have spread out from the harbor inocula. There are too many examples to list here (e.g., Hallegraeff et al. 1988), which is in itself proof that there is a distinct trend. However, eutrophication appears to play a direct role in the spread of only those species that are adapted to life in eutrophic bays. An indirect effect of eutrophication might also be postulated because red tide populations are likely to grow to a bigger size, both in terms of numbers as well as areal coverage. This will favor dispersal to new localities. Marshall (1986) reports an increase in the abundance of organisms typical of Chesapeake Bay on the outer shelf as a result of increasing nutrient load of the former.

The spread of all species cannot be attributed to human agency alone. Thus, *Gyrodinium aureolum*, one of the recent newcomers to Europe, now forms intense blooms in the North Sea and along the Norwegian coast, where it was first recorded in 1966, and has harmful but not catastrophic effects on aquaculture. However, it also forms extensive blooms on the outer shelf between Britain and France (Holligan 1979) far away from the sphere of direct human influence. Similarly, the first appearance of the toxic species *Ptychodiscus brevis* off the coast of North Carolina in 1988 could be related to transport of the bloom in a Gulf Stream filament that retained its integrity for an exceptionally long period (Tester et al. 1989). *P. brevis* blooms are normally restricted to the eastern Gulf of Mexico.

Another group of dinoflagellates that is implicated with eutrophication belong to the widespread genus *Ceratium*. Their biology differs considerably from the classic red tide organisms dealt with above. The species of this genus are large, long-lived, and robust; the marine species are not known to form resting spores. For unknown reasons, they are not relished by zooplankton although they do not appear to possess toxins. They are characteristic of late summer plankton and appear to build up their numbers through the summer. In some regions, such as the Belt Sea and the German Bight, they form large blooms that seem to undergo mass mortality (again for unknown reasons) and then sink out to the sea floor where their remains promote anoxia. Particularly massive blooms were recorded in 1976 in the New York Bight (*Ceratium tripos*) and in 1981 in the German Bight (*C. furca*). The former could only be marginally related to eutrophication (Walsh

1988), whereas the proximate causes of the latter could not be ascertained (Hickel 1982). Blooms of the same magnitude have not been reported since; however, in Laholm Bay (southern Swedish west coast), late summer anoxia has been recorded successively for several years and is attributed to an increase in the size of the regular ceratia blooms. As anoxia was never reported before the eighties, the excess biomass is believed to be the result of eutrophication (Edler 1984).

Unusual diatom blooms. Huge cloud-like aggregations of mucus, in which pennate diatoms seemed to flourish, appeared along the bathing beaches of the Adriatic at the height of the tourist season in 1989 and attracted considerable attention from the press. North European tourists, in particular, sensitized by the "killer algal" bloom of the previous year, abandoned the beaches in droves which plunged the flourishing tourist industry into a deep depression. This sudden misfortune required a culprit and the blame was quickly put on eutrophication. A massive "cleanup" program was hurriedly launched by the authorities, but it was forgotten, in all the flurry, that the mysterious clouds of algae were recorded and described already at the turn of the century (Steuer 1911). The phenomenon was known to local fishermen, whose nets were ripped by the weight of the slime, as "Mare sporco" (dirty sea). Although the exact mechanisms of "mucous cloud" formation have not yet been established, preliminary results suggest that the same sequence of events reported by Steuer in 1911 also lead to development of the peculiar bloom celebrated by the media in 1989. The reason why such a phenomenon has been reported only from the Adriatic is probably related to regional peculiarities of hydrography/topography and nutrient regime, the latter prevalent before the advent of eutrophication. A slime-producing diatom that has formed blooms large enough to rip fish nets is *Coscinodiscus wailesii* in the English Channel (Boalch 1987). This species is a recent introduction from the Pacific and has since established itself as a regular feature of species succession in the southern North Sea (Smayda 1989).

Evaluation of the Bloom Reports

Several tentative conclusions can be drawn from the above accounts. In the majority of cases, eutrophication cannot be considered the triggering mechanism, although there is some justification in regarding unusual blooms of noxious species as symptoms of advancing eutrophication. "Exceptional" blooms are actually part and parcel of the seasonal cycle of pelagic systems. In some cases these can be related to an unusual hydrographical event, but in many cases the causative mechanisms are not known. Growth of prymnesiophytes and *Gyrodinium aureolum* is probably enhanced by eutrophication. These algae are known as copious DMS producers (see

Wolfe et al., this volume) in contrast to diatoms and ceratia (Turner et al. 1988). The global implications of release of these unusual gases is discussed by Lovelock (this volume); Keller et al. (1989) suggest that eutrophication-induced blooms will play an important role in this scenario.

The large number of records of unusual blooms in the last decade are partly due to the increased monitoring effort in connection with the spread of aquaculture. In many cases it can be argued that the blooms were induced by the aquaculture operation. Indeed it is this growing sector of human endeavor that is most affected by its own actions. One might consider the noxious algae as the "weeds and pests" that arise whenever humans attempt to manipulate natural environments to their own specific needs. Combating them will be much more difficult than in the case of their terrestrial counterparts. For many areas, this factor alone might leave no other option than to depart radically from present techniques. Whether alternatives can be developed remains to be seen. An encouraging example comes from the Seto Inland Sea of Japan where incidences of red tides increased annually from ca. 40 to over 300 per year in the time span of 1965–1973 (Prakash 1987). Radical measures introduced in 1972, aimed at reducing organic inputs to the sea by more than half within a few years, appear to have met with success, as the incidence of red tides peaked in 1975 and has declined since then.

The fact that all algal classes commonly represented in the marine phytoplankton have contributed to the unusual blooms indicates that there can be no single factor responsible. Indeed, studies of the effects of prolonged eutrophication in large-volume experimental systems—the MERL mesocosms in Rhode Island—revealed that chain-forming diatoms dominated algal biomass throughout the year. Flagellate blooms appeared only transiently (Oviatt et al. 1989). However, the tanks were also supplied with silicon and hence not comparable to the high N/Si ratios postulated for eutrophied areas by Officer and Ryther (1980).

THE RISE OF ENVIRONMENTAL CONSCIOUSNESS

The populations of northwestern Europe and North America have been particularly sensitized by recent events, presumably because the higher demands on environmental quality in these affluent countries has lowered tolerance levels significantly. Environmental cleanup is a trend going back to the unsanitary, plague-ridden medieval cities. Comparison of our present-day expectations of what constitutes an acceptable environment to those of our not so removed ancestors, or even neighbors, indicates how environmental consciousness changes. The recent campaign against dog droppings on the streets is a case in point. The present attitude toward environmental pollution can be considered a new manifestation of this old trend. We, the scientific community, should accept it as such. Scientists, being part of the public, are not immune from the Zeitgeist of accepted cleanliness standards. Our community is now under pressure to prove the detrimental effects of eutrophication. The recent increase in the willingness to do so certainly contains (we are practicing self-criticism) elements of opportunism as it directly leads to increased funding squeezed out of the respective governments by public pressure. It also caters to the beliefs of the "scruffy generation" (from which many of us have been recruited), who emphasize environmental cleanliness at the cost of the detergent industry's whiter-than-white-shirt standards of personal "clean" appearance.

Although large-scale effects of eutrophication cannot be clearly demonstrated, its noxious smaller-scale effects, encompassing destruction of estuarine habitats, loss of underwater landscape, and accumulation of rotting algae along beaches, should suffice to convince us that prophylaxis rather than intensification of the search for new remedies should be advocated. On a global scale, extollment of the cleansing potential of the biological pump raises the temptation to attempt large-scale environmental manipulation. Past experience shows that such activity will merely bring new problems in its wake. Overstating the hazards of eutrophication, on the other hand, can also be detrimental to the cause of curbing wasteful use of restricted resources as overstimulation invariably results in desensitization or even apathy. Ecologists inciting a concerned public to environmental hysteria, through the medium of the sensationalist press, are tantamount to doctors encouraging hypochondria in their patients. We should simply endeavor to use resources more judiciously than in the past. Experience from lacustrine environments suggests that in most marine cases, the effected systems are likely to revert to the original state if the nutrient supply is drastically reduced. We are well on the road to making a global mess; it is time we sought for ways to curb the trend.

Acknowledgements. We thank G. Billen and U. Riebesell for helpful criticism and valuable suggestions. I. Lukait helped produce the manuscript. This is publ. nr. 306 of the Alfred-Wegener-Institut für Polar- und Meeresforschung, Bremerhaven.

REFERENCES

Ackefors, H., and M. Enell. 1990. Discharge of nutrients from Swedish fish farming to adjacent sea areas. *Ambio* **19**:28–35.

Balzer, W., and P. Kähler. 1988. Natürliche Entfernung von Stickstoff durch Denitrifikation aus dem System Kieler Bucht. *Abschlußbericht zum UBA-Projekt Wasser Teilvorhaben* **21**:1–22.

Barney, G.O., ed. 1980. Global 2000 report to the president. Council on environmental quality. Washington: U.S. Govt. Printing Office.

Bätje, M., and H. Michaelis. 1986. *Phaeocystis pouchetii* blooms in East Frisian coastal waters German Bight, North Sea. *Marine Biol.* **93**:21–27.

Billen, G., M. Somville, E. de Becker, and P. Servais. 1985. A nitrogen budget of the Scheldt hydrographical basin. *Neth. J. Sea Res.* **19**:223–230.

Boalch, G.J. 1987. Changes in the phytoplankton of the western English Channel in recent years. *Br. Phycol. J.* **22**:225–235.

Bodungen, B. von. 1986. Annual cycles of nutrients in a shallow inshore area, Kiel Bight: variability and trends. *Ophelia* **26**:91–107.

Brockmann, U., G. Billen, and W.W.C. Gieskes. 1988. North Sea nutrients and eutrophication. In: Pollution of the North Sea, an Assessment, ed. W. Salomons, B.L. Bayne, E.K. Duursma, and U. Förstner, pp. 348–389. Berlin: Springer.

Cadée, G.C. 1986. Organic carbon in the water column and its sedimentation, Fladen Ground North Sea, May 1983. *Neth. J. Sea Res.* **20**:347–358.

Cadée, G.C., and J. Hegeman. 1986. Seasonal and annual variation in *Phaeocystis pouchetii* (Haptophyceae) in the westernmost inlet of the Wadden Sea during 1973 to 1985 period. *Neth. J. Sea Res.* **20**:29–36.

Christensen, J.P., J.W. Murray, A.H. Devol, and L.A. Codispoti. 1987. Denitrification in continental shelf sediments has major impact on the oceanic nitrogen budget. *Global Biogeochemical Cycles* **1**:97–116.

Cosper, E.M., J.E. Carpenter, and M. Cottrell. 1989. Primary productivity and growth dynamics of the "brown tide" in Long Island embayments. In: Coastal Marine Studies, vol. **35** Novel Phytoplankton Blooms, ed. E.M. Cosper, V.M. Bricelj, and E.J. Carpenter, pp. 139–158. Berlin: Springer.

Davies, J.M., and R. Payne. 1984. Supply of organic matter in the northern North Sea during a spring phytoplankton bloom. *Marine Biol.* **78**:315–324.

Edler, L. 1984. A mass development of *Ceratium* species on the Swedish west coast. *Limnologica* **15**:353–357.

Elmgren, R. 1989. Man's impact on the ecosystem of the Baltic Sea: energy flows today and at the turn of the century. *Ambio* **18**:326–332.

Eversberg, U. 1990. Abbau und Akkumulation von organischer Substanz in den Sedimenten der Kieler Bucht. *Ber. Inst. Meer.* Kiel **193**:98.

Finck, A. 1979. Dünger und Düngung: Grundlagen, Anleitung zur Düngung der Kulturpflanzen. Weinheim: Chemie.

Gieskes, W.W.C., and G.W. Kraay. 1986. Analysis of phytoplankton pigments by HPLC before, during and after mass occurrence of the microflagellate *Corymbellus aureus* during the spring bloom in the open northern North Sea in 1983. *Marine Biol.* **92**:45–52.

Graf, G. 1987. Benthic response to the annual sedimentation pattern. In: Lecture notes on coastal and estuarine studies, Seawater–sediment interactions in coastal waters: an interdisciplinary approach, ed. J. Rumohr, E. Walger, B. Zeitzschel, vol. 13, pp. 84–92. Berlin: Springer.

Greve, W., and T.R. Parsons. 1977. Photosynthesis and fish production: Hypothetical effects of climatic change and pollution. *Helgoländer Wiss. Meer.* **30**:666–672.

Groom, S.B., and P.M. Holligan. 1986. Remote sensing of coccolithophore blooms. *Adv. Space Res.* **7**:73–78.

Hallegraeff, G.M., D.A. Steffensen, and R. Wetherbee. 1988. Three estuarine Australian dinoflagellates that can produce paralytic shellfish toxins. *J. Plank. Res.* **10**:533–541.

Hickel, W. 1982. *Ceratium* red tide in the German Bight in August, 1981: spatial distribution. *ICES* C. M. **1982/L**:8.

Holligan, P.M. 1979. Dinoflagellate associated with tidal fronts around the British

Isles. In: Toxic dinoflagellate blooms, ed. D.L. Taylor and H.H. Seliger, pp. 249–256. New York: Elsevier.

Honjo, S. 1982. Seasonality and interaction of biogenic and lithogenic particulate flux at the Panama Basin. *Science* **218**:883–884.

Jochem, F., and B. Babenerd. 1989. Naked *Dictyocha speculum*—a new type of phytoplankton bloom in the Western Baltic. *Marine Biol.* **103**:373–379.

Keller, M.D., W.K. Bellows, and R.R.L. Guillard. 1989. Dimethylsulfide production and marine phytoplankton: an additional impact of unusual blooms. In: Coastal Marine Studies, vol. **35**, Novel Phytoplankton Blooms, ed. E.M. Cosper, V.M. Bricelj, and E.J. Carpenter, pp. 101–115. Berlin: Springer.

Lancelot, C., G. Billen, A. Sournia, T. Weisse, F. Colijn, M.J.W. Veldhuis, A. Davies, and P. Wassmann. 1987. *Phaeocystis* blooms and nutrient enrichment in the continental coastal zones of the North Sea. *Ambio* **16**:38–46.

Marshall, H.G. 1986. The seasonal influence of Chesapeake Bay phytoplankton to the continental shelf. In: The Role of Freshwater Outflow in Coastal Marine Ecosystems, ed. S. Skreslet, pp. 319–327. Berlin: Springer.

Nelissen, P.H.M., and J. Stefels. eds. 1988. Eutrophication in the North Sea. *NIOZ-Rapp.* **4**:100.

Oeschger, R. 1990. Long-term anaerobiosis in sublittoral marine invertebrates from the western Baltic Sea: *Halicryptus spinulosus* (Priapulidae), *Astarte borealis* and *Arctica islandica* (Bivalvia). *Marine Ecol. Prog. Ser.* **59**:133–143.

Officer, C.B., and J.H. Ryther. 1980. The possible importance of silicon in marine eutrophication. *Marine Ecol.-Prog. Ser.* **3**:83–91.

Oviatt, C., P. Lane, F. French, and P. Donaghay. 1989. Phytoplankton species and abundance in response to eutrophication in coastal marine mesocosms. *J. Plank. Res.* **11**:1223–1244.

Prakash, A. 1987. Coastal organic pollution as a contributing factor to red-tide development. *Rapp. P.-v. Cons. Int. Explor. Mer.* **187**:61–65.

Radach, G., J. Berg, and E. Hagmeier. 1990. Long-term changes of the annual cycles of meteorological, hydrographic, nutrient, and phytoplankton time series at Heligoland and at LV Elbe 1 in the German Bight. *Cont. Shelf Res.* **10**:305–328.

Rönner, U. 1985. Nitrogen transformations in the Baltic Proper: denitrification counteracts eutrophication. *Ambio* **14**:1008–1013.

Rosenberg, R., O. Lindahl, and H. Blanck. 1988. Silent spring in the sea. *Ambio* **17**:289–290.

Skjoldal, H.R., and I. Dundas, eds. 1989. The *Chrysochromulina polylepis* bloom in the Skagerrak and the Kattegat in May–June 1988: environmental conditions, possible causes, and effects. Report ICES Workshop on the *Chrysochromulina polylepis* bloom in the Skagerak and Kattegat in May–June 1988. *JCES* C.M. **1989**:61.

Smayda, T.J. 1989. Primary production and the global epidemic of phytoplankton blooms in the sea: A linkage? In: Coastal Marine Studies, vol. **35**, Novel Phytoplankton Blooms, ed. E.M. Cosper, V.M. Bricelj, and E.J. Carpenter, pp. 449–483. Berlin: Springer.

Smetacek, V.S. 1985. Role of sinking in diatom life-history cycles: ecological, evolutionary and geological significance. *Marine Biol.* **84**:239–251.

Smetacek, V.S., B. Knoppers, R. Peinert, F. Pollehne, P. Stegmann, and B. Zeitzschel. 1984. Seasonal stages characterizing the annual cycle of an inshore pelagic system. *Rapp. P.-v. Cons. Int. Explor. Mer.* **183**:126–135.

Smith, S.V. 1984. Phosphorus versus nitrogen limitation in the marine environment. *Limnol. Ocean.* **29**:1149–1160.

Steuer, A. 1911. Leitfaden der Planktonkunde. Berlin: Teubner.

Stienen, C. 1986. Increased nutrient load and phytoplankton biomass in Kiel Fjord as compared with Kiel Bight, western Baltic. *Ophelia Suppl.* **4**:259–271.

Tester, P.A., P.K. Fowler, and J.T. Turner. 1989. Gulf Stream transport of the toxic red tide dinoflagellate *Ptychodiscus brevis* from Florida to North Carolina. In: Coastal Marine Studies, vol. **35**, Novel Phytoplankton Blooms, ed. E.M. Cosper, V.M. Bricelj, and E.J. Carpenter, pp. 349–358. Berlin: Springer.

Turner, S.M., G. Malin, P.S. Liss, D.S. Harbour, and P.M. Holligan. 1988. The seasonal variation of dimethylsulfide and dimethylsulfoniopropionate concentrations in nearshore waters. *Limnol. Ocean.* **33**:364–375.

Veldhuis, M.J.W., F. Colijn, and L.A.H. Venekamp. 1986. The spring bloom of *Phaeocystis pouchetii* (Haptophyceae) in Dutch coastal waters. *Neth. J. Sea Res.* **20**:37–48.

Vollenweider, R.A. 1982. Eutrophication of waters: monitoring, assessment and control. Paris: OECD.

Walsh, J.J. 1988. On the Nature of Continental Shelves. London: Academic.

Walsh, J.J., E.T. Premuzic, J.S. Gaffney, G.T. Rowe, G. Harbottle, R.W. Stoenner, W.L. Basalm, P.R. Betzer, and S.A. Mackoll. 1985. Organic storage of CO_2 on the continental slope off the mid-Atlantic bight, the southeastern Bering Sea, and the Peru coast. *Deep-Sea Res.* **32**:853–883.

Wulff, F., G. Aertebjerg, G. Nicolaus, Å. Niemi, P. Ciszewski, S. Schulz, and W. Kaiser. 1986. The changing pelagic ecosystem of the Baltic Sea. *Ophelia Suppl.* **4**:299–319.

Yoshida, N., H. Morimoto, M. Hirano, I. Koike, S. Matsuo, E. Wada, T. Saino, and A. Hattori. 1989. Nitrification rates and ^{15}N abundances of N_2O and NO_3- in the western North Pacific. *Nature* **342**:895–897.

Environmental Capacity of the Ocean Margins: Reality or Myth?

M.D. Krom[1] and Y. Cohen

National Institute of Oceanography
Israel Oceanographic and Limnological Research
POB 8030, Haifa, Israel

Abstract. The concept of environmental capacity (EC) has been put forward as a scientific basis for allowing controlled discharge of wastes into the environment rather than zero discharge. It is defined as a property of the environment and as its ability to accommodate a particular activity or rate of activity without unacceptable impact. A detailed description of the steps necessary to carry out an EC estimation is presented together with five case studies. Estimating EC using a predictive model is complex, expensive, and as yet unreliable. To date the endpoints which have most commonly been used for EC estimates are critical pathways to humankind, because unlike marine ecosystem effects, which are difficult to identify and quantify, it is relatively easy to assign values for the levels of pollutant exposure which are harmful to humans. Mariculture is expected to become important both as a means to develop more sophisticated endpoints for marine organisms and as an endpoint in itself. The actual endpoint chosen is a social decision albeit within a scientific framework. It is concluded that EC is a valid scientific concept which can be useful in defining the framework for management decisions; however, it is too complex and expensive to use as a basis for legislation.

INTRODUCTION

Ocean margins have been used throughout human history as a depository for waste material. Industrial society is producing increasing amounts of wastes which include not only those that have been produced throughout civilization such as building rubble and sewage but also new and potentially

[1] Presently on sabbatical leave at: University of Rhode Island, Graduate School of Oceanography, Narragansett Bay Campus, Narragansett, RI 02882-1197, U.S.A.

Ocean Margin Processes in Global Change
Edited by R.F.C. Mantoura, J.-M. Martin and R. Wollast

toxic materials such as artificial radionuclides and synthetic organic compounds (e.g., PCBs or plastics) which have never been present in the environment before. Even if direct ocean dumping of industrial and domestic wastes is banned, as has been legislated to occur off the coastline of the U.S.A. by Dec. 31, 1991, vast amounts of these wastes will continue to reach the ocean margins from rivers, point sources such as outfalls and storm sewage systems, nonpoint sources, and atmospheric inputs. In addition, one can expect the stress on the marine environment to increase in the coming years despite the growing impacts of environmental groups, since society is only just beginning to curtail the production of wastes within the industrial process and to reduce the amount of packaging material for the customer. With the growing quantities of waste material worldwide, waste disposal options will have to be selected on the basis of multimedia analysis. As our land area becomes more utilized and as fresh water becomes a scarcer resource, there will be renewed pressure to use the ocean margins for waste disposal. This process may be enhanced in small countries with limited freshwater resources, where waste disposal on land may endanger groundwater aquifers. However, it can also be expected to occur within the foreseeable future even in the industrialized countries of the northern hemisphere, where until recently unlimited quantities of unpolluted fresh water had seemed inexhaustible.

A number of scientists have examined the problem of waste disposal in the ocean with the aim of producing a rational basis for permitting controlled discharge of wastes rather than nominal zero discharge. One of the most frequent concepts brought forward in such discussions is that of "assimilative capacity." This concept was first proposed by Cairns (1977) to mean the ability of an ecosystem to cope with certain levels of waste discharges without suffering any significant deleterious biological effects. The concept was discussed in depth at the Crystal Mountain workshop on "The Assimilative Capacity of U.S. Coastal Waters for Pollutants," held in July, 1979 (Goldberg et al. 1979). The concept brought together a number of ideas which had been developed in separate fields of marine pollution, particularly the critical pathway approach developed to quantify the limits allowable for artificial radionuclide disposal such that public health is maintained (ICRP 1966; Preston 1971) as well as studies of bacteria, trace metals, and organic compounds such as PCBs. The ideas put forward during this workshop provoked lively debate within the scientific community (e.g., Stebbing 1981; Kamlet 1981). At that time, it was unclear, however, what the practical implications of the concept were. As a result, several authors set out to define the concept more precisely and to critique its use as a basis for management decisions (Kester et al. 1981; Krom 1986).

The authors of a recent report (GESAMP 1986) took this idea one stage further and provided a detailed discussion of how to put the concept of EC into practice. In this review we will use the GESAMP report and other work to define exactly what is presently meant by EC and will provide several case studies showing how it is applied in practice. Having reviewed the existing studies, we set out to put the concept of EC in a wider context in order to determine the reasons why the concept, though true in theory, has not been applied in practice. In this chapter we intend to present some of the practical problems which have been encountered in trying to carry out EC estimations and in defining what is an unacceptable level of pollutant in the system. In carrying out this review we have noted several areas in which the developing industry of mariculture interacts with and will provide relevant data that will help in solving marine pollution problems in general and EC estimations in particular. Although it is clear to most marine scientists that the oceans do have a finite capacity to accommodate wastes, the concept of EC has rarely been used either as a basis for scientific studies or for regulating waste disposal in the marine environment. We therefore ask whether for all practical purposes the concept of EC is a "Reality or Myth."

DEFINITION OF TERMS

The assimilative capacity of a body of water was defined by Goldberg (1979) as "the amount of material that could be contained within a body of seawater without producing unacceptable biological impacts." This amount was determined by a "titration" of the pollutant with the water body to an "endpoint," that is a point at which a definable endpoint is seen. Krom (1986) argued that the use of the word assimilation was unfortunate, since the term to assimilate means to "make like or similar" and it is implied that the receiving system ultimately benefits from the assimilation. That is obviously invalid if one is considering the input of pollutants into a body of water. As a result, the term "accommodate capacity" was proposed as an alternative to assimilative capacity. GESAMP (1986) also did not use the term assimilative capacity, choosing in its place the term environmental capacity. They defined EC as "a property of the environment and . . . as its ability to accommodate a particular activity or rate of activity without unacceptable impact." These terms as well as receiving capacity, are essentially interchangeable. We have chosen to use the term EC throughout for the reasons stated above and also because the report of the GESAMP group represents the most comprehensive and up to date review of the subject presently available.

STEPS INVOLVED IN AN EC ESTIMATE

GESAMP (1986) set out a set of guidelines for the scientific assessment of the impact of pollutants on the marine environment. The purpose of this summary is to indicate in outline what are the steps involved in a complete estimate of EC as described in the GESAMP report. If, however, an EC estimate is to be carried out in practice, the original (and more comprehensive) report should be consulted (GESAMP 1986).

Definition of the Project and/or Problem

As a first step it is necessary to specify the nature of the proposed project including the quantity and composition of the discharge. The rate of supply and any expected variability of the discharge should be known or approximated. The time scale of the proposed development should be stated together with expected changes in any other interacting activities likely to add additional stresses to the environment.

Estimation of the EC

Definition of the boundary conditions. It is necessary to define the boundary of the area being considered. This will depend not only on the geographic boundaries but also on the hydrodynamics of the system, taking into account the expected residence times of the pollutant components in the system.

Identification of possible targets. This could include human population and/or all or part of the marine ecosystem. Marine organism targets might include specific organisms such as adults, during sensitive stages of their life cycle or their reproductive potential. Rare, endangered species or critical habitats must be emphasised.

Identification of pathway from the pollutant to the target at risk. The possible pathways of the pollutant to the target organism or ecosystem should be identified and quantified. Where practical, a critical pathway approach should be used to identify the most probable route by which the pollutant(s) will reach and affect the target(s). It may be necessary to follow several pathways to several targets in order to establish which is the most sensitive.

Definition of the endpoint. Where appropriate existing effluent standards, water quality criteria, or other specific standards should be used directly. Where none exist these must be derived from similar situations elsewhere or generated from toxicity testing under appropriate conditions.

Estimation of EC. A model should be constructed to relate the inputs to the effects. This might be a simple conceptual model or a more complex model requiring numerical or probabalistic approaches.

Determination of acceptable discharge rates. The allowable input rate can now be calculated based on the EC estimate.

Design and treatment options. One must determine whether the proposed project is capable of reaching the target level or input rate specified, using available technology at a reasonable cost.

Decision phase. On the basis of the EC estimate, it will be possible to offer scientific advice to the regulatory agency and/or developers on what is needed to protect the marine environment. If the expected discharges cannot meet the concentrations or rates defined, alternatives will need to be considered.

Monitoring and Validation of the Estimation

After the proposed activity has begun, the effects on the environment should be assessed to ascertain whether the predictions were correct. The monitoring, which is thus goal oriented, should determine the flux of the pollutant in the discharge and how they vary with time. The concentration of the pollutant(s) in the biotic and abiotic components of the ecosystem should be measured. Specific components of the biota within the expected exposure plume and at control areas outside it should be sampled. Changes in the numbers, biomass, and species composition should be noted to determine whether there have been significant changes outside the expected natural level of variation.

A schematic illustration of EC was used by Kester et al. (1981) to identify the principle features of the concept (Fig. 1). Using this graphical representation, the EC for a given system is the intercept of the environmental response and the critical response value. Three possible response curves are shown for a given range of contaminant input flux. The delayed response (case 2) is the most insidious since inputs may occur with little effect initially (and appearing to be linear) and then after some threshold value a small further increase will result in a large effect. Clearly it is thus important to determine the response curve over the entire range of the pollutant flux to evaluate the nature of the curve.

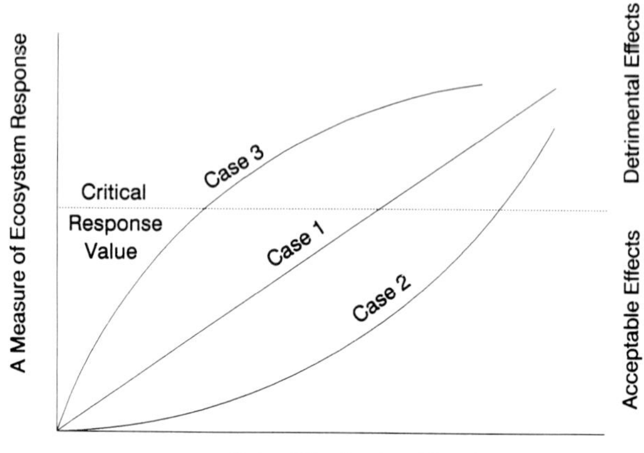

Fig. 1—A schematic illustration of the relationship between ecosystem response and the contaminant input rate. The critical response value corresponds to the endpoint in an EC estimation. After Kester et al. (1981).

CASE STUDIES WHICH USE THE EC APPROACH

Radioactive Waste Discharged into the Irish Sea from Sellafield

The disposal of radioactive effluent into the sea of Great Britain has been considered quantitatively using a "Critical Pathway to Man Approach" (Preston and Jeffries 1969a, b, Pentreath 1980). Generally, it is relatively simple to monitor the input of man-made radionuclides into a given body of water. There are no natural sources of these compounds and the input of these materials began only after 1950. They can be determined in the laboratory fairly easily at environmental concentrations. By far the largest source of liquid radioactive effluent in the U.K. (and in fact the world) is the Sellafield facility in northwest England. The flux of this material discharged into the Irish Sea was well known (Pentreath 1980). The endpoint considered was the increase in the risk of contracting cancer as the result of exposure to this effluent for a critical population.

In the EC estimation for radioactive input into the Irish Sea, the dose limit was set somewhat arbitrarily assuming an additional risk of 10^{-5} to 10^{-6} human years above the normal incidence of cancer was acceptable and the ingestion value was made to be 10 to 20% of this annual dose. In the original critical pathway calculation, a small group of Welshmen who consumed porphyra (a seaweed made into a pudding called laver bread) were taken as the critical population. Once the annual consumption of laver bread by the highest consumers was determined, it was possible to estimate whether this critical population was receiving an unacceptable dose of total radioactivity or not. On the basis of this calculation it was found that the EC of the Irish Sea had not been exceeded and no additional pollution abatement measures were required.

Radionuclide Disposal in the Deep Ocean

Thus far the only case where the concept of EC has been used as a basis for drawing up specific regulations for marine waste disposal is for the disposal of certain high level radionuclides in the deep ocean (IAEA 1986). Thus although strictly speaking this example does not involve the ocean margins, it is nevertheless worth presenting in brief summary since it shows what would be necessary to do to draw up regulations for the discharge of a pollutant to the ocean margins based on an EC estimation.

The oceanographic model used is a complex one as it attempts to include all the relevant biogeochemical processes (Fig. 2). It assumed that

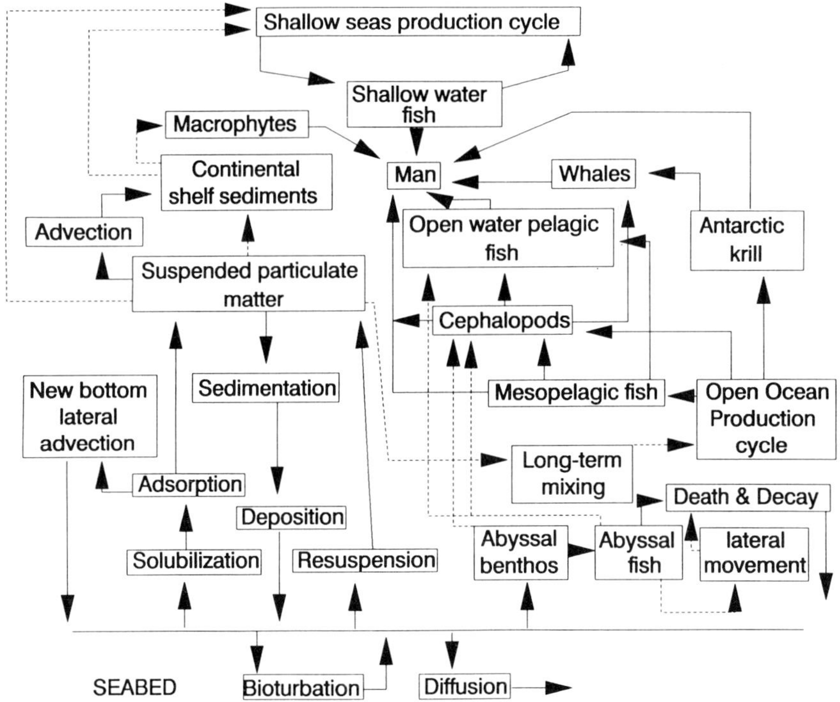

Fig. 2—A schematic illustration of the biogeochemical processes considered in constructing the oceanographic model used to determine the flux of radioactive waste which could be dumped into the deep sea. Unbroken arrows show known pathways and broken arrows pathways that possibly exist (IAEA 1978).

radionuclides would be disposed of in containers which would reach the deep-sea sediments intact and would instantaneously release all the pollutant. For each transfer process, an equation (or set of equations) was defined and a reasonable value (and range) taken for the appropriate coefficients (Table 1). Then for each of 106 radionuclides, the concentration changes for each step in a series of possible pathways was calculated. A value for the exposure rate to humans was assumed for each of these pathways (Table 2) and hence the amount received by the critical population from each exposure route could be calculated. Finally a dominant pathway(s) for each radionuclide was predicted and a limit suggested on the amount which could be dumped in the ocean in a given time. The details of the model are given in GESAMP (1983) and the results of the calculation presented in IAEA (1986).

This example is by far the most detailed attempt made so far to calculate the EC of a body of water. Despite the effort made and the regulations which were drawn up as a result, no subsequent deep-sea disposal of high

TABLE 1. The exposure pathways considered in the model used to define the critical pathways to humans for radionuclides in the deep sea (IAEA 1986).

Pathway	Intake rate or occupancy time
Actual Pathway	*Ingestion rate*
Surface fish consumption	600 $g.d.^{-1}$
Mid-depth fish consumption	600 $g.d.^{-1}$
Crustacea consumption	300 $g.d.^{-1}$
Mollusc consumption	100 $g.d.^{-1}$
Seaweed consumption	100 $g.d.^{-1}$
Salt consumption	3 $g.d.^{-1}$
Desalinated water consumption	2000 $g.d.^{-1}$
	Inhalation rate
Suspended airborne sediments	23 $m^3.d^{-1}$
Marine aerosols	23 $m^3.d^{-1}$
	Occupancy times
Boating	5000 $h.a^{-1}$
Swimming	300 $h.a^{-1}$
Beach sediments	2000 $h.a^{-1}$
Deep-sea mining	500 $h.a^{-1}$
Hypothetical Pathways	
Deep fish consumption	60 $g.d^{-1}$
Plankton consumption	3 $g.d^{-1}$

TABLE 2. Radionuclide independent parameters used in the oceanographic model in the deep-sea EC estimation.

	Parameter	Best estimated value	Values for sensitivity analysis
K_h	horizontal eddy diffusivity in water column	10^3 m^2.s^{-1}	10^2, 10^4
K_v	vertical eddy diffusivity in water column	10^{-4} m^2.s^{-1}	10^{-5}, 10^{-3}
V	ocean volume	10^{17} m^3	
H	ocean depth	4×10^3 m	
w	upwelling velocity	10^{-7} m.s^{-1}	3×10^{-8}, 3×10^{-7}
fw_p	volume flux of particles	10^{-13} m.s^{-1}	2×10^{-13}, 5×10^{-14}
w_p	vertical speed of particles	10^{-5} m.s^{-1}	
f	volume of particles	10^{-8}	
$f'w_s$	volume of accumulation of sediment	4×10 m.s^{-1} 8×10^{-14}	2×10^{-14}
f'	volume of sediments	0.4	0.2, 0.6
w_s	net sediment	10^{-13} m.s^{-1}	
K_b	efficient vertical diffusivity in bioturbation layer	3×10^{-13} m^2.s^{-1} 10^{-13}	10^{-12}
h	depth of bioturbated layer	0.1 m	0.05, 0.15
ρ_s	ρ_p density of sediments and particulates (dry)	1.5 t.m^{-3}	
K_{pw}	pore water vertical diffusivity	2×10^{-10} m^2.s^{-1} 1×10^{-10}	4×10^{-10}
r_s	radius of source	5×10^4 m	

level radioactive waste has been carried out. This is due at least in part to the lack of confidence scientists (and the general public) have in the reliability of such models to make predictions. It is generally agreed that the deep ocean and its floor are relatively homogeneous and stable areas. By contrast, the ocean margins are dynamic and highly variable. Predictive models which deal with pollution inputs into the ocean margins are thus even more difficult to develop and uncertain in their predictions.

Mercury in Liverpool Bay

One of the first attempts to apply the EC concept to a nonradioactive pollutant was applied to Liverpool Bay (and the Thames Estuary) by Preston and Portmann (1981). A critical pathway to man was considered in this calculation. In this case the mercury content in commercial flatfish (flounder and plaice), taken from the areas of the Irish Sea adjacent to Liverpool Bay, was used. The amounts of the various inputs of mercury to the relevant area of the Irish Sea were estimated. These were from the River Mersey flowing into Liverpool Bay (1 kg/d), direct sewage outfall (0.8 kg/d), dumping of sewage sludge (3.0 kg/d), dumping of dredge spoil from the Manchester ship canal (1.6 kg/d), and direct industrial discharges principally from chloralkali plants in the area which were 16 kg/d in 1972 but had been reduced to 5.0 kg/d by 1979. The average concentration of mercury in the flesh of flatfish collected in the area in 1972 was 0.44 mg/kg. Using data for the natural content of mercury in flatfish of 0.2 mg/kg, the excess due to pollution was calculated. By assuming that zero anthropogenic flux resulted in zero excess, a straight line could be drawn between the single point representing the 1972 data and the zero concentration–zero flux origin. No attempt was made in this study to estimate the different bioavailabilities and speciation of mercury from these different sources. Using this relationship, it was found that the level of mercury in flatfish taken in Liverpool Bay between 1976 and 1979 was predicted reasonably well and this simple relationship was therefore considered usable in estimating flatfish mercury concentrations from discharge rates.

In order to estimate the EC for Liverpool Bay, an acceptable level of mercury in fish flesh destined for human use was chosen. Originally a decision was made to set the Environmental Quality Standard (EQS) at 0.3 mg/kg because Preston and Portmann believed this was the lowest value which could be distinguished by analytical means to be above the ambient uncontaminated value of 0.2 mg/kg. It was also shown to be significantly lower than the 1 mg/kg concentration that was estimated to cause physical symptoms in 5% of the exposed population. Subsequently this limit for mercury in flatfish was confirmed as acceptable based on the FAO/WHO recommended provisional weekly intake of 0.3 mg, a survey of the dietary habits of extreme fish consumers in the area and the estimated mercury uptake rate from fish eaten into the blood of the consumer (GESAMP 1986). This study concluded that by 1980 the inputs of mercury to Liverpool Bay would be within the EC of the system. This was principally due to the reduction in inputs from the adjacent chloralkali plants, which had been instituted independent of the results of this study.

Mercury in Haifa Bay, Israel

More recently, the EC of Haifa Bay, on the Mediterranean coast of Israel has been estimated with respect to inputs of mercury. From 1956 until 1975, a chloralkali plant discharged effluent to the Bay without pollution abatement measures with a maximum discharge in 1975 of 5 kg/d. From 1976, the input flux was drastically reduced until in 1981 the present (1988) level of 0.1 kg/d was reached. Despite a considerable amount of monitoring data available from the area, it was not possible to relate any specific changes in the marine ecosystem structure to this particular pollutant input. This was even though it had been shown that the genetic variability of several species of invertebrates in the area had decreased as a direct result of the mercury input (Nevo et al. 1984). Thus a marine ecosystem endpoint could not be used for this calculation.

A critical pathway to man was used: the mercury content in the flesh of *Diplodus sargus*, an important commercial species. This species was chosen from a list of 28 species of fish sampled because it was the only species in which there was a significant accumulation of mercury in its flesh. As in other fish species, it has been shown that mercury accumulates in fish flesh as a function of weight (age). A series of linear regression lines of mercury content vs. weight were created for fish of the size range 40–175 g. The calculated concentration of mercury in the flesh of a "standard 150 g fish" was found to decrease with time in the polluted area, although the present (1988) level of 0.6 mg/kg was still significantly greater than the level of 0.15 mg/kg in adjacent unpolluted areas. This was attributed to elevated levels of mercury in the sediment and benthic food chain. A simple linear regression ($n = 6$) was used to relate the mercury content in the muscle of *Diplodus sargus* and the average mercury input to the Bay over the preceding three years which is the estimated age of a 150 g fish (Fig. 3).

As in the Preston and Portmann (1981) study, no data were available for the food consumption habits of a local critical population. The endpoints used were thus chosen based on such studies made elsewhere. If the endpoint is taken as 1 mg/kg (USFDA 1969), then the level of mercury in the bay is now considered acceptable. If, however, the action level was taken as 0.3 mg/kg which is the level chosen by GESAMP (1986) in their mercury pollution example, then despite the installation of pollution abatement measures in 1976, the present mercury levels in the bay would still be considered to be unacceptably high.

Environmental Capacity of the Krka River Estuary, Yugoslavia

All of the examples given thus far use a critical pathway to man as the endpoint for the EC estimation, although many examples exist where pollution abatement measures have been installed to reduce observed marine ecosystem effects. To our knowledge there is only one example published where a marine ecosystem limit has been used in an actual EC calculation. That was in the estimation of the EC of the Krka river estuary in Yugoslavia with respect to copper inputs (Pravdic and Juracic 1988). This estuary is

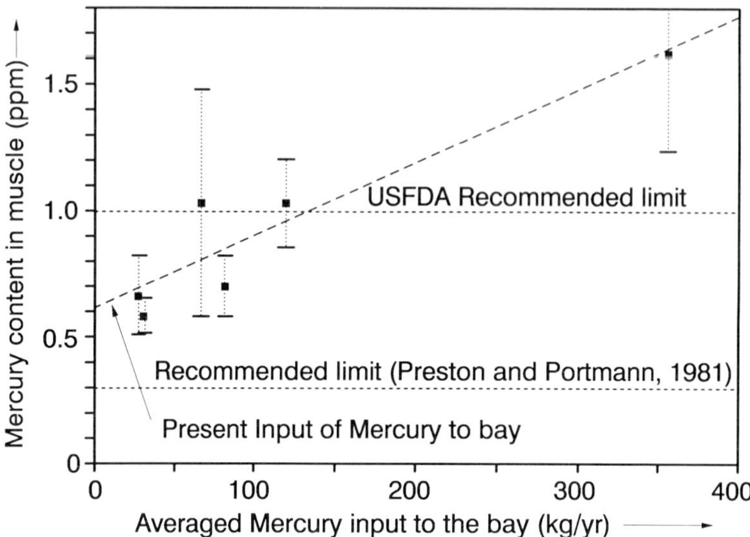

Fig. 3—The relationship determined between the ecosystem response (mercury content in the muscle of *Diplodus sargus*) and the contaminant input (average mercury input over the previous three years) used in the estimation of the EC of Haifa Bay, Israel. The error bars given represent 95% confidence limits.

situated on the eastern Middle Adriatic coast of Yugoslavia. The inputs of copper into the estuary were from a metal works at the upstream town of Knin and from copper sulfate which was used as a pesticide in the wine-growing regions in the river catchment area. Both of these inputs were considered for the purposes of the EC estimate to be unregulatable anthropogenic inputs. The third and potentially controllable input was copper leached from the antifouling paints on boats moored in the estuary.

The main activity in the area which required high quality water was the mariculture of salmonids. The juvenile stages of salmonids are sensitive to the presence of many toxicants, including copper. The choice of this sensitive target organism implies that other species constituting the natural ecological balance of the area would be protected as well. Water quality standards have been established in Yugoslavia based on primary scientific literature and the recommendations of WHO and FAO. This legislation specifies a concentration of copper of 0.01 mg/dm^3 in waters intended for salmonid aquaculture. This value was accepted as the endpoint in the estimate of the EC of the estuary. The aim of the calculation was to advise on how much copper released from antifouling paint could be accommodated within the system.

To determine this, a simple mass balance model was developed. The model included both the basic hydrology of the estuary and an estimate of the sedimentation rate of copper contaminated particles within the system. It did not, however, attempt to include all the biogeochemical processes which are known to be important in controlling the concentration of copper in an estuary or its chemical speciation (which is important in controlling its toxicity to salmonids). The total usable EC was estimated from the model as 4.7 kg/d. A safety factor of 0.2 was applied to allow for uncertainties in the calculation resulting in calculated EC for copper as the contaminant and salmonids as the critical target of 0.94 kg/d. Finally it was suggested that 20% of the unused capacity could be apportioned for the use of the boats, which resulted in a calculated maximum of 190 boats being allowed to moor in the estuary during any 24-hour period.

THE ENVIRONMENTAL CAPACITY ESTIMATION

In reviewing the case studies of EC estimations, both those given here and others considered have almost all been carried out a posteriori. The principal reason seems to be the problems involved in developing a valid predictive model to relate the input flux to the concentration in or at the target species. The models which need to be used are extremely complex if all the relevant major processes are to be taken into account (e.g., GESAMP 1983). It is an expensive and often lengthy procedure to carry out the relevant research needed to parametize all the relevant physical, chemical, biological, and geological variables needed in such a model. One of the only examples where such a procedure has been carried out is in the disposal of nuclear wastes in the deep ocean, which is thought to be a relatively stable and homogeneous environment compared to most of the ocean margins. Even when such models have been developed, they are often considered unreliable if used as predictive tools.

Thus most existing EC estimates involve an existing pollution input and typically the question asked is, can we change (generally increase) the existing inputs to the area and not have unacceptable effects? Often the input has already changed over a considerable range and the environmental consequences have been monitored (Preston and Portmann 1981; Krom et al. 1990). It is generally assumed that the existing (often simple empirical) relationship between the pollutant input and the concentration at the target will remain valid. The possible problems of a delayed abrupt response (Case 2, Fig. 1), such as has been suggested as being a potential long-term problem (Kester et al. 1981), are generally ignored. Clearly these constraints put severe limits on the usefulness of the EC concept.

CHOOSING THE TYPE OF ENDPOINT TO BE USED

Most case studies that have carried out EC estimates have used a critical pathway to man approach (Preston and Portmann 1981; GESAMP 1986; Krom et al. 1990). One major reason for this is that relatively speaking we have values for the levels of pollutant exposure which are harmful to humans. Often these values are the subject of specific regulations or recommendations of national and/or international committees (e.g. ICRP). However, as marine scientists we have to consider the protection of the marine environment and not just the health of a land mammal, *Homo sapiens*. Ideally we would like to quantify the effects of the entire suite of contaminants on the entire marine ecosystem which is under stress. That is generally a practical impossibility. It is often a problem to even be sure if an observed change in the marine ecosystem structure is due to pollution at all unless a large perturbation has occurred (Gray et al. 1980). Marine organisms typically have very high fecundity and may have large swings in population numbers as a result of natural predator–prey cycles. Even in the situation where a clear environmental degradation can be observed, it may be difficult to relate it quantitatively to a known input of a specific pollutant or suite of pollutants.

One approach that might be chosen to solve this problem is to choose a critical pathway approach to a particularly commercially important or sensitive organism and to target the estimation to that organism, as suggested by GESAMP (1986) and carried out by Pravdic and Juracic (1988). Most often the only quantified data available is the acute LC50 for one pollutant to the adult (or most easily cultured) stage. If, as is often the case in the environment, two or more pollutants are acting on the target organism simultaneously, the experimental design becomes rapidly more complex (and costly) to carry out. It is generally desirable to determine the level at which chronic effects appear rather than to wait for death. It is difficult to conduct such experiments even for single pollutants in part because the endpoint to be used is often not obvious. While overall growth rate, scope for growth, or feeding rate of the organism may be the most convenient to determine (Bayne et al. 1980), fecundity or the genetic integrity of the offspring might be the most important effect environmentally. Furthermore conducting a long-term chronic test may be extremely complicated if the pollutants involved or their interactions in nature have short-term variability such as diurnal changes in oxygen concentration or tidal variability.

This is an area in which mariculture may have an important part to play in the future. Mariculture, which is expected to become a major industry in the coming years, both generates waste products and is sensitive to exogenous pollution. As part of this commercial development, a detailed knowledge of the effects of a whole variety of pollutants to the cultured organisms is very important. An almost unlimited quantity of organisms is available to the toxicologist at little or no cost to conduct experimental tests. A detailed knowledge of the variability of control groups will also be available. The experiments will be carried out because of the almost immediate commercial benefits to the industry. The first sets of experiments will probably be carried out on the effects of pollutants such as ammonia, oxygen, and nitrite produced as a result of the mariculture operation itself. Others will involve the effects of therapeutic agents such as copper sulfate. There will also be a need to conduct experiments on more general pollutants which will limit mariculture operations in what would be otherwise suitable sites.

We also expect mariculture to contribute to more sophisticated measurements for the effects of low levels of pollutants on marine organisms. At present, most mariculture operations only see water quality problems when a mass mortality of the cultured organism occurs. Clearly measurements of environmental stress, prior to death, are desired. The modern trend in mariculture, particularly in the western countries, is for high technology, intensive culture, both in sea cages and in the potentially less polluting ponds and tanks. In such systems a more sophisticated knowledge of the biochemistry and physiology of the cultured organism is required if they are to be commercially successful. This trend is similar to that which occurred in the poultry industry over recent years. From such detailed knowledge of each organism it should be possible to develop biochemical indicators of stress based on some of the many modern advances in these areas of research. We predict that this will be the route that such advances will enter into marine pollution studies, much expanding the present range of biochemical indicators available. These effects can subsequently be used both as endpoints in experimental tests for the chronic effects of pollutants on marine organisms and eventually also in field studies on "wild" organisms.

Mariculture may also be important in providing the endpoint of EC estimates per se. In many locations such as Hong Kong (Holmes, pers. comm.), South Africa (Lord and Roberts, pers. comm.), and Scotland, the practice of sea cage mariculture is setting the limits for the discharge of a variety of effluents into marine waters. In the example from Yugoslavia given above (Pravdic and Juracic 1988), this relationship was used to carry out an EC estimate. In that study it was suggested that protecting the environment for salmonid culture would also protect the remainder of the marine ecosystem. Clearly, however, such an argument has its limits since the mariculture operation itself has an environmental effect.

IS THE ENDPOINT A SCIENTIFIC DECISION?

Thus far we have described some of the problems involved in making a scientific decision on determining the quantitative relationship between the flux of the contaminant to the environment and the measured effect. Determining the endpoint requires deciding where along this relationship the impact is no longer acceptable. This decision is not scientific but social, economic, and/or political, albeit within a scientific framework (Kester et al. 1981; GESAMP 1986). Science may be able to quantify the degree of environmental degradation caused by a given input of the contaminant, but it is unable to answer the question of what is unacceptable. Clearly the degree of environmental degradation that has been considered acceptable in the Hudson-Raritan estuary or the Thames estuary would be considered unacceptable for Puget Sound, Washington. While GESAMP (1986) suggested that the decision process might be aided by using probabilistic analysis derived from the procedures of decision analysis, the ultimate decision will not be that of scientists alone.

A worrying recent development is that despite all the work carried out in marine pollution research, the political decision-making process and resulting legislation is often totally at odds with the scientific understanding of the real environmental problems. A recent example was the ban on ocean dumping of wastes passed by the U.S. Congress in 1988. In the summer of 1986 and again in 1987 a small quantity of medical wastes, including hypodermic syringes and needles from New York clinics, washed up onto the New Jersey beaches. There was a great cry in the media about the danger of catching AIDS, and there was a drastic reduction in the numbers of people using the beaches, some of which were temporarily closed with a consequent loss in revenues for the local tradesmen (Walker and Paul 1989). Although it was shown by the relevant responsible scientist that this pollution incident was the result of overloading the storm sewer system in New York (which would cost billions of dollars to repair) and had nothing to do with the ocean dumping of sludge, the politicans felt they had to do something quickly and ocean dumping of sludge was banned, even though cost-effective alternatives to this practice may not exist. Another example is the pressure being exerted in the European Economic Community to pass legislation to control marine discharges based on a Uniform Emission Standard approach, because that would be "fair" to all polluters and all countries even though it is agreed by most environmental scientists that such an approach is not environmentally sound (B. Millemann, pers. comm.).

IS THE ENVIRONMENTAL CAPACITY A REALITY OR A MYTH?

In this context, there are two types of reality: one which represents a usable scientific theory or framework and one which involves the use of EC estimates directly in management planning and decisions.

It is our contention, as has been noted before, that EC is a scientific truth (Stebbing 1981). Clearly 10 g of mercury chloride could be poured into the center of the Pacific Gyre without causing an unacceptable impact, while all environmental scientists would say that the level of contamination of Minimata Bay, Japan was unacceptable. Several scientists have spent a considerable amount of time and effort defining all the parameters which need to be known if the EC of a given body of water is to be estimated (Goldberg et al. 1979; GESAMP 1983; GESAMP 1986). Indeed it could be argued that much of the research being carried out in marine pollution at present, as represented for example by the papers presented at the most recent International Ocean Disposal Symposium in Dubrovnik, Yugoslavia (IDOS 1989), are related in some way to the work which would be needed to carry out an EC estimate. To be a scientific reality, it is only necessary for the process to be valid and for the results to be in theory calculable.

By contrast, managers need specific answers to particular problems now. Managers need to set regulations which are economically reasonable and politically acceptable. In this context there are three major problems with EC estimations:

1. The cost and time involved in collecting sufficient data to make the estimation.
2. The large uncertainties involved particularly in those estimates which require complex predictive modeling.
3. The arbitrariness of the endpoint which is often chosen to justify economic or political reality and not scientific truth.

It is our contention that EC is a useful concept to design scientific studies. It can be used to quantify the effects of an existing discharge and to define areas where further detailed research is needed. It can often be used for designing monitoring studies by reducing wasteful undirected effort, the "if-it-moves-analyze-it" approach. Thus in a scientific sense it is a reality. For management purposes, the EC concept is at present a myth. While the ideas contained in the EC concept are useful in defining the framework for management decisions, it is too complex and expensive with our present knowledge to use to set regulations. Simpler site-specific concepts should and indeed are being used.

Acknowledgements. We would like to thank H. Hornung, D. Kester, H. Walker, K. Park, and M. Pilson for the useful discussions and suggestions made at various stages during the preparation of this manuscript and for the critical review provided by E. Goldberg. This work was supported by funds from Israel Oceanographic and Limnological Research and the Ministry of Energy and Infrastructure of Israel. M. Krom would also like to thank M. Pilson for the generous access given to his facilities granted during the later stages of completing this manuscript.

REFERENCES

Bayne, B.L. (chairman), J. Anderson, D. Engel, E. Gillfillan, D. Hoss, R. Lloyd, and F.D. Thunberg. 1980. Physiological techniques for measuring the biological effects of pollution in the sea. In: Biological Effects of Marine Pollution and the Problems of Monitoring, ed. A.D. McIntyre and J.B. Pearce, vol. 179, p. 88–99. International Council for the Exploration of the Sea, Rapports et Proces-Verbaux des Reunions. Copenhague K, Denmark.

Cairns, J. 1977. Quantification of biological integrity. In: The Integrity of Water, ed. R.K. Ballentine and L.J. Guarraia, pp. 171–187. Washington, D.C.: U.S.E.P.A.

GESAMP. 1983. IMO/FAO/UNESCO/IAEA/WHO/WMO/UN/UNEP Joint Group of Experts on the Scientific Aspects of Marine Pollution. An Oceanographic Model for the Dispersion of Wastes disposed of in the Deep Sea. Rep. Stud. GESAMP 19. Rome: FAO.

GESAMP. 1986. IMO/FAO/UNESCO/IAEA/WHO/WMO/UN/UNEP Joint Group of Experts on the Scientific Aspects of Marine Pollution. Environmental Capacity, An Approach to Marine Pollution Prevention. Rep. Stud. GEASAMP, 30. Vienna: IAEA.

Goldberg, E.D. 1979. Assimilative Capacity of U.S. Coastal Waters for Pollutants: Overview and Summary. In: Proc. of Workshop on Assimilative Capacity of U.S. Coastal Waters for pollutants at Crystal Mountain, WA, ed. E.D. Goldberg, pp. 1–7. U.S. Dept. of Commerce NOAA Working Paper No. 1.

Goldberg, E.D. et al. 1979. Proc. of Workshop on Assimilative Capacity of U.S. Coastal Waters for Pollutants at Crystal Mountain, WA, ed. E.D. Goldberg. U.S. Dept. of Commerce NOAA Working Paper No. 1.

Gray, J.S. (chairman), D. Boesch, C. Heip, A. Jones, J. Lassig, R. Vanderhorst, and D. Wolfe. 1980. Ecological working group report to monitoring of biological effects of pollution in the sea. In: Biological Effects of Marine Pollution and the Problems of Monitoring, ed. A.D. McIntyre and J.B. Pearce. International Council for the Exploration of the Sea, Rapports et Proces-Verbaux des Reunions, vol. 179, pp. 237–252. Copenhague K, Denmark.

IAEA. 1978. The oceanographic basis of the IAEA revised definition and recommendations concerning high-level radioactive waste unsuitable for dumping at sea. IAEA Technical Document No. 210. IAEA, Vienna, pp. 59.

IAEA. 1986. Definition and Recommendations for the Convention on the Prevention of Marine Pollution by Dumping of Wastes and Other Matter, 1972, pp. 73. Vienna: IAEA.

IDOS. 1989. Program and Abstract, Eighth International Ocean Disposal Symposium, at Dubrovnik, Yugoslavia, Oct. 9–13, 1989.

ICRP. 1966. Principles of environmental monitoring related to the handling of radioactive materials. In: Report of Committee of 4 of the International Commission on Radiological Protection. Oxford: Pergamon.

Kamlet, K.S. 1981. The Ocean as Waste Space: The Rebuttal. *Oceanus* **24**(1):10–18.

Kester, D.R., B.H. Ketchum, and P.K. Park. 1981. Future Prospects of Ocean Dumping. In: Ocean Dumping of Industrial Wastes, ed. B.H. Ketchum, D.R. Kester, and P.K. Park, pp. 505–517. New York: Plenum.

Krom, M.D. 1986. An evaluation of the concept of assimilative capacity as applied to marine waters. *Ambio* **15(4)**:208–214.

Krom, M.D., H. Hornung, and Y. Cohen. 1990. Determination of the environmental capacity of Haifa Bay with respect to mercury. *Marine Poll. Bull.* **27(7)**:349–354.

Nevo, E., R. Ben-Shlomo, and B. Lavie. 1984. Mercury selection of allozymes in marine organisms: prediction and verification in nature. Proceedings of the National Academy of Science, USA. **81**:1258–1259.

Pentreath, R.J. 1980. Nuclear Power, Man and the Environment. London: Taylor and Francis.

Pravdic, V. and M. Juracic. 1988. The environmental capacity approach to the control of marine pollution: the case of copper in the Krka river estuary. *Chem. Ecol.* **3**:105–117.

Preston, A. 1971. The United Kingdom approach to the application of ICRP standards to the controlled disposal of radioactive waste resulting from nuclear power programs. In: Environmental aspects of nuclear power stations, pp. 147–157. Vienna: IAEA.

Preston, A., and D.F. Jefferies. 1969a. The assessment of the principal public radiation exposure from, and the resulting control of, discharges of aqueous radioactive wastes from the U.K. Atomic Energy Authority Factory at Windscale, Cumberland. *Health Phys.* **16**:477–485.

Preston, A., and D.F. Jefferies. 1969b. The ICRP critical group concept in relation to the Windscale sea discharges. *Health Phys.* **16**:33–46.

Preston, A., and J.E. Portmann. 1981. Critical path analysis applied to the control of mercury inputs to U.K. coastal waters. *Environ. Poll.* Series (B) **2**:451–464.

Stebbing, A.R.D. 1981. Assimilative capacity. *Mar. Poll. Bull.* **12**(11):362–363.

USFDA. 1969. Action level for mercury in fish and shellfish. Federal Register, 39 (236, Pt. II), Dec. 1974: 42737–42739.

Walker, H.A., and J.F. Paul. 1989. Ocean dumping of sewage sludge. *Maritimes* **33**(2):15–17.

The Impact of Human Alterations of the Hydrological Cycle on Ocean Margins

Y. Halim

Department of Oceanography, Faculty of Science
Moharram Bey, Alexandria, Egypt

Abstract. Humans have manipulated the natural hydrological cycle of rivers both directly, by creating impoundments, and indirectly, through various land uses. The trend in river control for water resource recovery and the generation of hydropower has accelerated on all continents since the 1950s, regardless of the downstream impacts. Reduction in the upstream vegetation cover enhances soil erosion, leading to river siltation and increasing the danger of catastrophic floods. Reduced and stabilized flow downstream from reservoirs impairs contaminant flushing and directly affects biological cycles in the coastal zone. Where rivers are the major source of nutrients, productivity becomes drastically lowered and important fisheries are lost. Salt intrusion affects deltaic crops, damages mangrove ecosystems, causes stratification, and aggravates oxygen deficiency in estuaries. It also affects species such as shrimps, oysters, mussels, and anadromous fish that require brackish water for part or all of their life cycle. Trapping of sediments in reservoirs disrupts the coastal zone equilibrium, leading to land erosion and loss of valuable habitable space. Several cases are presented.

INTRODUCTION

Although human-induced alterations of river flow date back to ancient civilizations, the era of major dam-building activity did not begin until the early twentieth century. With the development of engineering technology, particularly concrete technology, the building of great multipurpose dams has accelerated since the 1950s (Fig. 1, Table 1). This worldwide trend was motivated by the pressing need for more water resources for industry and urban developments, the energy crisis, and the resulting need for alternative energy sources, particularly hydroelectric power. Successive years of severe drought in some developing countries, catastrophic floods in others, and the vital need for promoting agriculture and land reclamation induced these

Ocean Margin Processes in Global Change
Edited by R.F.C. Mantoura, J.-M. Martin and R. Wollast

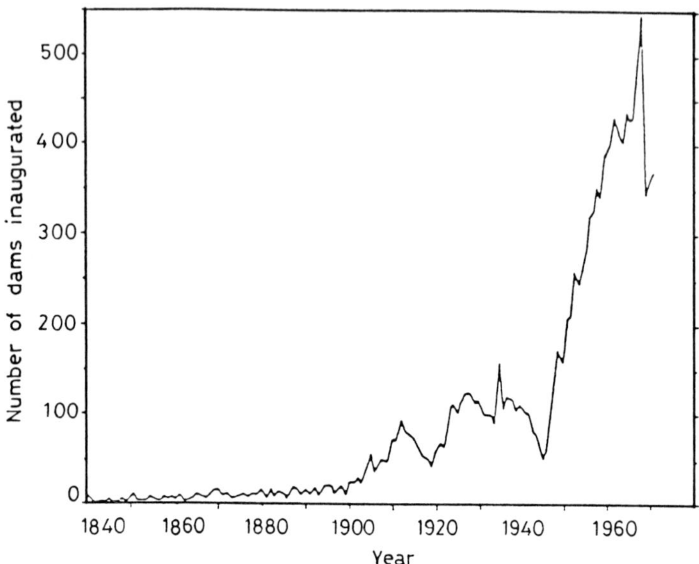

Fig. 1—World dam construction. Since 1971, the annual rate of dam building has not declined: for the countries represented in the 1973 Register, it has been maintained at about 400 per year. This figure is increased, however, to 700 per year if all countries are considered (Petts 1984).

countries to dam their rivers for water storage and regulation irrespective of downstream impacts. Almost all main rivers of Western and Central Africa have been or are planned to be dammed (UNEP 1985). The Nile waters are almost entirely withdrawn for various inland purposes. The Indus river is on its way to becoming entirely diverted for farmland irrigation. Rivers in developed countries are even more thoroughly harnessed; the Danube, Rhine, and Colorado are typical examples. Industrial countries developed hydropower to 30% of its potential in Europe, in North America to 20%. As a result of the increasing trend in river impoundment and artificial floodwater storage, the proportion of stable runoff has been augmented on every continent. In Africa and North America Petts (1984) estimates that about 20% of the stable runoff is contributed by impoundments. In Europe and Asia the figures are 15% and 14%, respectively, while rivers of South America (4.1%) and Austral-Asia (6.1%) are less affected. Croome et al. (1976) suggest that by the year 2000 about 66% of the world's total stream flow to the ocean margins will be controlled by dams.

Deforestation and various land uses have activated soil erosion. On a global scale the relation between erosion and human-induced soil erosion is not well known. According to some estimates (UNESCO 1982) the present

TABLE 1. The world's largest capacity reservoirs. Data from Mermel (1983) (Between brackets: under construction).

Name	Completion	Location	Capacity $10^9 \times m^3$
Owen Falls			204
Bratsk	1964	Angara, USSR	169
Aswan (High)	1970	Nile, Egypt	164
Kariba	1959	Zambezi, Zimbabwe	160
Akosombo	1966	Volta, Ghana	148
Daniel Johnson	1968	Manicouagan, Canada	141
Guri	(1986)	Caroni, Venezuela	136
Bennett	1967	Peace, Canada	74
Krasnoyarsk	1972	Yenisei, USSR	73
Zeya	1978	Zeya, USSR	68
Cabora Bassa	1974	Zambezi, Mozambique	64
La Grande 2	1982	La Grande, Canada	61
La Grande 3	1982	La Grande, Canada	60
Ust-Ilim	1977	Angara, USSR	59
Saratov	1967	Volga, USSR	58
Volga—V.I. Lenin	1955	Volga, USSR	58
Caniapiscau	1981	Caniapiscau, Canada	53
Chapeton	(1998)	Paraña, Argentina	53
Shintoyone	1973	Onyu, Japan	53
Upper Wainganga	(1987)	Wainganga, India	50
Bukhtarma	1960	Irtysh, USSR	49
Atatürk	(1990)	Euphrates, Turkey	48
Cerros	1973	Neuquen, Argentina	48
Irkutsk	1956	Angara, USSR	46
Tucurui	(1984)	Tocantins, Brazil	43

rate of erosion should be two and a half times the rate before humankind started to affect the landscape on a large scale. Although this figure is hypothetical, the increased rate of land erosion increases the sediment load of natural rivers and its subsequent deposition in river beds and deltas. The result is a reduced river gradient, increasing the danger of bank breaches and flooding, as happens with the rivers of China.

In the first part of this chapter, I will present an overview of some of the up- and downstream impacts brought about by human alterations of the natural hydrological cycles, both direct (e.g. impoundments) and indirect (changes in the vegetation cover). Several case studies from different continents illustrating the impacts on the ocean margins are then presented, followed by general conclusions.

IMPACTS OF HYDROLOGICAL ALTERATIONS

Among the most pervasive human effect on the hydrological regime and sediment load of rivers are the changes in vegetation cover and soil structure that have accompanied deforestation, intensive farming, and overgrazing. No less drastic and complex, however, are the impacts of river control by dams and reservoirs.

Alterations Resulting from Changes in the Vegetation Cover

Compared to natural conditions, most agricultural practices imply a reduction of vegetation cover at least during part of the year, including loss of surface litter and humus content. Humus content is related to temperature and humidity. Especially in tropical and subtropical regions cultivation will tend to reduce the humus content and increase erodibility. Due to the protection by the leaf canopy and ground cover of litter and vegetation, forests and woodlands are normally characterized by low surface runoff, high infiltration rates, and insignificant soil erosion. Forest soils often have a relatively porous structure which facilitates groundwater supply. When forest cover is removed, hydrological conditions change, and soil erosion, nutrient levels, and runoff become substantially increased. Bormann et al. (1974) reported on the export of particulate matter, erodibility, and the relative importance of dissolved substances to particulate matter in exported materials following deforestation at Hubbard Brook. Deforestation, accompanied by repression of growth by herbicides, increased export 15 times. The increase in export was exponential, slow in the first two years after cutting, rising sharply in the third year. The increases in particulate matter export are primarily due to increases in erodibility rather than in flow rates. The first response to deforestation is the mobilization of nutrients and leakage in stream water. While particulate matter rises sharply as biotic control of erodibility weakens, dissolved substance export declines, probably because of diminution of readily available nutrients stored in the system. The average ratio of annual net export of dissolved substance to particulate matter shifts from 2.3 before, to more than 8.0 after deforestation during the first two years.

Losses of nitrogen due to the clearing of tropical forests are higher than those for phosphorus and carbon (Van Bennekom and Salomons 1981). Lumbering of temperate forests leads to very high (more than 4000 μM) concentrations of inorganic nitrogen (nearly all in the form of nitrate) in stream water, corresponding to losses of 400 mM Nm^{-2} per year. This value, assumed to hold for two years after clearing, combined with annual deforestation of 0.5 to 1.6% of the 48.5 MKm^2 forested areas (Van Bennekom and Salomons 1981) gives an annual contribution from this source of 0.4 Tmol of dissolved nitrogen.

Alterations Resulting from River Impoundment

Dams interrupt the "natural" pattern of downstream transfers, the "river continuum." The discharges, the sediment and organic loads, and the water quality then become governed by the releases from the reservoirs.

Alterations of flow regime. Three types of flood regulation reservoirs are common (Petts 1984):

1. Retarding reservoirs (Fig. 2A) which are simple flood storage basins that release water through uncontrolled outlets. The flood peak will be reduced and displaced in time.
2. Detention reservoirs which have a storage volume larger than that of maximum flood. The flood peak is completely absorbed and the outflow is fully controlled (Fig. 2B).
3. Small capacity detention basins. An effective flow regulation is achieved by the prior evacuation of stored water. The flood peak is absorbed, the flow rate being slightly raised and spread over a longer duration (Fig. 2C).

Quality alterations of reservoir releases. The storage of water in open

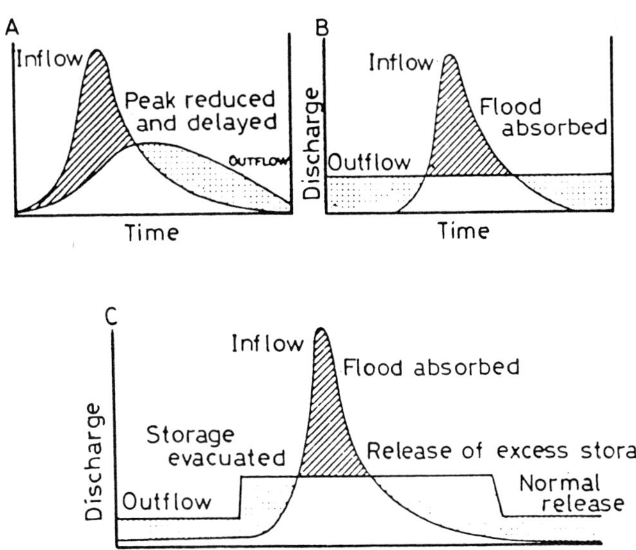

Fig. 2—Primary types of flow regulation: reservoir attenuation (A), reservoir storage (B), and release manipulation (C) (Petts 1984).

reservoirs induces physical, chemical, and biological changes within the stored water. Consequently, water discharged from impoundments can be of different composition and can show a different seasonal pattern from that of the natural river. A characteristic seasonal pattern of thermal behavior develops in reservoirs due to temperature–density differences in the water. The density gradient is poorly developed in spring but becomes increasingly well-defined as summer progresses.

The elevation of the outlet will determine the quality of releases from a stratified reservoir. Once a steady rate of outflow has been achieved, a relatively narrow layer of approximately constant density will be withdrawn so that the water quality of the outflow will vary considerably if releases are abstracted from the epilimnion or hypolimnion (Petts 1984).

During the summer, surface release from a stratified reservoir will discharge well-oxygenated, warm and nutrient-depleted water, while low-level outlets will produce relatively cold, oxygen-depleted and nutrient-rich releases which may contain high concentrations of iron, manganese, and hydrogen sulfide. Seasonal pulses are climatically induced and significant water quality changes below reservoirs can occur annually with the autumnal overturn. Sudden and extreme nutrient peaks have been reported from below Kariba Dam on the Zambezi River (Hall et al. 1977) and the Ucha Reservoir, USSR (Petts 1984).

Alterations of sediment transport. Reservoirs act as major sediment traps within the drainage basin of impounded rivers, although their trapping efficiency is variable. Early attempts to forecast a reservoir's trap efficiency (TE), the percentage of incoming sediment deposited within the reservoir, were based upon ratios between the reservoir storage capacity and the drainage area. The relationship (Brown 1944) is represented as:

$$\text{TE}\,(\%) = 100\,\{1 - 1/[1+0.1(\text{C/DA})\}$$

where C is the reservoir capacity and DA the drainage area.

The use of the ratio C/inflow (C/I) has been recognized as a more accurate index than C/DA (Brune 1953). The reservoir retention time has been identified as an important parameter in determining the trap efficiency. It has been estimated that short retention times (less than one day) commonly show trap efficiencies of between 40% and 90%. For longer retention times, trap efficiency usually exceeds 90%. Besides any sediment release from dams, the suspended sediment load for the river below a dam will also be derived from unimpounded tributaries, effluent outfalls, and the erosion of fine material from the river banks. The channel degradation and scour below dams also provide an important sediment supply for the river downstream.

CASE STUDIES

The Yellow River and the Yangtze in China

River control has long been a major issue in Chinese history and will remain so in the future. As a result of the uneven distribution of rainfall and of the disproportion between irrigation needs and water resources in China, there is a trend towards the excessive use of land and water resources at the expense of forests, pastures, rivers, and lakes. In return, flood and drought problems are aggravated, thus forming a vicious circle (Zhengying 1983). A tremendous amount of river engineering work has been carried out on numerous rivers, particularly on the seven major rivers (Table 2). Numerous storage projects of various sizes were constructed: 86,800 reservoirs, 6.4 million storage ponds of a total storage capacity of 410 km^3. Dikes along 16,000 km of rivers have been built or renovated (Zhengying 1983) (Table 3).

The Yellow River (The Huang Ho): a case of natural river displacement. Several radical changes have occurred in the Yellow River's course over the past 4000 years. At different times, the river has entered the Yellow Sea at points varying by as much as 500 miles. From 2278 B.C. to 602 B.C., it occupied its northernmost course, entering the Gulf of Bo Hai. From 602 B.C. to A.D. 70 both the river and its mouth shifted to the south of the Shantung Peninsula. From A.D. 70 to 1048 the Yellow River again shifted north, much along its present bed. In 1194 the river reoccupied its southernmost course. After two more shifts, it remained stable for more than 500 years until 1855, when it moved once more north of the Shantung Peninsula. Of the world's major rivers, the Yellow River carries by far the greatest load of suspended sediments (Fig. 3). Its annual runoff ranges from 20 km^3 (1960) to 86 km^3 (1964) (Zhengying 1983). Flowing through a loess plateau, the Yellow River erodes and transports an average annual sediment load of 1.6 billion tons. Much of this is deposited in the river channel below the loess plateau and in the estuarine zone. Aggradation of the channel and progradation of the delta reduce the river gradient, thus increasing deposition in the river bed, resulting in periodical breaches and subsequent displacements of the channel in history.

The Yangtze. The long-term average annual runoff of the Yangtze is 979 km^3. The main problems in this basin stem from the deterioration of soil and inadequate water conservation as a consequence of (a) deforestation in the upper reaches and (b) the narrowing of the river channel in its middle and lower reaches as a result of the reclamation of the floodplains with polders, intensifying the threat of flood volumes. The record large floods of 1931 and 1954 reached peak discharges exceeding 60,000 $m^3.S^{-1}$ and 3-

TABLE 2. Principal characteristics of seven major rivers in China (Zhengying 1983).

River	Catchment area ($10^3 km^2$)	Annual average runoff ($10^9 m^3$)	Annual average sediment load ($10^6 t$)	Water & soil loss area ($10^3 km^2$)	Population (10^6)	Cultivated land ($10^6 mu$)	Max. recorded annual runoff ($10^9 m^3$/year)	Min. annual recorded runoff ($10^9 m^3$/year)
The Yangtze	1,800	979	478	265.5	345	370	1360/1954	676/1978
The Yellow	750	56	1,640	391.3	82	196	86.1/1964	20.1/1960
The Huai	270	50	14	52.7	125	188	84.1/1954	6.3/1966
The Hai-Luan	320	29.2	81	123.8	98	170	45.8/1963	5.0/1920
The Pearl	450	341	69	35.6	76	78	529.2/1915	127.7/1969
The Liao	230	15.7	41	85.6	29	69	30.2/1954	4.7/1978
The Songhua	550	76		72.2	47	175	121.4/1960	45.1/1968
National total	9,600	2,600		1,203.4	1,031	1,500		

TABLE 3. The gross storage capacity of all large- and medium-size reservoirs as a percentage of annual average runoff of the respective river (Zhengying 1983).

84% Yellow river	11% Pearl river
72% Huai river	24% Songhua river
73% Hai-Luan river	82% Lia river
9% Yangtze river	

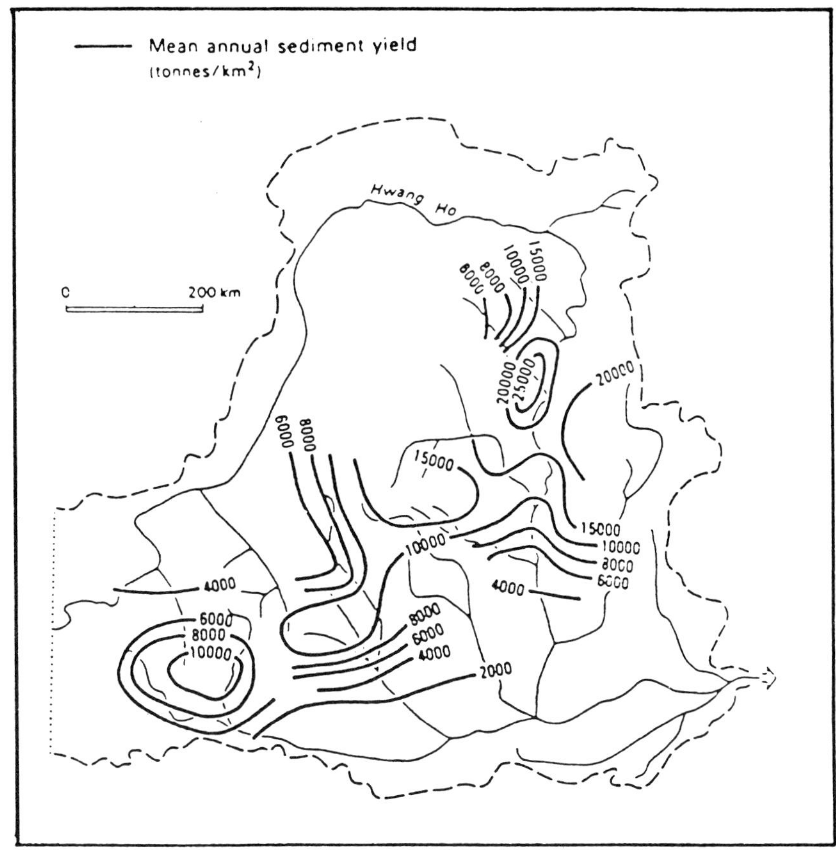

Fig. 3—Mean annual suspended sediment yields in the loess region of the middle reaches of the Yellow River (Hwang Ho), People's Republic of China (Hadley et al. 1985).

day flood volumes of over 20 km^3, largely exceeding the discharge capacity of the river and resulting in breach of dikes. The archives record 19 breaches of the Jingjiang dikes by the river from 1644 to 1949 (Zhengying 1983).

The Indus River Management

The Indus River is one of the world's largest rivers in terms of drainage area, river discharge, and sediment load. The Indus delta is located at the head of the Arabian Sea. Since the 1940s, engineering activities have greatly altered the discharge pattern of the Indus River (Milliman et al. 1984): two major dams have been built at Tarbela and Mangla; an extensive system of canals set up for farmland irrigation, barrages erected to control the river and divert its flow to channels (Fig. 4), and embankments constructed to restrict the river flow to the main channels.

Water resources development plans appear to be oriented towards the ultimate total capture and diversion of river water for irrigation. In some near future the Indus will no longer have a freshwater input to the delta area and Arabian Sea (Snedaker 1984). Such wide-scale alterations of the Indus have already had a drastic impact on both the river and its delta.

Before the human-made changes upstream, the Indus delta prograded annually at the rate of about 113 ft (Kazmi 1984). The active delta has now shrunk from approximately 1000 square miles to a small 100 square mile triangular zone.

Soon after completion of the dams, the sediment load of the river fell sharply. Though downdam erosion restored some of the load, it appears that sediment input to the estuary has decreased from a previous 300 million tons annually (according to Kazmi 1984) or 250 million tons (Milliman et al. 1984) to less than 100 mt in 1974–1975 (Milliman et al. 1984); however, marked yearly fluctuations persist. Whereas the water discharge in the 1950s exceeded 100 $km^3.y^{-1}$, it was frequently less than 60 $km^3.y^{-1}$ in the 1960–1970s. Embankments and levees, while preventing the flooding of the floodplain, resulted in an increased salinity in groundwater.

Flooded soils and saltwater intrusion appear to be a problem in the lower Indus delta. Much of the lower delta plain receives saltwater flooding from sea level elevation associated with the southwest monsoon winds in summer. Surface salt accumulation and effects of hypersalinity are presumably responsible for the partial deterioration of the rice and fishery industries (Snedaker 1984).

Observations of the mangrove forests of the Indus delta made in 1977 and in 1982 showed a progressive deterioration, undoubtedly associated with the reduction of the freshwater river flow (Snedaker 1984). Forested areas are now wholly restricted to the well-flushed banks of the tidal channels; the interiors of the deltaic islands appear bare. This process of deterioration

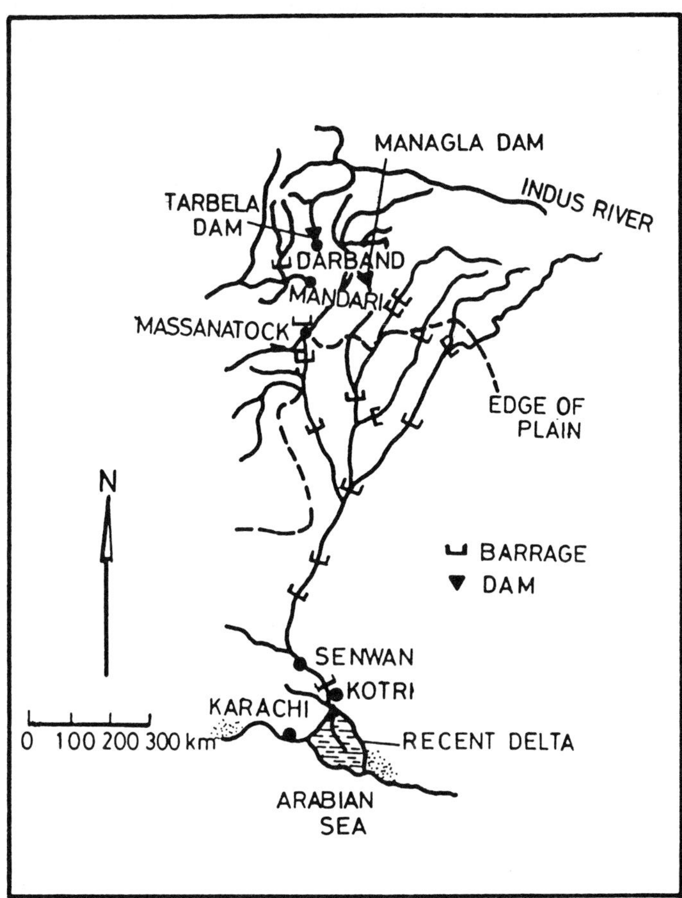

Fig. 4—The Indus River basin.

will eventually lead to the loss of the remnants of the mangrove ecosystems in the Indus delta. It should be noted, however, that the annual harvests of shrimp remain high in spite of the steady decrease in the area of mangrove forests (Snedaker 1984).

Effects of Rivers Management on the Black Sea and the Sea of Azov

River input. The Black Sea has a drainage area of 184,000 km^2. Most drainage comes from the eastern Russian platform; only 15% comes from the high mountain areas. Several large rivers empty into the Black Sea (Fig. 5 and Table 4).

Alterations in river inputs. The rivers draining into the northwestern and northeastern Black Sea (the Danube, Dneper, Dnester, Don, and Kuban rivers) have been largely manipulated since 1950. Human-induced changes were intended to serve several purposes: to generate hydropower; to provide inland water reserves for industrial, municipal, and agricultural needs; to accommodate river shipping. A series of hydropower stations were constructed creating large storage lakes.

The total area of the Dneper and Dnester storage lakes reached about 7160 km^2. The Don and Kuban reservoirs covered 2850 km^2. On the Danube, 28 hydroelectric plants have been built in Germany alone; however, the "Iron-Gates" plant in Yugoslavia/Romania is greater than all of them

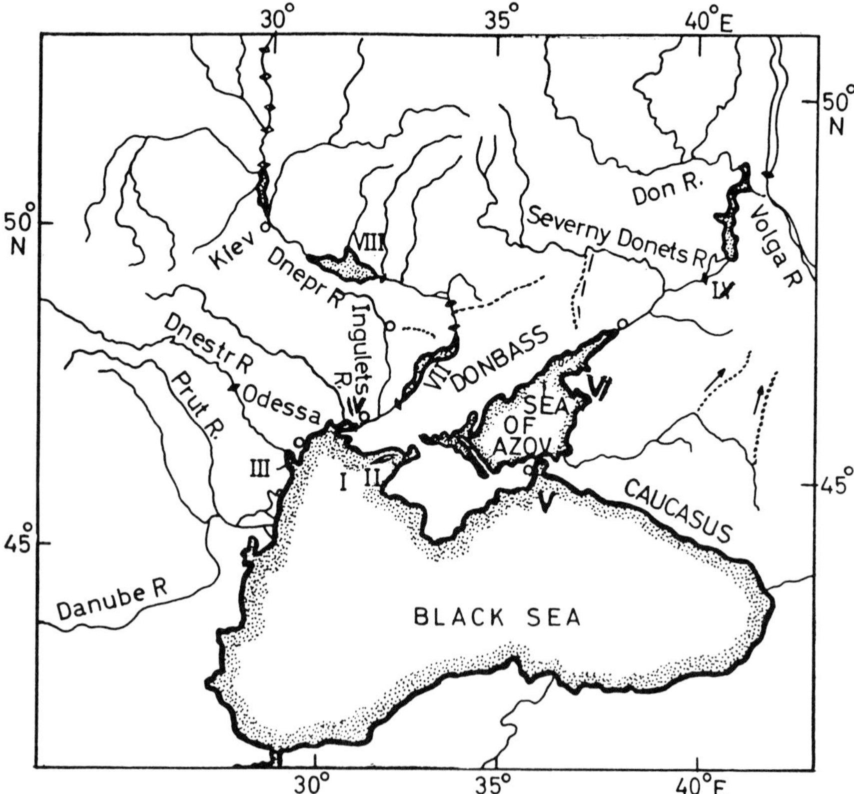

Fig. 5—Major rivers, estuarine regions, and associated geographical settings of the Black Sea. (I–IX) Water bodies: (I) Northwestern Black Sea, (II) Karkinitsky Zaliv (Bay), (III) Dnestr Estuary, (IV) Dneper Estuary, (V) Kerch Strait, (VI) Taganrogsky Zaliv (Bay), (VII) Kakhovskoye Vdkhr (Storage Lake), (VIII) Kremenchugskoye Vdkhr, (IX) Tsymlianskoye Vdkhr. Modified after Tolmazin 1985.

TABLE 4. Main rivers draining into the Black Sea: Basin areas, water and solids discharge and mechanical denudation (Ross 1977).

River	Basin Area (10^3km^2)	Water Discharge (km^3.y^{-1})	Solid Discharge (10^6 tons.y^{-1})	Denudation (t. km^2)
Danube	816	201	83	101
Dneper	503	53	2.02	5
Don	422	28	7.75	18.3
Rioni	13.4	13.5	8.5	633 (2000)
Dnester	72		2.5	31.5
Kuban				180

together. Projects to irrigate 2.6 million ha east of the Danube, if implemented, will require the withdrawal of about one third of the present river water flow.

Impact on the ecosystem of the Black Sea. The reduction in freshwater flow has induced salinity and density changes in the coastal areas of the Black Sea. In addition, the overloading of these river systems with industrial and sewage wastes has seriously depleted oxygen supplies.

Inland water management projects have produced a salinity increase of 0.19‰ in the Black Sea; the increase is more marked in the northwestern sector. In the immediate vicinity of the Dneper and Dnester estuaries, the increase is of 2.0–2.5‰. The size of the surface outflow to the Mediterranean Sea is decreased and the bottom inflow through the Bosphorus must be accentuated. The slow increase in the volume of the bottom anoxic layer in the Black Sea might, in time, have a drastic effect on this basin by gradually reducing the thickness of the productive and well-oxygenated surface layer.

Stratification and oxygen depletion. Oxygen depletion became a problem in the northwest Black Sea (NWBS) and the Sea of Azov as a result of the increased stratification in estuaries and of the input of organic materials from irrigated lands. Oxygen depletion in the NWBS was a serious problem during 1973–1975, following an intensified irrigation program in the Dneper and Dnester basins. As a result, mussel fields were destroyed (Tolmazin 1985) and bottom fish were driven toward the surface, where they eventually died washed up on the beaches.

Fish catches. The Black Sea has historically been one of the most biologically productive regions in the world. The productivity of this area, however, has been sharply curtailed by human-induced changes in the environment. Major fishing industries have suffered a drastic decline. The catch of bottom fish in NWBS, such as flounder, dropped from 80,000 tons before 1949 to only 4,000 tons in 1971–1975. On the whole, the total catch decreased from 300,000 tons to about 100,000 tons in 1971–1975. The decline of major fish stocks is attributable to a reduction in the brackish water environments where anadromous species reproduce (Tolmazin 1985).

Impact of water withdrawal on the Sea of Azov. The Sea of Azov—the most productive low salinity area of the Black Sea—provides strong evidence that freshwater inflow plays a major role in maintaining the biological productivity of the sea and its estuarine systems. Water withdrawals from the Sea of Azov have grown as high as 46% (Fig. 6). The total losses of freshwater supply between 1950 and 1975 account for almost 250 km^3 or about 11 $km^3.y^{-1}$. This sustained trend in declining water supply had a negative impact on physical properties and biological productivity (Rozengurt et al. 1985; Tolmazin 1985): increased salt intrusion resulting in a rise in mean salinity from 9 to 14–16 ppt; a 60% reduction in primary and secondary productivity, and over 95% reduction in catches of anadromous fish (Table 5); a massive intrusion of billions of medusae into the Sea of Azov and the Don River delta (Fig. 7).

Management of the Nile River and Its Effects on the East Mediterranean

Human intervention in the flow of the Nile dates back to pharaonic times. Modern intervention began with the construction of the "Delta Barrage" near Cairo in 1861.

The pre-damming Nile regime. The High Aswan Dam was completed in 1965, and thus the summer of 1964 saw the last "natural" flood discharge to the East Mediterranean. Until then, the Nile flow followed a rather regular two-phase pattern: the floodwave with its heavy load of suspended

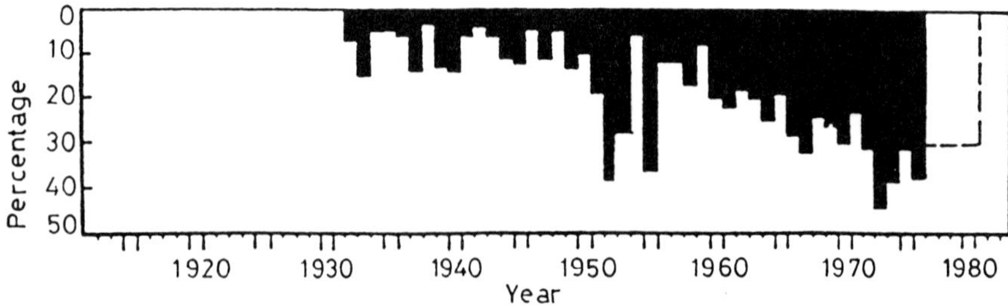

Fig. 6—Freshwater diversions from the Don–Kuban river system expressed as the percentage of annual natural river inflow to the Sea of Azov (Rozengurt et al. 1985).

TABLE 5. Annual catches in thousand hundred weights (10^5 kg) and extraction of caviar in the Sea of Azov (Tolmazin 1985).

Years	Pike-perch	Bream	Roach	Sturgeon Family	Vimba and Shamaya	Herring	Caviar
1921–1936	244.6	446.1	n/a	29.5	4.88	52.0	2.0
1948–1952	96.4	76.1	n/a	15.7	3.67	6.3	1.0
1954–1961	19.9	29.9	38.6 (in 1955)	4.3	0.90	3.8	0.1
1962–1963	13.1	19.5	n/a	4.4	0.45	0.49	n/a
1973–1975	n/a	3.2	n/a	1.5	n/a	occasional	n/a

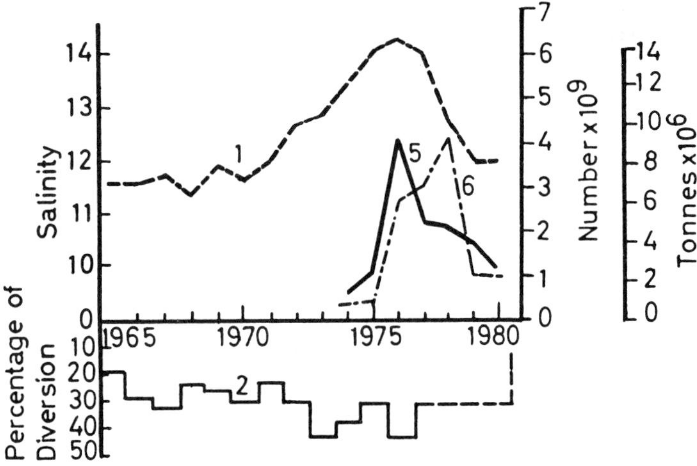

Fig. 7—Population explosion of the marine jellyfish (Aurelia) inside the formerly brackish Sea of Azov as a result of increased freshwater diversions and the resulting rise in salinity concentrations. (1) Annual average salinity, (2) combined average annual freshwater diversions expressed as a percentage of the natural runoff of the Sea of Azov, (5) raw weight of jellyfish (*Aurelia aurita* and *Rhizostoma*) in millions of tons, (6) combined number of jellyfish in billions. Modified after Rozengurt et al. (1985).

solids reaching southern Egypt in June–July, rose to a peak by the end of September and dropped to its low winter level by December. To prevent salt intrusion during the low-level phase, the two river outlets at Rosetta and Damietta were kept tightly closed by means of a weir-dam on the former and an earthen dam on the latter. Both dams were then opened in mid-August to allow the excess floodwater to flow to the sea.

Large yearly fluctuations in the total freshwater discharge have been recorded over the past centuries (Hsu, this volume) following fluctuations in the rainfall over the Ethiopian highlands and Ugandan plateau. The average discharge for five years (1959–1963) amounted to $42.9 \times 10^9 m^3$. In 1964, it was higher: $52.89 \times 10^9 m^3$. The Nile coastal current progressed rapidly eastward and then northward along the coast of Asia Minor, its northward extension depending upon the height of the floodwave. Large amounts of nutrient salts were brought down to the oligotrophic southeast Mediterranean basin with the floodwaters, becoming distributed over a wide area by the flood stream. Floodwater, 3 km upstream from the outlet (Halim 1960) contained 340 μM silicate and 6.4 μM phosphate. Assuming an average outflow of $43 \times 10^9 m^3$, the amount of Si and P brought down to the sea during the flood season would be about 410,000 tons of Si and 8,260 tons of P. These figures, however, appear to be an underestimate, as has been shown by Halim and Morcos (1966), since gradual dilution of the spreading stream waters was accompanied by the release of silt-adsorbed nutrients. The fertilizing effect of the floodwater, both on land and in the sea, was mostly due to its suspended material. The sudden release of the floodwater triggered a thick and continuous diatom bloom in the surrounding seawater; the floodwater itself remained devoid of plankton owing to its opacity. Soon after the onset of the bloom along the edge of the Nile delta, large shoals of sardinella (*S. eba*, *S. aurita*, and *S. granigera*) were attracted to the coastal belt to feed on the dense diatom crop. The stomach contents of the fish in this season were entirely composed of diatoms. The sardinella catch accounted for 30 to 40% of the total sea fish landings of the year. The shrimp fisheries were next in importance. Spawning for many benthic organisms such as echinoderms, polychaetes, and lamellibranchs appears to have been induced by (or at least to coincide with) the Nile bloom, their pelagic larvae contributing a large proportion to the zooplankton crop in this season.

The sediment load. The average annual suspended sediment load for the period 1904–1963 at Gaafra, 1100 km upstream from the outlets, is estimated to have been about 160 million ton yr^{-1}, ranging from 50 to 300 in 1913 and 1954 respectively. It was higher in the nineteenth century with an estimated average of 200 million tons yr^{-1} from 1825 to 1902 (UNESCO/UNDP 1978). Major deposition sites were the irrigated farmlands, the reservoirs, and the river bed. It appears, however, that some 130 to 140 million tons of sediment per year reached the Mediterranean through both outlets in the period 1903–1963.

The post-damming conditions. The gradual filling of the reservoir behind the dam and full regulation of the river became complete in 1968–1969. The new Lake Nasser behind the dam has a storage volume larger than that of the maximum flood and serves as a detention reservoir. The floodwave is completely absorbed and the outflow is now fully controlled (Fig. 8). Some surplus freshwater, however, about 2.5 to 4 × $10^9 m^3$, still reaches the Mediterranean, but the suspended sediment load in its totality is deposited on the lake bottom upstream from the dam (Fig. 9). Subsequent surveys of the continental shelf waters (Emara et al. 1973; Mostafa 1985), carried out during the autumn, show conditions in sharp contrast with the past (Fig. 10). The phytoplankton standing crop in autumn, 1965, dropped to about 5% of its corresponding value in 1964 (Guerguess 1970) and the total fish landings have been steadily decreasing since. By 1969 the yield was less than 20% of its 1962–1963 value (Fig. 11). The effect on the sardinella fisheries was more pronounced, since this fishery depended directly on the Nile bloom. The rise in fish landings since 1978 is only due to an increased fishing effort and to improved fishing methods.

Shoreline changes. On a longer timescale, the Nile Delta cone was building up with steady uniformity; on a shorter timescale, it was oscillating around a state of unstable equilibrium caused by fluctuations in sediment discharges. Human intervention changed this pattern into one of erosion. Changes in the coastline of the Nile Delta can be followed from topographic surveys since early 1800 supported by recent drilling data (UNESCO/UNDP 1978). Until about 1910 there was active progradation of the two subdeltas and

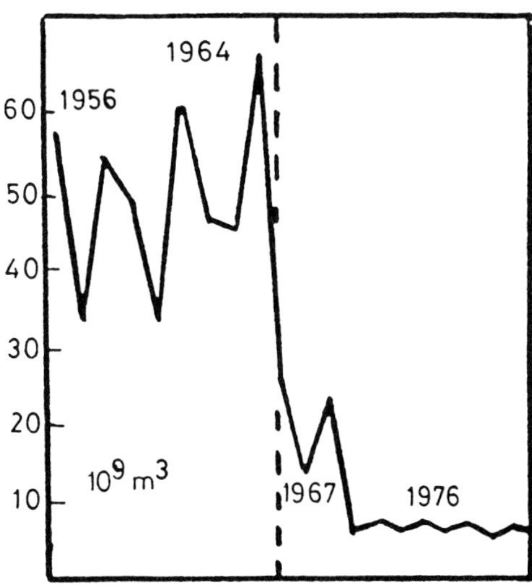

Fig. 8—Total water intake of the two Nile branches, pre- (1956–1965) and post-impoundment (1965–1976). Data from the Ministry of Irrigation, Cairo.

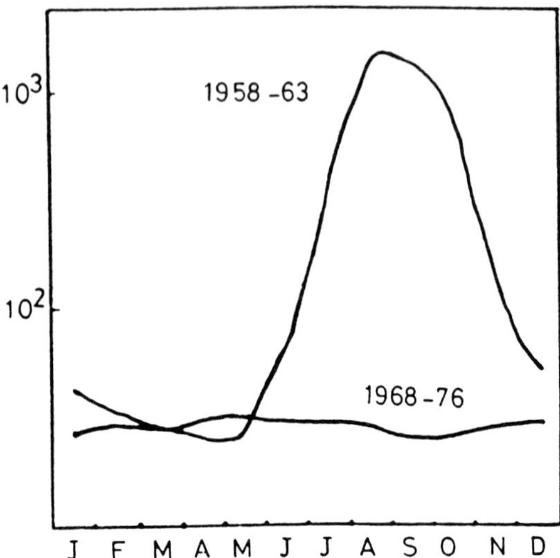

Fig. 9—Average monthly suspended load of the Nile, downstream of Aswan, pre- (1958–1963) and post-impoundment (1968–1976) in ppm. Data from the Ministry of Irrigation, Cairo.

their offshore silty-mud cones. An overall 25% decrease in suspended load since, due to the decrease in rainfall over Ethiopia, was followed by a general retreat. Erosion at Rosetta from 1915 to 1964 (UNESCO/UNDP 1978) was of the order of 30 m a year. With the absence of dynamic readjustments, promontories were fast retreating and the embayments advancing. Since 1964 the Rosetta promontory has eroded at a progressively accelerating rate, as shown by successive surveys, being 125 m y^{-1} and 211 m y^{-1}, respectively, in 1973 and 1982 (Fig. 12).

The High Aswan Dam: towards a balance sheet. Any balance sheet of the positive and negative aftereffects of the High Aswan Dam will have to take into account at least three major benefits: the magnitude of the hydropower generated, the doubling of the farmland area, and, no less crucial, the regulation of water supply to the country despite a decade of severe drought in the Sahel–Sudan–Ethiopian belt. Life in the Nile Valley and Delta is entirely dependent on the Nile River as a water artery, rainfall being insignificant. The deficiency in the floodwater supply necessitated a continuous emergency withdrawal of large volumes of water from the reservoir to protect the country from catastrophic effects of drought. From 1978–1979 to 1984–1985, the total deficit compensated from the reservoir reached 73.5 km^3 (Fig. 13).

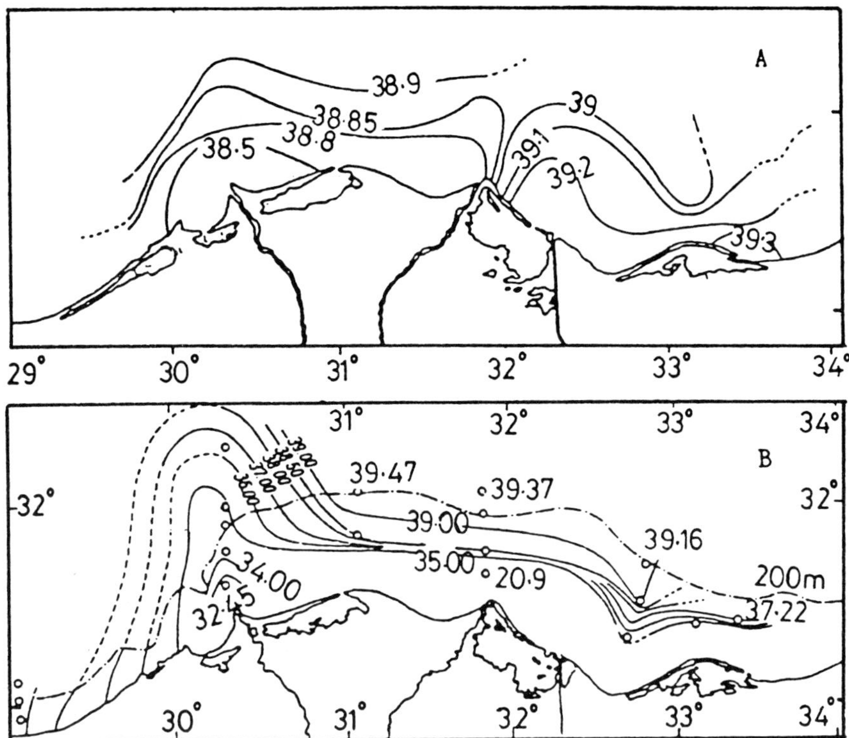

Fig. 10—Surface isohalines off the Nile delta before and after the High Aswan Dam became functional. A: November, 1982 (Mostafa 1985), B: October, 1964 (Halim et al. 1967).

Impacts of Water Diversions on the San Francisco Bay, Delta, and Estuary Ecosystems

Over the past 130 years, the San Francisco Bay and Delta, the largest inland estuary of the U.S.A. West Coast, has been altered greatly by land and water development. Upstream impoundments, diversions, and pumped exports from the Delta have reduced river flows by more than 50%. The long-term effects of the upstream impoundments, diversions, and removal are just beginning to be understood (Davoren and Ayres 1984; Rozengurt et al. 1985). Between 1950 and 1978, freshwater diversions amounted to a total of 286 km^3, 40 times the volume of the San Francisco Bay. For 28 years, therefore, an average of 10.0 $km^3\ yr^{-1}$ was withdrawn from river inflow to the Delta. This decline in water supply had a negative impact on both physical properties and biological productivity (Fig. 14):

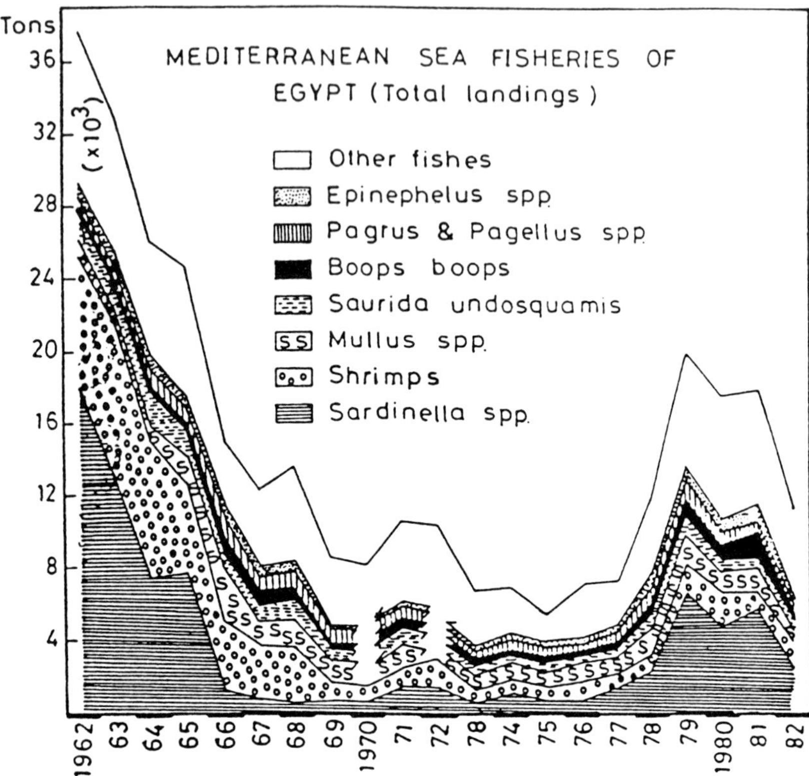

Fig. 11—Total fish and shrimp landings from Egyptian Mediterranean waters from 1962 to 1982 (Sh.K. Guerguess, pers. comm.).

1. Salt intrusion into the upper part of the Delta and Bay has increased, resulting in an increase of mean salinity in the Bay from about 20 to 27 ppt.
2. There has been massive reduction in the sediment load discharge to the delta–coastal zone ecosystem by 60–75% of the mean 8×10^6 tons discharged per year under natural runoff conditions.
3. Between 1967 and 1982, when reliable counts were made of winter salmon past Red Bluff Dam, the average annual volume of water diversion was approximately 12.2 km^3; cumulative withdrawals from the Sacramento–San Joaquin River supply to the estuarine system reached about 190 km^3 between 1967 and 1982. During the same period, the number of winter-run chinook salmon returning to spawn in the upper reaches of the river was reduced as much as 60 times, despite the release of millions of hatchery juveniles into the western Delta. Four factors have been proposed to explain the decline of the chinook salmon

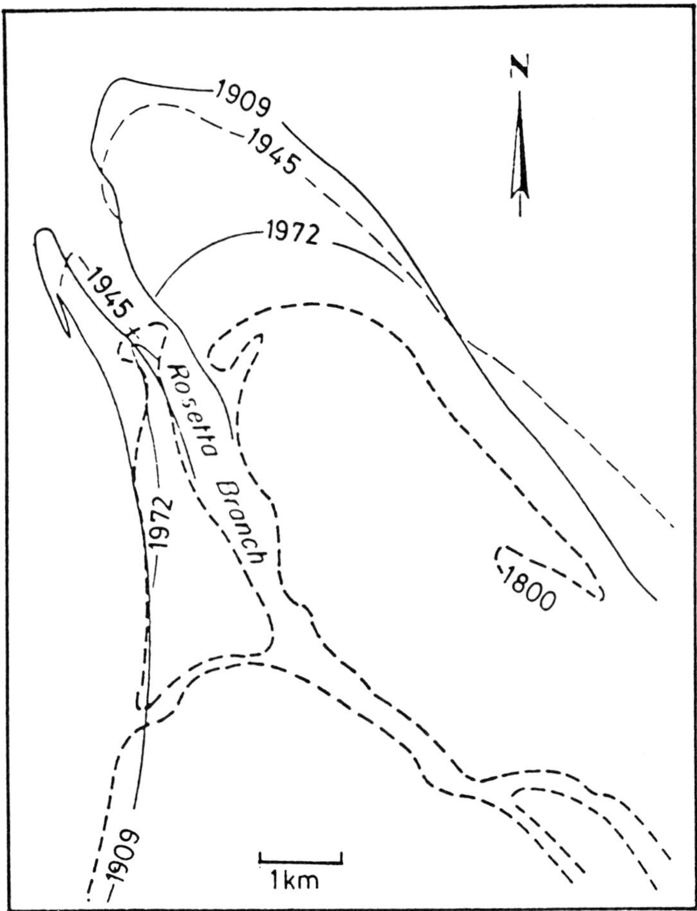

Fig. 12—Accretions and retreats of the Rosetta promontory since 1800 (UNESCO/UNDP 1978).

spawning population: dams, water diversions, pollutants, and the loss of 95% of their habitat. All of these factors may contribute to the reduction of the salmon population. Rozengurt et al. (1985) strongly suggest that the gradual increase of cumulative losses of water and nutrients resulting from diversions will continue to be the principal factor governing migration, spawning success, and recruitment in this stock for years to come. Nichols et al. (1986) conclude that the synergy between increased waste input and the reduction in freshwater inflow might be the critical factor in the deterioration of the estuary.

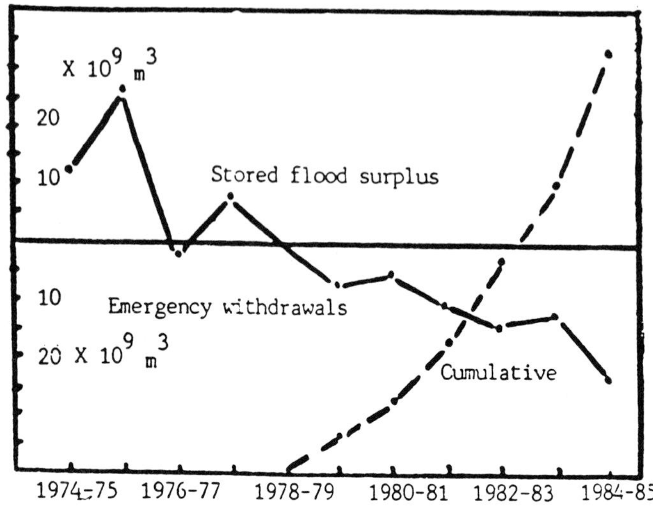

Fig. 13—Emergency withdrawals from lake Nasser reservoir to compensate for low flood input from 1976–1977 to 1984–1985. Data from the Ministry of Irrigation, Cairo.

CONCLUSIONS

Changes of considerable significance to the coastal zone can take place in a river regime due to natural causes. Wide displacements in the course of the Yellow River in China provide a striking example. Natural fluctuations of freshwater supply to estuaries within a given cycle can vary within 25% of 50 to 60 years average as shown for 126 rivers investigated by UNESCO (McMahon 1982). If diversions within a cycle do not exceed the natural average variability, the estuarine and coastal ecosystems with their natural resilience would survive the regulated water supply. In many parts of the world, however, massive river diversions have altered both the physical and biological environments. The periodicity and timing of the flow and the water quality are also major factors beside the flow volume.

The formation of river deltas and the dynamic stability of the coastline are determined by the higher rate of river sediment input versus wave and current erosion. Human intervention disrupts this pattern, enhancing erosion and consequently the loss of valuable habitable space. The Nile and the Indus deltas are obvious examples.

Flood control and reduced flow rate downstream from reservoirs may combine with the release of oxygen-poor water to exacerbate pollution problems. The flushing capacity for domestic and industrial effluents is greatly reduced.

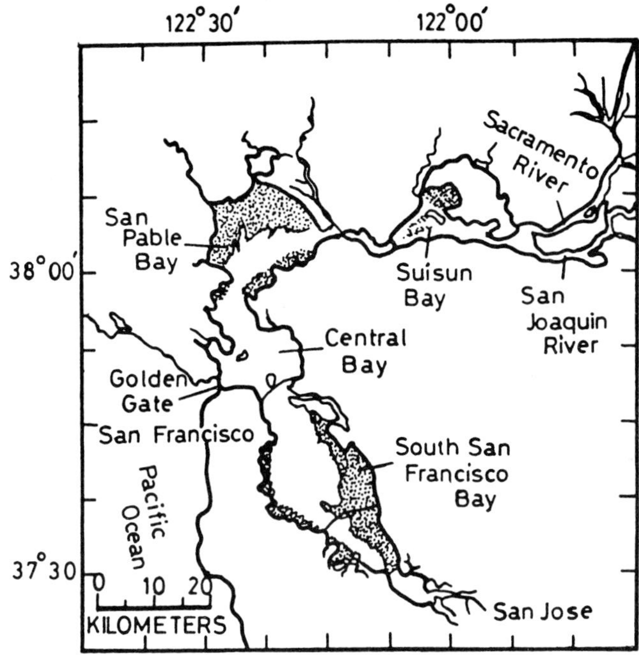

Fig. 14—San Francisco Bay and western Delta. Modified from Davoren and Ayres (1984).

The rise in salinity impacts deltaic, estuarine, and coastal ecosystems in various direct and indirect ways. Intrusion of the saltwater wedge upriver and the corresponding stratification of waters aggravate the effect of oxygen depletion in the subsurface. As a result mortality of bottom fish and the destruction of mussel fields occur in the Black Sea. Increased groundwater salinity is responsible for the deterioration of mangroves. Massive freshwater diversions combined with pollution lead to the decline of spawning salmon, striped bass, and shad populations in the San Francisco and the Sacramento–San Joaquin bays by more than 50%. With the rise of salinity, exotic species extend to the coastal zone. Massive invasions by jellyfish occurred in the Sea of Azov following freshwater withdrawal. The increase in predation, competition, and disease with increased salinity causes many euryhaline species to avoid higher salinity which they could otherwise tolerate. The occasional flushing of oyster beds with fresh water helps control predators such as *Thais haemastoma* and pathogens such as the fungus *Dermatocystidium marinum* (Copeland 1966). Alterations in the flow pattern of a river are followed by changes in the life cycle of coastal zone organisms, even when the total outflow is unchanged. Since the Cabora

Bassa Dam became functional in 1974, the peak runoff of the Zambeze River lasts for the period of December to July instead of January to April (Fig. 15). The longer duration of the flood is thought to shorten the residence period of juvenile shrimps in the estuary, inducing an earlier seaward migration. Consequently, the age at first capture and the catchable biomass are decreased (Da Silva 1986).

Productivity is enhanced in river plumes and in fronts between freshwater-driven coastal currents and the ocean. Salinity stratification in the Norwegian coastal current driven by freshwater runoff permits an earlier initiation of the spring bloom and prolongs the growth season by 20% in comparison to the adjoining oceanic water (Smetacek 1986). Major rivers influence large oceanic areas. During peak runoff, the influence of the Amazon plume, in terms of higher plankton biomass, is felt as far as Barbados (Kidd and Sanders 1979). Extensive regulation of river flow deprives estuaries and oceanic margins of the organic detrital material, nutrients, and growth-promoting factors usually brought down with runoff, depressing productivity

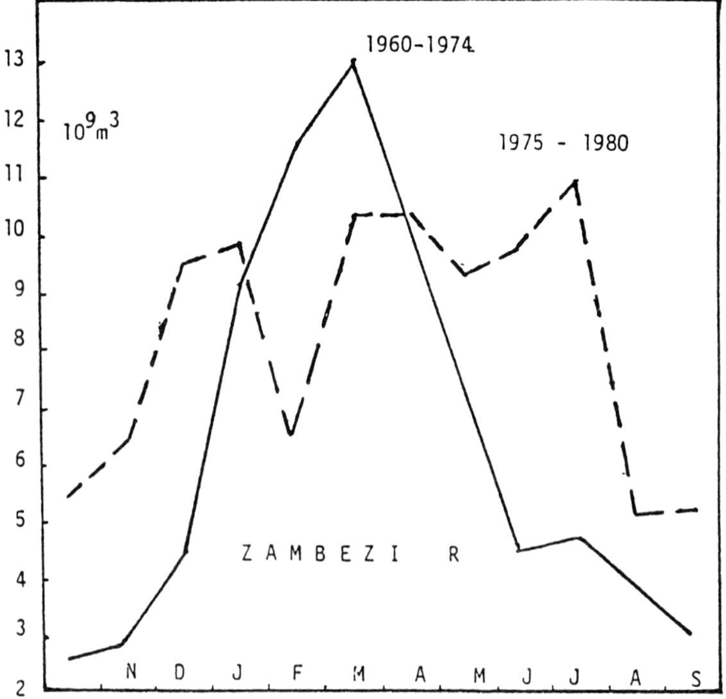

Fig. 15—Average monthly outflow of the Zambezi river past Matundo, below the Cabora Bassa Dam, before (1960–1974) and after damming (1975–1980) (Halim 1984).

at all levels. The case of the Nile River provides a clear-cut example. The response to runoff alterations, however, is complex and not everywhere linear. While the general pattern appears to confirm the "freshwater-driven food chain" of Sutcliff (as quoted by Sinclair et al. 1986), response will depend upon the relative importance of the runoff alterations to natural variability and to the prevailing oceanographic conditions in the receiving basin. In oligotrophic seas such as the Mediterranean, river runoff is the major source of nutrients. In the Gulf of St. Lawrence and the adjacent shelf, increased winter runoff due to regulation is believed to prevent nutrient convection from the underlying nutrient-rich layer. The regulation of the St. Lawrence system rivers is thought not to have had a measurable impact on fisheries production. If there has been an impact, it has been masked by the natural interannual variability of runoff (Sinclair et al. 1986).

In spite of the variability and the complexity of the impacts, there is diverse compelling evidence that inland manipulations of river hydrological cycles have adverse effects on oceanic margins. Since dams and water diversions are planned to benefit humankind in various ways, the downstream losses, usually overlooked, must be taken into account.

REFERENCES

Bormann, F.H., G.E. Likens, T.G. Siccama, R.S. Pierce, and J.S. Eaton. 1974. The export of nutrients and recovery of stable conditions following deforestation at Hubbard Brook. *Ecological Monographs* **44**:255–277.

Brown, C.B. 1944. Sedimentation in reservoirs. *Trans. Am. Soc. Civil Eng.* **109**:1085.

Brune, G.M. 1953. Trap efficiency of reservoirs. *Trans. Geophys. Union* **34**(3):407–18.

Copeland, B.J. 1966. Effects of decreased river flow on estuarine ecology. *J. Water Poll. Control Fed.* **38**(11):1831–1837.

Croome, R.L., P.A. Tyler, K.F. Walker, and W.D. Williams. 1976. A limnological survey of the River Murray in the Alburg-Wodonga area. *Search* **7**(1):14–17.

Da Silva, A.J. 1986. River runoff and shrimp abundance in a tropical coastal ecosystem. The example of Sofala Bank (Central Mozambique). In: The Role of Freshwater Outflow in Coastal Marine Ecosystems, ed. S. Skreslet. pp. 329–334, NATO ASI Series, vol. 67. Berlin, Heidelberg: Springer-Verlag.

Davoren, W.T., and J.E. Ayres. 1984. Fast and pending decisions controlling San Francisco Bay and Delta. *Wat. Sci. Tech.* **16**:667–676.

Emara, H.I., Y. Halim, and S.A. Morcos. 1973. Oxygen, phosphate and oxidizable organic matter in the Mediterranean waters along the Egyptian coast. *Rapp. Comm. Int. Mer Medit.* **21**(7):345–347.

Guergess, Sh.K. 1970. Zooplankton Studies in the UAR Mediterranean Waters with Special Reference to the Chaetognatha. M.Sc. Thesis, Faculty of Science, Alexandria, 263 pp., unpublished.

Hadley, R.F., R. Lal, C.A. Onstad, D.E. Walling, and A. Yair. 1985. Recent developments in erosion and sediment yield studies. Technical Documents in Hydrology, IHP, 127 pp. Paris: UNESCO.

Halim, Y. 1960. Observations on the Nile bloom of phytoplankton in the Mediterranean. *J. Conseil* **26**(1):57–67.

Halim, Y. 1984. Marine Sciences. People's Republic of Mozambique. Technical Report, 91 pp. Paris: UNESCO.

Halim, Y., Sh.K. Guerguess, and H.H. Saleh. 1967. Hydrographic conditions and plankton in the southeast Mediterranean during the last normal Nile flood (1964). *Int. Revue. Ges. Hydrobiol.* **52**(3):401–425.

Halim, Y., and S.A. Morcos. 1966. Le role des particules en suspension dans l'eau du Nil en crue dans la repartition des sels nutritifs au large de ses embouchures. Rapp. et Proc.-verb. *Comm. Inter. Explor. Mer Medit.* **18**(3):733–736.

Hall, A., I. Valente, and B.R. Davies. 1977. The Zambezi River in Mozambique: the physico-chemical status of the middle and lower Zambezi prior to the closure of the Cabora Bassa Dam. *Freshwater Biol.* **7**:187–206.

Kazmi, A.H. 1984. Geology of the Indus Delta. In: Marine Geology and Oceanography of Arabian Sea and Coastal Pakistan, ed. B.U. Haq and J.D. Milliman, pp. 71–84. New York: Van Nostrand Reinhold Co.

Kidd, R., and F. Sander. 1979. Influence of the Amazon River Discharge on the Marine Production System off Barbados, West Indies. *J. Marine Res.* **37**(4):669–681.

McMahon, T.A. 1982. Hydrological characteristics of selected rivers of the world. Technical Documents in Hydrology. I.H.P., 23 p. Paris: UNESCO.

Mermel, T.W. 1983. Major dams of the world, 1983. *Water, Power and Dam Const.* **35**(8):43.

Milliman, J.D., G.S. Quraishee, and M. Beg. 1984. Sediment discharge from the Indus River to the ocean: past, present and future. In: Marine Geology and Oceanography of Arabian Sea and Coastal Pakistan, ed. B.U. Haq and J.D. Milliman, pp. 65–70. New York: Van Nostrand Reinhold Co.

Mostafa, Hesham M.M. 1985. Phytoplankton production and biomass in the South East Mediterranean Waters off the Egyptian Coast. M.Sc. Thesis, Faculty of Science, Alexandria, 215 pp., unpublished.

Nichols, F.H., J.E. Cloern, S.N. Luoma, and D.H. Peterson. 1986. The modification of an estuary. *Science* **231**:567–73.

Petts, G.E. 1984. Impounded Rivers. Perspectives for Ecological Management, Chichester: John Wiley and Sons.

Ross, D.A. 1977. The Black Sea and the Sea of Azov. In: The Ocean Basins and Margins, ed. A.E.M. Nairn, W.H. Kanes, and F.G. Stehli, vol. 4A, pp. 445–481.

Rozengurt, M., M. Josslyn, and M. Hertz. 1985. The Impact of Water Diversions on the River–Delta–Estuary Sea Ecosystems of San Francisco Bay and the Sea of Azov. Estuary of the Month Seminar. San Francisco Bay. Washington, D.C.: NOAA Estuarine Programs Office.

Sinclair, M., G.L. Bugden, G.L. Tang, J.C. Therriault, and P.A. Yeats. 1986. Assessment of effects of freshwater runoff variations on fisheries production in coastal waters. In: The Role of Freshwater Outflow in Coastal Marine Ecosystems, ed. S. Skreslet, pp. 139–160, NATO ASI Series vol. 67. Berlin, Heidelberg: Springer-Verlag.

Smetacek, V.S. 1986. Impact of freshwater discharge on production and transfer of materials in the marine environment. In: The Role of Freshwater Outflow in Coastal Marine Ecosystems, ed. S. Skreslet, pp. 85–106, NATO ASI Series, vol. 67. Berlin, Heidelberg: Springer-Verlag.

Snedaker, S.C. 1984. Mangroves: a summary of knowledge with emphasis on Pakistan. In: Marine Geology and Oceanography of Arabian Sea and Coastal Pakistan, ed. B.U. Haq and J.D. Milliman, pp. 225–262. New York: Van Nostrand Reinhold Co.

Tolmazin, D. 1985. Economic impact on the riverine–estuarine environment of the USSR: the Black Sea Basin. *Geojournal* **11**:137–152.

UNEP. 1985. Coastal erosion in West and Central Africa. UNEP Regional Seas Reports and Studies. No. 67, 242 p. Geneva: UN.

UNESCO. 1982. Sedimentation problems in river basins. Report IHP Working Group 5.3 under the chairmanship of A. Sundberg, ed. W.R. White. Paris: UNESCO.

UNESCO/UNDP. 1978. Coastal Protection Studies. UNDP/EGY/73/063. Project Final Technical Report, vol. 1, 205 p. Paris: UNESCO.

Van Bennekom, A.J., and W. Salomons. 1981. Pathways of nutrients and organic matter from land to ocean through rivers. In: River Input to Ocean System, ed. J.M. Martin, J.D. Burton, and D. Eisma, pp. 35–51. Paris: UNEP/UNESCO.

Zhengying, Qian. 1983. The problems of river control in China. In: Proc. Second Int. Symp. on River Sedimentation. Nanjing, China, 11–16 October 1983, pp. 8–19. Chinese Soc. Hydraulic Eng. Beijing: Water Resources and Electric Power Press.

Morphological and Ecological Effects of Sea Level Rise: An Evaluation for the Western Wadden Sea

E.B. Peerbolte[1], W.D. Eysink[1], and P. Ruardij[2]

[1]*Delft Hydraulics*
8300 AD Emmeloord, The Netherlands

[2]*Netherlands Institute for Sea Research*
1790 AB Den Burg, The Netherlands

Abstract. The ecology of shallow coastal waters is partly determined by hydraulic and morphological parameters such as flow velocity, water depth, tidal amplitude, and seabed configuration. Increasing sea level rise will possibly lead to changes in the coastal morphology which may, in turn, influence certain ecological characteristics. This chapter deals with an evaluation of the sensitiveness of the ecosystem in tidal basins to morphological changes induced by sea level rise. Because the study was carried out in the context of the problem of climate change, the consequence of temperature rise on the ecosystem was considered as well. The study results are indicative and do not pretend to present detailed quantitative information in relation to the above changes. They rather show the possibilities of combining available scientific tools and the gaps in knowledge in this type of integrated analysis.

INTRODUCTION

On November 11, 1987, a memorandum of understanding between UNEP (United Nations Environment Programme) and the government of the Netherlands, marked the onset of a global program of studies related to the consequences of sea level rise and possible management or policy options needed to counteract it. This program is called ISOS (Impact of Sea level On Society) and is actually the continuation of an international workshop on the economic and social implications of sea level rise, held by Delft Hydraulics in August, 1986 (Wind 1987). Today the greenhouse issue is

Ocean Margin Processes in Global Change
Edited by R.F.C. Mantoura, J.-M. Martin and R. Wollast

being discussed widely in many national and international platforms, with a major emphasis on the problem of climate change.

Within the framework of the above memorandum, various studies and activities were undertaken. One regional project focused on the Netherlands, in cooperation with and partly sponsored by the Department of Public Works. The aim was to develop an ISOS model for the Netherlands, to be used as a tool for decision-making in the context of the sea level rise problem. The aspect of safety against flooding and coastal erosion, in relation to the cost of protection, formed the major subject of study. However, other effects of sea level rise and related climate effects have also been evaluated. The resulting model is now being used to assess a wide range of impacts and to develop optimal dike-raising programs. This includes, apart from dike-raising costs and coastal protection costs, impacts on activities in flood prone areas (industry, harbors, farming), impact on the water management system (pumping cost, crop damage due to increasing levels of salinity, drought damage due to temperature rise), and ecological impacts on the fresh- and saltwater systems. This chapter reflects the study which was undertaken to evaluate the impact of temperature rise and sea level rise on the ecosystem to estuaries, one of the physical subsystems of the ISOS model (Peerbolte et al. 1988). A principle diagram of the ISOS model is given in Fig. 1.

The subsystem "estuaries and coastal seas" consists of two major areas, viz. the Wadden Sea in the north of the country and the Delta Area in the southwest. These areas are shown on Fig. 1. The aim of the study was to analyze the effects of a 4°C temperature rise and 0.4 to 0.6 m sea level rise, representing a scenario for a change to evolve during the next 100 years.

It is emphasized that the results, although expressed in a quantitative sense, are mainly tentative or indicative. The morphological and ecological system is complex, and the present state of knowledge does not allow for accurate predictions over such a long time horizon. The morphological computations are in fact extrapolations of an evaluation of a small tidal basin in the western part of the Wadden Sea. In addition, the ecological results are based on a model for the western Wadden Sea that was originally developed for an area in the eastern part. In fact, the ecological results are not applicable to other areas in the subsystem of the ISOS model.

The analysis of the impacts on the ecosystem basically serves to assess the impacts on the elements "nature" and "fishery" of the ISOS model. Two possibly major effects have been evaluated: changes in the hydrodynamic and morphological system of the estuaries induced by sea level rise and changes in temperature. The first effect comprises the working through of sea level rise on the hydrodynamic and morphological characteristics of the estuary, including the feedback effect on waves. These effects are expressed in terms of siltation, leading to changes of the salt marshes, the intertidal

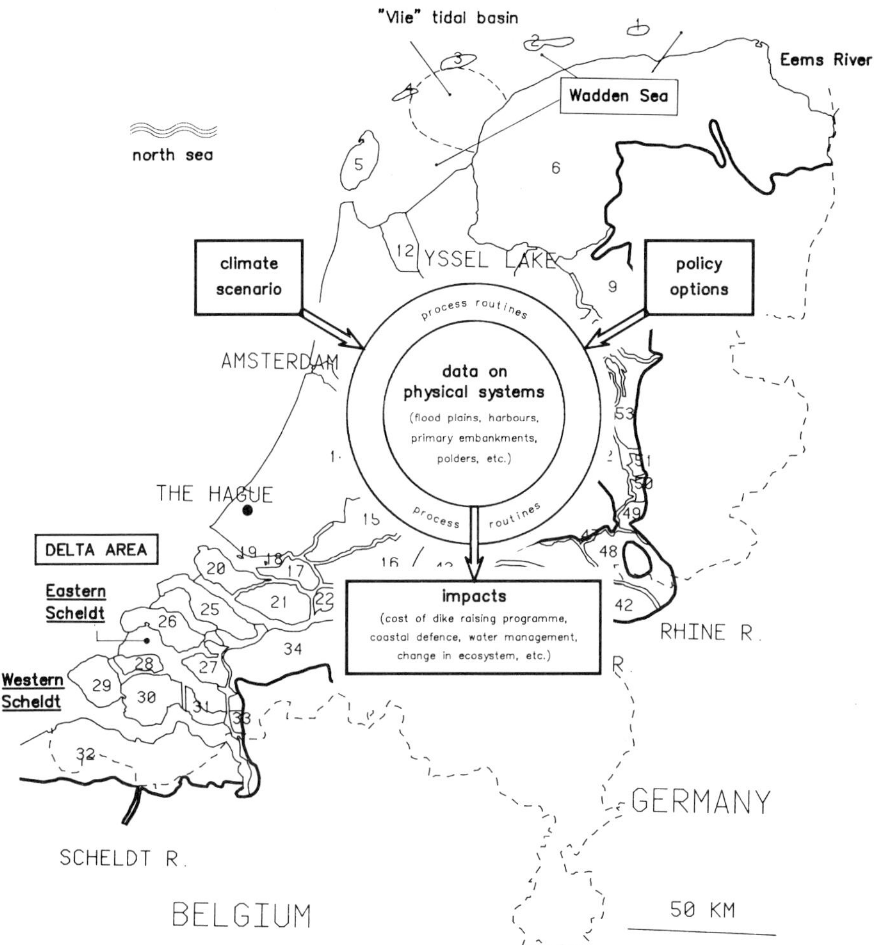

Fig. 1—Principle of ISOS model system.

zones, and subtidal zones, which are supposed to be characteristic zones in the estuary from the ecological point.

Second, the above morphological changes, combined with an increase of the water temperature, have been used as a starting point for ecological computations. The output of these computations are in units of production (carbon) and biomass, which are indicators for possible changes in the ecosystem.

One measure that interferes in the morphological process of an estuarine system is considered, viz. regulations with respect to sand mining. Presently sand is reclaimed from the estuaries and the Wadden Sea to meet the market demands. In the ISOS model the effect of sand mining is approximated by averaging the reclaimed volumes over the relevant tidal basin, which leads to changes in the average depth. Obviously, this is a simple approximation of the real physical process and represents only the fact that sand is extracted from the morphological system and forms a sink in the sand budget. However, this approach agrees with the simplicity of the schematization of the autonomous effect of sea level rise on the estuary morphology. In the following section, the morphological effect is discussed, followed by an analysis of the impact on the ecosystem.

EFFECT ON MORPHOLOGY

The estuary is divided in three different zones: salt marshes, intertidal areas, and subtidal areas. The boundaries between these zones are horizontal: between marshes and intertidal areas at a level of high water +0.6 m and between intertidal and subtidal areas at low water. Based on historical data related to the morphological development of the tidal basin of Vlie and the boundary conditions pertaining to this development, the increase in average water depth in this area is computed for different scenarios of sea level rise. The basic principle of the computations is that before the estuary can react by increasing siltation, average flow velocities have to reduce to effect decreasing sediment transport capacities. This is realized by increasing water depths, i.e., by a delayed response of the seabed to the mean water level rise. Thus the actual increase of water depth is less than the sea level rise. Because the siltation will lag behind sea level rise, marshes, intertidal and subtidal areas will change in size. The principle is shown in Fig. 2.

Based upon the results for the Vlie tidal basin and the geometric properties of the other basins, a prediction is made for the following estuarine areas in this subsystem: western Wadden Sea, eastern Wadden Sea, Eastern Scheldt and Western Scheldt. The results are expressed in areal changes, specified for the above typical zones. From the point of sensitivity of the ecosystem to morphological changes, it is assumed that areal changes of less than 10% will have no significant impacts.

To account for the effect of sand mining, two different levels of sand mining have been considered: (a) termination of sand mining and (b) doubling sand mining w.r.t. present levels. As we explained before, this results respectively in a decrease and increase in the average depth of the relevant tidal basin.

The next analysis starts with a fixed-seabed situation. Next the natural adaptation mechanism is discussed, followed by some remarks on shoal formation. Finally the time lag the sea bed reacts with on the rising sea level is approximated.

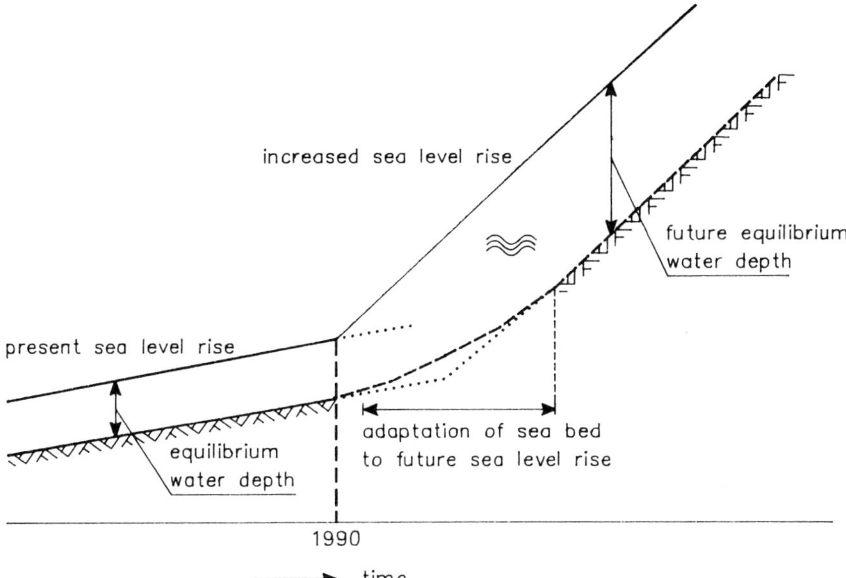

Fig. 2—Principle of delayed seabed response on sea level rise.

Fixed Seabed

Sea level rise will cause the mean low water (MLW) and the mean high water (MHW) to rise. If the seabed does not react on this increase of the tidal water levels, subtidal areas, intertidal areas, and marshes will shift (marshes are defined as the areas above HW +0.6 m). The changes in these areas can be derived from the storage relationships (i.e., storage volume vs. elevation) for the relevant tidal basins. This is summarized in Table 1.

The boundary between the western Wadden Sea and the eastern Wadden Sea is located at the tidal basin south of the island of Terschelling. This is shown in Fig. 2. Further it is noted that the areal change in marshes is given by $dA_m/dh = -(dA/dh)_{MHW + 0.6\ m}$, where A represents area (km^2)

TABLE 1. Area–elevation relationship expressed in km^2 per 0.1 m sea level rise.

Area	$(dA/dh)_{MHW + 0.6\ m}$	$(dA/dh)_{MLW}$	dA_s/dh
East Wadden Sea	9.20	38.60	−29.40
West Wadden Sea	1.25	61.50	−60.25
Eastern Scheldt	0.80	2.70	−1.90
Western Scheldt	2.20	2.40	−0.20

and h elevation (dm). The subscript m denotes marshes. The intertidal area will change following

$$dA_s/dh = (dA/dh)_{MHW + 0.6\,m} - (dA/dh)_{MLW},$$

which is given in the right-most column. The subscript s represents the intertidal area (shoals).

From Table 1 it can be seen that a considerable loss of intertidal area will occur in the western Wadden Sea (more than 10% areal loss per 0.1 m sea level rise) if the shoals do not follow the sea level rise. This holds for the marshes as well. Figure 3 shows the storage relationship for the distinct areas.

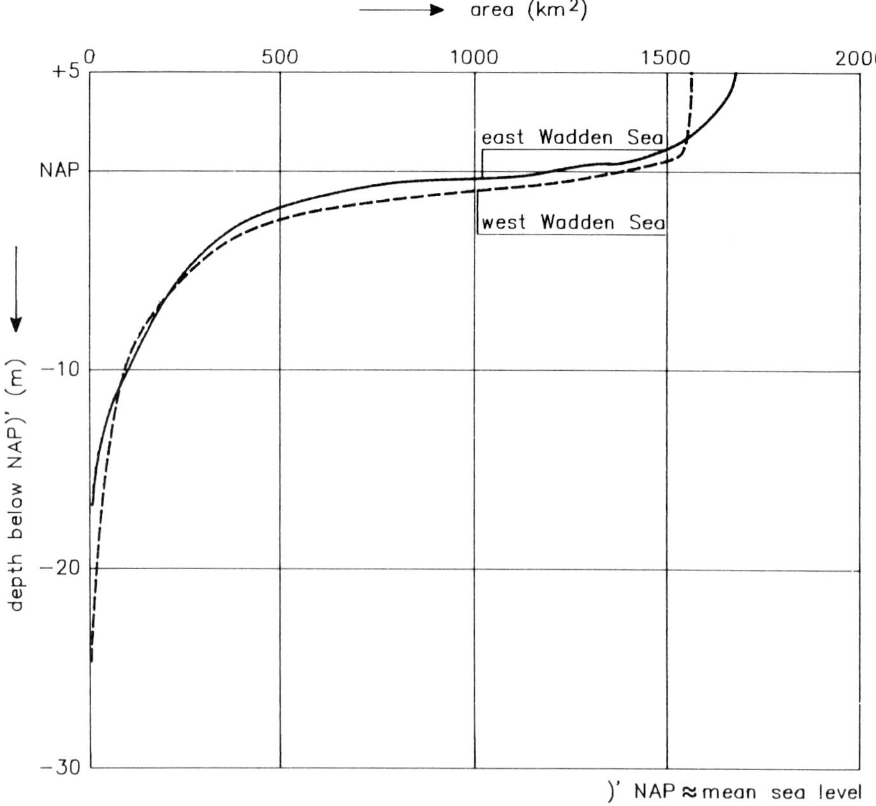

Fig. 3a—Area–depth relationship of the Wadden Sea.

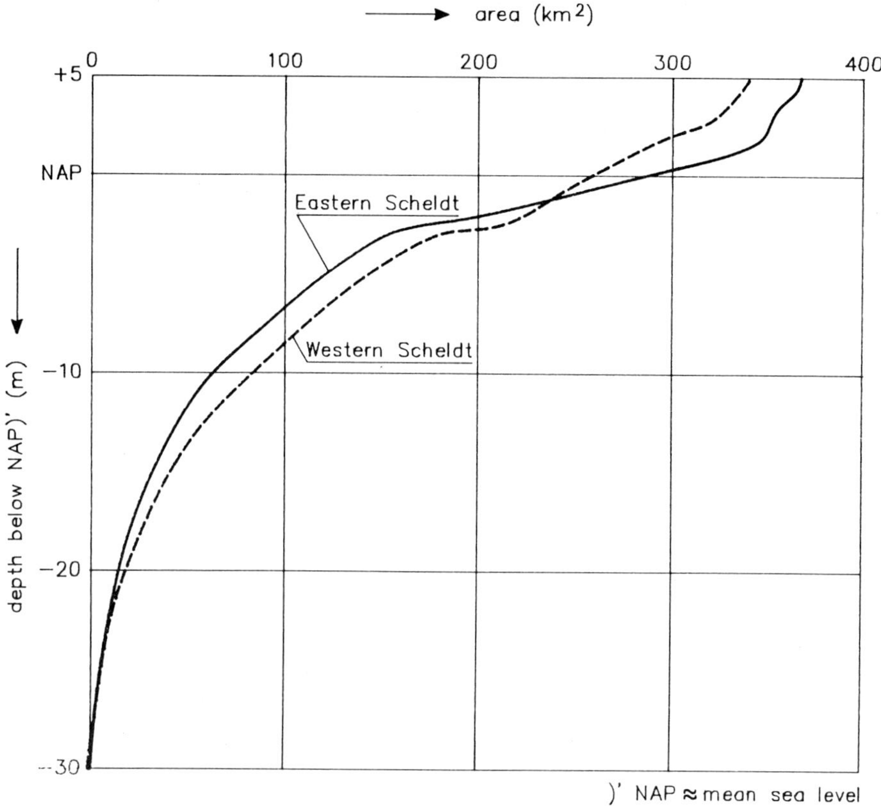

Fig. 3b—Area–depth relationship of Eastern Scheldt and Western Scheldt.

Natural Adaptation Mechanism

It is expected that the shoals and marshes are able to follow sea level rise to a considerable extent. As far as marshes are concerned, data from De Glopper (De Glopper 1981) indicate that in the past decades sedimentation in marshes of the Provinces of Friesland and Groningen, which are actually reclamation areas with very limited maintenance, developed with a much higher rate than present sea level rise, as is shown in Table 2.

The tendency shown in Fig. 2 can be observed to a lesser extent in the salt marsh of Nieuwlandsrijd on the Island of Ameland, where sedimentation rates of 0.4 cm/yr at NAP +1.3 m, 0.25 cm/y at NAP +1.5 m, and 0.05 cm/y at NAP +1.7 m have been observed (Eysink et al. 1987; NAP refers to approximately mean sea level). A relation was found between the cumulative flooding depth per year and the sedimentation. Thus the growth (vertical)

TABLE 2. Sedimentation rates in the marshes along Friesland and Groningen coasts.

Elevation w.r.t. Mean Sea Level	Sedimentation Rate
0.9 m	1.7 to 3.6 cm/y
1.2 m	0.7 to 2.7 cm/y
1.5 m	0.4 to 1.3 cm/y
1.7 m	0.3 to 1.0 cm/y

of a marsh depends on the elevation with respect to MHW and will, consequently, change due to increasing sea level rise. Land subsidence induced by gas extraction will obviously have the same effect. In case of the Island of Ameland, a land subsidence of 0.1 m will "produce" 1.5 times higher sedimentation rates.

Marshes will probably not or only slightly reduce at rates of sea level rise less than 1 m per century. A number of marshes may flatten because the lower parts will rise quicker than the higher parts; this is because of the above-observed dependence of flooding frequency.

Unpublished data from Dijkema and De Glopper (1988) confirm that the marshes along the coasts of Friesland and Groningen may keep pace with increasing sea level rise and may even silt up with a higher rate. Between MHW and MHW −0.6 m, the observed sedimentation ranges from 0 to only 0.35 cm per year. However, seaward of this zone, sedimentation rates of 0.75 to 0.85 cm per year are found, concluding that marshes are expected to grow with sea level rise up to the above scenario of 1 m per century.

The shoals, forming the intertidal areas, will probably follow the increasing sea level rise as well. From the observations at a shoal (Grienderwaard) west of the Islet of Griend, it is concluded that the sudden increase of MHW due to the enclosure of the Zuiderzee, which was about 0.15 m, was compensated for in the subsequent period of 40 to 50 years, including the relative sea level rise. The shoal has raised in that period with 0.5 to 0.6 cm per year, which is approximately 0.25 m in the above period. The sediment was carried from the western extremity of the Grienderwaard by the combined effects of tides, waves, currents, and wind.

Shoal Formation

The Wadden Sea is composed of a complicated pattern of tidal channels and shoals that are in a state of dynamic equilibrium, subject to the forces of the tides, waves, and wind. On the whole, the Wadden Sea is a sedimentation basin, i.e., it has a resulting inward sediment transport, carried from the North Sea and adjacent coastal areas through the channels between the Wadden Islands. Inside the Wadden Sea, the above-mentioned equilibrium exists between sediment transport towards the shoals during rising water (flood) and the downward transport towards the tidal channels during falling water (ebb).

An increase of sea level rise will disturb this equilibrium. The tidal volume tends to increase, resulting in higher flow velocities in the channels. Consequently, sediment transport capacity will increase, leading to more sediment transport towards the shoals during flood. On the shoals, the water depth increase (due to sea level rise) is relatively large because of the limited water depths. This means that the condition for sedimentation tends to become more favorable, possibly leading to reduction of the sand transport towards the channel during ebb. Thus the shoal will tend to follow the rising sea, in first instance with deepening tidal channels. However, due to the higher shoals, the tidal volume decreases again, causing adaptation of the channel bed, i.e., the channel bed follows the rising shoals with some time lag.

Summarizing, the shoals tend to follow an increase in the sea level rise due to adaptation of the pattern of sediment transport in the Wadden Sea. The channels will follow with some time lag. The total volume of sediment imported in the Wadden Sea has to increase to realize the above adaptation. Before this additional sediment can be trapped, the transport capacity in the Wadden Sea must first be reduced. This is realized by a time lag between sea level rise and the adaptation. The following section deals with the estimation of this time lag.

Time Lag in Adaptation of the Seabed

From the available information it is concluded that before 1932 sedimentation in the tidal storage basin of Vlie compensated for the sea level rise of 0.17 m/century and a land reclamation rate of 0.3‰ per year. This sedimentation was about 1.8×10^6 m^3 per year, which was approximately 10% of the annual sand influx. Corresponding theoretical lagging of the water depth $L_{17} = 0.12$ m represents the overdepth to achieve a velocity reduction of 3%, necessary for this sedimentation rate (based upon the relation $S \sim v^{3.5}$).

Further, the outer delta of the tidal inlets, forming the door step to the tidal basin, will adapt more or less simultaneously with the increasing sea level rise. Thus the hydraulic and sedimentological conditions will remain unaltered in this zone.

If in the year 1990 the sea level rise has changed to 0.40 m/century, then the lagging in water depth has to increase as well in order to adjust the sedimentation rate to a level of approximately 4.1×10^6 m^3 per year ($0.4/0.17 \times 1.8 \times 10^6$). The required lagging water depth L_{40} is 0.291 m.

As explained earlier, the lagging water depth is a measure (theoretical) to allow for sufficient velocity reduction to arrive at the sedimentation required to follow the increased sea level rise.

It is expected that the seabed will adapt gradually to the new lagging depth (see Fig. 4). This is approximated as follows:

1. At time t, after 1990, the water level z_w with respect to the water level z_{wo}, which would occur with the original (pre-1990) value of the sea level rise, is equal to: $z_w = t \times (0.40 - 0.17)/100$ [m], where t is expressed in years and 0.40 and 0.17 are the values of the sea level rise after and before 1990, respectively.

Sand mining has effect on the resulting water depth. To account for this effect, a yearly volume of 0.75×10^6 m^3 is supposed to be dredged in the Vlie tidal basin.

2. On time t the actual (average) increase in water depth is $z_{wa} = z_w - x$, where x = seabed adaptation on time t.

In the above formula it is implicitly assumed that in the period before 1990 the seabed keeps pace with sea level rise z_{wo}.

3. Adaptation x is the result of additional sedimentation in the period between 1990 and time t with respect to the sedimentation due to the pre-1990 sea level rise, which was equal to 1.8×10^6 m^3 per year.
4. The resulting increase in mean water depth is computed iteratively and summarized in Table 3.

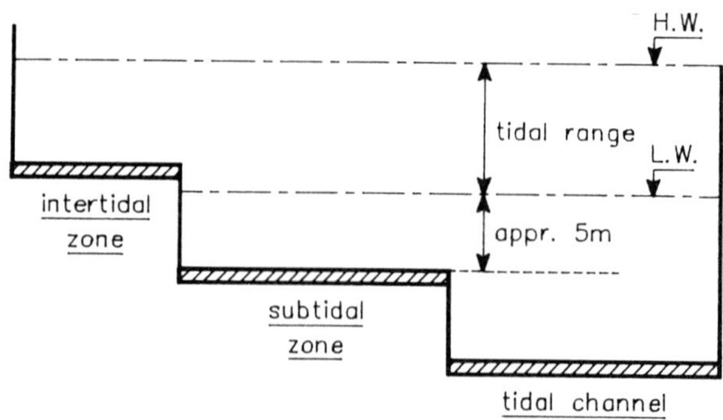

Fig. 4—Schematization of the EMOWAD model.

The values of the "lagging water depth" are for slr = 0.4 m/100 y 0.29 m and for slr = 0.6 m/100 y 0.41 m. The present value of the lagging water depth is 0.12 m.

It is assumed that the results shown in Table 3 can be applied for all tidal basins. First the effect of the past sand mining operations in the tidal basin of the Vlie has been eliminated from the above results for x. For this purpose an annual volume of 750,000 m^3 has been taken into account, yielding a depth reduction of 0.11 cm per year. The resulting depth changes are exclusively the possible effect of sand mining. Based on available predictions of the level of sand mining, the scenarios without sand mining, predicted sand mining, and predicted sand mining can be evaluated. The basic data are given in Table 4.

Combined with the elevation/area relations for the salt marshes and the intertidal zones in the Wadden Sea and the Eastern and Western Scheldt, the area reduction can be computed. The results are presented in Table 5. The areal changes are expressed respectively in absolute (km^2) and relative (%) units. Apart from the areal changes of marshes, intertidal areas, and subtidal areas, the change in average depth has been computed. Because of the rise of HW and LW marks, the interface between the above zones will shift landward. Depending on the shape of the estuarine depth profile or on the shape of the elevation/area relationship, the average depth of the marshes, intertidal and subtidal zones will change. This is a factor of importance to ecological computations in connection with light extinction,

TABLE 3. Adaptation of mean water depth (in cm) to sea level rise.

Year	Sea level rise	
	40	60 (cm/100 y)
0	0	0
10	3.1	5.0
20	5.7	9.1
30	7.7	12.7
40	9.5	15.6
50	10.9	17.9
60	12.0	20.0
70	12.9	21.5
80	13.7	22.9
90	14.3	
100	14.8	25.1
120	15.6	26.3
150	16.3	27.7
200	17.0	28.8

TABLE 4. Effect of sand mining.

Tidal basin	Predict. m^3/y	area km^2	dh$_{sm}$ cm/y
East W'Sea			
Borndiep	10,000	315	0.003
Pinkegat	0	65	0
Zoutkamperlaag	50,000	130	0.039
Eilanderbalg	0	55	0
Lauwers	220,000	145	0.152
Schild	0	50	0
W.Eems/NL	500,000	520	0.096
Average:			0.061
West W'Sea			
Marsdiep	500,000	710	0.070
Eyerl.Gat	0	160	0
Vlie	750,000	660	0.114
Average:			0.082
Eastern Scheldt	no sand mining		
Western Scheldt	2,000,000	330	0.606

turbidity, sedimentation rates, etc. The resulting depth changes are given in the tables below as well.

IMPACT ON THE ECOSYSTEM

The model EMOWAD has been used to carry out a number of simulations of the ecosystem in the Western Wadden Sea to assess the effect of morphological changes and temperature rise. Basically, the model produces outputs related to the carbon balance and biomass on lower levels in the ecosystem. The processes simulated by the model include production and consumption functions and transportation, exchange, and sedimentation processes of plankton. The parameters which cause possible effects in the ecosystem are changes of the intertidal and subtidal zones, the average depth of these zones, and the temperature. The biological submodels of the EMOWAD model have originally been developed for the estuary of Eems river. The total sublittoral zone, i.e., the zone between LW and LW −5 m, in this estuary is limited. In the western Wadden Sea the sublittoral zone,

TABLE 5. Summary results of morphological computations. The results, expressed in km² and %, represent the areal changes in marshes and intertidal areas.

Morphological changes after a period of 100 years

Sea level rise	0.4 m/100 y				0.6 m/100 y			
Zone	marsh		intertidal		marsh		intertidal	
Tidal basin:	km²	%	km²	%	km²	%	km²	%
east Wadden Sea	6.7	11%	31.7	3	13.3	22%	63.3	7%
west Wadden Sea	0.0	0%	61.1	11%	0.1	1%	122.0	21%
Eastern Scheldt	1.5	23%	4.1	3%	3.0	43%	8.6	5%
Western Scheldt	0.3	2%	2.1	2%	0.6	4%	4.2	4%

Without sand mining

Morphological changes after a period of 100 years

Sea level rise	0.4 m/100 y				0.6 m/100 y			
Zone	marsh		intertidal		marsh		intertidal	
Tidal basin:	km²	%	km²	%	km²	%	km²	%
east Wadden Sea	10.7	18%	51.2	3%	17.4	28%	82.7	9%
west Wadden Sea	0.7	1%	111.4	11%	0.1	1%	172.2	30%
Eastern Scheldt								
Western Scheldt	2.12	14%	14.8	13%	16	16%	16.9	15%

1* predicted future sand mining

Morphological changes after a period of 100 years

Sea level rise	0.4 m/100 y				0.6 m/100 y			
Zone	marsh		intertidal		marsh		intertidal	
Tidal basin:	km²	%	km²	%	km²	%	km²	%
east Wadden Sea	14.8	24%	70.6	8%	21.5	35%	102.1	11%
west Wadden Sea	0.1	1%	161.5	28%	0.1	1%	222.5	39%
Eastern Scheldt								
Western Scheldt	3.9	26%	27.5	24%	4.2	28%	29.6	26%

2* predicted future sand mining

or actually the subtidal zone excluding the tidal channels, is very extensive. Therefore, a separate sublittoral submodel was developed [EON 1988], based on the benthic model of the intertidal zone. The resulting model for the western Wadden Sea is described in NIOZ (1988). Here, information on the development, applicability, and limitations of the model can be found. The EMOWAD model schematizes the western Wadden Sea in 12 compartments. Figure 4 shows the cross section of a compartment. Each compartment is composed by an intertidal zone, a sublittoral zone, and a tidal channel zone.

The predicted morphological changes resulting from sea level rise have been disaggregated on the compartment level. To reproduce field conditions as accurate as possible, 1985 and 1986 data have been used to define the baseline for the runs which include the temperature rise and the sea level rise. The initial conditions for the 1986 run were derived from the boundary values of a 1985 run. The 1986 run delivered the boundary conditions for the standard simulation (baseline).

Next, four runs were made with the model to evaluate the effect of a 4°C temperature rise (run 1), 21% reduction of intertidal areas due to sea level rise (run 2), 39% reduction of intertidal areas due to sea level rise and an increased level of sand mining (run 3), and a combination of run 1 and 3 (run 4). Table 6 presents an overview of the EMOWAD runs. Areal changes of marshes, intertidal and subtidal areas lead to changes in the average depth of these zones, as explained in the previous section. These effects are included in the overview as well.

The results of the computations are presented in Tables 7 and 8. Table 7 gives the carbon flow budgets in the three subsystems of the model, i.e., the pelagic subsystem, the intertidal benthic subsystem, and the sublittoral benthic subsystem. The results are expressed in terms of carbon mass per unit area per year ($gCm^{-1}\ y^{-1}$). The pelagic subsystem extends over the whole model area (western Wadden Sea), the benthic subsystem over respectively the intertidal and sublittoral zones. Table 8 presents the resulting

TABLE 6. Overview of EMOWAD runs.

Run no.	1	2	3	4
temperature rise	4°C	0	0	4°C
loss intertidal area	0%	21%	39%	39%
depth change intert. area	0 m	−0.32 m	−0.36 m	−0.36 m
depth change subl. zone	0 m	−0.51 m	−0.74 m	−0.74 m
depth change channels	0 m	+0.20 m	+0.36 m	+0.36 m
tidal amplitude	1.8 m	10.8 m	10.9 m	10.9 m

TABLE 7. Summary resulting carbon budgets (gC/m²/y).

Run	0	1	2	3	4
PELAGIC					
Inputs:					
input IJ'lake	95	95	94	94	94
input Vlie	37	33	31	29	21
Gross prim. production	488	599	484	480	587
Total	620	727	609	603	702
Outputs:					
export Marsdiep	142	150	147	149	159
consumption	297	414	292	293	410
sedimentation in i.z.	33	44	20	16	22
sedimentation in s.z.	14	20	11 11	11	16
filtering in i.z.	11	6.20	14	11	7.70
filtering in s.z.	114	82	116	116	79
export EWS	6	6	6	6.10	6.30
increase biomass	2.50	2	2.40	2.40	2.40
Total	619.50	724.20	608.40	604.50	702.40
INTERTIDAL ZONE					
Inputs:					
sedimentation	129	173	98	104	141
filtering	44	24	69	68	50
gross prim. production	150	157	223	230	233
decrease biomass	68	39	94	92	65
Total	391	393	484	494	489
Outputs:					
consumption	304	302	400	408	399
burying	86	90	82	82	86
birds	1.90	2.20	2.40	3.20	3.60
export to s.z.	0.17	−1.10	0.64	0.66	−0.08
Total	392.07	393.10	485.04	493.86	488.52
SUBLITTORAL ZONE					
Inputs:					
sedimentation	23	34	17	16	24
filtering	192	138	179	168	115
import from i.z.	0.07	−0.45	0.20	0.15	−0.02
decrease biomass	3.70	7.50	7.70	8.80	10.62
Total	218.77	179.05	203.90	192.95	149.60
Outputs:					
consumption	218	179	205	193	149
Total	218	179	205	193	149

i.z. = intertidal zone; s.z. = sublittoral zone

TABLE 8. Average biomasses and some other state variables.

Run		1	2	3	4
Pelagic					
Oxygen concentration (mg C.l^{-1})	9.49	8.89	9.53	9.53	8.92
Primary production (mg C.m^{-3})	401.95	382.56	430.59	421.67	409.06
Micro & macroplankton	76.95	109.66	73.55	76.10	113.87
Bacteria	44.66	44.91	51.15	50.20	52.91
Intertidal Zone					
Primary production (mg C.m^{-2})	1172.95	1211.15	1286.00	1320.85	1322.95
Filter feeders	3102.55	1581.00	6908.90	7055.30	4289.95
Deposit feeders	638.74	443.43	1306.22	1422.05	1075.13
Meiobenthos	304.42	202.27	343.35	342.70	429.19
Bacteria	2469.60	2798.90	2410.11	2409.95	2653.40
Aerobic layer	3.14	2.13	3.85	3.95	2.79
Subtidal Zone					
Filter feeders (mg C.m^{-2})	3895.25	2510.65	3481.15	3072.70	1858.00
Deposit feeders	1095.55	654.22	859.25	690.65	386.00
Meiobenthos	6.56	27.35	5.04	3.52	19.18
Bacteria	2630.70	2069.43	2494.81	2237.23	1529.31
Aerobic layer	1.12	1.12	1.04	1.05	1.07
Epibenthos					
Mesoepibenthos (mg C.m^{-2})	201.98	273.85	125.73	107.24	155.57
Macroepibenthos	466.95	534.96	543.17	455.41	498.73

biomasses in mg carbon per liter or cubic meter in the pelagic subsystem and mg carbon per square meter in the benthic subsystem. The results of the computations are summarized as follows.

Run 1

The temperature increase yields two major effects: (a) a (approximately) 20% increase of the pelagic primary production and (b) decreasing filter activity of the filter feeders in the intertidal and sublittoral zones. This holds for the filter feeders as well. Consequently their biomass reduces with about 40%.

These results have some indirect effects. The production increase will cause the consumption (mineralization) in the water column to increase as well. The lower level of filtering will contribute to this effect as well. Thus the biomass of micro- and zooplankton will increase, as can be seen from

Table 8. This increase is about 30%. This leads, in turn, to higher sedimentation rates, related to detritus production. Finally, an indirect effect is the increase of mesoepibenthos, colonizing on benthic primary producers which become available for consumption due to the lower activity of the filter feeders.

Runs 2 and 3

There appear no significant differences between the outcomes of runs 2 and 3. Two major effects are observed from the computational runs: (a) the benthic primary production in the intertidal zone increases approximately 50% because the phase of the tidal cycle the shoals emerge has become longer; (b) the sedimentation rate has been reduced because of the decrease in average water depth in the intertidal zone.

The increase of benthic primary production in the intertidal areas has a number of consequences. First, the filter feeders react on the increased food supply; their mass grows with more than 100%. Consequently, more food is extracted from the pelagic subsystem by filtering. Second, the deposit feeders increase in number, feeding on the increased mass of benthic producers. Third, carbon flux from the pelagic subsystem to the benthic subsystem in the sublittoral zone (maintained by filter feeders) is not changing. Thus an increase of the sublittoral zone (due to the decrease of the intertidal zone) does not influence the growth of filter feeders.

Run 4

The results of run 4 can be reproduced by adding up the results of run 1 and run 2/3. No effects have been observed which are beyond the above outcome.

Based upon the above results and the characteristics of the model (limitations), the following is observed. The temperature dependency of the macrobenthos is based upon the species living in the Eems–Dollard estuary (east Wadden Sea), i.e., Macona Balthica and Cerastoderma Edule. These species have a boreal origin and are in fact not able to withstand a strong temperature rise. However, in the western Wadden Sea the mussel is an important representative of the filter feeders; this species lives in southern areas as well (Gironde) and will probably not be affected by temperature rise. Therefore, it is expected that the model overestimates the decrease of filter feeders due to temperature rise. In conclusion, although temperature rise may lead to a disrupture of succession of the boreal species, more southern species will, as far as not already present, replace them. Thus temperature rise may have an impact on the diversity of the filter feeders. Such a replacement effect may occur, in a more general sense, with other species as well. On the other hand, the observed increase in pelagic primary production is a more realistic effect.

From the results of run 2/3, it has been observed that the increased benthic primary production is due to the increasing emerging of shoals in the intertidal area (the average depth has decreased due to landward shifting of this zone). This is based upon the results of the model for the Eems–Dollard estuary, where the assumption was adopted that benthic primary production only takes place when, e.g., the shoals are above the water level. However, some researchers (e.g., Cadeé 1978) suggest that this requirement for benthic production is not that strong. In that case the real increase of production will be less than predicted by the model.

CONCLUSIONS

The following conclusions are drawn from the morphological and ecological computations.

1. Due to a delayed response of the seabed to sea level rise, the average water depth will tend to increase. This results in changes of salt marsh and intertidal and subtidal areas, in terms of areal surface. Moreover, the average water depth in intertidal area may change due to the concave shape of the intertidal zone.
2. The applied scenarios for sea level rise, being 0.4 m and 0.6 m per century, do not result in significant differences in the resulting impacts on the ecosystem. The most important factor which determines possible ecological impacts is the change in average water depth in the intertidal and sublittoral (= subtidal to a depth of approximately 5 m) zones.
3. The most significant impact of the morphological change is the increase in benthic primary production, leading to an increase of the biomass of macrobenthos per unit area. If the reduction of the habitat is taken into account, i.e., decreasing intertidal areas, then the above biomass increase is limited. In the sublittoral zone it is observed that the areal increase does not result in a higher macrobenthos growth, over the total sublittoral area. The mussel is the most important representative species in the macrobenthos group.
4. Temperature rise leads to increasing primary production (20%) in the pelagic subsystem. Basically this has a positive effect on higher trophical levels. However, the EMOWAD model does not predict diversity changes of the functional species.
5. Both in the morphological and ecological evaluation, rather strong schematizations are made of the natural processes at work. It is recommended to explore in more detail the desired methodology for impact assessment. In regard to morphology, the long time scale and

spatial variability of the processes form a special difficulty. The ecological computations disregard the potential effects of temperature rise on the functioning of the different species and possible changes in succession. Also, changing boundary conditions with respect to nutrient supply, irradiation, and transport processes need to be taken into account.

REFERENCES

Cadeé, G.C. 1978. Voedselproduktie in de Waddenzee (in dutch). *Natuur en Techniek* **46-11**:730–749.

De Glopper, R.J. 1981. Siltation and erosion of the marshes at the Friesland and Groningen coast (in dutch). In: 50 Jaar Onderzoek Rijksdienst for IJsselmeerpolders. Lelystad, The Netherlands.

Dijkema, K.S., and De Glopper, R.J. Workshop on Expected Effects of Climatic Changes on Marine Ecosystems, 1988. Netherlands: Research Institute for Nature Management.

Eysink, W.D. et al. 1987. Gas extraction on the island of Ameland, effects of bottom subsidence (in dutch). Delft Hydraulics: H114 (report). The Netherlands.

NIOZ. 1988. The ecosystem of the western Wadden Sea: field research and mathematical modelling. Netherlands Institute for Sea Research. EMOWAD II, NIOZ report 1988-11.

Peerbolte, E.B. et al. 1988. Impact of sea level rise on society, a case study for the Netherlands. Phase I. Delft Hydraulics: H750 (report). The Netherlands.

Wind, H.G., ed. 1987. Impact of sea level rise on society. Delft Hydraulics. University of Twente. Rotterdam: A.A. Balkema.

Standing, left to right:
Michael Krom, Edward Goldberg, Philip O'Kane, Thomas Church, Ken Hsü, Bart Peerbolte, Franca Frascari, Youssef Halim
Seated, left to right:
Patrick Holligan, Victor Smetacek, John Steel, Dagmar Hainbucher

Group Report: What Is the Response of Ocean Margin Systems to Natural and Anthropogenic Perturbations?

J.P. O'Kane, Rapporteur
T.M. Church
F. Frascari
E.D. Goldberg
D. Hainbucher
Y. Halim
P.M. Holligan
K.J. Hsü
M.D. Krom
E.B. Peerbolte
V.S. Smetacek
J.H. Steele

INTRODUCTION

The working group addressed the question, "What is the response of ocean margin systems to natural and anthropogenic perturbations?" The background papers by Halim, Hsü, Krom, Peerbolte,and Smetacek (this volume) were taken as read and the group ordered its deliberations according to a number of more detailed questions:

1. How can we separate natural and anthropogenic changes and trends?
2. What are the time scales of changes (e.g., events and trends)?
3. Can we use variability in space to separate the historical effects of natural and anthropogenic changes?
4. How do we detect change? What do we need to measure, how and where?
5. What specific topics have a worldwide relevance for the coastal zone (e.g., eutrophication, land use, sea level, etc.)?
6. What future issues may have such relevance (e.g., mariculture)?
7. How do we construct good management schemes for the next century?

The rapporteur has attempted to record in the following paragraphs the

Ocean Margin Processes in Global Change
Edited by R.F.C. Mantoura, J.-M. Martin and R. Wollast

fruits of a very lively discussion, the disagreements and recommendations concerning what we do not know, and what we should like to know.

The societal and scientific problems of the coastal margins are worldwide. While there are many textbooks and monographs which treat different aspects of this (e.g., Bowden 1983; Davis et al. 1985; Dyer 1986; Csanady 1982; Walsh 1988; Clark 1977), there is no worldwide encyclopedic review of (a) its hydrography, physics, chemistry, biology, and geology or (b) its exchanges of mass, energy, and momentum with the land, atmosphere, open ocean, and seabed, or (c) its value to and use by people. In addition, the resources which are available for studying it are unevenly distributed.

We can attempt to separate natural and anthropogenic changes in the ocean margins only if we know the variability of the processes of interest. Consequently, our report begins with time and space scales of processes in oceanic margins, including their treatment in mathematical models. References to text books and monographs are included to help the reader who is unfamiliar with the terminology.

The report then addresses the problem of separating natural and anthropogenic changes using geochronology, statistical method, modeling, and environmental data banks. The third section of the report considers specific world wide problems. The report ends with a discussion of the management of the coastal zone in the next century. Recommendations for future tasks are made at the end of each section.

TIME AND SPACE SCALES OF PROCESSES IN OCEANIC MARGINS

One reason for the complexity of the oceanic margins is the very wide diversity of scales and processes. We can regard the ocean margin as a fractal object separating land from the open ocean (Mandlebrot 1977). Thus the ocean margin as a biogeochemical environment is best described by patterns in the space and time scales of its state variables and in its boundary forcing functions.

Spectral analysis, in both space and time, and other techniques may be used to identify these scales in the physical, chemical, and trophic environments of the oceanic margin (Platt and Denman 1975; Steele 1989; Stommel 1963; Kisiel 1969; Bath 1974; Bendat and Piersol 1966; Press et al. 1986). We may quote Nihoul (1985, p. 27) on the current state of knowledge: "Although the variability of the atmosphere and the ocean is universally recognized, nothing like a unanimous spectral description of geohydrodynamic phenomena appears to be available in the literature which often offers conflicting views based on bits and parts of energy spectra plotted in irreconcilable units." The spectra from the oceanic margins of the world may be more complicated than those from the open ocean, and this presents us with a considerable challenge for the future.

The extension of existing spectra to geological time scales is not complete, and a gap exists at frequencies between decades and millenia. Whether this portion of the spectrum follows a white noise or a power law is of very considerable importance for the management of the coastal zone over the next century. Numerical experiments (Solow and Steele 1990) suggest that a record at least ten times longer than the critical time scale is required. A proxy geological interpolation may be the only alternative to distinguish these two possibilities.

Attempts have been made to relate biological to physical spectra and to build holistic system models of the open ocean in spectral space (Platt and Denman 1978). The paradigm for this work is "a close connection of physical and biological scales and, through evolution and adaptation, a dependence of ecological processes on the details of the physical dynamics." In contrast to this, the Gaia Hypothesis states that on geological time scales there is global feedback from ecological to physical and chemical processes. For example, Lovelock has speculated that algae in the coastal margins may play a role on these geological scales.

Spectra have peaks which are associated with particular processes and scales. Models which attempt to resolve processes at widely differing scales are "stiff" and very difficult to solve numerically (Press et al. 1986, 572); e.g., prey–predator interactions between small and large organisms. Perturbation analysis is a suitable technique for treating processes with two widely different time scales (see, e.g., Lin et al. [1974] vol. I, section 2.2 of part A and chapters 6 to 11 of part B).

The theory of scaling for process models is well developed in physical oceanography (Pedlosky 1977). For an introduction to the theory, see Lin et al. (1974) chapter 6, volume I. The function of scaling is to reduce the dimensionality of the model parameter space as much as possible in order to comprehend the dynamics of several interacting processes. The extension of these concepts and techniques to coupled physical/biogeochemical models of the coastal zone is a challenging interdisciplinary task for the future (see, e.g., section 3.2 of Platt et al. [1981]).

In making an environmental model, a spectral window is usually chosen containing a single peak in the spectrum (Nihoul 1977, 1985). Processes at higher temporal frequencies are averaged or filtered out and are parameterized in terms of the state variables in order to "close" the system of equations which govern the model. A similar parameterization is required for site-specific numerical models, which fail to resolve sub-grid-scale spatial processes. For an introduction to this field, see Bedford (1981) and Abbott et al. (1985a, b). Process parameterization is a source of considerable uncertainty in current models and will remain active as an area of research in the future. The dependence of parametrization on scale is poorly known.

Future Research Tasks

1. Encyclopedic review of existing data and knowledge of the coastal margins of the world.
2. Measurement of temporal events covering a wide range of physical, chemical, and biological variables at selected long-term sites and transects on the oceanic margins of the world. Autonomous vehicles or "smart" ROVs, in addition to more conventional techniques such as survey ships and moored instruments, are indicated in order to describe the scales of variability and to identify significant processes at various scales.
3. Further development and application of remote imaging systems, to provide data at the spatial and temporal frequencies which resolve important processes in the oceanic margin.
4. Extension of coastal margin spectra to historical and geological time scales using, e.g., geochronology to understand the role of coastal margins in global climate at decadal and longer time scales.
5. Application of the theory of scaling to coupled, physico-biogeochemical processes (a) to understand their dynamics, (b) to make better site-specific numerical models, and (c) to provide a theoretical basis for the classification of the oceanic margin.

SEPARATION OF NATURAL AND ANTHROPOGENIC CHANGES

Having examined the diversity of scales and variability of processes in the oceanic margin, the working group proceeded to address the question of separating natural from anthropogenic changes.

The terms "top down" and "bottom up" are used in ecology to denote control of ecosystem structure by grazers and carnivores, or by physicochemical properties of the environment, respectively. These terms can also be applied to the effects of anthropogenic perturbation: thus, "top down perturbation" would be engendered by removal of a key grazer or carnivore, by overfishing or extermination by some specific toxin, whereas "bottom up perturbation" would be due to changing environmental conditions. The latter can be tested if a suspicion arises, whereas the former can only be inferred.

The preferred approach among some scientists who study the environment is (a) to find a place where one process dominates, (b) to understand the principles which govern it, in as general a way as possible, and (c) to apply their understanding to other situations, which may or may not be perturbed by humans.

Differentiating "anthropogenic" from "natural" change is also a matter of scale, as "anthropogenic" impact impinges on a variety of processes ranging from the proximate introduction of a given element or compound, to changing weather and climate. The problem spans a vast scale of time and space, and requires appropriate measuring systems from remote sensing to point sampling.

Except for the few sites where long-term data sets have been collected, change as deviation from the "normal" has to be assessed by the yardstick of individual intuition. Due to sampling constraints, data sets covering biogeochemical processes cannot be comprehensive and hence tend to be ambiguous.

The group discussed a number of approaches to the question of separating "anthropogenic" and "natural" change under the following headings: geochronology; statistical method; modeling; and environmental data banks.

Geochronology

Geochronology is the most important approach to the problem of separating natural and anthropogenic changes on longer time scales.

The advent of agrarian and industrial man resulted in changes in land use, clearing of forests, and various agricultural and industrial practices that directly and indirectly affect the mobilization and depositional records of chemical elements. Changes in the delivery to the ocean margin of excess quantities of substances due to human activity can be documented in coastal sediment records. In some places, the horizon at the start of anthropogenic influence coincides with an horizon of active diagenesis. In this case, we must discriminate between temporal changes due to depositional processes and those due to diagenetic changes.

During the past hundreds to thousands of years, it has been estimated that suspended loads and deposition of silt may have increased by as much as tenfold. It is one of the most important anthropogenic trends in the coastal zone. On a shorter time scale, the construction of dams reduces the sediment load temporarily.

The increased input of sediment is the result of the conversion of land to agriculture in many parts of the world, e.g., northwest Europe, southeast Asia, and the Mediterranean. It may also be seen in the accretion of salt marshes in North America after the clearing of primeval forests. Associated with an increase in suspended silt, there may be (a) a seaward shift in primary production and (b) a decrease in the stock of macrophytes, increasing bottom instability, with an associated loss of benthic filter feeders.

Sediment cores taken from appropriate sites in the ocean margin may record the introduction in the last century and subsequent accumulation of excess amounts of trace elements. Records of anthropogenic organic compounds and of excess metals in banded corals and ice cores have been observed. In principle, there should be evidence of similar increases in the accumulation of excess organic carbon, produced in oceanic margins from increased inputs of nitrate and nitrogen and associated eutrophication.

Changes in the extent, frequency, and magnitude of coastal noxious blooms, associated anoxia, and the appearance of nuisance and opportunistic species appear to be linked to increased eutrophication of estuarine and coastal waters. Also, increased coastal eutrophication can lead to changes in relative phytoplankton abundances, such as the ratio of flagellates to diatoms.

Anthropogenically induced perturbations of the inputs of nitrogen and phosphorous relative to silicon could lead not only to anomalies in biomass production, but also to changes in the composition of biological communities. Diatom blooms in spring could be replaced by dinoflagellate blooms in summer, with indirect consequences on carbon cycling and eutrophication. Finally, such stress may result in unique biomarkers such as genetic or enzymatic materials in the sediments.

Direct excess inputs can be carried into sediment records by worldwide eolian and water transport. Thus it is important to determine when sedimentary perturbations in trace elements became global. Physical, chemical, and biological perturbation is possible.

The analytical ability to document such change in ocean margins will depend on statistically meaningful sampling and on a high precision and accuracy of analytical measurements. A simple geochronological record of an element concentration may display a trend that can be due to either natural or anthropogenic perturbation. Appropriate controls from the preindustrial sediment record should be used to provide the natural physical conditions of the sedimentary regime.

Statistical Method

The theory of the design of experiments, the testing of hypotheses, and the theory of sampling are used routinely in disciplines such as agronomy, demography, pharmacology, and epidemiology, where natural variability is also a problem (see e.g., Bradford Hill 1965; Anderson et al. 1980; Cochran 1977). However, these techniques are rarely used in environmental science. Why is this?

The absence of statistical theory from environmental science is partly due to conflicting objectives among individual scientists. Sharing a research vessel makes it extremely difficult to define a unified sampling design. The absence is also due to a preference for formulating hypotheses *after* the data have been collected, rather than beforehand. This approach, which is necessary initially, appeals to the scientist as environmental detective and avoids the need to state objectives in a quantitative way.

But there may be a more fundamental reason. If we do not admit the distinction between the mathematical concepts of "signal" and "noise," "trend" and "scatter," or "mean" and "variance," believing that every bit of information is always significant, statistical theory has nothing to say or offer in separating cause from cause or anthropogenic effect from natural variation.

Controlled and replicated experiments can always be used to separate anthropogenic and natural variation on short time and space scales. Classical statistical theory relies on the "vital requirement that the actual and physical conduct of the experiments should govern the statistical procedure of their interpretation" (Fisher 1966, p.34). Consequently, hypotheses should be stated in quantitative terms *before* the data is collected.

Consider, for example, the problem of making a mass balance. The methodological issue is always: What conclusion can we draw from a given closing error? No conclusion is possible unless we have some quantity with which to compare the error. Speculation on "missing fluxes" is groundless since the closing error may be an artefact of the sampling. An honest closing error of size zero may also be an artefact of the sampling.

Statistical theory can overcome these difficulties when the mass balance is expressed as a null hypothesis: the expected value of the closing error is zero. In order to carry out the test, independent estimates of the mass fluxes and their sampling variances are required during a fixed interval of time. We first calculate the probability distribution of the closing error on the assumption that sampling variability is the only source of error. If the closing error lies "outside" this distribution, we reject the hypothesis at the chosen level of probability (O'Kane 1986).

Similarly, differences between spatial or temporal averages or trends in space or time can only be tested statistically if the sampling variances are also estimated by an appropriate probability sampling method (Cochran 1977).

Mathematical Modeling

Mathematical modeling is the third approach for separating natural and anthropogenic changes. Models are well-established instruments in environmental management where anthropogenic changes must be estimated *before* some human intervention is made in the biosphere. In this case, models are used simply because they provide a basis for making uncertain decisions. Models are often based on dynamic mass balances, which can, in principle, be tested as null hypotheses.

The problem posed or the questions to be answered define the model: its time and space scales, its content of processes, and its boundary. The model, in turn, specifies the data to be collected.

The resolution of site-specific, numerical models will improve as better process models and numerical schemes are discovered and as computers grow in power. Their predictive capability will remain low until major advances are made in instrumentation systems that provide real-time data on the state variables and boundary fluxes. Real-time data assimilation in meteorological models for weather forecasting is an indication of possible future developments in the monitoring and control of the environment.

In the meantime, models of the physical and biogeochemical environment of coastal margins will play an important role in (a) the integration of data from multidisciplinary projects at specific sites, (b) the elaboration of management plans for coastal margins that address specific problems and questions, and (c) in separating natural and anthropogenic changes.

Environmental Data Banks

Environmental data are stored in computers at both national and international levels from a wide variety of sources. One of the reasons given for investing in environmental data bases is the possibility of using that data to separate natural from anthropogenic effects.

However, "both valid and invalid data are entered into storage. Both are retrieved from storage. Since they cannot be distinguished from one another, they are essentially useless except for bibliographic or historical purposes" (Goldberg and Taylor 1985). Consequently, they are unreliable guides in the separation of anthropogenic and natural change.

In their editorial, Goldberg and Taylor (1985) argue for the routine application of data validation, quality control and quality assurance procedures in the archiving of environmental data, especially when the data may be used for decision making.

Future Research Tasks

1. Development of new methodologies to measure the temporal and spatial extent of cultural eutrophication. Some possibilities are:
   (a) measurement of the accumulation of biogenic material in coastal sediments as a function of time in relation to industrial activity (soot, magnetite) deforestation (pollen),
   (b) direct measurement of areas of anoxia in sediments as a function of time and space using redox sensitive elements such as Rhenium,
   (c) measurement of the frequency of anoxic events in sediments.

2. Measurement of the fluxes of organic carbon in time-dated sediments with the following characteristics:
   (a) Deposition under anoxic conditions to minimize effects of bioturbation. Strata should be dated by ^{210}Pb and ^{137}Cs techniques to ensure particle by particle desposition.
   (b) Very rapid deposition of sediments under aerobic conditions. Strata should be dated as in (a).
   (c) Analysis of the organic phases for total organic carbon, isotopic composition of the organic carbon, and selected organic molecules.
   (d) Diagenetic models based on both natural and anthropogenic carbon. The group was not unanimous on whether the flux-ratio of anthropogenic to natural carbon could be determined in this way.
3. Examination of surrogate elements such as barium for silicon, cadmium for phosphate, and molybdinum for nitrate, as a means of extracting the eutrophication signal from cores subject to diagenesis.
4. Discovery and assessment of indicators and/or response organisms to global change and determination of when and how local changes became global.
5. Development and application of new biomolecular tools for inferring the biological response to stress such as enzyme reduction.
6. Development and deployment of new instrumentation systems to provide real-time data on environmental state variables and boundary fluxes.
7. Development and application of new statistical techniques for the design of sampling surveys, impact experiments, mass balances, and models as dynamic mass balances, which take account of practical constraints.
8. Application of data validation, quality control, and quality assurance procedures to the archiving of environmental data, especially when the data may be used for decision making.

SPECIFIC WORLDWIDE PROBLEMS IN THE OCEANIC MARGIN

Appendix A contains a list of different types of perturbation to the oceanic margin, followed by the various types of response that may occur. The following research tasks were identified during the discussion of this list.

Future Research Tasks

1. Study of the boundary conditions (physics) of the coastal zone and its relation to biogeochemical processes:
   (a) Changes in and impact of fresh water (and associated material) fluxes from land.

   (b) Dynamics of and exchange of material across estuary/shelf sea fronts in response to changes in sea level and freshwater inputs.
   (c) Combined effects of shelf (wind/tide mixing, freshwater buoyancy) and ocean (upwelling, boundary current, etc.) processes in controlling the exchange across the shelf break.
2. Coastal zone processes:
   (a) Study of the chain of relationships: Land use changes → eutrophication → carbon cycle → anoxia → trace gas (N_2O, CH_4) emissions → deposition (including calcite).
   (b) Study of the worldwide impact of toxic compounds (industrial wastes and particular compounds such as chlorinated hydrocarbons, TBT, etc.) and ecological changes (species, genetic diversity),
   (c) Study of the overall effects of changes in boundary conditions, climate and shelf processes on living resources. Predictive modeling (interdisciplinary).
3. Future problems:
   (a) Effects of temperature changes at high latitudes on coastal ecosystems.
   (b) Mariculture and fisheries—sustainability and ecological impacts (also other uses of coastal zone).
   (c) Sea level impacts (assuming change in present trend).

SPECIFIC WORLDWIDE PROBLEMS IN THE OCEANIC MARGIN

An important statement of objectives for managing the coastal zone is Clark's (1977) manual for coastal ecosystem management. His overall objective is maximum "carrying capacity," understood as "the potential of the coastal ecosystem to provide products useful to human society." From this objective, Clark derives sets of principles, standards, and development guidelines, which constrain economic and social activity, so that the vital areas and essential functions of the coastal ecosystem are conserved.

Living resources are a source of food, fuel, fiber, feedstock, pharmaceuticals, research material, and recreation. "They have two important properties, the combination of which distinguishes them from non-living resources: they are renewable if conserved; and they are destructible if not. Conservation is that aspect of management which ensures that the fullest sustainable advantage is derived from the living resource base and that activities are so located and conducted that the resource base is maintained" (World Conservation Strategy 1980).

The World Conservation Strategy (1980) represents a second step in the refinement of objectives for environmental management. Its goal is living resource conservation, which has three interrelated and specific objectives: (a) maintenance of essential ecological processes and life-support systems, (b) preservation of genetic diversity, and (c) sustainable utilization of species and ecosystems. Several governments have made national conservation strategies based on its recommendations.

The concept of the "sustainable development of planet earth" is now common currency, but it is by no means clear what it ultimately means. On the one hand, the "technological optimists" risking nemesis believe that a highly malleable technology can solve its environmental problems by decoupling the supply of people's material needs from the biosphere. While some of the features of this mode of production may be under discussion and partly in operation, e.g., clean technologies, closed-cycle production, recycling of waste, benign energy vectors such as hydrogen, etc., the final form is not. On the other hand, the "environmental pessimists" believe that the path to a sustainable planet requires a global decrease in human population, following the Chinese example, with or without a matching reduction or redistribution of material wealth. This debate will continue into the next century.

In the short term, however, environmental management will be concerned with the resolution of conflicting claims on the oceanic margin as a common property resource. This conflict is likely to persist longest in those countries which are least well endowed economically.

The most important uses of the coastal zone are (Goldberg 1990):

1. waste disposal,
2. transportation and port activities,
3. extraction of oil, gas, and materials,
4. mariculture—farming and ranching in sea cages and ponds,
5. coastal engineering,
6. recreation and tourism,
7. energy production from wind, waves, and thermal differences.

Waste disposal has been placed at the top because there is evidence of surprisingly high rates of child mortality from contact with enteric pathogens in the coastal zones of developing countries. This is not the unanimous view of the working group since there are at least three other possible routes of infection: polluted drinking water, contaminated fruit and vegetables, and contaminated shellfish (Bradford Hill 1965).

The fully engineered and managed coastal margin involving most of these uses is a possible response to population pressure and rising sea level. Like King Canute, people may have to step into the advancing sea and colonize it in the next century.

Decision makers will continue to ask specialists in coastal zone science for advice on the implications of different courses of action for resolving conflicting claims on the resource. This will require problem-orientated research projects which attempt to describe, understand, predict, and control aspects of particular parts of the oceanic margin. In addition, various international initiatives have been taken by ICSU and various UN agencies, e.g., GESAMP, to address common problems and deficiencies in scientific information and know-how in the countries of the world.

Future Task

A Dahlem conference on the sustainable development of the biosphere and the coastal margin.

References

Abbott M.B., J. Larsen, and Jianhua Tao. 1985a. Modelling circulations in depth-integrated flows. Part 1: The accumulation of evidence. *J. Hydraulic Res.* **23**(4):309–326.

Abbott, M.B., J. Larsen, Jianhua Tao 1985b. Modelling circulations in depth-integrated flows. Part 2: A reconciliation. *J. Hydraulic Res.* **23**(5):397–420.

Anderson, S. et al. 1980. Statistical Methods for Comparative Studies, Techniques for Bias Reduction. New York: Wiley.

Bath, M. 1974. Spectral Analysis in Geophysics. Amsterdam: Elsevier.

Bedford, K. 1981. Spectra preservation capabilities of Great Lakes transport models. In: Transport Models For Inland and Coastal Waters, ed. H.B. Fischer, pp. 172–221, New York: Academic.

Bendat, J.S., and A.G. Piersol. 1966. Measurement and Analysis of Random Data. New York: Wiley.

Bowden, K.F. 1983. Physical Oceanography of Coastal Waters. Ellis Horwood.

Bradford Hill, A. 1965. The environment and disease: association or causation. *Proc. Royal Soc. Med.* **58**:295–300.

Clark, J.R. 1977. Coastal Ecosystem Management. A Technical Manual for the Conservation of Coastal Zone Resources. New York: Wiley.

Cochran, W.G. 1977. Sampling Techniques, 3rd edition. New York: Wiley.

Csanady, G.T. 1982. Circulation in the Coastal Ocean. Dordrecht: D. Reidel.

Davis, R.A., ed. (1985) Coastal Sedimentary Environments. Berlin: Springer-Verlag.

Dyer, K.R. 1986. Coastal and Estuarine Sediment Dynamics. New York: Wiley.

Fisher, R.A. 1966. The Design of Experiments, 8th edition. Edinburgh: Oliver and Boyd.

Goldberg, E.D., and J.K. Taylor. 1985. The VD Conspiracy (editorial). *Water Pollution Bull.* **16**(1):1.

Goldberg, E.D. 1990. Protecting the wet commons. *Environ. Sci. Tech.* **24**:450–454.

Kisiel, C.C. 1969. Time series analysis of hydrologic data. In: Advances in Hydroscience, ed. Ven Te Chow, vol. 5, pp. 1–120. New York: Academic Press.

Lin, C.C., L.A. Segel, and G.H. Handelman. 1974. Mathematics Applied to deterministic problems in the natural sciences (2 volumes). London: Macmillan.

Mandlebrot, B. 1977. The Fractal Geometry of Nature. San Francisco: Freeman and Co.

Nihoul, J.C.J. 1977. Modèles mathématiques et dynamique de l'environnement. Liège: Ele.

Nihoul, J.C.J. 1985. Marine variability and Mathematical Models of the Marine System. In: Progress in Belgian Oceanographic Research, ed. R.Van Grieken and R. Wollast. Brussels: The Belgian Academy of Sciences, Committee of Oceanology.

O'Kane, J.P.J. 1986. Statistical mass-balances for estuaries. In: Estuarine Processes: An application to the Tagus Estuary, Proc. Int. Sci. Workshop organized by UNESCO/IOC/ICSU-SCOR/UNDP and the National Commission for the Environment (CNA), Lisbon, Portugal, pp. 505–520. Published by UNESCO, Paris and DGQA (CNA), Lisbon.

Pedlosky, J. 1977. Geophysical Fluid Dynamics, 2nd Edition. Berlin: Springer-Verlag.

Platt, T., and K.L. Denman. 1975. Spectral analysis in ecology. *Ann. Rev. Ecol. Syst.* **6**:189–210.

Platt, T., and K.L. Denman. 1978. The structure of pelagic marine ecosystems. *Rapp. P.-V. Reun. Cons. Int. Explor. Mer* **173**:60–65.

Platt, T., K.H. Mann, and R.E. Ulanowicz. 1981. Mathematical models in biological oceanography. Paris: UNESCO Press.

Press, W.H. et al. 1986. Fourier transform spectral methods. In: Numerical Recipes – the Art of Scientific Computing. Cambridge: Cambridge Univ. Press.

Solow, A.R., and J.R. Steele. 1990. On sample size, statistical power, and the detection of density dependence. *J. Animal Ecology*, in press.

Steele, J. 1989. The ocean "landscape." *Landscape Ecology* **3**(3/4):185–192.

Stommel, H. 1963. Varieties of oceanographic experience. *Science* **139**:572–576.

Walsh, J.J. 1988. On the Nature of Continental Shelves. New York: Academic Press.

World Conservation Strategy 1980. Int. Union for Conservation of Natural Resources. Gland, Switzerland.

APPENDIX A: Specific Worldwide Problems in the Oceanic Margin

A list of different types of perturbation to the oceanic margin is given below, followed by the various types of response that may occur. References may be found in the background papers.

1. Land/ocean boundary modification:
   (a) Physical
      - coastal engineering and development activities
      - freshwater management and changes in river flows
      - changes in the natural input/output balance of sediments in delta and river systems
      - dredging for harbor maintenance and hydraulic shell fishing

   (b) Biological
      - indirect effects related to physical modification (habitat, disturbance, and loss)
      - direct effects due to people: cutting of mangrove forests, draining of marshes and associated effect on sediment dynamics,

  - coastal subsidence due to extraction of coal, oil, gas, or water from the ground
2. Chemical modification (land use changes, waste disposal)
   (a) Nutrients → eutrophication (total nutrient load, nutrient balance)
   (b) Toxic substances (hydrocarbons, metals, synthetic goods, etc.)
3. Direct effects of climate
   (a) Physical processes—wind mixing, seasonal stratification, ocean boundary effects, land boundary effects (storm frequency)
   (b) Temperature
      - rates of growth and productivity
      - biogeographic population change
   (c) Precipitation
      - delivery of dissolved and suspended materials from land
      - delivery of water → coastal ocean circulation
4. Indirect effects of climate
   (a) Sea level rise 14 mm/yr in early postglacial → 0.4mm/yr in late postglacial → 1.5 mm/yr today → 5–10 mm/yr projected
   (b) Ocean circulation effects (also see 3a), e.g., El Niño

The various types of response to these perturbations are listed below. The response of the boundary dynamics is an open question; that of living resources not well known and of ecological viability, excluding the effects of eutrophication, largely unknown. The response of the carbon cycle in the coastal zone remains controversial.

1. Boundary dynamics
   (a) Land
      - movement of land/sea boundary due to sea level change
      - general decline in freshwater discharges and estuarine and groundwater salt intrusions
      - decline in sediment input from land—subsidence of deltas, instability of salt marches, etc.
      - changes in properties of estuarine/shelf sea boundary
   (b) Ocean
      - climatic effects on boundary currents and shelf break processes
      - effects of shelf processes (brine formation, freshwater exchange) on deep water formation
      - impact of ocean on coastal waters (e.g., El Niño)
   (c) Interaction between land (material inputs) and ocean (energy inputs) influences, relative sources of variability, etc.
2. Carbon cycle in the coastal zone
   (a) Nature of carbon fluxes, source of variability in global ocean carbon budgets, implications of eutrophication (land input of NO_3 increased

from 1% to 10% contributing to coastal "new production") and associated shifts in phytoplankton species (including toxic species), CO_2 exchange

   (b) Associated changes in anoxia in inshore water, CH_4 production
   (c) Dependent geochemical processes affecting the cycling of N,P,S, especially in relation to exchange of other trace gases (N_2O, DMS)
3. Living resources
   (a) Fish production—combined effects of climate, overfishing, and pollution
   (b) Mariculture—impact on coastal waters—eutrophication, addition of synthetic compounds
4. Ecological viability (excluding effects of eutrophication)
   (a) Addition of toxic materials
      - additive impact of various compounds, environmental capacity
      - new compounds (e.g., TBT)
      - chlorinated hydrocarbons (especially in tropics)
   (b) Species and genetic diversity (little known about this problem).
5. Human health and recreation
   (a) Enteric viruses and bacteria (strong disagreement among the working group on the risk to health in tropical coastal waters that are contaminated with human sewage)
   (b) Recreational resources—beaches, clean water

The Coastal Organic Carbon Cycle: Fluxes, Sources, and Sinks

R. Wollast

Univ. of Brussels, Laboratory of Chemical Oceanography
Campus de la Plaine, CP 208
Bd. du Triomphe, Brussels, Belgium

Abstract. Characteristic features of the carbon and nitrogen cycles in the coastal zone are presented. A tentative mass balance on a global scale indicates that export of organic C to the intermediate and deep water on the continental slope may constitute a significant sink. Preservation of organic C in the coastal sediments is of less importance although this flux has been strongly affected recently by the increase of the river particulate load and enhanced sedimentation rates on the continental shelves. On the other hand, the characteristics of the nitrogen cycle in the coastal zone indicate that the increase of the riverine nitrogen flux has only small effects on the coastal productivity on a global scale.

INTRODUCTION

The ability of the ocean to take up carbon dioxide injected into the atmosphere due to human activities is theoretically considerable. On a long-term basis, there is no doubt that the oceanic system will be the main sink for the anthropogenic carbon presently released. In addition to the direct transfer of CO_2 between the atmosphere and the ocean, carbon can be sequestered in the marine sediments as organic carbon or calcium carbonate. If reliable projections of future CO_2 changes are to be made, exchange rates between the atmosphere and the ocean, as well as the influence of physical, chemical, and biological factors on the exchange processes, must be thoroughly understood.

Most of the models of the global carbon cycle developed to evaluate the CO_2 fluxes between reservoirs use properties of the open ocean to represent the global ocean. The special role of the continental shelf and slope has

Ocean Margin Processes in Global Change
Edited by R.F.C. Mantoura, J.-M. Martin and R. Wollast

not usually been considered explicitly, although this environment is distinguishable from the open sea by several characteristic features.

The coastal zone directly receives the flux of dissolved and particulate material carried by rivers. Most of the solid phase is deposited and accumulated on the continental shelf and slope, including detrital terrestrial organic carbon. However, the most prominent feature is certainly the relatively high biological productivity of the coastal area due to the river input of nutrients, upwelling of fertile deep waters, the absence of losses of nutrients below a compensation depth, and the close coupling of the benthic and pelagic systems.

Due to the shallow depths, characteristic of continental shelves, a significant fraction of the primary productivity rapidly reaches the sea floor yielding organic rich sediments where intensive diagenetic processes occur. Fluxes of dissolved species from sediments in the coastal zone, particularly that of nutrients but also of trace gases of climatic significance (N_2O, CH_4, etc.), play a major role in biogeochemical processes (for references, see Wollast and Mackenzie 1989). Furthermore, since these areas are also characterized by high sedimentation rates, the coastal sediments constitute efficient sinks for organic carbon of terrestrial or coastal origin and associated elements.

All these distinctive features justify that more detailed attention be paid to the carbon cycle in the coastal area when global cycles are considered. This becomes even more imperative when the influence of perturbations for societal activities on geochemical cycles are discussed. The coastal and nearshore oceanic realm is particularly disturbed by human activities that may furthermore induce significant positive or negative feedback processes on a global scale. For instance, eutrophication can lead to an increased preservation of detrital organic carbon in coastal sediments and thus reduce the accumulation of anthropogenic CO_2 in the atmosphere. On the other hand, an increase of the organic carbon content in the surficial marine sediments will favor the denitrification process, which is responsible for the release of N_2O, an efficient greenhouse gas. This process, however, also removes some of the excess nitrogen responsible for eutrophication.

Considering the complexity of these interactions, it is not surprising that the relative significance of coastal process on a global scale is in many cases controversial. Even the fact that the continental shelf and slope acts as a source or sink for certain constituents is open to debate. Certainly the export fluxes of materials to the open ocean or to the atmosphere from the coastal zone on a global scale is poorly understood. In this paper I will present simplified biogeochemical cycles in order to evaluate the relative importance of processes associated with the carbon cycle in the coastal zone, using selected sets of values which in my opinion represent the best reality. I will mainly discuss primary production and the degradation of organic carbon and its preservation in the sediments. The influence of human activities on the fluxes resulting from the processes occurring in the coastal zone will also be discussed.

THE ORGANIC CARBON CYCLE IN THE WATER COLUMN

There are numerous measurements of primary productivity in coastal zones, and several evaluations of annual productions have been performed on a global basis. Platt and Subba Rao (1975) have estimated the annual productions of the marine microphytes divided into inshore and offshore zones. Their estimate of the average annual net primary production of shelf regions (<200 m depth) is 183 g C m^{-2} y^{-1}. However, the early evaluations of primary production were generally too low, partly because extracellular products and picoplanktonic production were not taken into account. Based on more detailed studies of the organic carbon cycle in the North Sea, Wollast and Billen (1981) suggested a value of 220 g C m^{-2} y^{-1} for the mean global productivity of phytoplankton in the coastal area. Similarly, Martin et al. (1987) estimated a mean value of 250 g C m^{-2} y^{-1} based on their data from the California coast. Based on a larger set of recent data, especially from mesotrophic areas, Walsh (1984) suggested a mean net primary production equal to 200 g C m^{-2} y^{-1} on a global scale. If we adopt the mean value of Table 1, it represents a net primary production of 6.9×10^{15} g C y^{-1} for the total surface area of the continental shelves (30×10^6 km^2).

Besides the primary production related to the activity of the phytoplankton, there is also an intensive production mainly due to macrophytes in coastal marshes and open water regions of estuaries. Woodwell et al. (1973) calculated that the primary production of these ecosystems was close to 1.4×10^{15} g C y^{-1}. The total production of the continental shelves can thus be estimated to be 8.3×10^{15} g C y^{-1}. Although the surface area of the coastal region represents only 8% of the global ocean, its contribution to the global production of the ocean is comprised between 18% and 33% if the highest estimate of 45×10^{15} g C y^{-1} (de Vooys 1979; Wollast 1981; Martin et al. 1987) or the lowest estimate of 25×10^{15} g C y^{-1} (Eppley and Peterson 1979; Berger et al. 1987) is considered.

The coastal area distinguishes itself not only by its high fertility but also by the subsequent utilization or fate of the fixed carbon. To demonstrate the special characteristics of the carbon cycle in the continental shelfs, I have summarized in Table 1 some of the main annual fluxes of the organic carbon cycle reported in a few detailed studies performed in these environments. These fluxes are related to the net primary productivity of the phytoplankton, the grazing of the algal production by zooplankton and microplankton, the production of detrital organic matter due to natural

TABLE 1. Annual carbon fluxes in the water column of coastal environments (in g C m^{-2} y^{-1}). The variance (in brackets) is calculated for the results of the six environments listed in the table.

	Net production	Planktonic grazing	Detritus production	Total respiration	Export plus sedimentation	Ref.
Bering Sea						
Outer shelf	162	68	114	83	79	(1)
Middle shelf	166	36	138	138	28	(1)
Anadyr Strait	285	87	198	167	118	(1)
Mid-Atlantic						
New York shelf	300	100	240	120	180	(1)
Gulf of Mexico						
Texas–Louisiana	100	23	84	44	56	(1)
North Sea						
Southern bight	370	70	233	200	170	(2)
Mean (variance)	230(94)	64(27)	168(59)	125(51)	105(56)	
% of net primary production	100%	28%	73%	54%	46%	

(1) Detailed references and discussion in Walsh (1988a)
(2) Joiris et al. (1982)

decay of phytoplankton or to grazing, the total respiration due to the activity of all the heterotrophic organisms, and finally the flux of detrital organic carbon exported to the continental slope or buried in the sediments.

As shown in Table 1, only a small fraction of the primary production is grazed. Zooplankton populations are relatively small and unable to efficiently utilize the algal production. Most of the production ends up as detrital organic carbon. This detrital organic matter can be oxidized by heterotrophic bacteria in the water column and the nutrients released during this process are further available to maintain recycled primary productivity. However, a significant fraction of the detrital organic matter escapes oxidation and is either deposited at the water–sediment interface or exported offshore. The ratio of the heterotrophic respiration in the water column to the net primary production allows us to evaluate the fraction of the recycled production in these environments. From the data in Table 1, it can be seen that this value varies between 40% and 83%, with a mean of 54%. This is in contrast with open ocean food webs, where grazing is usually extremely efficient and where up to 90% of the nitrogen demand of primary productivity is met by recycled forms of nitrogen (Eppley and Peterson 1979). The low values of the fraction of the primary productivity recycled in the water column are confirmed by measurements of the nitrogen fluxes and budgets carried out in the case of Georges Bank and the New York bight (cited in Walsh 1988b) and of the North Sea (Billen 1978).

This tendency can also be observed when the carbon cycles in inner and outer shelves are compared. The results of a study of the food web within these two regions for the Bering Sea (detailed references in Walsh 1988b) are shown in Table 1. On the outer shelf, primary production supports an essentially pelagic food web, with a suite of grazers and predators active in the water column. The situation is reversed in the inner shelf where most of the primary production settles instead to the bottom, favoring the development of a benthic food web. Furthermore, in the case of the outer shelf, a significant amount of the detrital material is exported to the continental slopes. In the case of the inner shelf, the main sink of organic carbon is the burial in the sediments. This distinction is important with respect to possible sinks of carbon in the ocean and will be discussed later in more detail.

Despite these differences, there are striking similarities between the cycles of the various coastal areas presented in Table 1, which allow us to propose a general scheme typical for the continental shelf. Mean values for the various fluxes are presented in Table 1, where I have also indicated the percentage of each flux with respect to the mean net primary productivity.

If we accept a mean value of 230 g C m^{-2} y^{-1} (Table 1 and discussion above) for the productivity of the global continental shelf, we can recalculate the relative importance of the individual C fluxes in the global carbon cycles in the coastal areas. These are represented in Figure 1. I have also distinguished in this cycle the production of detrital organic C of the herbivores based on the observations reported by the various authors listed in Table 1. The existing data show consistently that approximately 2/3 of the phytoplankton uptake by the herbivores is respired and 1/3 is excreted.

The most important feature of this cycle, in the context of the global carbon budget, is the large fraction of organic C exported from the water column either to the sediment or to the slope and open ocean. The fraction of the C sedimented that is buried and preserved in the sedimentary column escapes the ocean system and represents a sink on a geological time scale. On the other hand, the fraction exported to the deeper waters will be further respired in the water column during settling. If the respiration occurs below the thermocline, the organic carbon mineralized will increase the bicarbonate reservoir of the deep waters. This may be considered as a sink for organic C if short time scales in relation with anthropogenic activities are considered.

Sedimentation and respiration of organic C in the water column are however closely related. Figure 2 represents the fraction of the primary production collected in sediment traps as a function of depth, the remaining fraction being of course respired. Note that the decrease in organic C flux is roughly exponential. This figure confirms that in shallow areas about 50% of the primary production may reach the bottom in contrast with the pelagic

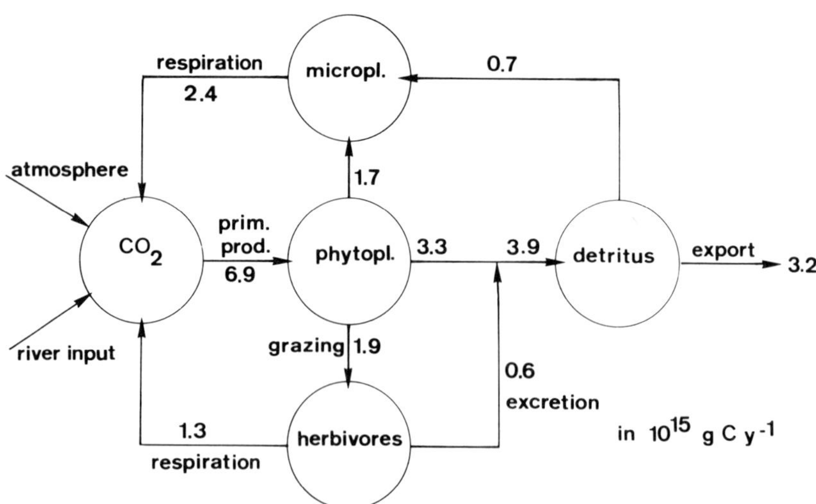

Fig. 1—Global annual fluxes of C associated with the biological activity in coastal environments (values based on data of Table 1). Fluxes in 10^{15} g C y^{-1}.

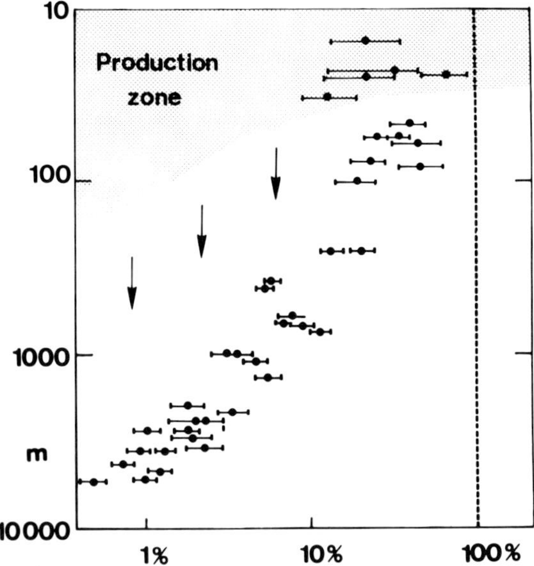

Fig. 2—Vertical fluxes of detrital organic matter collected in sediment traps, relative to the primary productivity, as a function of depth. After Suess (1980).

zone where typically only 1–2% of the productivity is deposited at depths from 2000 to 5000 meters.

The detrital organic carbon reaching the bottom of shallow areas is, however, not necessarily buried and preserved in the coastal sediments. The shelf is also a zone of high turbulence under the influence of vigorous currents where fine material is unable to accumulate. Storms, of course, are particularly efficient processes to resuspend the sediments there. On narrow shelves, dominated by coarse sand deposits, the tendency to export the primary production prevails and thus only a small fraction of the detrital organic C is accumulated in the coastal sediments.

The export of particulate C is also demonstrated by several sediment trap studies on or along the continental slope. These studies indicate that the particulate flux significantly increases with depth due to lateral transport at mid-water depths. Figure 3 shows the fluxes of organic C along the Middle Atlantic bight reported by Biscaye et al. (1988) during the SEEP I program.

A maximum concentration of organic C is often found in the sediments of the continental slope rather than in the coastal sediments along shelf areas with strong currents and high turbulence. Using the same data of SEEP I study, Walsh (1988a) estimated that the seasonally averaged vertical flux of organic carbon which eventually arrives on the continental slopes was 7.7 g C m^{-2} y^{-1}. It is, however, very difficult to extrapolate this flux

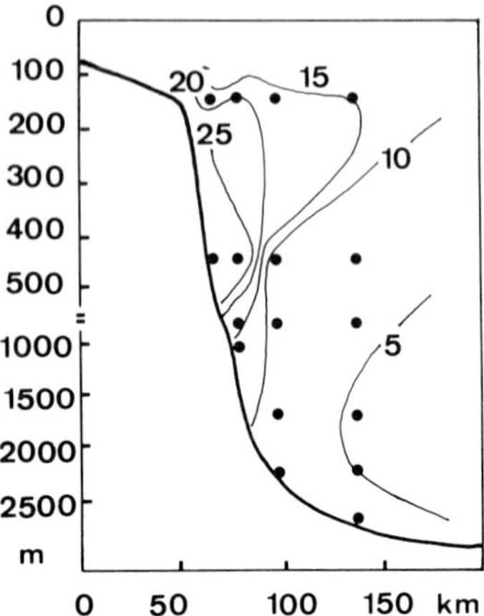

Fig. 3—Fluxes of organic carbon on the continental slope observed during the SEEP I experiment along the Mid-Atlantic shelf. After Biscaye et al. (1988). Fluxes in mg C m^{-2} d^{-1}.

on a global basis. Taking into account a surface area of the continental slope of 32.5×10^6 km^2, defined as the area extending from the 200 m to the 2000 m isobaths, Walsh (1988a) estimated the deposition of detrital organic carbon to be 0.25 10^{15}g C y^{-1}.

It must be remembered that this flux represents only a small fraction of the organic carbon exported from the shelf. Most of the carbon has been mineralized in the water column during transport and settling. According to Figs. 2 and 3 we can estimate that the total export flux from the near shore area is probably 4-5 times larger or about equivalent to 1×10^{15} g C y^{-1}. According to Fig. 1, this flux accounts only for one third of the organic matter which is removed from the water column in the coastal zone. This would suggest that by difference, approximately 2×10^{15} g C y^{-1} of organic carbon is deposited on the shelf.

I will now discuss in more detail the carbon cycle in the sediments and try to estimate how much of the deposited organic carbon is preserved.

PRESERVATION OF ORGANIC CARBON IN THE SEDIMENTS

It is of interest to evaluate first the burial of organic carbon on a long-term geological time scale, without the perturbations induced by humans. This evaluation can be done using an estimate of the relative proportions of shelf, hemipelagic, and pelagic sediments in the oceans, combined with average sedimentation rates on a geological time scale, and areas for these regions (see Wollast and Mackenzie 1983 for references in data selection).

Assuming sedimentation rates of 5, 50, and 125 g m^{-2} y^{-1} and areas of 268, 63, and 30 $\times$ 10^{12} m^2, respectively for pelagic, hemipelagic, and shelf sediments, we have total deposition rates of 1.3, 3.2, and 3.8 $\times$ 10^{15} g y^{-1} for these three oceanic provinces. These proportions correspond well to the observed abundance in the marine sedimentary record. I have, on the other hand, selected mean values for the organic carbon content of these three categories of sediments mainly based on the detailed discussion of this subject by Berner (1982), but also on recent data in relation to fluxes given in Table 1 (see discussion and references in Walsh 1988b). Assuming a mean organic carbon content of 0.3%, 0.7%, and 1.5% for the pelagic, hemipelagic, and shelf sediments, the mean rate of burial of organic carbon on a geological time scale would be respectively 4, 21, and 57 $\times$ 10^{12} g C y^{-1}. This gives a mean total rate of preservation on geological time scale of 82 $\times$ 10^{12} g C y^{-1}.

These estimates underline once again the particular importance of the coastal and nearshore area in the C budget of the ocean. If we include one half of the hemipelagic zone in the nearshore environment to represent the continental slope (32.5 $\times$ 10^6 km^2) (see Walsh 1988a), almost 80% of the past detrital organic C yearly buried in the marine sediments has accumulated there. This situation has certainly been intensified since the last 15,000 years due to the sea level rise, which has strongly reduced the export of particulate matter to the open ocean. It has been suggested (Gibbs 1981) that sediments carried by rivers are now being deposited entirely on the continental shelves with little being transported along the continental rise. If we consider, according to Gibbs that 90% of the sediments is now accumulating on the shelf and 10% on the continental slope, then the rate of accumulation of organic C can be estimated to reach 157 $\times$ 10^{12} g C y^{-1} for a similar sediment load and the same organic content in the various categories of deposits. Thus changes in the sedimentation pattern in the oceans might have affected strongly the preservation of organic matter in the past.

In order to evaluate and predict the effects of human activities, it is necessary to analyze in more detail the accumulation of organic matter in the sediments.

The organic matter freshly deposited at the sediment–water interface undergoes rapid degradation by aerobic bacterial respiration. In the oxic layer of surficial sediments, most of the metabolizable organic C is mineralized. If the supply of organic matter of the sediments is high enough, then anoxic conditions can be established, and less efficient and slower degradation processes like sulfate reduction or methane fermentation will take place. A significant fraction of the organic matter could be preserved if high rates of sedimentation were realized at the same time. They tend to increase the rate of burial of organic matter and prevent the downward diffusion of oxidants in the sediments. Furthermore, less efficient metabolic processes are involved under the reducing conditions involved (see Blackburn, this volume).

Figure 4 shows, for example, the rate of carbon burial as a function of the rate of sedimentation for a large variety of marine environments. According to this figure, the amount of carbon preserved in regions of high sedimentation rate (10^{-1} to 10^{-2} cm y^{-1}) would be about 10% of the organic matter delivered to the surface of the sediment.

Going back to the fluxes of organic C in the coastal zone presented in Fig. 2, of the 3.2×10^{15} g C y^{-1} detrital organic matter removed from the coastal zone, I have considered here that roughly 1×10^{15} g C y^{-1} is exported to the adjacent continental slope; the remaining 2.2×10^{15} g C y^{-1} must thus be deposited on the continental shelf. If only 10% of that is preserved in the sediments, the sink of organic matter of marine origin in the coastal deposits would represent 0.22×10^{15} g C y^{-1}.

The present-day particulate flux of the rivers is probably 15.5×10^{15} g y^{-1} (Meybeck 1982) and if 90% of this input accumulates in the coastal zone (Gibbs 1981), the rate of preservation of organic C estimated here would lead to a mean concentration of organic C of 1.6% in very good agreement with the observed range which fluctuates from 0.75% for deltaic-shelf sediments to 3% C for productive coastal areas far from the mouths of major rivers (Berner 1982).

The same processes occur on the continental slope. Previously I estimated that 1×10^{15} g C y^{-1} is exported from the nearshore area to the continental

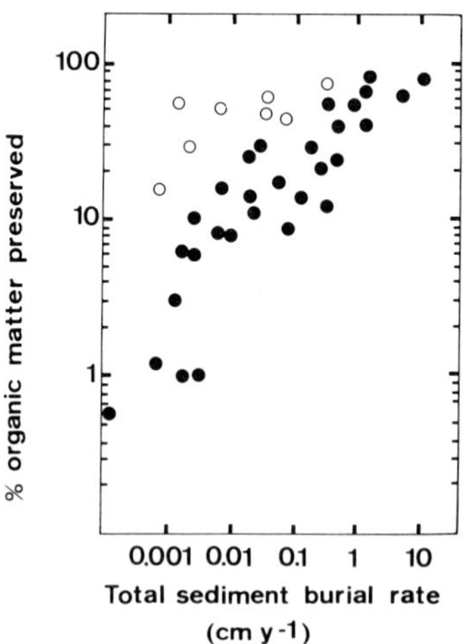

Fig. 4—Percentage of the organic matter preserved in the sediments as a function of the total rate of sedimentation. Open circles correspond to sediments in anoxic bassins. After Canfield (1989).

slope and we may assume that 80% of this flux is respired in the water column. Of the 0.2×10^{15} g C y^{-1} which is estimated to settle on the continental shelf, a large proportion is probably subjected to further degradation. The slower sedimentation rate, which would favor the respiration of the detrital organic matter, is however compensated by an increase of refractive nature of the residual organic substrate, which has already been intensively metabolized in the water column. If we take, as above, a burial ratio of about 10% and a sedimentation flux of 1.55×10^{15} g y^{-1}, representing 10% of the particulate river input, as representative for the present-day accumulation of sediments on the continental slope, we obtain a burial rate of 0.02×10^{15} g C y^{-1} and a mean organic C content of 1.3% by weight. This is similar to the value obtained on the geological time scale and represents probably an upper limit, indicating that the sediments of the continental slope are probably not significant sinks for C compared to the coastal sediments.

There is, however, another significant difference between the shelf and slope environments. In the case of the continental slope, the organic matter sedimented or deposited is mostly respired below the thermocline, and on short time scales the carbon released will increase the dissolved inorganic reservoir of the deeper waters, which may be considered as a temporary sink. In the coastal waters, the inorganic carbon resulting from the respiration of organic C in the surficial layers of the sediments is seasonally mixed with the surface waters and may thus ventilate with the atmosphere.

Figure 5 summarizes the mean fluxes of organic C on the continental shelves and slopes as discussed here.

INFLUENCE OF HUMAN ACTIVITIES

The increase of nutrient fluxes from rivers due to human activities has potentially enhanced the productivity of the coastal zone. In order to evaluate the importance of this perturbation, it is necessary to define the relative contribution of the riverine input with respect to other sources of nutrients necessary to satisfy the requirements of the new production. These sources must compensate for the losses due to burial of nutrients or their exportation to the deep pelagic waters.

The nitrogen cycle, which is most likely controlling the primary productivity, will be used to illustrate these calculations. The amount of nitrogen associated with the global primary production of the coastal zone, estimated to be 6.9×10^{15} g C y^{-1}, must be of the order of 1.4×10^{15} g N y^{-1}, if a C:N ratio by weight close to 5 is assumed for phytoplankton in this environment. A large fraction of this nitrogen (54% or 740×10^{12} g N y^{-1}) is provided by respiration of organic matter and recycling of the nitrogen released in the water column. Another large fraction of recycled nitrogen

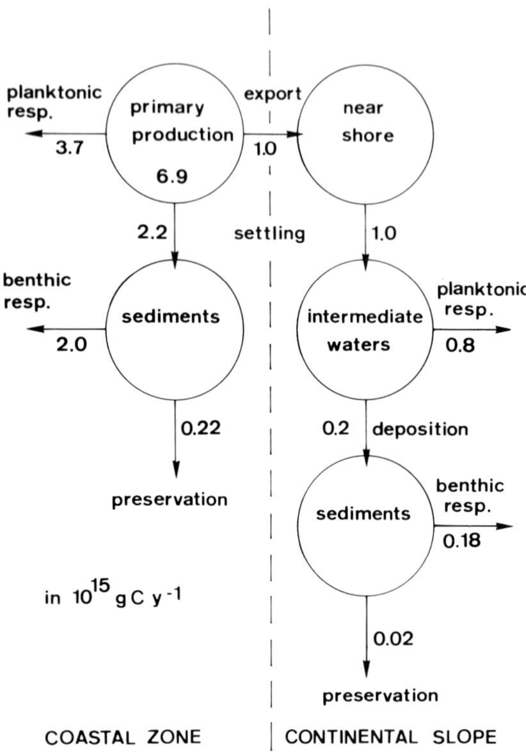

Fig. 5—Tentative global mass balance of the organic C fluxes on the continental shelf and slope. Fluxes in 10^{15} g C y^{-1}.

is supplied from the respiration of organic matter in the superficial layer of the coastal sediments. Some of the deposited organic nitrogen is, however, preserved in the sediments and some is lost due to denitrification under anoxic conditions. The C:N ratio of the organic matter preserved in the sediment is rather close to 12, and the amount of N buried in the coastal zone is on the order of 20×10^{12} g N y^{-1} for the 0.22×10^{15} g C y^{-1} of organic matter preserved. Denitrification affects more significantly the nitrogen budget in sediments (Christensen et al. 1987). According to an extensive study of the nitrogen cycle in the North Sea, Billen (1978) suggests a loss of 17% of the nitrogen recycled by denitrification in the sediments. This would represent a total loss of 70×10^{12} g N y^{-1} for the global coastal zone on the basis of the amount of organic matter mineralized in the sediments (2×10^{15} g C y^{-1}) (see Fig. 5). Christensen et al. (1987) calculated a mean flux based on measurements performed on continental shelf sediments of a wide range of areas and found 50×10^{12} g N y^{-1} for the coastal area or 62 × 10^{12} g N y^{-1} if the estuaries are included. Note that the same authors suggest an organic nitrogen burial of 21 × 10^{12} g N y^{-1}, very similar to our estimation. By difference, the N recycled as ammonia or nitrate from the sediments to the surface waters amounts to 370 × 10^{12} g N y^{-1}.

From this we can estimate that the total amount of N recycled in the coastal zone is equal to 1110 × 10^{12} g N y^{-1}, which represents 80% of the nitrogen necessary for the estimated net primary productivity. The new production, by difference, is thus equal to 20% of the total productivity, in very good agreement with the values suggested by Martin et al. (1987). In terms of nitrogen fluxes, the new production requires an input of 290 × 10^{12} g N y^{-1}. The present-day river input has been estimated to be 25 × 10^{12} g N y^{-1} (Wollast 1983), in contrast with the pre-human flux of 4.3 × 10^{12} g N y^{-1} of inorganic dissolved nitrogen prior to human activities given by Meybeck (1982). This represents only a small fraction of the nitrogen required for the new primary production and the main source of nitrogen must thus be from the upwelling of deep, rich waters. If we neglect the flux due to nitrogen fixation, which is usually considered to be small in the oceans, the upwelling flux in the coastal zone may be estimated by difference to be 265 × 10^{12} g N y^{-1}. This value is close to a previous estimate of the total upwelling flux for the entire ocean, equal to 340 × 10^{12} g N y^{-1} (Wollast 1981). A large proportion of this upwelling flux must indeed take place in the coastal zones. Even disturbed, the river input represents only 10% of the upwelling flux of nitrogen. A tentative summary of the nitrogen cycle associated with the carbon cycle on the continental shelf is given in Fig. 6.

CONCLUSIONS

Analysis of the organic carbon cycle in the coastal zone indicates that the fluxes encountered in this environment are similar to those of C resulting from anthropogenic activities. A well-balanced budget of the present-day C cycle must take into account potential sinks of this element on the continental shelf and slope.

A coherent set of fluxes can be obtained by considering either the primary production and processes occurring in the water column, or the sedimentation rates and composition of coastal and hemi-pelagic sediments. These estimations indicate that the preservation or export of carbon to the deep water may be presently on the order of one gigaton of C per year. It may thus represent a significant sink for some of the "missing CO_2." Field measurements with sediment traps and surveys of the composition of sediments indicate that the main scavenging process of organic C in the ocean is export to the deep waters of the open ocean rather than accumulation and preservation in the sediments. This represents, however, only a

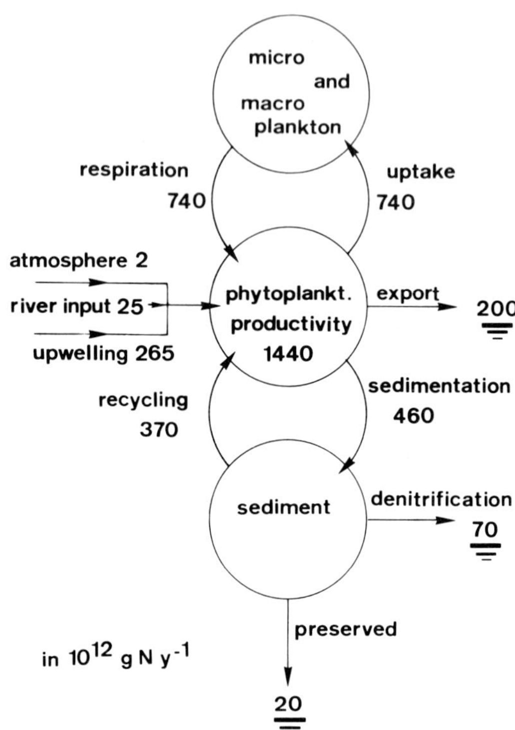

Fig. 6—Fluxes of nitrogen in the coastal zone associated with the carbon cycle given in Figs. 1 and 5. Fluxes in 10^{12} g N y^{-1}.

temporary sink for anthropogenic C, on the order of a few hundred years, which is the time necessary to mix the oceanic water column.

The coastal system is furthermore very sensitive to small perturbations. One of the most important factors affecting the present rate of preservation of organic C in the sediments is the rate of sedimentation of terrestrial particulate matter. This sedimentation rate has been increased in the last thousand years due to sea level rise and higher erosion rates of the continents related to human activities. Crude calculations indicate that the present-day accumulation rate of organic C is probably three times higher than its mean rate recorded during the geological time.

Another interesting conclusion drawn from these calculations is that the river input of nutrients related to human activities represents, on a global scale, only a small amount of the nitrogen used in the primary production of the coastal zone. The present-day river input of N accounts for only 10% of the new production of this fertile environment. Eutrophication may certainly constitute a significant perturbation locally, in areas directly influenced by the freshwater input of highly polluted rivers, as in the case of enclosed seas such as the Baltic and the North Sea. But it has probably not changed significantly the global productivity of the coastal zone.

One can argue, on the other hand, that the areas in the vicinity of the river mouth affected by an increase of nutrients are also submitted presently to higher particulate loads, which are obviously favoring the accumulation of organic matter in the sediments. If the 21×10^{12} g N y^{-1} of nitrogen nutrients added to the coastal zone due to the anthropogenic activities are preserved in the sediments as detrital organic matter with a C:N ratio of 12, this would correspond to a maximum burial of 0.25×10^{15} g C y^{-1}. Note that this value is similar to the one shown in Fig. 5, based on the composition of coastal sediments and on the present-day rate of sedimentation in that area. This indicates that the increased nutrient supply can theoretically compensate the enhanced burial of organic matter without impoverishing the ocean reservoir.

Several other interesting aspects of the carbon cycle have not been discussed here. The river input of particulate and dissolved organic carbon corresponds to 0.18×10^{15} g C y^{-1} and of 0.22×10^{15} g C y^{-1}, respectively (Meybeck 1982). The fate of this rather refractory material is not well known, but we may assume that their rate of supply to the ocean has been increased due to human activities and their rate of burial in the coastal zone has been increased due to higher sedimentation rates. This may represent another sink for C.

The active biological activity characteristic of the coastal zone can have a pronounced effect on the partial pressure of CO_2 in the surface water and thus on the transfer of carbon at the air–sea interface. The existing evaluations of the exchange of CO_2 between surface waters and the atmosphere have not taken the coastal zone into account. The problem, however, is complicated because of strong local and seasonal fluctuations of the partial pressure of CO_2 in the surface layer of the water column. Upwelling of deep waters in the coastal zone provides water masses oversaturated in CO_2 with respect to the atmosphere. On the other hand, the intensive productivity occurring in these nutrient-rich waters results in a drastic decrease of the dissolved CO_2, which in turn produces undersaturated water masses. The departures from equilibrium in the coastal zone are usually much higher than those observed in the open ocean and the exchange of CO_2 at the air–sea interface cannot be ignored there.

Finally, the carbon cycle associated with the precipitation and dissolution of carbonate minerals has not been included in our discussion. It is, however, intimately linked with the organic carbon cycle. The production of calcium carbonate skeletons is of course dependent on the productivity of the ecosystem and there is a positive correlation between these two factors. However, in areas characterized by high productivities and sedimentation rates of organic matter, the carbonate minerals may undergo severe dissolution in the surficial layers of the sediments by carbonic acid due to bacterial respiration. Sediment trap studies along the continental slope have furthermore shown that calcium carbonate produced on the continental shelf may also be exported to a large extent to the intermediate and deep waters (Biscaye et al. 1988).

All the processes affecting the organic and inorganic C related cycle in the coastal zone evoked here are in fact intimately intricated. An evaluation of net fluxes occurring on the continental shelf and slope requires the development of specific models and the acquisition of more field data. The importance of the coastal zone in global oceanic models has been underestimated in the past; it is fortunate that several international programs have more recently begun to study this complicated and changing environment.

Acknowledgements. The author thanks L. Chou and M. Whitfield for their helpful comments and discussions which helped to improve the presentation of this paper.

REFERENCES

Berger, W.H., K. Fisher, C. Lai, and G. Wu. 1987. Ocean productivity and organic carbon flux. Part I. Overview and map of primary production and export production. *Scripps Inst. Ocean.* 87–30.

Berner, R.A. 1982. Burial of organic carbon and pyrite sulfur in the modern ocean: its geochemical and environmental significance. *Am. J. Sci.* **282**:451–473.

Billen, G. 1978. A budget of nitrogen recycling in North Sea sediment off the Belgian coast. *Est. Coast. Marine Sci.* **7**:127–146.

Biscaye, P.E., R.F. Anderson, and B.L. Deck. 1988. Fluxes of particles and constituents to the eastern U.S. continental slope and rise: SEEP-1. *Cont. Shelf Res.* **8**:811–840.

Canfield, D.E. 1989. Sulfate reduction and oxic respiration in marine sediments: implications for organic carbon preservation in euxinix environments. *Deep-Sea Res.* **36**:121–138.

Christensen, J.P., J.W. Murray, A.H. Devol, and L.A. Codispoti. 1987. Denitrification in continental shelf sediments has major impact on the oceanic nitrogen budget. *Glob. Biogeochem. Cyc.* **1**:97–116.

de Vooys, C.G. 1979. Primary production in aquatic environments. In: The Global Carbon Cycle, ed. B. Bolin, E.T. Degens, S. Kempe and P. Ketner, pp. 259–292. New York: Wiley.

Eppley, R.W., and B.J. Peterson, 1979. Particulate organic matter flux and planktonic new production in the deep ocean. *Nature* **282**:677–680.

Gibbs, R.J. 1981. Sites of river derived sedimentation in the ocean. *Geology* **9**:77–80.

Joiris, C., G. Billen, C. Lancelot, M.H. Daro. J.P. Mommaerts, J.H. Hecq, A. Bertels, H. Bossicart, and J. Nijs. 1982. A budget of carbon cycling in the Belgian coastal zone: relative roles of zooplankton, bacterioplankton and benthos in the utilization of primary production. *Neth. J. Sea Res.* **16**:260–275.

Martin, J.H., G.A. Knauer, D.M. Karl, and W.W. Broenkow. 1987. VERTEX: Carbon cycling in the northeast Pacific. *Deep-Sea Res.* **34/2**:267–285.

Meybeck. M. 1982. Carbon, nitrogen and phosphorus transport by world rivers. *Am. J. Sci.* **282**:401–450.

Platt, T., and D.V. Subba Rao. 1975. Primary production of marine macrophytes. In: Photosynthesis and Productivity in Different Environments, ed. J.P. Cooper, pp. 249–280. London and New York: Cambridge.

Suess, E. 1980. Particulate organic carbon flux in the ocean-surface productivity and oxygen utilization. *Nature* **288**:260–263.

Walsh, J.J. 1984. The role of the ocean biota in accelerate ecological cycles: a temporal view. *Bioscience* **34**:499–507.

Walsh, J. J. 1988a. How much shelf production reaches the deep sea. In: Productivity of the ocean: present and past. Dahlem Workshop 53. Berlin.

Walsh, J.J. 1988b. On the Nature of Continental Shelves. London: Academic Press.

Wollast, R. 1981. Interactions between major biogeochemical cycles: marine cycles. In: Some Perspectives of the Major Biogeochemical Cycles, ed. G.E. Likens, SCOPE Report 17, pp. 125–141. Chichester: Wiley.

Wollast, R. 1983. Interactions in estuaries and coastal waters. In: The Major Biogeochemical Cycles and their Interactions, ed. B. Bolin and R.B. Cook. SCOPE Report 21, pp. 385–410. Chichester: Wiley.

Wollast, R., and G. Billen. 1981. The fate of terrestrial organic carbon in the coastal area. In: Carbon Dioxide Effects Research and Assessment Program, ed. G.E. Likens, F.T. Mackenzie, J.E. Richey, J.R. Sedell, and K.K. Turekian, CONF-8009140, pp. 331–359. Washington: U.S. Dept. of Energy.

Wollast, R., and F.T. Mackenzie. 1983. The global cycle of silica. In: Silicon Geochemistry and Biogeochemistry, ed. S.R. Aston, pp. 39–76. London: Academic.

Wollast, R., and F.T. Mackenzie. 1989. Global biogeochemical cycles and climate. In: Climate and Geo-Sciences, ed. A. Berger, S. Schneider, and J. Cl. Duplessy, NATO ASI series, vol. 285, pp. 453–510. Dordrecht: Kluwer.

Woodwell, G.M., P.H. Rich, and C.A.S. Hall. 1973. Carbon in estuaries. In: Carbon and Biosphere, ed. G.M. Woodwell and E.V. Pecan, pp. 223–240. Springfield, Virginia: U.S. Atomic Energy Commission.

Climatic and Environmental Implications of Biogas Exchange at the Sea Surface: Modeling DMS and the Marine Biologic Sulfur Cycle

G.V. Wolfe[1], T.S. Bates[2], R.J. Charlson[3]

[1]*University of Washington and Joint Institute for the Study of Atmosphere and Oceans (JISAO)*
Seattle, WA 98195, U.S.A.

[2]*NOAA/PMEL/OCRD*
7600 Sandpoint Way
Seattle, WA 98115, U.S.A.

[3]*Dept. of Atmospheric Sciences, University of Washington*
Seattle, WA 98195, U.S.A.

Abstract. The direct and indirect climatic effects of trace gas emission by oceanic biota are considered; it is shown that the highest potential sensitivities derive from the change in remote marine stratus cloud albedo due to dimethyl sulfide (DMS) emissions. As a paradigm for modeling trace biogas emissions, a hierarchy of box models is employed to explore the controlling variables on the biogenic production of this gas in the global ocean. This remote marine scenario is contrasted with the climatic and environmental impacts of coastal regions, and the coupling of the sulfur cycle with those of other trace gases is considered. The need to understand controlling variables and processes is stressed as a means to understand the stability of the system.

INTRODUCTION

In recent years there has been increasing recognition of the importance of biogenic trace gas emissions on climate. The modulation of carbon, sulfur, nitrogen, and halogen containing gases by biota has important potential impacts on the radiative balance of the Earth's atmosphere, both directly

Ocean Margin Processes in Global Change
Edited by R.F.C. Mantoura, J.-M. Martin and R. Wollast

TABLE 1. Climatically important marine biogases.

Gas[1]	Origin	Net flux to atmosphere[2] (mole y^{-1})		Atmospheric residence time w.r.t. air–sea exchange (y)[3]	Climatically important atmospheric species[4]	Climatic effect[5]
I. Carbon Species:						
CO_2	respiration	−2 E13	(a)	500	CO_2	a
CO	microbial photooxidation of organic matter	1–3 E12	(b)	1	CO, CO_2	a, b
CH_4	biogenic	5–10 E11	(c)	100	CH_4, CO, CO_2 HCHO	a, b
NMHC (C2–C5)	biogenic	1 E11	(d)	0.03	NMHC	b
II. Sulfur Species:						
DMS[6]	enzymatic cleavage of biologically produced DMSP	5–10 E11	(e)	0.006	H_2SO_4, CH_3SO_3H	c
OCS	photooxidation of organic sulfur	1–2 E10	(f)	2	H_2SO_4	d

III. Nitrogen Species:						
N_2O	microbial	5–50 E10	(g)	40	N_2O	a, e
NH_3	biogenic	1 E12	(h)	0.007	$(NH_4)_2SO_4$	c
IV. Halogen Species:						
CH_3I	biogenic	9 E9	(i)	0.004	IO°	b
CH_3Cl	CH_3I	ND[(7)]		ND	ClO°	e
CH_3Br	biogenic	ND		ND	BrO	e

[1]Species which exchange between the ocean and atmosphere.
[2]Negative fluxes indicate the ocean is a net sink.
[3]Residence time with respect to air–sea exchange: calculated from flux and average atmospheric concentration at sea level over the remote oceans assuming a mixing height of 2 km. This is a lower estimate since the long-lived gases are obviously mixed to higher altitudes.
[4]Includes the species itself or its atmospheric oxidation products.
[5]Effect which influences the Earth's climate: (a) tropospheric greenhouse gas, (b) major influence on oxidative properties of troposphere (i.e., O_3 OH^-), (c) source of tropospheric aerosols and cloud condensation nuclei,
(d) source of stratospheric aerosols, (e) major influence on oxidative properties of stratosphere.
[6]Related sulfur gases (CH_3SH, DMDS, H_2S) can account for as much as 10–20% of the total reduced sulfur flux, and are thought to have similar atmospheric effects as DMS.
[7]ND implies that no data exist.
[8]NMHC = nonmethane hydrocarbons.
Sources: a. Tans et al. (1990); b. Logan et al. (1981); c. Ehhalt (1974); d. Rudolph and Ehhalt (1981); e. Bates et al. (1987); f. Ferek and Andreae (1984); g. Butler et al. (1989); h. Placet et al. (1989); i. Rasmussen et al. (1982).

and via aerosol/cloud formation. In turn, aerosol particles from the atmospheric oxidation of biogases have both a direct effect of scattering of solar radiation into space and a more important indirect effect by acting as cloud condensation nuclei (CCN), particularly for marine stratus clouds. Table 1 lists the climatically important biogases and reaction products along with their net fluxes, atmospheric residence times, and climatic effect mechanisms. It is clear from this table that the ocean–atmosphere fluxes of DMS, NMHC, CH_3I, and NH_3 are extremely important in controlling the atmospheric concentration of these species (short atmospheric residence time with respect to air–sea exchange). Table 2 explores the quantitative effects of these biogases on the Earth's radiation balance. Included in the table is a simple index of the climate sensitivity, $\Delta T_{+30\%}$, illustrating in principle that DMS emissions and the resulting sulfate–aerosol effects are indeed the most sensitive species in spite of the exceedingly low mean concentrations.

While other components may (and probably will) appear in this picture, this list serves to focus attention on the individual chemical species and, more importantly, on the processes that control their concentrations. Clearly all of them deserve to be fully understood; however, due to the large uncertainties regarding CH_4 and N_2O and due to recent progress in understanding the formation of CCN from the atmospheric oxidation of $(CH_3)_2S$ (DMS), we have selected the latter as an initial focus. Our purpose in doing so is to present a developing strategy for research on such complex biological–chemical–physical systems. Figure 1 describes the atmospheric/climatic side of the remote marine system comprising DMS emission by phytoplankton, gas transfer to the atmosphere, reaction to non-sea-salt sulfate (nss-SO_4^{2-}) aerosol particles acting as CCN and their effect on cloud albedo. This figure has served as a useful paradigm, particularly in the design of field experiments and data analysis.

In this chapter we first consider microbiologic DMS production processes which occur globally, and whose CCN/cloud albedo climatic sensitivities are thought to be highest in the remote marine atmosphere. We will build a hierarchy of box models to explore the possible controls on the flux of this gas to the atmosphere and use these models to suggest important measurements which future studies might consider. We will also stress the importance of coastal zones in other environmental and climatic effects, such as haze production and acid rain, as well as the impact of human activity on natural trace gas emissions.

TABLE 2. Possible physical climatic effects of marine biogases and their reaction products.

Component[1]	Concentration[2] (mole m^{-3})		Current trend 1/C dC/dt (% y^{-1})		$\Delta T_{+30\%}$[3]
I. Gaseous Species					
CO_2	15 E–3	(a)	0.5	(a)	0.5–0.7
CH_4	74 E–6	(a)	1.1	(a)	0.1–0.2
N_2O	14 E–6	(b)	0.35	(d)	0.1–0.2
O_3	0–2 E–6	(c)	1.4	(d)	0.1–0.2
II. Aerosol Species					
SO_4^{2-}	2–7 E–9	(e)	ND[5]		−1.3[4]
$CH_3SO_3^-$	5–10 E–10	(f)	ND		ND
NH_4^+	2–7 E–9	(f)	ND		ND

[1]Climatically important species are taken from Table 1. Although O_3 is not a direct product of marine biogases, its concentration over the remote oceans is strongly influenced by marine biogas emissions (Johnson et al. 1989).
[2]Approximate current remote mean atmospheric concentrations (C) at sea level based on: (a) Ramanathan et al. (1985); (b) Butler et al. (1989); (c) Johnson et al. (1990); (d) Oltmans and Komhyr (1986); (e) Savoie and Prospero (1989); (f) Quinn et al. (1989).
[3]From Ramanathan et al. (1985). $\Delta T_{+30\%}$ is the estimated change in surface temperature in response to increasing the concentration of component C by 30% from present levels, all other factors held constant.
[4]From Charlson et al. (1987). Based on a change of CCN concentration in remote marine stratus clouds assumed proportional to H_2SO_4 concentration (a weak assumption).
[5]ND implies no data exist.

THE MARINE SULFUR CYCLE AND IMPLICATIONS FOR DMS PRODUCTION

Direct Production Model

To understand the production of DMS and what controls the flux to the atmosphere, it is necessary to consider the cycling of sulfur in the ocean photic zone by biota. In order to gain some insight into DMS flux to the atmosphere, we consider a hierarchy of models which progress from simple to complex. This model is given for historical reference and assumes no other processes exist, other than those that are shown.

What we term the "standard" model of DMS production may be summarized in Fig. 2, which considers two global boxes, a marine photic zone 100 m deep and an atmospheric marine boundary layer 1 km deep (note the units used are moles rather than mass). Although extrapolation

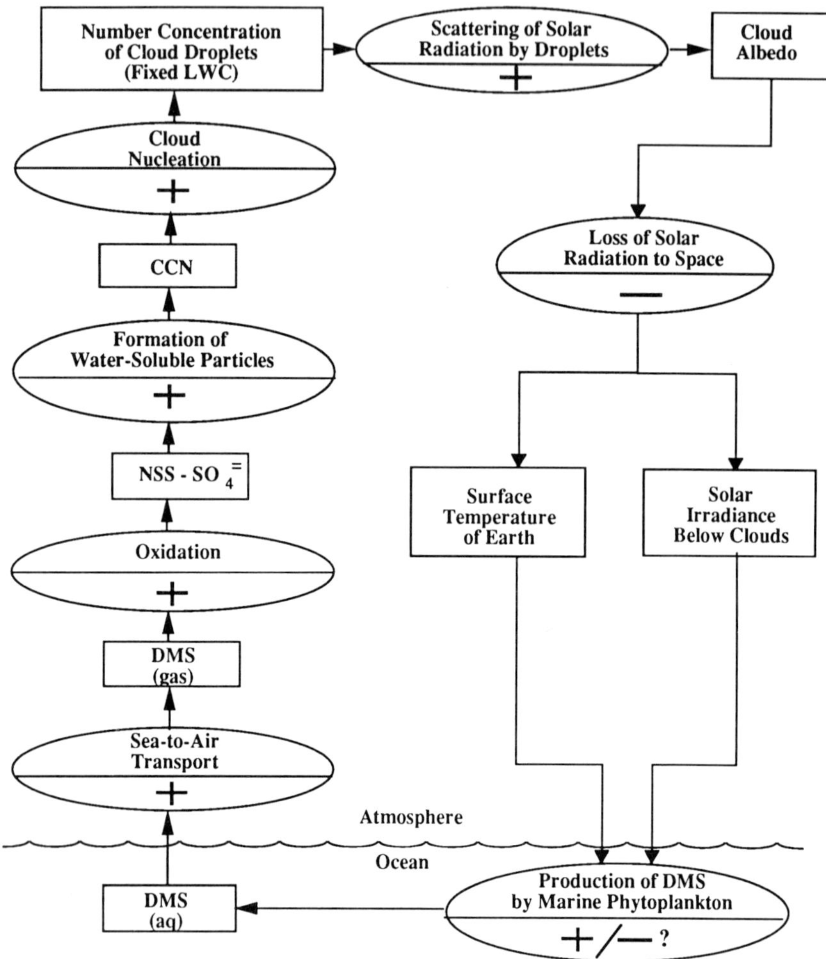

Fig. 1—Hypothesized DMS/CCN/cloud albedo feedback loop (from Charlson et al. 1987). Ellipses describe processes (with sign of feedback); boxes describe quantities or states.

of regional DMS measurements to global boxes is risky, we choose to do so (a) to stress the global nature of the marine sulfur cycle and (b) because no good regional data sets exist for the other sulfur compounds we will address below. Here we see direct emission of DMS into the water column by marine phytoplankton (a highly species-specific phenomenon, dominated in lab culture studies by such groups as the dinoflagellates, coccolithophores, and prymnesiophytes [Keller et al. 1989]), followed by loss to the atmosphere. The magnitude of DMS production by phytoplankton has not yet been

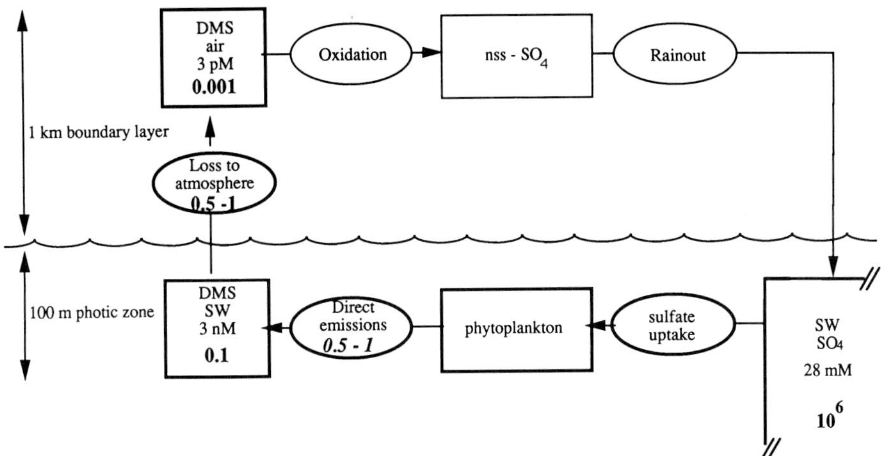

Fig. 2—"Standard" DMS production model. Boxes assume global oceanic area (3.6 E14 m^2). Bold face units are Tmoles (10^{12}moles) or Tmoles/yr; italicized fluxes are inferred from steady-state assumption.

measured directly in the natural environment, so this flux has been set artificially to balance the loss to the atmosphere (deduced from sea/air DMS concentration measurements and gas transfer velocity models, which is a gross oversimplification) under assumptions of steady state and no other major oceanic sources or sinks. The cycle is completed by oxidation of DMS to sulfate in the marine atmosphere followed by rainout back to the sea surface (and also via transport to land areas followed by rainout and river transport back to the sea). Numerous studies have shown only poor correlation of seawater (DMS) with total biotic activity as measured by chlorophyll *a*, primary productivity, or nutrient or salt concentrations (Andreae et al. 1983; Turner et al. 1988; Iverson et al. 1989); rather, these studies suggest that changes in phytoplankton speciation might be a controlling factor in DMS flux to the atmosphere in the standard model.

The ultimate source of DMS-S is the huge inorganic sulfate pool, which as the second-most abundant anion in the sea (28 mM) is essentially infinite compared to other compounds we are considering. Therefore, the first limit to consider on DMS production is the portion of this sulfate which is taken up or assimilated by phytoplankton. What is the magnitude of sulfate assimilation by phytoplankton in the world oceans? This number has not been measured, but a first estimate may be derived from carbon fixation and consideration of C:S stoichiometries. We do so with the following crude assumptions:

1. Phytoplankton are the sole primary producers (fixers of carbon).
2. There is a "Redfield" ratio for C:S for a given group of organisms such as phytoplankton.

These assumptions are justified insofar as (a) phytoplankton production constitutes a majority of oceanic carbon fixation, and (b) sulfur plays an important structural role in protein conformation and is also present in universal biochemical compounds (for example, the FeS reaction centers in the electron transport chain, and sulfoquinovosyl diglyceride is present in all plant chloroplast membranes).

What are typical molar C:S ratios? Data are sparse, but Table 3 presents some of the measurements for marine macroalgae, phytoplankton, bacteria, and particulate matter. Note that the average ratio of approximately 100:1–200:1 yields an average dry weight sulfur composition of about 0.7–1.5%, assuming that carbon is 50% of the dry mass and a wet:dry mass ratio of 4:1. These values are considerably higher than the 0.5% S typical of terrestrial producers (Bowen 1979). In a sulfur-rich environment this is not unreasonable, but many of the data are old and uncertain and this estimate is probably an upper limit.

Taking recent primary productivity and biomass data, Table 4 shows the calculated global sulfate assimilation to be 8–25 Tmol S y^{-1} and bio-sulfur burden to be 0.4–2.5 Tmol S. A recent study of total particulate carbon, nitrogen, and sulfur in the highly productive southern California bight (Matrai and Eppley 1989), extrapolated to a global box, yields sulfate uptake of 20 Tmol y^{-1} and sulfur biomass (both living plus dead material) of 0.6 Tmol, suggesting that the calculated values are at least reasonable as a first

TABLE 3. C:S (molar) in marine biota.

C:S (molar)	Source	Origin	Notes
25–50	(a)	red macroalgae	
100	(a)	green/brown macroalgae	
100–200	(a)	marine phytoplankton	
119	(b)	ocean plants	
250	(c)	marine bacteria	uptake ratios similar
280–340	(d)	marine bacteria	
120	(e)	sinking particulates	living + dead
225	(e)	suspended particulates	living + dead
100–200 average (mass ratio 40–80)			

Sources: (a) Bowen (1979); (b) Likens et al. (1981); (c) Cuhel (1982); (d) Jordan and Peterson (1978); (e) Matrai and Eppley (1989).

TABLE 4. Global carbon fluxes and biomass and derived sulfur.

Carbon[1]:		
Ocean productivity:	20–30 Gt C y^{-1}	1650–2500 Tmol C y^{-1}
Ocean biomass:	1–3 Gt C	80–250 Tmol C
Sulfur derived from 100–200 C:S ratio:		
Sulfate assimilation:		8–25 Tmol S y^{-1}
Ocean S biomass:		0.4–2.5 Tmol S

[1]De Vooys (1979)

approximation. Figure 3 incorporates these numbers into the direct production model. This figure also includes the one potentially important abiotic sink for DMS which has been suggested (Brimblecombe and Shooter 1986): photooxidation in the presence of organic photosensitizers to dimethyl sulfoxide (DMSO) (see below).

We note that the flow of DMS out of phytoplankton is a relatively small

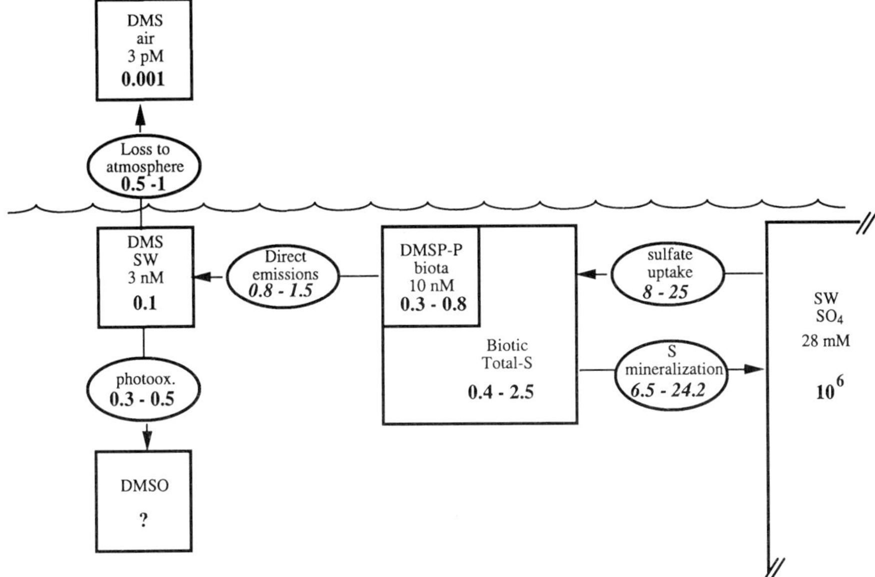

Fig. 3—Direct DMS production model. Boxes assume global oceanic area (3.6 E14 m^2). Boldface units are Tmoles (10^{12} moles) or Tmoles/yr; italicized fluxes are inferred from steady-state assumption.

leak in the uptake of sulfate, suggesting that DMS emissions might potentially be considerably larger without seriously straining the marine biologic sulfur cycle, as for example given changes in phytoplankton speciation. However, although the sensitivity of climate to a change in sea–air DMS flux is not yet known, it is thought that an increase of even a small amount might lead to large albedo changes, raising the paradox of why such an intrinsically unbalanced model is so seemingly stable.

Direct and Indirect Production Pathways

Next we consider more closely the process of DMS production. Although DMS is derived biochemically from a variety of sulfonium compounds, including S-adenosyl methionine (SAM), S-methyl methionine (SMM), and dimethyl sulfonio propionate (DMSP), the latter seems to be nearly the sole DMS precursor in marine phytoplankton (Tocher et al. 1966). This compound has been shown to function as an osmolyte in marine macrophytes (Blunden and Gordon 1986); however, evidence for similar activity in phytoplankton is less convincing (Vaivaramurthy et al. 1985), and its presence in phytoplankton, sometimes at extremely high concentrations (Keller et al. 1989), remains something of a functional mystery.

What fraction of marine total sulfur is present as DMSP? This number is exceedingly hard to estimate, since it depends on species distribution as well as species DMSP concentrations, and neither number is well known. However, even if some species are high accumulators of this compound, overall its presence is likely still a minority of biogenic sulfur, which has prominent protein-structure functions. The few studies which have measured DMSP (Turner et al. 1988; Iverson et al. 1989; and Wolfe et al., unpublished) show that it is generally most abundant in particulate material and occurs in concentrations 5–50 times those of DMS. Using a very rough estimate of an oceanic mean concentration of 10 nM, and an upper limit of 25% of total bio-S, we tentatively assign 0.3–0.8 Tmoles of DMSP-S in oceanic phytoplankton. This number completes the cycle in Fig. 3, the direct production model.

However, very little evidence actually exists to confirm direct production of DMS; rather, increasing evidence points to an indirect process by which phytoplankton cells lyse during senescence or grazing, releasing DMSP into the water column (we use $DMSP_p$ to denote the particulate, *in vitro* cell DMSP and $DMSP_d$ the dissolved pool of this compound). Particulate and dissolved DMSP, though stable to physical degradation at marine pH (Dacey and Blough 1987), are acted upon enzymatically by bacteria, or by bacteria in the guts of zooplankton, to produce DMS (Dacey and Wakeham 1986; Turner et al. 1988).

Figure 4 incorporates the indirect emission process. We note that although the overall picture presented in the previous model is unchanged, the two methods of production "open the door" for participation of more of the marine microbial food loop in DMS production. Such a feature makes the production of this gas much less likely to be highly variable and points to a stability to which we will later return.

Other Photic Zone Sources and Sinks

The loss of DMS to the atmosphere is but one of many possible sinks for this molecule, and its magnitude depends on the number and strength of other water-column sinks as well as production.

Several abiotic processes have been suggested for DMS loss in the water column, including photooxidation in the presence of photosensitizers (Shooter and Brimblecombe 1989) and adsorption onto clay colloids (Brimblecombe and Shooter 1986). Of these, only photooxidation is thought to present any major sink, and current estimates are perhaps up to half the atmospheric loss, as indicated in Fig. 4.

However, biotic sinks are much more likely to be important in the water

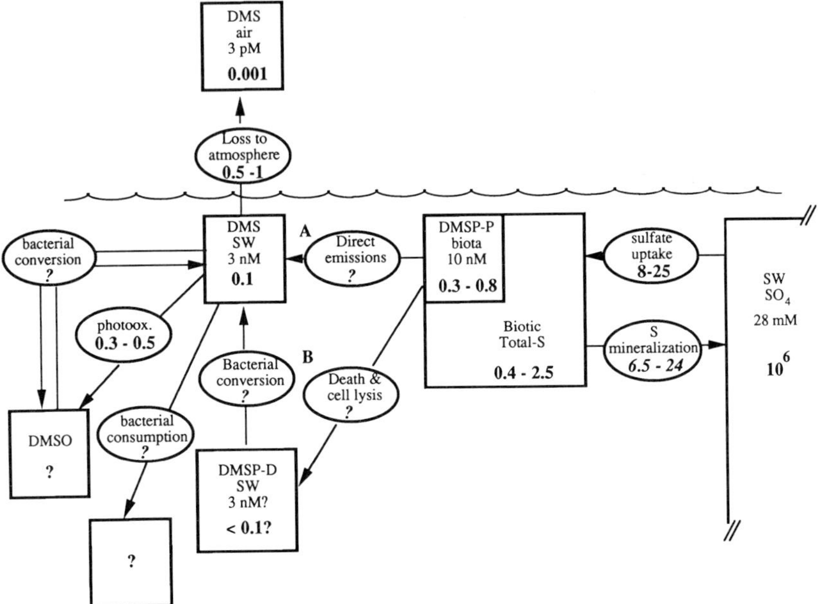

Fig. 4—Direct and indirect DMS production model. Boxes assume global oceanic area (3.6 E14 m^2). Boldface units are Tmoles (10^{12} moles) or Tmoles/y; italicized fluxes are inferred from steady-state assumption.

column. Consumption of the compound by methylotrophic bacteria (bacteria which can strip methyl groups from compounds) has been demonstrated in other environments, and both thiobacilli and purple photosynthetic bacteria have been isolated which can utilize DMS as a sole carbon source (Suylen and Kuenen 1986). Recent work has determined that microbial consumption of DMS in seawater can be much more rapid (turnover times as low as two days [Leck et al. 1989]) than loss to the atmosphere. Although work in this area is only beginning, it is clear that there is potentially a large biologic sink in the water column.

Stability and Feedbacks: Coupling with Other Trace Gas Cycles

We therefore propose the possibility of a very active biologic sulfur cycle in the ocean photic zone, limited only by sulfate incorporation as mediated by primary productivity and C:S stoichiometries, and which features DMS loss to the atmosphere as possibly only a small leak of an active cycle. This leads to an interesting irony, for if the climate sensitivity to cloud modulation is as high as proposed (−1.3 K for a 30% increase of DMS-derived CCN; see Table 2), it seems extraordinary that such a system could be so stable. What links with other elemental cycles might stabilize this loop? What feedbacks are present, what are their magnitudes, and are there possibly multiple stable states to the system?

The linkage of the marine sulfur cycle with other elemental cycles is likely to be important. We have already implied such a linkage with the carbon cycle in our estimate of sulfur pools and assimilation by using C:S stoichiometries. In this case, DMS production is linked to photosynthetic CO_2 uptake. It should be noted that if the proposed DMS-climate loop has a negative feedback, such a linkage would be doubly important for climate change since an increase in both fluxes would force temperature in the same direction; e.g. an increase of DMS production coupled with an increased drawdown of CO_2 would both tend to cool the atmosphere. Figure 5 sketches a possible framework for the carbon–sulfur connection.

The marine sulfur cycle may also be linked to the nitrogen cycle. Sulfuric acid is the dominant acid in marine rain, and its presence accounts for the acidic nature of even the most remote, pristine marine rain (5.0–5.5 [Charlson and Rodhe 1982]). Ammonia is the dominant base in marine rain (Quinn et al. 1987). Many studies have determined that remote rain pH is surprisingly consistent (Charlson and Rodhe 1982). We have already pointed out the potential for a large variation in DMS flux, which would lead to variation in rain pH. Is the observed constancy therefore due to a linkage between DMS and NH_3 production? Since exceedingly little is known about NH_3 production from the oceans, we will not speculate on biotic or atmospheric mechanisms but point out that ice cores show evidence that over ice-age

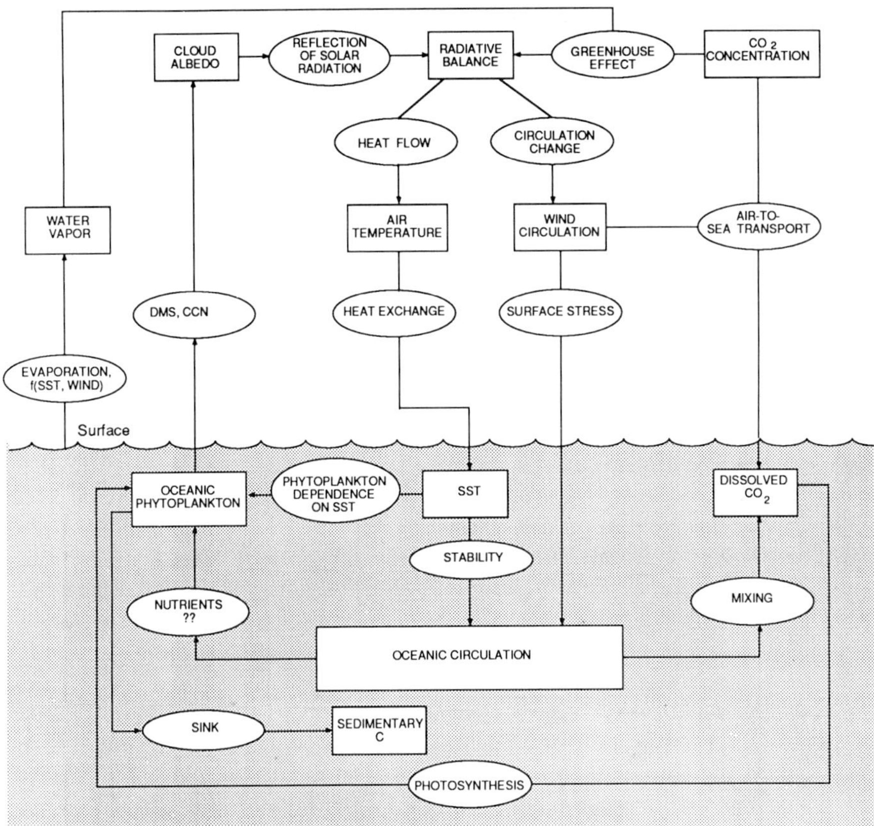

Fig. 5—A schematic diagram of possible linkages between the carbon and sulfur cycles in the marine atmosphere and ocean.

time scales, atmospheric sulfate and methane sulfonate did change by a factor of 2 or more (Legrand et al. 1988). A useful test of this proposed DMS–NH_3 linkage would be an ice-core chronology for NH_4^+.

COASTAL ZONES

Although their areal extent is negligible, coastal margins may have an important impact on trace gas production and are noteworthy in any case for several reasons. With respect to trace gas emissions (and DMS emissions in particular), coastal biomes display:

1. the highest primary productivity,
2. the highest biogas fluxes/unit area,

3. the highest degree of macrophyte activity,
4. the greatest terrestrial nutrient inputs,
5. the greatest physical (temperature, salinity, nutrient composition) variability,
6. the greatest diurnal cycles,
7. the greatest change in area during ice ages,
8. the greatest impact from human activity.

The regions of sea-ice margins have also been found to be important in marine DMS production, where intensely seasonal production displays some of the highest measured DMS emissions. Therefore, we must not dismiss these regions as unimportant simply because of their small area.

Besides the global effects of biogas exchange on climate via infrared absorption (emission) of gases and cloud effects of aerosol, other regional environmental effects exist which are probably most important in coastal zones and possibly in some upwelling open-ocean areas. These effects are not uniformly distributed over the globe due to the short lifetime of the chemical species that are involved, typically one to a few days. Horizontal transport rates in the marine atmospheric boundary layer of a few m/sec dictate a geographic scale of ca. 10^3 km, comparable to that of the coastal zone. The outstanding environmental effects of biogases include:

1. The contribution of DMS oxidation and NH_3 to the acid-base balance of remote marine rainwater (Charlson and Rodhe 1982). Recent studies suggest that the emission of sulfur gases from coastal margins may play some role in the acidification of rain which falls on terrestrial regions; the extent of this acidity likely varies from insignificant on the western coast of the United States (Bates and Cline 1985) to important in the summertime North Sea (Turner et al. 1988; Leck and Rodhe 1990).
2. The role of marine sulfate aerosol in controlling atmospheric light scattering (Shaw 1987).
3. The role of CH_4, nonmethane hydrocarbons and possibly the methyl halides as they influence tropospheric O_3.

The climatic and environmental consequences of marine biogas fluxes in coastal ocean areas are inherently difficult to assess due to terrestrial sources (both natural and anthropogenic) which often dwarf those from coastal biomes. In the case of DMS, one useful comparison might be between two contrasting coastal regions which have been intensively studied: the northeastern Pacific and the North Sea. These regions exemplify coastal zones which have great differences in the amount of influence upon terrestrial regions, and also in the amount of anthropogenic influence on the coastal biome.

CONTROVERSIES

The hypothesis connecting oceanic DMS emissions to climate has received considerable comment and some objection. Notably, Schwartz (1988), claiming that industrial emissions of SO_2 have had no effect on the midlatitude cloud component of albedo in the Northern versus the Southern hemispheres and also have had no effect on global temperature or the differences in warming rates of the two hemispheres, suggested that these key climatic variables were not controlled by anthropogenic sulfates and "by extension are not controlled by marine DMS emissions." Wigley (1989) considered possible effects of CCN from anthropogenic SO_2 emissions and concluded that the hemispheric temperature data indeed do allow that the Northern hemisphere may be warming more slowly than the Southern, contrary to Schwartz. Savoie and Prospero (1989) suggested that the sulfate aerosol of the remote marine, North Pacific areas are still dominated by natural DMS emissions. Coakley et al. (1987) showed that the ship-stack emissions do cause an expected increase in stratus cloud albedo. Several other arguments were put forward in correspondence (*Nature* 340:436–438 and 515–516). No test of the feedback hypothesis has yet been attempted, and no formal objection has yet appeared to the hypothesis that most natural marine CCN are from the oxidation of biogenic DMS. However, the ice-core record of both nss-SO_4^{2-} and methane sulfonate show that values were elevated during the ice age (Legrand et al. 1988). Thus if a feedback was present, it may have acted more as an amplifier of the ice age (and the switch to interglacial climate) than as a thermostat.

CONCLUSIONS

One useful method of assessing our understanding of the cycling of the climatically important biogas species is through the use of simple models as outlined here for the marine sulfur cycle and DMS production. Through the use of these models we are able to explore some of the potential controlling variables on the fluxes of trace gas and the processes which drive the cycles. Such an exercise also helps us define some of the major uncertainties in our understanding of this cycle: in this case, the amount of total oceanic bio-sulfur apportioned as DMSP, the major DMS precursor, the competing processes which transform this compound into the volatile gas DMS, and additional sources and especially microbial sinks in the water column. The overall magnitude of the biologic sulfur cycle in the marine photic zone is also a key to our understanding the present-day flux of DMS to the atmosphere, as well as long-term stability or variability.

In this paper we have concentrated on water-column biologic processes; a complete linkage between sulfur trace gas emission and climate must also consider atmospheric processes controlling the concentrations of aerosol components over the oceans. Included among these are the oxidation reactions between DMS and other biogenic trace gases, or between competing biogases and the oxidation agents. The physical climate interaction of aerosol/cloud effects and radiatively active species such as CO_2 are also crucial to an overall understanding of trace gases and climate. Therefore we must be careful to include in our analysis some consideration of the linkages between the different species.

We stress the need to understand the controlling variables, processes, and stability of this system. It is especially interesting that in the unpolluted marine atmosphere, the effects of the sulfur cycle on both rain and on atmospheric optics are currently only of marginal importance. However, even small increases of nss-SO_4^{-2} aerosol concentration have effects that are disproportionate to the increase. In the case of rainwater, the existing near-balance of NH_4^+ and SO_4^{2-} yields a system analogous to a titration close to the endpoint; small additions of H_2SO_4 can lower the pH dramatically. Similarly, the present-day optical scattering coefficient of this same aerosol is a small but finite fraction of the Rayleigh scattering due to the air molecules, such that relatively small increase in the amount of aerosol causes it to dominate over the optical effects of the gas. Such a system seems inherently unstable and suggests possible feedbacks and interlinked cycles which we do not yet understand. Clearly, this potential for environmental effects close to the marine source region provides an additional rationale for studying factors controlling biogas exchange.

Regardless of the particular environmental or climatic effect of concern, for perhaps 10^4 years the world has come to depend on current climatic conditions which in many instances appear to have been in a steady state prior to industrialization. Therefore it is necessary to understand the factors and potential feedbacks in these systems if we are to be able to predict the response to perturbations.

Acknowledgements. This work was funded in part by NSF through the atmosphere sciences program, NOAA through the Radiatively Important Trace Species division, the University of Washington, the Joint Institute for the Study of the Atmosphere and Oceans, and the Ford Motor Corporation Research Institute.

REFERENCES

Andreae, M.O., W.R. Barnard, and J.M. Ammons. 1983. The biological production of dimethylsulfide in the ocean and its role in the global atmospheric sulfur budget. *Ecol. Bull.* **25**:167–177.

Bates, T.S., and J.D. Cline. 1985. The role of the ocean in a regional sulfur cycle. *J. Geophys. Res.* **90**(C5):9168–9172.

Bates, T.S., J.D. Cline, R.G. Gammon, and S.R. Kelly-Hansen. 1987. Regional and seasonal variations in the flux of oceanic dimethylsulfide to the atmosphere. *J. Geophys. Res.* **92**(C3):2930–2938.

Blunden, G., and S.M. Gordon. 1986. Betaines and their sulphonium analogues in marine algae. In: Progress in Phycological Research, ed. E. Round, vol. 4, pp. 39–80. Amsterdam: Elsevier Biomedical Press.

Bowen, H.J.M. 1979. Environmental Chemistry of the Elements. London: Academic.

Brimblecombe, P., and D. Shooter. 1986. Photo-oxidation of dimethylsulphide in aqueous solution. *Marine Chem.* **19**:343–353.

Butler, J.H., J.W. Elkins, and T.M. Thompson. 1989. Tropospheric and dissolved N_2O of the West Pacific and East Indian oceans during the El Niño southern oscillation event of 1987. *J. Geophys. Res.* **94**(D12):14865–14877.

Charlson, R.J., J.E. Lovelock, M.O. Andreae, and S.G. Warren. 1987. Oceanic phytoplankton, atmospheric sulphur, cloud albedo and climate. *Nature* **326**:655–661.

Charlson, R.J., and H. Rodhe. 1982. Factors controlling the acidity of rainwater. *Nature* **295**:683–685.

Coakley, J.A. Jr., R.L. Bernstein, and P.A. Durkee. 1987. Effect of ship-stack effluents on cloud reflectivity. *Science* **237**:1020–1022.

Cuhel, R.L., C.D. Taylor, and H.W. Jannasch. 1982. Assimilatory sulfur metabolism in marine microorganisms: considerations for the application of sulfate incorporation into protein as a measurement of natural population protein synthesis. *Appl. Env. Microbiol.* **43**:160–168.

Dacey, J.W.H., and N.V. Blough. 1987. Hydroxide decomposition of dimethlysulfoniopropionate to form dimethylsulfide. *Geophys. Res. Let.* **14**(12):1246–1249.

Dacey, J.W.H., and S.G. Wakeham. 1986. Oceanic dimethylsulfide: production during zooplankton grazing on phytoplankton. *Science* **233**:1314–1316.

De Vooys, C.G.N. 1979. Primary production in aquatic environments. In: The Global Carbon Cycle, ed. B. Bolin, E.T. Degens, S. Kempe, and P. Ketner, SCOPE 13. Chichester: Wiley.

Ehhalt, D.H. 1974. The atmospheric cycle of methane. *Tellus* **26**:58–70.

Ferek, R.J., and M.O. Andreae. 1984. Photochemical production of carbonyl sulfide in marine surface waters. *Nature* **307**:148–150.

Iverson, R.L., F.L. Nearhoff, and M.P. Andreae. 1989. Production of dimethylsulfonium propionate and dimethylsulfide by phytoplankton in estuarine and coastal waters. *Limnol. Ocean.* **34**:53–67.

Johnson, J.E., R.H. Gammon, J. Larson, T.S. Bates, S.J. Oltmans, and J.C. Farmer. 1990. Ozone in the marine boundary layer over the Pacific and Indian oceans: latitudinal gradients and diurnal cycles. *J. Geophys. Res.* **95**:11847–11856.

Jordan, M.J., and B.J. Peterson. 1978. Sulfate uptake as a measure of bacterial production. *Limnol. Ocean.* **23**:146–150.

Keller, M.D., W.K. Bellows, and R.L. Guillard. 1989. Dimethyl sulfide production in marine phytoplankton. In: Biogenic Sulfur in the Environment, ed. E.S. Saltzman & W.C. Cooper, pp. 167–183. Washington, D.C.: Am. Chemical Soc.

Leck, C., U. Larsson, L.E. Bagander, S. Johansson, and S. Hajdu. 1989. DMS in the Baltic Sea—annual variability in relation to biological activity. *J. Geophys. Res.* **95**:3353–3363.

Leck, C., and H. Rodhe. 1990. Emissions of marine biogenic sulfur to the atmosphere of northern Europe. *J. Atmos. Chem.*, in press.

Legrand, M.R., R.J. Delmas, and R.J. Charlson. 1988. Climate forcing implications from Vostok ice-core sulphate data. *Nature* **334**:418–420.

Likens, G.E., F.H. Bormann, and N.M. Johnson. 1981. Interactions between major biogeochemical cycles in terrestrial ecosystems. In: Some Perspectives of the Major Biogeochemical Cycles, ed. G.E. Likens, SCOPE 17. Chichester: Wiley.

Logan, J.A., M.J. Prather, S.C. Wofsky, and M.B. McElroy. 1981. Tropospheric chemistry: a global perspective. *J. Geophys. Res.* **86**:7210–7254.

Matrai, P.A., and R.W. Eppley. 1989. Particulate organic sulfur in the waters of the southern California bight. *Glob. Biogeochem. Cycl.* **3**(1):89–104.

Oltmans, S.J., and W.D. Komhyr. 1986. Surface ozone distributions and variations from 1973–1984 measurements at the NOAA geophysical monitoring for climatic change baseline observatories. *J. Geophys. Res.* **91**:5229–5236.

Placet, M., R. Bahye, F. Fehsenfeld, and G. Basseet. 1989. Emissions involved in acidic deposition processes. NAPAP State of Science Technology Report 1.

Quinn, P.K., T.S. Bates, J.E. Johnson, D.S. Covert, and R.J. Charlson. 1990. Interactions between the sulfur and reduced nitrogen cycles over the central Pacific Ocean. *J. Geophys. Res.* **95**:16405–16416.

Quinn, P.K., R.J. Charlson, and W.H. Zoller. 1987. Ammonia, the dominant base in the remote marine troposphere: a review. *Tellus* **39B**:413–425.

Ramanathan, V., R.J. Cicerone, H.B. Singh, and J.T. Kiehl. 1985. Trace gas trends and their potential role in climate change. *J. Geophys. Res.* **90**(D3):5547–5566.

Rasmussen, R.A., M.A.K. Khalil, R. Gunawardena, and S.D. Hoyt. 1982. Atmospheric methyl iodide (CH_3I). *J. Geophys. Res.* **87**:3086–3090.

Rudolph, J. and D.H. Ehhalt. 1981. Measurements of C2–C5 hydrocarbons over the North Atlantic. *J. Geophys. Res.* **86**:11,959–11,964.

Savoie, D.L., and J.M. Prospero. 1989. Comparison of oceanic and continental sources of non-sea-salt sulphate over the Pacific Ocean. *Nature* **339**:685–687.

Schwartz, S.E. 1988. Are global cloud albedo and climate controlled by marine phytoplankton? *Nature* **336**:441–445.

Shaw, G.E. 1987. Aerosols as climate regulators: a climate–biosphere linkage? *Atmos. Envir.* **21**(4):985–986.

Shooter, D., and P. Brimblecombe. 1989. Dimethylsulphide oxidation in the ocean. *Deep-Sea Res.* **36**(4):577–585.

Suylen, G.M.H., and J.G. Kuenen. 1986. Chemostat enrichment and isolation of *Hyphomicrobium* EG: a dimethyl-sulphide oxidizing methylotroph and reevaluation of *Thiobacillus* MS1. Antonie van Leeuwenhoek **52**:281–293.

Tans, P.P., I.Y. Fung, and T. Takahashi. 1990. Observational constraints on the atmospheric carbon dioxide budget. *Science* **247**:1431–1438.

Tocher, C.S., R.G. Ackman, and J. McLachlan. 1966. The identification of dimethyl-β-propiothetin in the algae *Syracosphaera carterae* and *Ulva lactuca*. *Can. J. Biochem.* **44**:519–522.

Turner, S.M., G. Malin, P.S. Liss, D.S. Harbour, and P.M. Holligan. 1988. The seasonal variation of dimethyl sulfide and dimethylsulfoniopropionate concentrations in nearshore waters. *Limnol. Ocean.* **33**(3):364–375.

Vaivaramurthy, A., M.O. Andreae, and R.L. Iverson. 1985. Biosynthesis of dimethylsulfide and dimethylpropiothetin by *Hymenomonas carterae* in relation to sulfur source and salinity variations. *Limnol. Ocean.* **30**(1):59–70.

Wigley, T.M.L. 1989. Possible climate change due to SO_2-derived cloud condensation nuclei. *Nature* **339**:365–367.

How Do We Generalize Coastal Models to Global Scale?

E.E. Hofmann

Department of Oceanography, Old Dominion University
Norfolk, VA 23529, U.S.A.

Abstract. It is now appropriate to begin the development of physical–biological models that treat processes which occur at ocean basin or global scales. Of particular interest is the development of models that can be used to quantify the flux of material from the ocean margins to the open ocean. Hence, it is worthwhile to review the types of models that could be extended to larger scales and/or used to quantify boundary fluxes and to indicate some of the problems that are associated with extending and using these models. A brief overview of the types of models developed to consider physical–biological interactions in limited coastal regions is presented. This is followed by a discussion of the issues and problems that must be addressed if these regional models are to be extended to larger scales or used to estimate material fluxes across boundaries, such as the shelf-slope boundary. Two issues are of primary concern: mismatches of space and time scales between regional and larger-scale systems, and the degree of complexity that needs to be transferred from regional to larger scale models. Some suggestions are given for research directions that might be pursued in order to develop physical–biological models for investigating basin or global scale processes and quantifying exchanges between ocean margins and the open ocean.

INTRODUCTION

Recently there has been much interest in quantifying the exchange of biogenic material between continental shelf and open ocean waters. However, for some environments, biological distributions and their associated fluxes are difficult to measure from ships, moored instruments, and even from satellites, such as the Coastal Zone Color Scanner (CZCS), because biological distributions are the result of interactions between circulation and biological processes. Additional complexity is introduced by the considerable space and time variability that often characterizes biological distributions.

Ocean Margin Processes in Global Change
Edited by R.F.C. Mantoura, J.-M. Martin and R. Wollast

Physical–biological models represent one approach for investigating the processes producing observed biological distributions and for quantifying the exchange of biogenic material between different marine environments, such as the continental margin and open ocean. Physical–biological models, if calibrated and verified, provide a mechanism for (a) producing biological distributions at space and time scales with finer resolution than that of observations, (b) synthesizing diverse data sets, (c) interpolating between observations, (d) separating the role of circulation and biological processes in producing observed distributions, and (e) making predictions of future conditions given present conditions. Physical–biological models that treat one or some of these points have been constructed for specific coastal environments; Wroblewski and Hofmann (1989) present a review of these models.

In general, the physical–biological models that are available were formulated and implemented to study processes in specific coastal regions. These regional models have gone through an evolution from a steady-state, two-dimensional framework (e.g., Walsh 1975) to a time-dependent, two-dimensional framework (e.g., Wroblewski 1977; Hofmann 1988; Ishizaka 1990b), to a time-dependent, three-dimensional framework (Walsh et al. 1988). These models use a variety of techniques encompassing theoretical approaches, laboratory measurements, and field observations to specify the circulation and biological processes. However, even the models that include sophisticated circulation dynamics (e.g., Wroblewski 1977; Walsh et al. 1988) or complex biological processes (e.g., Hofmann 1988) are limited in the space and time scales over which the model solutions are valid: typically time scales of a few days to a few weeks and space scales of a few tens of kilometers. The departure of the model solutions from reality provides a guide to the limitations in our understanding, the need to measure particular physical and biological processes, and the need to include space and time varying processes and exchanges.

The recent attention on global processes (e.g., Committee on Global Change 1988) has resulted in the first attempts at constructing physical–biological models that treat processes that occur on larger scales, such as those associated with ocean basins. The first of these models (Wroblewski et al. 1988) considered the effect of variable mixed layer depth on primary production in the North Atlantic. The model consists of a three component ecosystem (nitrate, phytoplankton, zooplankton), in which interactions among the ecosystem components are parameterized with relatively simple formulations. The biological processes included in the model were parameterized using average values and these were assumed to apply over the whole of the North Atlantic. The physical portion of the model consists of climatological distributions of mixed layer depth. The circulation is not explicitly parameterized and in particular excludes convergences and divergences (e.g., upwelling) which are important to carbon and nutrient fluxes. The model is used to obtain the steady-state nutrient, chlorophyll, and zooplankton distributions across the North Atlantic in May on a 1° latitude by 1° longitude grid. However, even with the coarse resolution and simple physics and biology, the simulated chlorophyll distributions show the gross features observed in phytoplankton distributions obtained from climatological and composite CZCS data sets.

The awareness that processes on continental shelves are continuous with those in the open ocean prompts the question of how do we interface the physical–biological models developed for continental shelf regions with those developed for ocean basins. Given the importance of continental shelves as a source of organic carbon, it is desirable to quantify the flux of this material from the ocean margins to the open ocean. The implementation of physical–biological models to estimate exchanges between the margins and open ocean has a significant problem associated with it in that the time and space scales associated with physical and biological processes, as well as the processes themselves, can be quite different in coastal and open ocean environments. Also, physical–biological models developed to investigate processes and consequences associated with global change issues will need to interface ocean margin and open ocean systems.

The following section presents a description of approaches that have been used to model physical–biological interactions in coastal environments. The types of results that are obtained from these models are also discussed. This is followed by a discussion of the points that must be considered if these regional physical–biological models are to be extended to larger space and time scales or coupled to larger (e.g., basin) scale models. Finally, some comments and recommendations for extending coastal models to basin or global scales are presented.

TYPES OF MODELS

The types of models that have been used to consider physical–biological processes can be partitioned into three broad categories. The first of these categories consists of Lagrangian particle tracing models, which consider primarily the effects of the circulation on the distribution of conservative properties. These models provide information on particle residence time and general particle transport patterns. The second category of models considers the role of biological interactions in producing observed biological distributions. These models typically consist of systems of coupled ordinary differential equations that describe the time-dependent behavior of the components of an ecosystem. These models provide information on the time scales associated with biological processes and allow investigation of biological processes independent of circulation effects. The third category consists of models in which circulation and biological effects are combined and, thus, consider the role of physical–biological interactions in determining observed biological distributions. These models are composed of systems of coupled partial differential equations that describe the space and time evolution of biological distributions. Below are given brief descriptions and examples of results that are obtained with each of these types of models.

Lagrangian Models

A Lagrangian model that describes the horizontal displacement of a particle is of the general form

$$X_n = X_o + u\Delta t \tag{1}$$

$$Y_n = Y_o + v\Delta t \tag{2}$$

where X_n and Y_n represent the new location of a particle that results from the displacement of the particle from its previous position (X_o, Y_o) by the horizontal velocity, u and v, in a time interval, Δt. If vertical velocities are important in altering the location of a particle, then an additional equation of the form

$$Z_n = Z_o + w\Delta t + w_b\Delta t \tag{3}$$

can be included. The effect of the vertical velocity on particle location is given by $w\Delta t$. Since many plankton species can alter their vertical location, a biological vertical velocity is included through $w_b\Delta t$.

A few examples of the use of Lagrangian models to investigate the transport patterns of planktonic populations are available (e.g., Rothlisberg et al. 1983; Smith et al. 1983; Ishizaka and Hofmann 1988; Bucklin et al. 1989; Ishizaka 1990a). These models have been developed to consider plankton transport patterns in specific coastal regions. To date there has not been a Lagrangian model developed to consider basin-scale plankton transport patterns.

One example of the application of a Lagrangian particle tracing model is from the southeastern U.S. continental shelf (Fig. 1), where Lagrangian trajectories were used to investigate the general transport patterns of plankton populations (Fig. 2). The trajectories show that, for the region included in the model domain, different particle transport patterns occur on the southeastern U.S. continental shelf during different times of the year. These patterns are determined by the combined effect of winds, shelf stratification, and the location of the Gulf Stream relative to the shelf break (Ishizaka and Hofmann 1988). The implication of these simulated trajectories

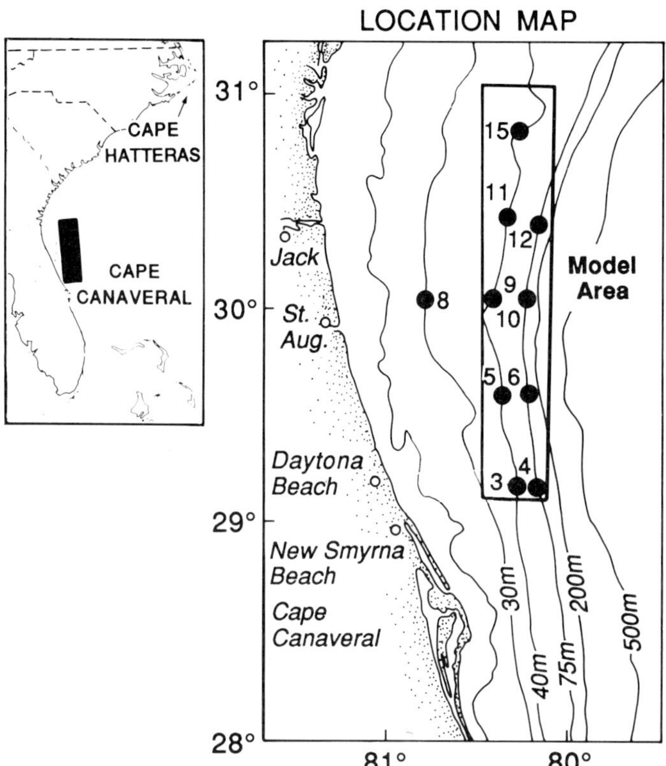

Fig. 1—Location map for the southeastern U.S. continental shelf. Larger map shows the portion of the southeastern U.S. continental shelf that is included in the model domain. Circles indicate the location of current meter moorings and the box indicates the model domain.

is that the exchange of material between this shelf and the open ocean can have a strong seasonal dependence.

The implementation of a Lagrangian model is straightforward once a velocity field is available. The velocity distribution can be specified either from observations, as done by Smith et al. (1983) and Ishizaka and Hofmann (1988), or from a theoretical circulation model, as done by Rothlisberg et al. (1983). However specified, it is important that the space and time scales associated with the flow field be compatible with those that are influencing the transport of the plankton populations. If biological velocities are to be included in the model, it is necessary to have observations and measurements of sufficient breadth to specify these velocities. For example, observations of the diurnal migration pattern at different stages of animal development may be needed (e.g., Rothlisberg et al. 1983).

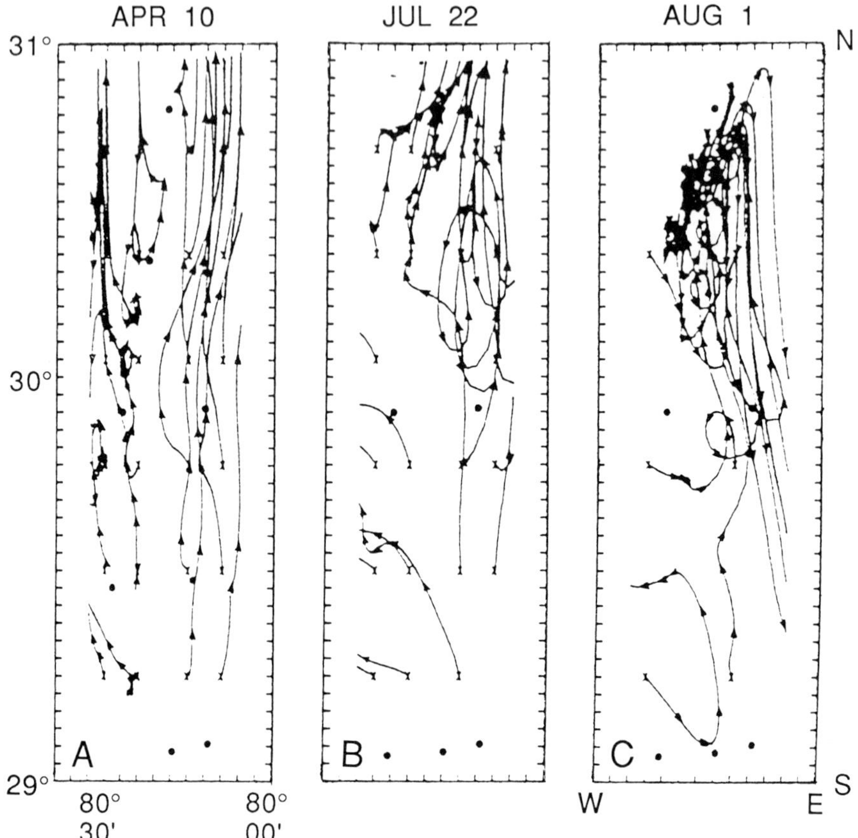

Fig. 2—Simulated trajectories at a depth of 37 m on the southeastern U.S. continental shelf for particles started on (A) April 10, 1980, (B) July 22, 1981, and (C) August 1, 1981. The arrows on the trajectories indicate the particle position in days and the particle flow direction. Figure reprinted from Ishizaka and Hofmann (1988).

Lagrangian models are relatively inexpensive to run and for most applications do not require extensive computer resources. These models are useful in that they yield information on the space scale over which planktonic organisms can be transported in their lifetime or during a particular circulation scenario. This in turn indicates the minimum space scale that needs to be resolved if exchanges between shelf and open ocean environments are to be quantified.

Time-dependent Biological Models

Time-dependent models that describe biological interactions are of the general form

$$\frac{dB}{dt} = f(t, B, P, Z, \ldots) \tag{4}$$

where B is the biological quantity of interest and $f(t, B, P, Z, \ldots)$ is a general function that describes how the biological component changes with time in response to other components of the ecosystem. There are numerous examples of time-dependent biological models that have been formulated for a diverse set of marine environments to consider a variety of hypotheses.

A schematic of the interactions that can be included in a time-dependent biological model is shown in Fig. 3. This particular time-dependent biological

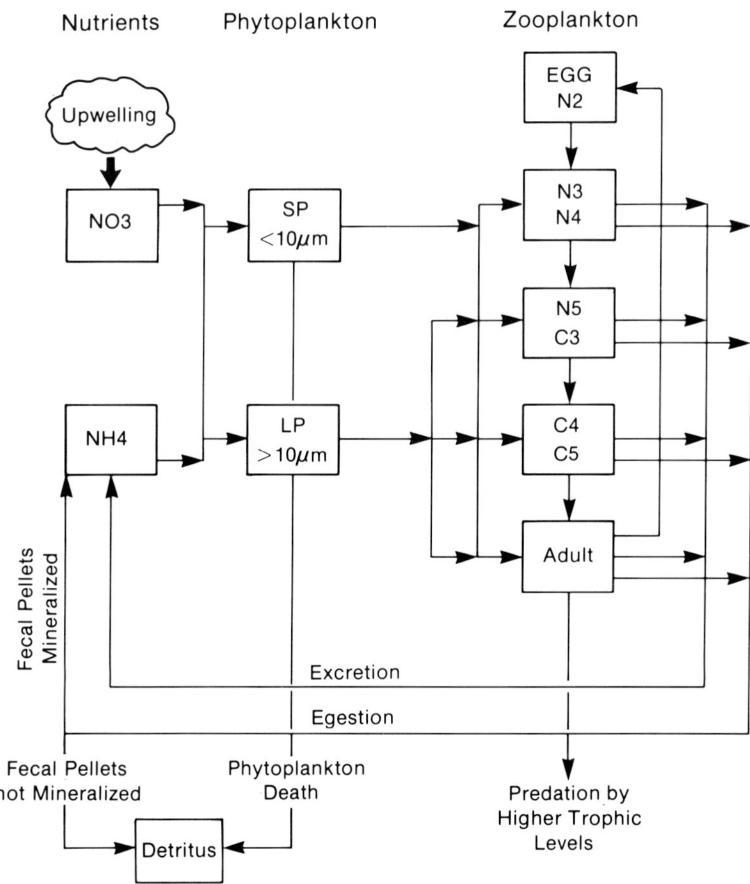

Fig. 3—Schematic of the components included in a time-dependent biological model developed for the southeastern U.S. shelf ecosystem. Figure reprinted from Hofmann and Ambler (1988).

model was used to investigate the effects of Gulf Stream-induced upwelling on the southeastern U.S. continental shelf (cf. Fig. 1). The components chosen for inclusion in the model were determined from field observations. The biological model shown in Fig. 3 appears complex, but it represents the minimum ecosystem that could be modeled and still expect to reproduce the biological interactions that occur on the southeastern U.S. continental shelf. An example of the simulated time-dependent biological distributions that were obtained with this model is shown in Fig. 4. An important piece of information that can be extracted from these simulated distributions is the time scale associated with the biological processes. This then determines the biological time scale that must be resolved in more complex models, such as a physical-biological model.

Time-dependent biological models are useful for integrating diverse data sets and for identifying biological processes that need to be studied. The implementation of these models, however, does require a certain level of information about a particular system. For example, measurements of the rates of processes, such as growth, need to be available. The biggest problem with these models is that many important biological processes, such as *in situ* mortality, are difficult or impossible to measure. These processes represent the reservoir of removed material and are necessary for the biological material to be conserved. Often the model solutions are sensitive to the values chosen for the parameterizations of these processes. Consequently, sensitivity analysis is an important part of analyzing the solutions obtained with time-dependent biological models. Once formulated, however, time-dependent biological models are not difficult to implement, even for complex systems. These models usually do not require sophisticated numerical techniques or extensive computer resources. One shortcoming of time-dependent models is that processes such as upwelling are important in determining biological production and these processes cannot be explicitly included in these models. Thus, accurate representation of these processes forces one to include spatial as well as temporal dependence.

Physical–biological Models

The role of physical–biological interactions in determining observed biological distributions can be described by the equation for the time-dependent advection and diffusion of a nonconservative substance, which is of the form

$$\frac{\partial B}{\partial t} + \frac{\partial(uB)}{\partial x} + \frac{\partial(vB)}{\partial y} + \frac{\partial(wB)}{\partial z} \tag{5}$$

$$= \frac{\partial}{\partial x}\left(K_x \frac{\partial B}{\partial x}\right) + \frac{\partial}{\partial y}\left(K_y \frac{\partial B}{\partial y}\right) + \frac{\partial}{\partial z}\left(K_z \frac{\partial B}{\partial z}\right) + \mathit{Biological\ Source/Sink\ Terms}$$

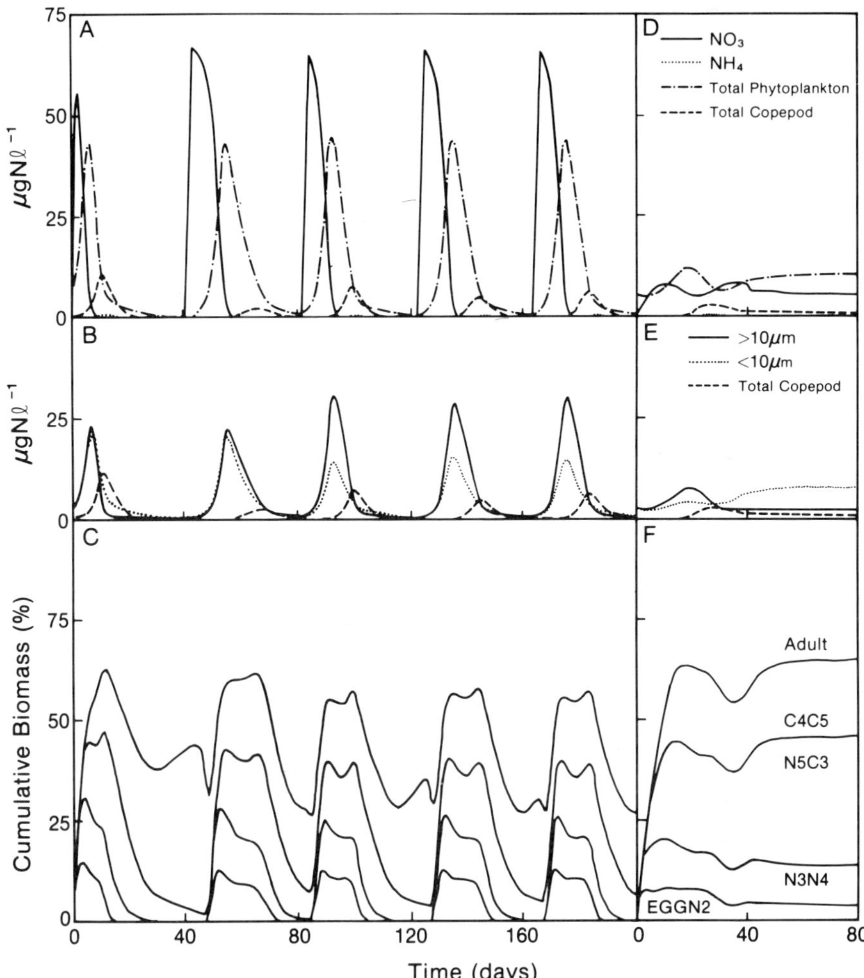

Fig. 4—Time-dependent simulated biological distributions obtained for the southeastern U.S. shelf ecosystem. Panels A–C show the distributions obtained with episodic inputs of nitrate. Panels D–F show the distributions obtained with a constant low level input of nitrate. Figure reprinted from Hofmann and Ambler (1988).

where the terms in Eq. 5 represent the time-dependent changes in B, horizontal and vertical advective effects on B, horizontal and vertical diffusive effects on B, and biological effects on B, respectively. The model given by Eq. 5 allows the circulation to effect the biological distributions; however, the biological processes have no effect on the circulation. In this sense the physical–biological model is not truly coupled.

The physical–biological models that have been developed to date have been designed to investigate processes in coastal upwelling or shelf systems. These models are relatively few in number, which is not surprising given the considerable amount of data required to formulate, calibrate, and verify such models. Consequently, the models that have been developed are the result of multidisciplinary research programs.

An example of the simulated chlorophyll distributions that were obtained from a physical–biological model constructed for the southeastern U.S. continental shelf (cf. Fig. 1) ecosystem is given in Fig. 5. The simulated fields provide information on the space and time evolution of features that are observed in chlorophyll distributions. As with the Lagrangian and time-dependent models, the model solutions give estimates of the space and time scales over which features change. The model-derived distributions are available with a resolution that is usually finer than that associated with field observations. Thus, one use for the simulated distributions is to interpolate between field observations.

Physical–biological models require sophisticated numerical procedures for solution and thus are more difficult to implement than the previously discussed models. These models also require all of the same laboratory and field measurements as do the others. It should be noted that implementation

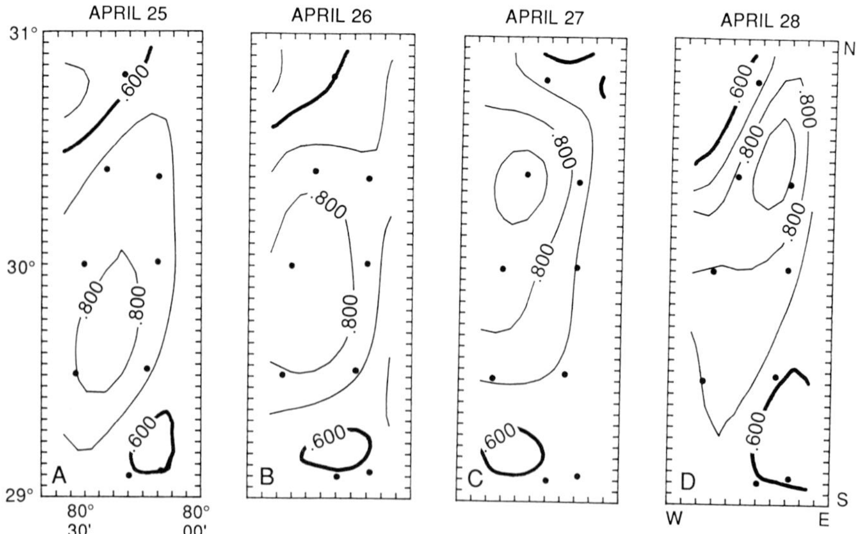

Fig. 5—Simulated chlorophyll distributions at a depth of 37 m obtained from a physical–biological model developed for the southeastern U.S. continental shelf ecosystem. These distributions correspond to April 25 to 28, 1980. Contour interval is 0.1 μg chl a l^{-1}.

of a physical–biological model presupposes the existence of a circulation model, since the advective velocities enter into the model as variable coefficients. Thus, for some applications obtaining a solution to a physical–biological model may require considerable computer resources, such as a supercomputer. This is especially true if the biological model is solved simultaneously with the circulation model. This is a point that will be returned to later.

Simulated phytoplankton and nutrient distributions, such as those shown in Fig. 5, can be used to provide estimates of the amount of biogenic material, in terms of carbon or nitrogen, that is exported from a given region of the continental shelf. For example, the results of simulations from the southeastern U.S. continental shelf were combined to produce an estimate of the average phytoplankton and nitrogen flux that occurred in April 1980 (Ishizaka 1990c). This type of output from regional models could then be used to provide estimates of the exchanges across continental margins that then could be used to specify boundary conditions for larger basin-scale physical–biological models.

DISCUSSION

The above models were constructed for a limited specific region of a particular continental shelf (cf. Fig. 1). Such regional models, focused on specific questions, are typical of the physical–biological models that have been developed in the last fifteen or so years. Regional physical–biological models are starting to have a history of use and as a result the sensitivity, dynamics, and limitations of these models are becoming better understood. The models that consider physical–biological processes in coastal and shelf environments often require fine space and time resolution and considerable detail in the specification of the biological processes. The question then arises as to the extension of the knowledge gained with regional physical–biological models to investigate processes that occur on basin and global scales. Two issues must be addressed: how to match the space and time scale requirements of coastal processes with those of larger scale systems, and how much of the complexity of the biological systems that characterize coastal systems needs to be transferred to larger scale models. Resolution of these issues is important especially if regional models are to be used to specify exchanges across continental margins that are then used to specify the boundary conditions for basin-scale physical–biological models.

The spatial resolution of the Lagrangian and physical–biological models developed for the southeastern U.S. continental shelf was on the order of five kilometers in the across- and along-shelf directions. This horizontal spacing is at the upper end of the spatial resolution needed to resolve the upwelling events that are primarily responsible for the biological production and distributions observed on this shelf. Thus, correct representation of the structure of these features is necessary to simulate the structures associated with the biological distributions. Further, the time scales associated with upwelling on the southeastern U.S. continental shelf are short, one to five days, and the doubling time of the phytoplankton populations associated with the upwelling events is on the order of one day. Thus, a physical–biological model designed to investigate the exchange of nutrients and carbon between this shelf and the open ocean must have comparable time scales.

As a comparison, the recent simulation of the global circulation (Semtner and Chervin 1988) was done with a horizontal spatial resolution of one-half of a degree of latitude and longitude and a vertical resolution of twenty levels. The horizontal resolution of this general circulation model is approximately fifty to sixty kilometers. In such a global model, the region represented in the physical–biological model developed for the southeastern U.S. continental shelf would be represented by a single across-shelf grid box and four to five grid boxes in the along-shelf direction. Thus, all of the biological interactions that regulate the magnitude of the flux of carbon, for example, from this shelf would be reduced to a few grid boxes. If the biological model developed for the southeastern U.S. shelf was included in the global scale model, the implicit assumption would then have to be made that these processes apply over the larger space scales. A similar assumption must be made for the circulation processes. Given the scales associated with the physical and biological processes that regulate the structure and magnitude of the biological distributions and production on the southeastern U.S. continental shelf, the horizontal spatial resolution of the global circulation model used by Semtner and Chervin (1988) would be inadequate to correctly represent the horizontal flux of carbon from this shelf.

Vertical gradients of biological quantities are often quite important in determining the vertical flux of biogenic material. Most biological processes that regulate vertical fluxes occur in the upper 100 to 200 m of the water column. The critical vertical scale associated with these processes can be estimated by $\frac{1}{k}$, where k is the light attenuation coefficient. In circulation models the upper 100 to 200 m of the water column may be represented by only two or three grid boxes, which is not sufficient to resolve biologically interesting features such as the subsurface chlorophyll maximum. Even in circulation models that use functional expressions for vertical processes (Haidvogel et al. 1990), many functions may be required to obtain fine resolution in the upper water column. These considerations point to a problem in that models designed to simulate circulation processes do not necessarily include the time and space scales that are of interest and importance for simulating biological distributions.

The models developed for coastal regions often include complex biology (cf. Fig. 3), but the biological processes are generic in nature, i.e., phytoplankton growth, zooplankton grazing. However, inclusion of these generic processes in models developed for specific regions requires that the rates and functional relationships associated with the biological processes be measured specifically for the region of interest. Moreover, rates and relationships measured for a given region do not necessarily apply to other regions, or to the same region if environmental conditions change. Also, the level of complexity needed in the biological model is often regional and question specific. The question is then how much of the biological complexity of the coastal models is required in basin or global scale models. Or equivalently, how can the parameterization of processes be transferred between regional and larger scale models. These are important considerations because species differ over regions such as the North Atlantic and the processes controlling phytoplankton production (e.g., nutrient availability and light) differ over such large regions. A biological model that is appropriate for the southeastern U.S. continental shelf may be totally inappropriate for an open ocean system or a higher latitude shelf system. Thus, the inclusion of realistic biology in a basin scale model may require several different biological models. Further, models that consider biological effects in global change will likely need to parameterize changing species composition at a specific location over time. It is not apparent how this can be handled in large or regional scale models designed to study global change processes.

An additional problem is the space and time scales over which the parameters used in the functional relationships for the biological processes are valid. For example, the coefficient values used in the Michaelis–Menten formulation for nutrient removal by phytoplankton measured in one area may not apply in another or may not apply in the same area except under specific conditions. The alternative is to use relationships for processes such as phytoplankton growth (e.g., Eppley 1972) that apply over many different species and environments. This approach has the effect of integrating over many details.

If realistic biology is to be included in a basin-scale model, then the available computer resources become a real constraint. The global circulation model done by Semtner and Chervin (1988) calculated six prognostic variables (u, v, w, P, S, T) on a relatively coarse resolution grid. This model required many CPU hours on a CRAY-XMP to obtain a single steady state solution. A minimal biological model would be on the order of five to six prognostic variables; hence, it is easy to conceive of a global biological model that could rapidly overwhelm the available computer resources, especially if the biological model is solved simultaneously with the circulation model, as would be the case for a coupled physical–biological model in which feedbacks between the biology and circulation are allowed. For example, it is known that the concentration of phytoplankton can affect solar irradiance adsorption in the upper part of the water column. Vertical changes in the water column heating rate due to the vertical gradient of phytoplankton can potentially alter vertical mixing intensity and the depth of the mixed layer (Lewis et al. 1983). Inclusion of processes such as this could be important in models of global change since it provides a feedback between biological and physical processes.

The above discussion has attempted to point to some of the problems that must be addressed if physical–biological models adequate for studying basin or global scale processes are to be developed. Such models have been included as integral parts of the global programs proposed for the coming decade. Coupled models will undoubtedly be important in investigating the processes contributing to and predicting the consequences of global change. Each subsystem (e.g., atmosphere, ocean) of the models developed for the Earth System has associated with it specific processes, length scales, and time scales which impose specific demands and constraints on the models. The issues discussed above are compounded as system components are added or existing components are expanded. Progress has been made in coupling portions of the Earth System in models. For example, results are now becoming available from linked atmosphere and ocean models and these solutions show behaviors that are different from the uncoupled models. The challenge to modelers and observationists is to add to these coupled atmosphere–ocean models the processes that control biogeochemical cycling.

The first attempts at developing physical–biological models for basin scale studies are now underway (Wroblewski et al. 1988) and the first theoretical studies (Bennett and Denman 1989) that consider factors that could produce large scale (basin) patchiness in phytoplankton distributions are becoming available. However, as mentioned, the most progress in developing coupled global/basin scale models has been in the areas of ocean–atmosphere (e.g., Washington et al. 1980) and ocean–geochemical (e.g., Toggweiler et al. 1989a,b; Sarmiento et al. 1989) models. These models must also consider mismatches in the scales of processes and the nonhomogeneity of model parameters. In theory, the techniques developed for these models can be applied to physical–biological models. For example, inverse modeling is a promising direction for using available data to identify important processes, constrain values for the parameters to which the models are most sensitive, and to produce in a quantitative manner error bars on measurements and models (e.g., Platt et al. 1981). Further, techniques for the assimilation of observations into ocean circulation models (Haidvogel and Robinson 1989) and regional physical–biological models (Ishizaka 1990c) are now being developed. The assimilation of data corrects errors in model parameterization and keeps the model solutions from drifting away from observations or climatology. This is also a promising direction for the development of physical–biological models, but it may be that these models require that data for assimilation be available at frequent (e.g., one day) intervals (Ishizaka 1990c).

Perhaps it is incorrect to conceive of a single physical–biological model, in a manner similar to the community models that now exist for atmospheric and ocean circulation studies. Rather it may be more appropriate to consider the development of many fine resolution regional physical–biological models that include all the complexity and dynamics that are needed to simulate biological distributions in these regions. For example, a biological model, combined with a regional primitive equation circulation model, like the one described in Haidvogel et al. (1990), could provide estimates of the export of biogenic material from a continental shelf. The output from this type of model could then be used to provide input to larger scale open ocean models and *vice versa*. One example of this approach for regional circulation modeling is given by Wang et al. (1984), in which the output from a global tidal model was used to specify the offshore boundary conditions of a shelf tidal model. Also, some of the techniques used in atmospheric modeling in which fine scale phenomena are embedded in larger scale models (e.g., Hill 1968) may be applicable to physical–biological models.

The real question then is not *how* we generalize coastal models to global scales, but rather *can* we generalize coastal models to global scales? Obtaining an answer to this question will require simultaneous investigation with both coupled regional and global models. It is likely that the two types of models will be run together and coupled only through boundary exchanges.

Acknowledgements. This research was supported by the National Aeronautics and Space Administration, the Office of Naval Research, and the National Science Foundation of the United States. I thank J. Steele for his constructive comments on this manuscript.

REFERENCES

Bennett, A.F., and K.L. Denman. 1989. Large-scale patchiness due to an annual plankton cycle. *J. Geophys. Res.* **94**:823–829.

Bucklin, A., M.M. Rienecker, and C.N.K. Mooers. 1989. Genetic tracers of zooplankton transport in coastal filaments off northern California. *J. Geophys. Res.* **94**:12555–12569.

Committee on Global Change. 1988. Toward an Understanding of Global Change. Initial Priorities for U.S. Contributions to the International Geosphere-Biosphere Program. Washington, D.C.: National Academy.

Eppley, R.W. 1972. Temperature and phytoplankton growth in the sea. *Fish. Bull.* **70**:1063–1085.

Haidvogel, D.B. and A. Robinson (eds.). 1989. Special issue: data assimilation. *Dynam. Atmos. Oceans* **13**:171–513.

Haidvogel, D.B., J. Wilkin, and R. Young. 1990. A semi-spectral primitive equation ocean circulation model using vertical sigma and orthogonal curvilinear coordinates. *J. Comput. Ph.*, in press.

Hill, G.E. 1968. Grid telescoping in numerical weather prediction. *J. Appl. Met.* **7**:29–38.

Hofmann, E.E. 1988. Plankton dynamics on the outer southeastern U.S. continental shelf. Part III: A coupled physical–biological model. *J. Marine Res.* **46**:919–946.

Hofmann, E.E., and J.W. Ambler. 1988. Plankton dynamics on the outer southeastern U.S. continental shelf. Part II: A time-dependent biological model. *J. Marine Res.* **46**:883–917.

Ishizaka, J. 1990a. Coupling of Coastal Zone Scanner data to a physical–biological model of the southeastern U.S. continental shelf ecosystem. 1. CZCS data description and Lagrangian particle tracing experiments. *J. Geophys. Res.* **95**:20167–20181.

Ishizaka, J. 1990b. Coupling of Coastal Zone Color Scanner data to a physical–biological model of the southeastern U.S. continental shelf ecosystem. 2. An Eulerian model. *J. Geophys. Res.* **95**:20183–20199.

Ishizaka, J. 1990c. Coupling of Coastal Zone Color Scanner data to a physical–biological model of the southeastern U.S. continental shelf ecosystem. 3. Nutrient and phytoplankton flux estimates and CZCS data assimilation. *J. Geophys. Res.* **95**:20201–20212.

Ishizaka, J., and E.E. Hofmann. 1988. Plankton dynamics on the outer southeastern U.S. continental shelf. Part I: Lagrangian particle tracing experiments. *J. Marine Res.* **46**:853–882.

Lewis, M.R., J.J. Cullen, and T. Platt. 1983. Phytoplankton and thermal structure in the upper ocean: Consequences of nonuniformity in chlorophyll profile. *J. Geophys. Res.* **88**:2565–2570.

Platt, T., K.H. Mann, and R.E. Ulanowicz. 1981. Mathematical Models in Biological Oceanography. France: UNESCO.

Rothlisberg, P.C., J.A. Church, and A.M.G. Forbes. 1983. Modelling the advection of vertically migrating shrimp larvae. *J. Marine Res.* **41**:511–538.

Sarmiento, J., M.J.R. Fasham, U. Siegenthaler, R. Najjar, and J.R. Toggweiler. 1989. Models of chemical cycling in the ocean: Progress Report II. Ocean Tracers Laboratory Technical Report No. 6, Princeton University, Princeton.

Semtner, A.J., Jr., and R.M. Chervin. 1988. A simulation of the global ocean circulation with resolved eddies. *J. Geophys. Res.* **93**:15502–15522.

Smith, W.O., G.W. Heburn, R.T. Barber, and J.J. O'Brien. 1983. Regulation of phytoplankton comunities by physical processes in upwelling ecosystems. *J. Marine Res.* **41**:539–556.

Toggweiler, J.R., K. Dixon, and K. Bryan. 1989a. Simulations of radiocarbon in a coarse-resolution world ocean model. 1. Steady state prebomb distributions. *J. Geophys. Res.* **94**:8217–8242.

Toggweiler, J.R., K. Dixon, and K. Bryan. 1989b. Simulations of radiocarbon in a coarse-resolution world ocean model. 2. Distribution of bomb-produced carbon 14. *J. Geophys. Res.* **94**:8243–8264.

Walsh, J.J. 1975. A spatial simulation model of the Peru upwelling ecosystem. *Deep-Sea Res.* **22**(A):201–236.

Walsh, J.J., D.A. Dieterle, and M.B. Meyers. 1988. A simulation analysis of the fate of phytoplankton within the Mid-Atlantic Bight. *Cont. Shelf Res.* **8**:757–787.

Wang, D.P., V. Kourafalou, and T.N. Lee. 1984. Circulation on the continental shelf of the southeastern United States. Part II: Model development and application to tidal flow. *J. Phys. Ocean.* **14**:1013–1021.

Washington, W.M., A.J. Semtner, Jr., G.A. Meehl, D.J. Knight, and T.A. Mayer. 1980. A general circulation experiment with a coupled atmosphere, ocean, and sea ice model. *J. Phys. Ocean.* **10**:1887–1908.

Wroblewski, J.S. 1977. A model of phytoplankton plume formation during variable Oregon upwelling. *J. Marine Res.* **35**:357–393.

Wroblewski, J.S. and E.E. Hofmann. 1989. U.S. interdisciplinary modeling studies of coastal-offshore exchange processes: past and future. *Prog. Ocean* **23**:65–99.

Wroblewski, J.S., J.L. Sarmiento, and G.R. Flierl. 1988. An ocean basin model of plankton dynamics in the North Atlantic, 1, Solutions for the climatological oceanographic conditions in May. *Glob. Biogeochem. Cycl.* **3**(3):199–218.

Geophysiology of the Oceans

J.E. Lovelock

Coombe Mill
St. Giles on the Heath
Launceston, Cornwall, PL15 9RY, U.K.

Abstract. In geophysiology, the Earth is taken to be an evolving system in which the organisms and their material environment are so tightly coupled together that they constitute a single entity. This postulated entity, or super-organism, has emergent properties like the self-regulation of climate and chemical composition. This chapter is about the particular function of the oceans in the physiology of the Earth.

INTRODUCTION

The Gaia hypothesis supposed that life maintained the material conditions of the Earth at a state favorable for its survival (Lovelock 1972; Margulis and Lovelock 1974). Stated like this the hypothesis was rather poetic and probably misleading. It arose while I was working for NASA on an experiment to detect life on Mars. The idea behind the experiment was that organisms present on a planet would inevitably change the chemical composition of the atmosphere in a way that would be recognizably different from that of the near equilibrium chemistry of a lifeless planet. This experiment forced a top-down view of Mars, one that saw the planet as a whole entity. This same top-down view when applied to the Earth saw an atmosphere so profoundly departed from chemical equilibrium, and inherently unstable, that the presence of some powerful means of regulation seemed needed to maintain constancy. At the time this thought was not much more than an intuition and I was not able to express it rigorously. The few attempts I made to do so were unsuccessful, most scientists at that time were deep in the security of their disciplines. Apart from the enlightenment of Sillen, Hutchinson, and Redfield, few were prepared to consider the atmosphere to be a biological product. They were not even prepared to entertain interdisciplinary ideas. Methane, for example, was

Ocean Margin Processes in Global Change
Edited by R.F.C. Mantoura, J.-M. Martin and R. Wollast

seriously considered to come from primeval reservoirs by outgassing, and the source of oxygen was the upper atmospheric photolysis of water vapor.

Over the past twenty years our knowledge of the Earth has greatly enlarged, but our understanding of it as a system is still hampered by the fragmentation of the sciences. The greatest enlargement has been in atmospheric science, especially atmospheric chemistry. The new discoveries about the atmosphere led us to realize that all but a few percent of atmospheric gases are the direct products of living organisms. By contrast, an immense ignorance still prevails about the oceans and the crustal rocks. We know that the bulk of the ocean, mainly salt and water, comes directly from inorganic sources. The possibility that organisms living in the ocean could regulate their own composition was not seriously considered until quite recently. The Gaia hypothesis was originally conceived from atmospheric evidence and has continued to be supported by recent advances in atmospheric chemistry. It went further and implied that the composition of the oceans might also be regulated by the ocean ecosystems; however, few were prepared to argue for or against this contention. The involvement of marine organisms in the system of control for seawater composition was first touched upon by Sillen in 1966 and later discussed by Broecker (1971) and by Garrels and Mackenzie (1971). Whitfield published his seminal paper, "World Ocean–Mechanism or Machination," in 1981. However, even today the quantitative implications of active regulation remain unclear.

This chapter will briefly explain how a systems science, geophysiology, arose from the Gaia hypothesis. It will go on to consider how this new science might be applied to the understanding of the oceans.

GEOPHYSIOLOGY AND THE GAIA HYPOTHESIS

In the 1970s the Gaia hypothesis was frequently stated as: "The climate and the chemical composition of the Earth's surface and atmosphere are regulated, at a state favorable for life, by and for the biosphere." I now acknowledge that the hypothesis, if so stated, is incorrect. It is not life, or the biosphere, that regulates the atmosphere, or the oceans; regulation, if it occurs, is a property of the whole tightly coupled system that comprises life and its material environment.

The mistake made by suggesting that life, rather than the system, is the regulator laid it open to heavy criticism from biologists and geochemists. The biologists saw life as a property of individual living organisms. They knew that large collections of organisms, like the biota or the biosphere, were not in their terms alive, nor could they evolve by natural selection; only organisms can evolve. Geochemists were critical because they thought that the Earth's composition could adequately be explained by their science alone.

Biologists, like Ford Doolittle (1981) and R. Dawkins (1982) expressed their objections succinctly and forcibly. The biota, they said, would need foresight and would have to plan if they were to regulate the Earth. There was no evidence that committees of the species met at annual conferences and agreed on the climate. There is just no way that natural selection can lead to altruism on a global scale. Richard Dawkins also argued that even if there was a planetary organism, Gaia, it could never have evolved by natural selection; it has no capacity for reproduction.

The eminent geochemist H.D. Holland in his book "The Chemical Evolution of the Atmosphere and Oceans," (1984) said that the Gaia hypothesis was not needed to account for the regulation of the Earth's chemistry and climate; the changing state of the Earth could fully be described by geochemistry and geophysics.

Gaia theory now sees regulation as a systems' property and the objections to the original hypothesis, although understandable, are, in the context of a system theory, wrong. Moreover, the critics failed to see that their own explanations of the Earth were contradictory. Biologists and geochemists were both circumscribed by the limitations of the disciplines and unable to see the larger system of which they were a part.

The geochemists' criticism, that Gaia is not needed to explain regulation, was challenged in two papers (Lovelock and Whitfield 1982; Lovelock and Watson 1982). Here it was shown that the abiological rate of weathering of the rocks was, by itself, insufficient to account for the low contemporary level of CO_2. In reality, the system comprising the organisms and their environment acts to pump CO_2 from the atmosphere and, as a consequence, sustains a much cooler and for them more comfortable climate than would otherwise be possible.

The biological criticism was answered by means of a simple model (Lovelock 1983; Watson and Lovelock 1983) called "daisyworld." The model was of the growth of a single plant species, daisies, on a planet that was subjected to a steadily increasing input of heat from its star. The growth and spread of daisies over the planetary surface altered the planetary albedo and hence the heat balance of the planet. The model illustrated that when the color of the daisies is different from that of the bare rocks, their growth and spread over the planetary surface powerfully regulates planetary temperature. I agree with the biologists who say that life cannot regulate the Earth, but observe that they have failed to recognize the existence of the system, composed of organisms and their material environment: a system that seems to have properties that include homeostasis and the capacity for self-regulation.

Another frequent, but easily answered, criticism of Gaia was that it had all been said before by Huxley in his explanations of the "Balance of Nature." For example, Postgate (1988) used the balance set between the source, consumers, and the sink, photosynthesizers, to explain the abundance of gases such as CO_2. The error of this explanation lies in its neglect of the geochemical cycle of volcanism and weathering. Those who believe in the balance of Nature seem to think that the exchange of CO_2 and oxygen between animals and plants sets the atmospheric abundance of these gases. Geochemists know that the cycle of CO_2 between plants and animals, although large, is a "do nothing" cycle. Organisms alone do not set the level of CO_2 in the air, they only accept what is there. Plants and animals must always, at steady state, exchange equal quantities of CO_2. The geochemical cycle alone, it would seem, sets the level of CO_2 in the air.

In fact, neither biology nor geochemistry can account for the current low level of CO_2 in the atmosphere. Even if there were an interdisciplinary gathering of geochemists and biologists, they could do little more than disagree about what regulated CO_2. There is a parallel here between the Earth as a system and an animal, such as a human. Neither anatomy nor biochemistry can alone explain why our core temperatures are constant at 37°C. Temperature regulation is a property within the remit of physiology, the systems' science of the body. Gaia theory, unlike biology or geochemistry, sees the Earth as a physiological system.

Regulation, if it occurs, is an emergent property of the system that comprises the living organisms of the Earth and their material environment. This system, when operating, is indivisible, and like a living organism, the whole is more than an anthology of its parts. As James Hutton said in 1789, "its proper study is by physiology."

In my spare time during the 1980s I was able to apply the geophysiological approach, the basis of the daisyworld model, more generally. I made models of the Archean and Proterozoic periods of the Earth's history. In these new models I extended the idea, first used in daisyworld, of regulating climate by coupling together the evolution of the growth of organisms and the planetary radiation balance. Climate, in these new models, was determined by the greenhouse gases of the atmosphere, water vapor, carbon dioxide, and methane. The growth of photosynthetic bacteria had, like white daisies, a cooling effect. This was due to the removal of CO_2 from the air, as they used it to sustain growth and also because their growth on surface rocks would increase the rate of removal of CO_2 by weathering. Warming came from the release to the atmosphere, by anaerobic bacteria, of methane, a fairly potent greenhouse gas. However, in addition to the coupling of the climate with growth, as in daisyworld, these new models coupled also the supply of nutrients released by rock weathering and the effects of oxygen gas as an environmental variable. An excess of oxygen was taken to be toxic, but a little oxygen was considered to be beneficial through its effect of increasing the rate of rock weathering, and consequently on the supply of nutrient elements. Oxygen also increases the efficiency of carbon turnover by making possible the direct consumption of plants rather than the slow fermentation of their debris by methanogens.

These geophysiological models differed considerably from the more conventional biogeochemical models. In biogeochemistry, the science of Vernadsky (1945) and Hutchinson (1954), the biota is assumed to adapt passively to changes in the chemical or physical environment. In geophysiology (Lovelock 1989), the primacy of adaptation no longer rules; organisms can change the environment as well as adapt to it. This seemingly small change in model structure profoundly affects mathematical stability. Biogeochemical box models are notoriously sensitive to the right choice of initial conditions and are often unstable and easily perturbed. Geophysiological models, by contrast, are insensitive to the choice of initial conditions; they maintain a stable state, homeostasis, and quickly return to it if perturbed. In addition, the geophysiological models predict levels for the environmental variables of climate and atmospheric gas composition that are chemically realistic and which would be acceptable to the organisms in the ecosystems of the model.

The existence of these well-behaved models that can predict climate and chemical composition was what changed the Gaia hypothesis into Gaia theory. It is a theory of evolution that grows from, and acknowledges, Darwin's great vision. It differs from the biological theory of evolution by including the evolution of the organisms and the evolution of their geochemical and geophysical environment together as a single system. In the new theory, these previously separated evolutionary processes are seen as one single tightly coupled process. This physiological approach to a combined Earth and life science is not new. It has been touched on before (Hutton 1788).

However, as the theory has developed, the Gaia hypothesis and the general ideas around it have been increasingly rejected by the scientific community. The rejection was often emotional rather than rational and rarely ever touched on the real weakness of the hypothesis: its failure to state that a system, Gaia, did the regulating, not just life. So strong were these feelings that it became impossible to publish papers on Gaia theory in peer reviewed journals. Consequently, most statements of Gaia theory have appeared either in my book, "The Ages of Gaia" (Lovelock 1988) or in conference proceedings (Lovelock 1989). Some indication of the heat aroused by the idea of Gaia is illustrated by a short exchange that occurred at the 1988 Dahlem Conference "The Changing Atmosphere." Here I was asked what the Gaia theory had to say about the effect of algal growth at the ocean surface on clouds and *climate*. Before I could reply, a participant objected on the grounds that the Dahlem Conferences were serious and respected scientific gatherings and their credibility would diminish if they were used to tell fairy stories about Greek goddesses.

WHAT IS GAIA?

Gaia is a system open to the exchange of radiation to and from space and to the exchange of matter to and from the Earth's interior. Its bounds are space on the outside and the base of the crustal rocks on the inside. The system Gaia is therefore a thin skin of organisms and of their controlled environment, spread circumferentially around the Earth. The Earth's interior is an "inner space" that is effectively an infinite source and sink for materials during the life span of the Earth. Scientists balk at the idea of the Earth as a living organism of any kind, mainly, I think, because all unconsciously have a definition of life that is hierarchical. We tend to classify mammals as more alive than reptiles, and they more alive than invertebrates. Plants are still less alive while bacteria, fungi and lichens are hardly alive at all. As a quasi-living system, the postulated entity Gaia is different from the various forms of life. I find it helpful to think of it as like a large, old tree. The bulk of the interior wood of a tree is dead matter, as is the bark, the living tissue is a thin layer around the circumference of the wood. Only a few percent of a tree is alive in a literal sense, but few would see a whole tree in full leaf as dead. The system Gaia is rather like the tree. The thin layer of organisms on the land surfaces, or dispersed in the oceans, are like the living cells of the tree, they exist below the nonliving air and above the now dead rocks. The resemblance is strengthened when it is recalled that the air, like the bark, is the direct product of the organisms, and the crustal rocks, like the wood, have extensively passed through living organisms during the Earth's evolution.

OCEAN GEOPHYSIOLOGY

For the purpose of this section, let us accept Gaia theory as a working description of the Earth and then consider to what extent the ocean participates as a subsystem in the process of planetary homeostasis. The ocean is composed of materials that, for the most part, seem to have an abiological origin. Water and salts have come from the interior by outgassing and by the weathering of igneous rocks. This makes it more difficult to distinguish the physiological role of Gaia in the ocean from the passive biogeochemistry of conventional science. I shall limit myself in this paper to considering the small body of knowledge so far gathered about the part played by ocean organisms in regulating the climate and chemical composition of the atmosphere.

Carbon Dioxide and Climate

Geophysiologists accept the geochemical view of the Earth and agree that the source and sink for carbon dioxide are, respectively, tectonic processes and weathering; geophysiologists, however, couple in the biota by recognizing that the rate of weathering is strongly determined by the organisms living in contact with the rocks. Volk and Schwartzman (1989) confirm this conclusion and claim that the weathering of basalt under sterile conditions is 1000 times less than when organisms are present. I used, in the model mentioned earlier, a system for climate regulation by CO_2 pumping. It involved the well-known reaction of atmospheric CO_2 and rainwater with calcium silicate rock (Walker et al. 1981). Since the growth of organisms follows a parabolic relationship with temperature, where growth is negligible below 5°C and above 50°C and is best at intermediate temperatures, there will be a strong link between temperature and the rate of the rock weathering reaction. It is easy to model such a system, and the model predicts stable regulation close to the current levels of temperature and CO_2. If we look on this system as an engineering construction and apply control theory, the open loop gain of the model is close to 50, amply large enough when the loop is closed to keep the system stable and well regulated.

There is much more, however, to the climate and CO_2 system than rock weathering on the land surfaces. My colleagues, Andrew Watson and David Turner (Pers. Comm.) of the Plymouth Marine Laboratory, have recently investigated the exchanges of CO_2 between the air and waters of the open Atlantic. They discovered a large scale absorption of CO_2 from the atmosphere at the time of the bloom of diatoms. Later, this almost eutrophicated growth of diatoms was succeeded by blooms of cocolithophores and other algae, this confirmed and extended the work of Holligan et al. (1983). During the second phase CO_2, dissolved in the ocean, is moved a step further towards burial in the sediments by its inclusion in the calcium carbonate skeletons of the algae.

These intricate geochemical and biological interactions have previously been ignored. Ocean investigators recognized that photosynthesis plays some part in reducing the partial pressure of CO_2 in the surface layers of the ocean but saw no pressing need for more than physical diffusion and physical chemistry to explain the exchange of CO_2 between the atmosphere and ocean. It was not clear that the biological signal could ever dominate the physical (solubility) signal on a wide enough scale to be noticed. There was an absence of any direct measurements of chlorophyll and pCO_2 taken simultaneously over a significant area of ocean. The real impact of Watson and Turner's research was to confirm the dominance of the biological signal over a large area of the North Atlantic during spring and summer. In particular, they showed a strong local coherence between chlorophyll and pCO_2 and a strong regional coherence between pCO_2 and temperature with a coefficient whose sign was opposite to that predicted by purely physical control.

The complete story is likely to be slow in unfolding. When it does, an important part will surely be the other connection between ocean algae and the atmosphere, their role as the main source of sulfur gases in the marine environment.

Algae in the Oceans and Clouds

A global rise in temperature is a reasonable prediction as *greenhouse gases*, like CO_2, accumulate in the air. However the extent of the temperature rise requires, among other things, an understanding of the effects and responses of clouds on climate and climate change. In a recent paper, Mitchell et al. 1989 suggest that the predictions based on the effects of the greenhouse gases alone may be in error by a factor of 100%. They included in their general circulation model not merely clouds, but also the effect of warming on the distribution of ice in the water of clouds.

The failure to take sufficient note of the liquid and solid phases of the atmosphere was responsible for the inability of the complex stratospheric models to predict the ozone hole over Antarctica.

These small misunderstandings of a disciplinary kind that hindered communication between cloud physicists and atmospheric chemists have twice led to large errors of prediction. Consider the possible errors that can come from the great misunderstanding that exists between atmospheric scientists and biologists. A single top-down physiological view of the system may provide the means to avoid this type of error.

Charlson et al. (1987) proposed that clouds over the open oceans may depend upon the emissions of sulfur gases from ocean surface ecosystems for their presence. This argument was based on the well-established observation (Andreae 1985) that dimethyl sulphide (DMS) is ubiquitously present in the ocean surface and atmosphere. DMS oxidizes rapidly in the air to form sulfuric and methane sulfonic acids. Microdroplets of these acids are potent cloud condensation nuclei (CCN). There are few other CCNs over the open oceans; sea salt spray particles, previously thought to be the predominant CCNs, are now known to be scarce at cloud level. A recent and definitive paper by Savoie and Prospero (1989) reports that 80% of the sulfate-bearing particles over the Pacific ocean are biological in origin. Their observations strongly suggest the gaseous intermediate DMS, as the source of particulate sulfur with a biological signature. Savoie and Prospero also note that sulphur gases from industrial and domestic sources, even though their emissions are much larger than those of DMS, do not reach the principal areas of the open oceans. It also seem unlikely that any process of sea salt spraying could selectively segregate biological sulfate from the abundant sulfate ion of the ocean.

Although there may be little doubt about the connection between CCNs over the ocean and the marine biota, we are still a long way from establishing the connection, if any, between climate and marine organisms.

The relevant facts are these: The presence of clouds in the troposphere does have a net cooling effect. The affair is quite complex and certainly all clouds do not cool. Denser clouds, and low lying clouds like marine stratus, are good at cooling, but high level clouds, like cirrus, tend to warm by reflecting outgoing infrared light back to the ground more than they cool by returning visible light to space. DMS is an end product of the metabolism of marine algae of many species. It does diffuse from the ocean surface to the atmosphere, and it does oxidize at the right rate to be a candidate source of CCNs. Calculations suggest that a doubling of CCNs from the oceans could produce a cooling effect comparable with the greenhouse effect of a doubling of CO_2. Such a doubling is not beyond the capacity of the ocean organisms. Ice core samples (Legrand et al. 1988) from Antarctica suggest that DMS emissions were twice as great during the glaciation as now.

The missing link is the connection between climate and the emission of DMS by the algae. It is not easy to imagine a direct connection between sea surface temperature and the algal emission of sulfur gases. A small rise in ocean surface temperature represents a vast quantity of stored heat but observations at different latitudes by Andreae (1985) show no consistent change of DMS emission with temperature. He did find that hot, oligotrophic, ocean regions emitted more DMS than did cool and more nearly eutrophic temperate waters. A great deal more ocean investigation is needed before we can be sure of the influence of speciation, season, nutrient supply, and salinity as well as climate on DMS emissions.

A tantalizing indication of an indirect link between DMS and climate comes from observations that suggest an increase in cloud cover, algal density, and wind velocity has occurred over the oceans in the past decade. The emission of DMS from an algal bloom could lead directly to a local increase in sea surface wind velocity. The energy driving the wind would come from the release of latent heat when water condensed on the CCNs to form the clouds. Clouds, where there was an updraft of damp air fueling this energy release, would draw in surface air and so generate wind at the sea surface. Becalmed sailors know the value of a cloud as marking a place on the ocean surface where wind can be expected. The corresponding source of biochemical energy would be the wind-induced circulation of nutrient-rich water from below the thermocline. Perhaps these local systems link in with the large scale geophysical weather systems in a coherent manner and so lead to a global system response. To add speculation to supposition, I wonder if the wind increase also serves to bring nutrient elements like iron from the deserts to the global oceans.

The first search for DMS emissions from the oceans was made in 1972 for quite different reasons. Gaia theory predicted the need for a mechanism to return nutritious elements that were in scarce supply on the land surfaces, but abundant in the oceans. DMS was known to be emitted by the large algae that grew in coastal waters. It was therefore postulated to be the candidate sulfur carrier in the marine environment. It was sought and found at all points during a voyage of the RS Shackelton from the U.K. to Antarctica and back in 1971 and 1972. As a bonus, *methyl iodide* and carbon disulfide were also found to be present in the air and in the surface waters of the ocean. Iodine is a rare element on the land and nutritional deficiencies involving a lack of iodine are well known. The biosynthesis of the necessary but highly toxic substance, methyl iodide, was unexpected but intriguing. Just as the role of DMS as a precursor for CCNs enlarged our interest in its emissions, so the suggestion Barnes et al. (1987) that iodine is involved in the tropospheric chemistry of DMS oxidation enlarges interest in the ocean ecosystems that emit both of these gases simultaneously.

Other Thoughts on Ocean Geophysiology

I will conclude by discussing briefly some thoughts that arise from a geophysiological view of oxygen, water, salinity, and the emissions of trace gases from the oceans.

Kump (1988) proposed a geophysiological system for the regulation of atmospheric oxygen. In this system the source of oxygen was, as usually believed, the burial of residual organic matter that had avoided aerobic digestion. The major site for carbon burial is usually taken to be the sediments of the continental shelves. In his paper Kump proposes a partition of carbon burial between the land surfaces and the sea, mediated by the prevalence of forest fires. The probability of fires following lightning strikes is a steeply rising function of oxygen abundance, nearly doubling for each 1% increase of oxygen at 21%. In Kump's model, fires change the cycle of phosphorus in such a way that carbon burial on the land surface declines and so reduces the total carbon burial and the rate of oxygen production. The tight link between fires and oxygen makes the land surface ecosystems sensitive to small variations of atmospheric oxygen. If Kump's model is correct it places the regulation of oxygen on the land surface rather than in the coastal waters.

The abundance of water on the Earth is difficult to explain by geochemistry. Mars and Venus are dry, yet these planets probably outgassed water in the same proportion to other volatiles as did the Earth. The small terrestrial planets readily lose hydrogen to space, because the temperature and the gravitational field of their exospheres are inconsistent with the retention of hydrogen atoms. Ordinarily the presence of atmospheric oxygen keeps the hydrogen abundance low in the upper atmosphere and the escape rate is low, but there is a continuous production of hydrogen from the reaction of ocean water with hot basalt at the seafloor spreading zones. If this hydrogen were to accumulate it would yield a hydrogen-rich reducing atmosphere and there would be a continuous loss of hydrogen to space, and hence a loss of water from the oceans. We need to know the rate of seawater reduction over the Earth's history. It seems possible that living organisms, by their production of oxygen and their removal of hydrogen, keep the Earth moist. If so then without life the Earth would by now be dry like the other terrestrial planets. If true it would mean that abundant water, like abundant oxygen, is a characteristic of a life-bearing planet. Not only does water enable life, but life sustains water.

In a similar, and possibly connected, way the constancy of the salinity of the oceans is not yet properly explained. One of the early arguments for the Gaia hypothesis was that marine life has a critical upper limit of salinity that it can tolerate and that the present ocean was close to this limit. How could the geochemical sources and sinks of salt have remained constant enough during the Earth's history to avoid exceeding the critical upper limit of 0.8 Molar?

The sources of the ions in the ocean are the weathering of continental rocks and releases of volatiles from the seafloor and other tectonic sites. The sinks for salt are reactions with the hot rocks of the seafloor spreading zones and the deposition of salt in evaporite beds. Was it just chance that kept these processes right for ocean life or does geophysiology operate to regulate salt? Hinkle and Margulis (1988) report that the answer is still unknown; however, they have found examples of biological intervention in the formation and survival of evaporite beds. For example, organisms present in algal mats at the surface of evaporite lagoons coat salt crystals with a varnish that protects them from dissolution by rain.

We have seen how the emission of a minor trace gas, DMS, may have profound effects on the Earth's climate. Do the other trace gases COS, CS2, methyl chloride also have roles larger than their mere abundances would suggest? Shaw (1983) proposed a climate regulator based on COS. This gas is the source of the background stratospheric aerosol of sulfuric acid. Shaw argued that the source of COS is biological and the aerosol of sulfuric acid has a cooling effect at the surface of the Earth.

The geophysiology of methyl chloride, another trace gas to be emitted from the oceans, remains obscure. Methyl chloride has an abundance not far from one part per billion and might well repay further investigation. It would be intriguing to think of it as a natural ozone depleter; too much stratospheric ozone might be as harmful as too little.

CONCLUSIONS

So great was the impact of the theory of evolution by natural selection among organisms that the older scientific tradition of a planetary physiology was displaced and has been forgotton for more than a hundred years. It took the top-down view of the Earth from space, both as an image and as a source of scientific information, to enlarge and change our understanding of the Earth, and to make it worth our while to re-visit the older systems way of Earth Science. Whether or not Gaia theory is correct, or offers a more complete account of the Earth, is just now much less important than the new insight it provides into the machinations and mechanisms of the Earth.

REFERENCES

Andreae, M.O. 1985. The emission of sulphur to the remote atmosphere. In: The Biogeochemical Cycling of Sulphur and Nitrogen in the Remote Atmosphere, ed. J. Galloway et al., pp. 5–25. Dordrecht: Reidel.

Barnes, I., V. Bastian, K.H. Becker, and H. Niki. 1987. FTIR spectroscopic studies of the methylthio and nitrogen dioxide reaction under atmospheric conditions. *Chem. Phys. Lett.* **140**(5):451-457.

Broecker, W.S. A kinetic model for the chemical composition of sea water. 1971. *Quatern. Res.* **1**:188-207.

Charlson, R.J., J.E. Lovelock, M.O. Andreae, and S.J. Warren. 1987. *Nature* **326**:655–661.

Dawkins, R. 1982. The Extended Phenotype. Oxford: W.H. Freeman.

Doolittle, W.F. 1981. *Co-Evol. Quart.* **29**:58–63.

Garrels, R.M., and F.T. Mackenzie. 1971. Evolution of Sedimentary Rocks. New York: W.W. Norton.

Hinkle, G.J. 1988. Marine salinity: a Gaian phenomenon. In: Gaia, the Thesis the Mechanisms and the Implications, ed. P. Bunyard and E. Goldsmith. Wadebridge: Quintrell.

Holland, H.D. 1984. The Chemical Evolution of the Atmosphere and the Oceans. Princeton: Princeton University Press.

Holligan, P.M., M. Viollier, D.S. Harbour, P. Camus, and M. Champagne-Philippe. 1983. *Nature* **304**:339–342.

Hutchinson, G.E. 1954. The biogeochemistry of the terrestrial atmosphere. In: The Solar System: The Earth as a Planet, ed. G.P. Kuiper, vol. 2, chapter 8. Chicago: University of Chicago Press.

Hutton, J. 1788. Theory of the Earth: or an investigation of the laws observable in the composition, dissolution, and restoration of the land upon the globe. *Trans. R. Soc. Edinburgh* **1**:209–304.

Kump, L.R. 1988. Terrestrial feedback in atmospheric oxygen regulation by fire and phosphorus. *Nature* **335**:152–154.

Legrand, M.R., R.J. Delman, and R.J. Charlson. 1988. Climate Forcing implications from Vostok ice core sulfate data. *Nature* **334**:418–420.

Lovelock, J.E. 1972. Gaia as seen through the Atmosphere. *Atmos. Envir.* **6**:579–580.

Lovelock, J.E. 1983. Gaia as seen through the Atmosphere. In: Biomineralisation and

Biological Metal Accumulation, ed. P. Westbroek and E.W. deJong. Dordrecht: Reidel.
Lovelock, J.E. 1988. The Ages of Gaia. New York: W.W. Norton.
Lovelock, J.E. 1989. Geophysiology. *Trans. R. Soc. Edinburgh* **80**:169–175.
Lovelock, J.E., and A.J. Watson. 1982. *Planet. Space. Sci.* **30**:795–802.
Lovelock, J.E., and M. Whitfield. 1982. Life span of the biosphere. *Nature* **296**:561–563.
Margulis, L., and J.E. Lovelock. 1974. Biological modulation of the Earth's atmosphere. *Icarus* **21**:205–221.
Mitchell, J.F.B., C.A. Senior, and W.J. Ingram. 1989. *Nature* **341**:132.
Postgate, J. 1988. *New Sci.* 60.
Savoie, D.L., and J.M. Prospero. 1989. *Nature* **339**:685–687.
Sillen, L.G. 1966. Regulaton of O_2, N_2, and CO_2 in the atmosphere: thoughts of a laboratory chemist. *Tellus* **18**:198–206.
Shaw, G. 1983. *Clim. Change* **5**:297–303.
Vernadsky, V. 1945. The biosphere and the noosphere. *Amer. Sci.* **33**:1–12.
Volk, T., and D.W. Schwartzman. 1989. *Nature* **340**:457–459.
Walker, J.C.G., P. Hayes, and J.F. Kasting. 1981. *J. Geophys. Res.* **86**:9776–9782.
Watson, A.J., and J.E. Lovelock. 1983. Biological homeostasis of the global environment. *Tellus* **35B**:284–289.
Whitfield, M. 1981. World ocean-mechanism or machination. *Interd. Sci. Rev.* **6**:12.

Standing, left to right:
George Knauer, Roland Wollast, Evi Nöthig, John Walsh, Robert Charlson, Birgit Quack, Michael Whitfield
Seated, left to right:
Eileen Hofmann, John Kraft, Fred Mackenzie, Michael Bewers

Group Report: What Is the Importance of Ocean Margin Processes in Global Change?

F.T. Mackenzie, Rapporteur
J.M. Bewers
R.J. Charlson
E.E. Hofmann
G.A. Knauer
J.C. Kraft
E.-M. Nöthig
B. Quack
J.J. Walsh
M. Whitfield
R. Wollast

INTRODUCTION

Undoubtedly, humankind will be faced with global (worldwide) environmental change in the future because of natural and anthropogenic activities/trends. Predicting these changes is fraught with difficulties. These difficulties include:

1. lack of definition of the state of the global system prior to substantial human interference—the "quasi-steady state";
2. the nature of the existing system—its material reservoirs, fluxes, and biogeochemical cycling dynamics; and
3. our frustration with defining and recognizing change and distinguishing between natural responses and anthropogenic influences.

There is no doubt that the biogeochemical cycles of elements in the global system, including the coastal margins, have been modified by the deforestation and agricultural practices of humankind spanning a time scale of several thousands of years. In the last century, industrial, fossil fuel burning, and intensive crop cultivation activities have further perturbed these cycles. Some of these anthropogenic perturbations have coinicided temporally with

Ocean Margin Processes in Global Change
Edited by R.F.C. Mantoura, J.-M. Martin and R. Wollast

changes in Earth's surface system owing to natural causes, such as the recovery of sea level since the last glacial maximum.

Most of the substances that have been added to the global environment are elements and compounds that have circulated freely throughout Earth's surface environment during geologic time; that is, they represent additions to natural biogeochemical cycles existing prior to human activities. Others are synthetic compounds that are new to the environment. It may even be the case that the increased fluxes of material and energy across global system boundaries, including ocean-margin boundaries, may reach rates in the future such as those that prevailed during catastrophisms in Earth's history (Hsü, this volume). There can be no controversy that these modifications to biogeochemical cycles and dynamics, if they continue, will modify coastal margin fluxes and energy transfers and interactions with the open ocean, regardless of the direction of change induced by natural causes. The questions revolve around the magnitude and direction of change, the time and spatial scales of change, and the possibility of abrupt shifts in coastal margin and global biogeochemical dynamics.

In this report, we first consider the dynamics of global change emphasizing that change is the nonlinear product of many natural interactions, superimposed on which are factors related to the activities of humankind. The coastal margin is a part of this global interactive system. We then consider, for a number of components of the global system, the likely direction of change during recent geological time and their future direction, and where possible, rates of change. The inorganic and organic carbon cycle and human intervention in this cycle are discussed as an example of potential worldwide change in the coastal margin. Finally, we consider briefly some future directions of research in the coastal margin that may enable us to perceive and eventually predict change in this critical region because of natural and anthropogenic causes.

THE DYNAMICS OF GLOBAL CHANGE

In this section, we consider the major forcing functions and biogeochemical dynamics of global change that can affect the coastal margin and its interaction with the atmosphere, land, and open ocean. Feedbacks in global and coastal margin systems are complex, but they must be understood if we anticipate making reasonable predictions of the future of coastal margin biogeochemical and physical dynamics. In the final section, we present, for some components of the global and coastal margin systems, past and present conditions and likely future trends as we see them.

Forcing Functions and Processes of Change

In order to understand the future of the coastal margin it must be viewed in the context of the major forcing functions and dynamics of global change. These are schematically shown in Fig. 1. The internal heat of the planet drives the endogenic cycle leading to movement of the great lithospheric plates of the planet. The exogenic, or surface cycle, is driven by incident solar radiation. Both of these forcing functions can affect, for example, the size and distribution of continental shelves. Plate tectonic processes can lead to an increase or decrease in the volume of mid-ocean ridges and displacement of water onto shelves or withdrawal of water from shelves. Changing solar insolation can produce temperature changes leading to changes in the storage of water in continental glaciers and hence sea level changes.

As another example of interacting dynamics, consider the temperature of the planet. The radiation balance determines Earth's surface temperature; in turn, the hydrologic cycle is driven by the temperature. The higher the mean surface temperature, the greater the evaporation rate and the gaseous and liquid water content of the atmosphere, and for a constant land area, the greater the discharge of water from the continents to coastal margins. Furthermore, this increase in temperature and enhanced circulation in the water cycle may lead to increased weathering rates, erosion, and riverine transport of dissolved and suspended materials to the ocean. Further interactions between the major forcing functions of change and the biogeochemical and physical dynamics of change are represented in Fig. 1.

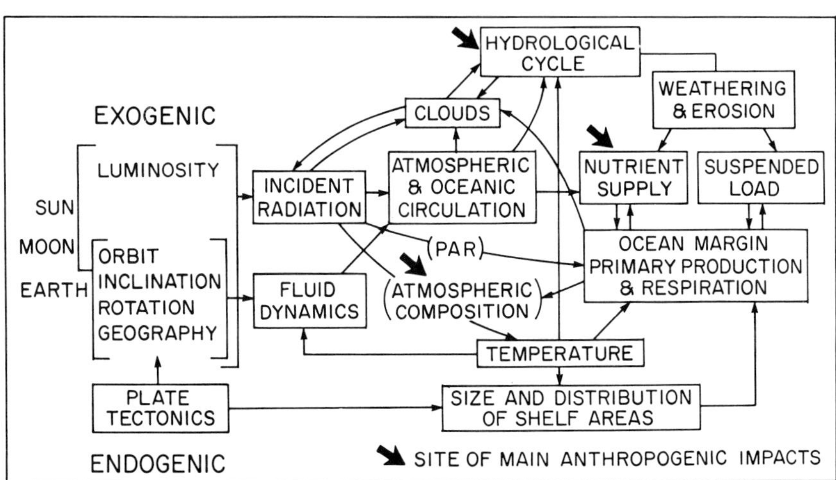

Fig. 1—The coastal margin viewed in the context of the major forcing functions and dynamics of global change. PAR is photosynthetically available radiation.

Consideration of future change in the global and coastal margin systems necessitates recognition of a complexly interwoven set of processes and feedbacks. The coastal margin is not an isolated system with only internal interactions, but is strongly influenced by a number of processes involving the lithosphere, atmosphere, biosphere, cryosphere, and the main body of the ocean. Some of the internal and external processes and the sea level and temperature consequences of changes in the magnitude of these processes are shown in Fig. 2. It is not our intention to discuss in detail the complexity of Fig. 2 but to emphasize that any consideration of future change in the coastal margin system must be viewed in the context of this interactive web with all of its complexity. It can be easily seen that a change in the flux related to one process, e.g., emission rates of greenhouse gases to the atmosphere, can result in changing fluxes related to other processes and to positive and negative feedbacks.

As a specific illustrative case, consider the question of global change and how primary production and phytoplankton biomass may change. On one hand, global change could result in a change in biomass accumulation rates because of higher or lower productivity or, on the other hand, in a change in species composition. Both biological changes can have an influence on the vertical particle flux as well as on atmospheric gas exchange in coastal margins and elsewhere in the oceanic realm.

One example of a biomass and species change probably induced by cultural eutrophication has been provided by Smetacek et al. (this volume). In this case, an increase has been observed in the biomass of a postdiatom bloom species in the North Sea. These species belong to the group Prymnesiophytes and include the species *Phaeocystis pouchetti*, which has been implicated in DMS production. Increase in the biomass of this species in the North Sea during the recent past is most likely due to higher nitrate and phosphate and lower silica inputs to this region, resulting in production of an algae that requires no silica frustule.

Such shifts in species composition can have important consequences for the structure of the food web such as:

1. The algae may not be eaten by secondary producers. This situation could lead to a higher sedimentation rate of the dead phytoplankton and a higher degradation of organic matter by microheterotrophs in the water column and on the seafloor.
2. However, the more plentiful and "new" species can be eaten by secondary producers, which ultimately may result in higher productivity as well as in a higher vertical particle flux of the feces of the grazers.

Both scenarios result in changes in the amount and pathways of the primary produced organic carbon and its fate. Furthermore, these changes

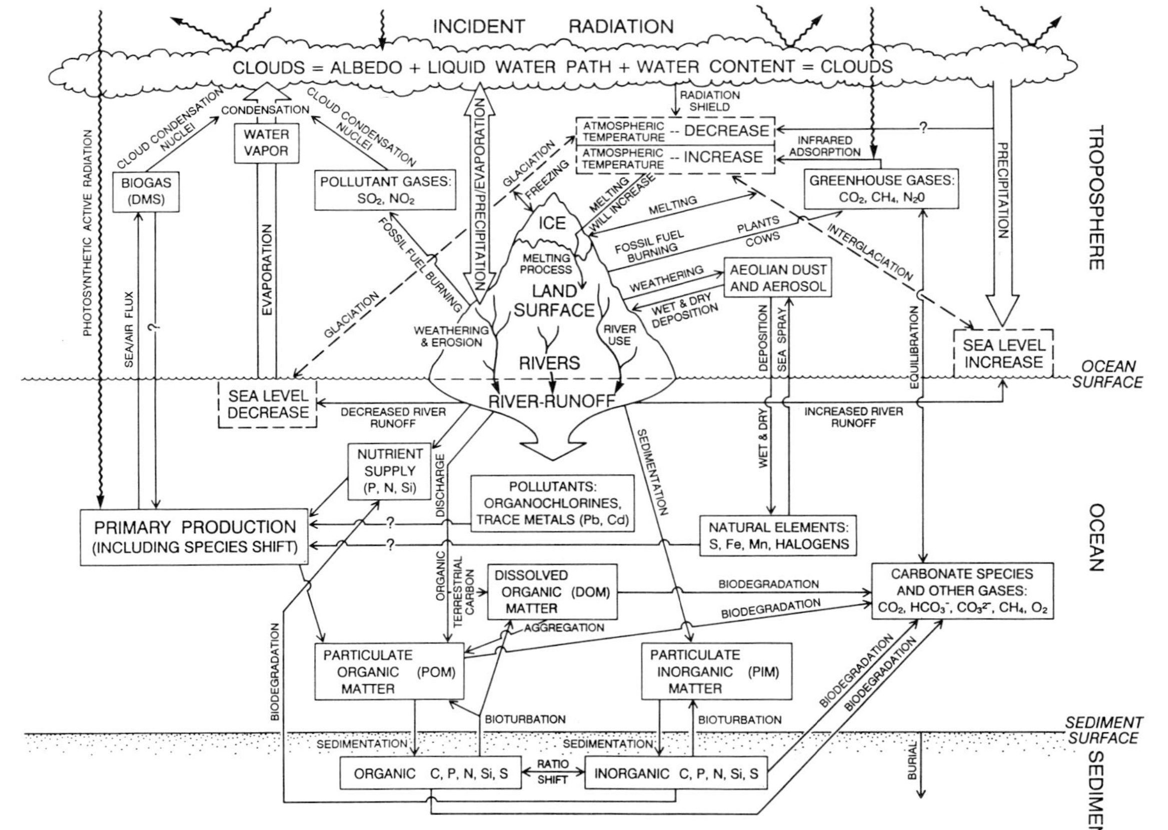

Fig. 2—Schematic diagram of internal and external interactive processes affecting coastal margin biogeochemical and physical dynamics.

can result in changes in O_2 and CO_2 concentrations, and perhaps in the case of *Phaeocystis*, DMS content of seawater on a short time scale. Thus, the biogeochemical dynamics of a coastal margin may be affected by a process occurring upstream in the estuarine system. The coupled and complex interactions shown in Fig. 2 must not be overlooked when considering the coastal margin in the context of global change.

Stability and Rates of Change

The ocean margins exert a global influence through their impact on the biological and chemical cycling of the elements. Two contrasting examples are provided by (a) the role of the shelf seas as a temporary repository for sedimentary material (mineral and organic phases with their associated trace metal, carbon, nitrogen, and phosphorus loads) and (b) the influence that biological processes in ocean margins can exert on the air–sea exchange of climatically important volatiles via the carbon, nitrogen, and sulfur cycles. These cycles are in turn driven by physical forcing functions (Fig. 1) with characteristic scales ranging from milliseconds and millimeters (micro-turbulence) to tens of thousands of years and tens of thousands of kilometers (orbital shifts). Through dimensional analysis, it is possible to ascribe the time and space scales appropriate to the physical measurements needed to characterize these processes. The picture is complicated by the presence of cyclical variations in the intensity of physical forcing related to the Earth's rotation, to periodicities in the luminosity of the sun and in the interactions between the Earth and the other bodies in the solar system. In the ocean margins, these periodicities may lead, for example, to diurnal and seasonal variations in the response of biological systems to tidal oscillations, and to seasonal fluctuations in river flow and in coastal upwelling. Such variations may also drive longer-term changes in ocean currents (e.g., ENSO events) and the cycle of glacial and interglacial episodes experienced in the recent past. For periodic events, the characteristic time and space scales can often be calculated with some precision from the fundamental physics (e.g., tidal levels and periodicities) or can be determined from the analysis of long-term records (e.g., temperature periodicities from ice core data). However, their quantitative impact on biologically and chemically significant processes (e.g., advection of water, sediment transport) will be significantly modified by processes characterized by chaotic dynamics (e.g., turbulence and wind stress). Nevertheless, the periodicity of such systems will play a central role in defining the appropriate time and space scales for the observations.

The chaotic behavior of systems subjected to fluctuating forcing functions provides a further category of variability in space and time. The periodic behavior of systems is in principle predictable given sufficient information on the component frequencies and amplitudes. In contrast, the behavior of chaotic systems is deterministically unpredictable, although the possible states that can be attained are constrained within certain bounds. A fundamental characteristic of such systems is that their variance increases with time scale—whatever can happen, will happen. In coastal systems, this kind of behavior is most clearly apparent in the atmospheric meteorology and in the turbulent dynamics of the water column. The consequences of increasing variance with time are apparent in the infrequent occurrence of storm events. They can also be demonstrated in the dynamics of sediment transport. Where sediment accumulation occurs, the longer the buildup continues, the more likely it is that a threshold flushing event will remove the sediment en masse. Such events occur in estuaries where the bulk of the annual sediment transport to the coast can occur over a few days of exceptional river flow and in shelf sediments where massive turbidite events can transport accumulated sediment thousands of kilometers into the deep ocean on a ten thousand-year time scale. In the case of chaotic systems, the most robust approach is to sample as frequently as is possible. Consequently, meteorological measurements are typically made every few hours over periods of years to provide adequate forcing for models of air–sea interactions. Structured sampling is also possible where the frequency of the measurements is altered in response to the significance of the event. Such systems are used in stream hydrology studies to optimize sampling rates during flash floods.

Besides the sampling regimes demanded by the physical forcing functions, problems arise in correlating the measurements of the resultant behavior of biological and chemical cycles. The time scales implied by the internal dynamics of biological systems (reproduction rates, growth rates, fluxes of components between organisms and their environment) and the space scales implied by their distribution and community structures do not necessarily match with those of the physical forcing functions. In establishing experimental designs, it is therefore essential that the time and space scales of the observations should be harmonized to ensure coherence in the measurements and to enable the chemical and biological processes to be embedded into appropriate physical models.

In studying the response and impact of the ocean margins on a global scale, we must also consider the problem of scaling up the observations made at particular sites and particular times to provide an integrated picture. This again implies some coherence in the experimental design applied at the individual sites in terms of the nature of the observations made and the time and space scales over which they are taken. In addition, a consistent approach must be employed in the selection of sites that takes into account the relative importance of the physical forcing functions.

Within the context of time and space scales is the concept of feedback. One class of feedback known to occur in biogeochemical systems arises when the source and/or sink processes for a chemical species have a dependence on the concentration of that same species. The classical example is that of oxygen in the atmosphere, which is thought to be in a geologically controlled steady state, and the consequence is that O_2 concentration is nearly constant over geological time scales. Numerous trace atmospheric species which are radiatively important or produce radiatively important aerosol particles (e.g., CO_2, CH_4, N_2O, $(CH_3)_2S$, OCS and perhaps the halogenated species, such as methyl chloride, bromide, and iodide) exhibit a degree of constancy from such records as their content in ice cores and from sediment samples. In particular, preindustrial CO_2, CH_4, N_2O, non-sea-salt SO_4^{2-} and $CH_3SO_3^-$ show constancy over periods of time that are much greater than their turnover times in the atmosphere. This constancy seems to imply the existence of active feedback processes, especially given the large reservoirs and fluxes within those systems. Small shifts in oceanic/coastal cycling, therefore, could induce large changes in atmospheric composition. For example, a small increase in $(CH_3)_2S$ emission would do little to the SO_4^{2-} content of seawater but could theoretically produce a climatic cooling and certainly would produce acidification of rainwater near the coastal ocean. It is therefore worthwhile to attempt to understand the processes that control the atmospheric exchange fluxes with the coastal ocean and to ascertain whether feedback mechanisms exist.

Given that the *net* fluxes of several of those climatically relevant species into and/or out of the ocean are relatively small compared to the oceanic cycling fluxes within the system, it will not be sufficient to treat this problem only in a mechanistic manner. Rather, direct observations of fluxes, comparing and contrasting open ocean versus coastal regions, will be required to quantify the respective fluxes to the atmosphere. It will be generally difficult to estimate net fluxes to the atmosphere by measurements of fluxes within oceanic cycles. Again, the example of $(CH_3)_2S$ emission to the atmosphere is best approached by measuring its concentrations in seawater and in air to yield an estimated flux rather than to measure other quantities such as production and destruction rates of its biochemical precursors. Internal consistency should be sought mainly to demonstrate a degree of understanding of the processes that are involved.

If feedbacks do exist (as may be indicated), it will be very difficult to produce estimates of changes in fluxes because of global change unless the feedback processes themselves can be quantified. It must be kept in mind that complex feedbacks may exist in which numerous biogeochemical cycles interact and feedback occurs from one cycle to another (Fig. 2).

In any case, we can conceive of a hierarchy of components that are needed for modeling of biogeochemical cycles:

1. mass balance relations for all the essential elements (C, N, P, S, Fe, ...),

2. linear or nonlinear relationships of all variables,
3. factors or processes that may act to produce either regulatory or amplifying feedbacks.

Interdisciplinary mathematical models, in general, provide one approach for addressing questions related to the conditions that may result in stability of global budgets of quantities such as carbon, and questions related to the conditions that produce changes (and rate of change) in these budgets. Interdisciplinary models can range from simple analytical models to complex numerical simulation models. In particular, these models provide a framework for testing the design of experiments, synthesizing observations, and making predictions about the effect of perturbations on global budgets. However, even with the knowledge that budgets of quantities such as carbon are global in nature (cf. Fig. 3), it is likely that interdisciplinary mathematical models will be developed, at least initially, for specific shelf regions. Each of these regional models will no doubt be useful in synthesizing the understanding of regional systems and in making predictions about changes that can occur in these systems as a result of natural or anthropogenic variations. The issue then arises as to how these regional models can be linked to provide estimates of natural fluxes between regions and estimates of fluxes between shelf and open ocean systems (i.e., boundary fluxes). The linking of regional models with one another and with global models represents an important and critical area of research that must be addressed.

A second critical area that needs to be addressed is the mismatch in space and time scales that arises when embedding a biological model within a circulation model. Frequently the relevant scales inherent (or needed) in

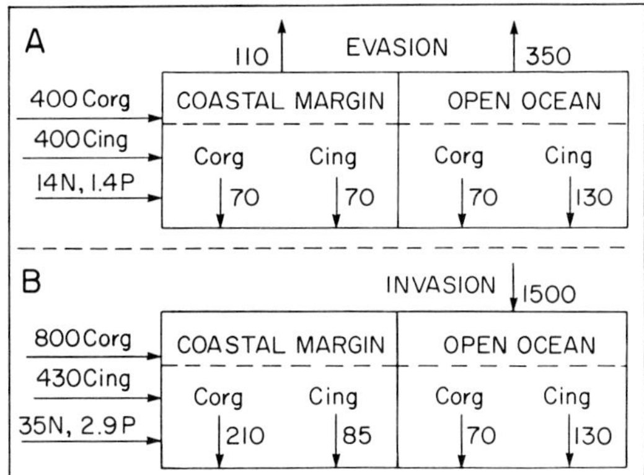

Fig. 3—Long-term, geological cycle (A) and present-day cycle (B) of carbon. Ing. = inorganic carbon; org. = organic carbon. Fluxes are in units of $MTCY^{-1}$. Modified from Wollast and Mackenzie (1989).

the two models are quite different. A related issue concerns the parameterization of biological and chemical processes, so that the processes can be included in circulation models. Resolving these problems requires input from experimentalists/observationalists, as well as input from modelers. In particular, it is important that interactions occur among physicists, biologists, and chemists at the onset of model development (i.e., interdisciplinary research), so that models will resolve the relevant physical as well as biological and chemical scales. Such interdisciplinary models for regional and ocean basin systems are currently being developed. The nature and frequency of measurements, physical, chemical, and biological, must be guided by the need to quantify the relevant processes.

The development and implementation of interdisciplinary models should precede, if possible, observational experiments that are designed to study processes relating to changes in the carbon (or any other) budget in specific regions. For regions for which historical data exist, the development of interdisciplinary models will synthesize these data and help clarify the relevant research questions. Once interdisciplinary models are available, simulations can be used to provide guidance as to what and where critical measurements are needed. An accelerated input of resources will be needed for global change studies in the context of the coastal margin. It is important that the development of regional interdisciplinary models begin now.

In concluding this section, it is appropriate to consider some of the changes in the components of the global system shown in Fig. 2 that can impact coastal margin ecosystems. These are summarized in Table 1, where the conditions of the glacial, interglacial, and present Earth's surface environment are given. The historical conditions provide a comparison with those of today and show that today's global environment for the parameters shown is substantially different from that of the interglacial and glacial climatic extremes. The surface environment of today is a system modified by the activities of humankind. There is little doubt that increasing fossil fuel burning and agricultural activities, as well as others, will continue for decades into the future. Such activities will continue to modify Earth's surface environment. The recent rates of change of the components of Earth's surface system are substantial, and although the future direction and rate of change are difficult to ascertain, qualitative estimates can be made (Table 1).

It can be seen from Table 1 that we anticipate future change in a number of the components shown in Fig. 2. These changes will certainly modify the physical and biogeochemical conditions of coastal margin ecosystems. Analysis of these systems will necessitate an expanded observational and modeling program to detect modifications and to discern the linkages and feedbacks that involve coastal margin systems and other global reservoirs. Because of its proximity to sources, the coastal margin system is the most

TABLE 1. Historical and present global environmental conditions of Earth's surface environment and recent and future changes in these conditions.

Component	Glacial	Conditions Interglacial	Present	Last 100 yrs (avg)	Future
CO_2	180 ppm	280 ppm	350 ppm	↑ 0.3% y^{-1}	↑ Acceler[a]
CH_4	0.3 ppm	0.8 ppm	1.6 ppm	↑ 0.6% y^{-1}	↑ Acceler[a]
N_2O	?	?	305 ppb	↑ 0.02% y^{-1}	↑ Acceler[a]
SO_2	—	—	90×10^9 kg y^{-1}	1% y^{-1}	↑
DMS	$60–400 \times 10^9$ kg y^{-1}	40×10^9 kg y^{-1}	40×10^9 kg y^{-1}	?	↑ ?
Temperature	284°K	288°K	288°K	↑	↑ ?
Mineral aerosol	5–10 x(y)	(y)	2×10^{12} kg y^{-1}	↑	?
Land runoff (H_2O)	2×10^{16} kg y^{-1}	3.7×10^{16} kg y^{-1}	3.7×10^{16} kg y^{-1}	↓ 0.1% y^{-1}	↑ Acceler[ab] ?
Particulate erosion products in runoff	~1×10^{13} kg y^{-1}	~7×10^{12} kg y^{-1}	1.5×10^{13} kg y^{-1}	↓	↓
Dissolved salts in runoff	—	—	4×10^{12} kg y^{-1}	↑	?
N_{tot} Riverine	—	1.4×10^{10} kg N y^{-1}	3.5×10^{10} kg N y^{-1}	↑	↑ Acceler[a]
P_{tot} Riverine	—	1.4×10^9 kg P y^{-1}	3×10^9 kg P y^{-1}	↑	↑
C_{org} Riverine	—	4×10^{11} kg C y^{-1}	8×10^{11} kg C y^{-1}	↑	?
Total marine net primary production	—	—	3.8×10^{13} kg y^{-1}	↑	↑ Acceler[a]

[a]Accelerating trend: ↑ increase; ↓ decrease.
[b]Damming will decrease runoff; temperature increase will increase runoff.

likely marine ecosystem to be affected by global and worldwide environmental change, and probably the most susceptible to environmental degradation. The projected trends of Table 1 provide some idea of the components of the global system that need to be considered in any future analysis of the coastal margins. In the next section, the biogeochemical cycle of carbon is presented as an example of how perturbations induced by human activities can affect the coastal margin system.

THE ROLE OF THE COASTAL MARGIN IN THE CARBON CYCLE: AN EXAMPLE OF CHANGE

In recent years, knowledge of the carbon cycle has expanded greatly. It is probably the best known cycle in terms of processes, reservoir magnitudes, and fluxes. However, even this knowledge is insufficient for an understanding of the past, present, and future behavior of the cycle. The role of coastal margins in sequestering organic and inorganic carbon, in exchange of CO_2 across the air–sea interface, and in carbon exchange with the open ocean is controversial. Thus, it is difficult to develop a scenario of future change for carbon in the coastal ocean ecosystem. However, one approach that may have some general application is to evaluate the long-term geological cycle of carbon and to compare it with the present cycle (see Wollast and Mackenzie 1989). Such an evaluation may provide us with a measure of the direction of carbon change in the coastal margin and a baseline against which to evaluate future change. The generic model presented below has application to the biogeochemical cycles of other elements in the coastal margin.

The comparison of the long-term geological cycling behavior of this element with its present-day biogeochemical cycle is illustrated in Fig. 3. The long-term geological cycle is that for the behavior of carbon prior to the industrial revolution and prior to extensive land use activities that led to deforestation and increased erosion rates. The coastal margin includes shallow water areas and the proximal regions of the slope, whereas the open ocean encompasses the distal slope, the rise, and the deep-sea floor.

Prior to human activities substantially affecting the carbon cycle, there was a *net* flux of CO_2 from the ocean through the atmosphere to the land, where the CO_2 was used in net production of organic matter and in weathering of minerals in continental rocks. The organic carbon produced and the dissolved inorganic carbon resulting from weathering were carried by streams to the ocean, where they were respired and precipitated, respectively. The oceans were a net heterotrophic system.

Because of fossil fuel burning and land use activities, the net transfer of CO_2 has been reversed, and the ocean is now an important sink of CO_2. However, the magnitude of this flux has been estimated only for the open ocean. There are no global estimations of CO_2 transfer in the coastal area where large departures in the partial pressure of CO_2 from equilibrium with the atmosphere are observed in relation to river input, upwelling, and intensive biological activity. Changes in the river input of materials have also strongly modified the fluxes of carbon, and this perturbation affects mainly the coastal zone. The organic carbon load of river waters has been increased by a factor of two, and the input of nutrients has more than doubled. Herein lies an important link between the carbon cycle and those of the nutrients N and P. These increased fluxes of N and P have increased primary production in the coastal area and thus the carbon flux to the sediments. The increase in total sedimentation rate because of increased erosion of the land has also enhanced the rate of burial of organic carbon on the continental shelf and slope. These increased fluxes of carbon all qualify as potential sinks of anthropogenic CO_2.

Another possible sink for CO_2 on a short time scale is the exportation of organic matter and calcium carbonate produced in the coastal area to the open ocean and sedimentation of this material to deep waters. The carbonate mineral phases of coastal zone sediments differ from those of the pelagic realm. The former are highly chemically reactive and can dissolve at oceanic depths shallower than the calcite chemical lysocline, or be sedimented on the seafloor. On the seafloor and within shallowly buried sediments, these phases may dissolve because of aerobic and anaerobic reactions involving organic matter. If they dissolve and produce alkalinity, their flux qualifies as a sink of fossil fuel CO_2. Furthermore, any deposition or respiration of organic carbon below the thermocline may also qualify as a sink. The uncertainties concerning these fluxes are considerable, ranging from a few percent of the fossil fuel flux to as much as 15%. Whatever the case, these fluxes may represent a significant fraction of the anthropogenic flux of carbon to the atmosphere.

This comparison between the long-term geological cycle of carbon and its modern cycle demonstrates how human activities can modify the coastal margin. It also shows how biogeochemical cycles are inextricably linked, in this case the C, N, P cycles. The demonstrated feedback to CO_2 accumulation in the atmosphere might not have been apparent if this analysis were not made. Future changes in the coastal zone carbon cycle depend, in particular, on changes in the riverine input of dissolved and particulate organic carbon, nutrients, and suspended material. Continued eutrophication of coastal margins will certainly enhance the differences between the geological cycle of carbon and that of the future. Possible future increased riverine runoff and decreased suspended load will further modify the biogeochemical cycling dynamics of carbon in the coastal zone and its interactions with the open ocean. Climate changes will affect not only the coastal margin carbon cycle but that of the whole globe. It is difficult to say specifically how all these interacting variables will modify the coastal zone carbon cycle because of our lack of knowledge of the operation of this cycle today.

FUTURE DIRECTIONS

Although it is generally accepted that coastal and marginal seas play a significant role in global biogeochemical cycles, it is less clear what quantitative role ocean margins play as sources/sinks for biogenic and other inorganic constituents and how they mediate fluxes of dissolved and particulate materials with respect to the ocean interior. However, because of the complexity and wide variability of marginal regimes, it has not as yet been possible to develop an appropriate rationale for selecting experimental sites within the global context. There have been a few attempts (e.g., Knauer 1987). In the Global Ocean Flux Studies report (Knauer 1987), ocean margins were classified within a matrix involving ocean circulation, river discharge, morphology, and latitude (Table 2). While this classification may at first seem tractable, amplification of any one of these factors operating in concert with corresponding factors soon produces a multidimensional matrix far beyond the resources of the most affluent nations to fund,

TABLE 2. Global coastal margin classification scheme (modified from Knauer 1987).

Order	Factor
1	Ocean Circulation a. Eastern boundary current b. Western boundary current c. Marginal sea–loop current
2	River Sediment Discharge[a] a. High, $>10^8$ T y^{-1} b. Low, $<10^6$ T y^{-1}
3	Margin Geometry (Shelf Width) a. Broad, >100 km b. Narrow, <50 km
4	Latitude a. Polar b. Mid-Latitude c. Tropic

[a]River water discharge must also be considered in this factor. Combinations of these factors produce a matrix of 36 types, not all of which occur on Earth.

if not the global scientific community to investigate. For example, the major physical processes thought to operate on a given margin can be broadly divided into Eastern boundary processes, Western boundary processes, and those processes common to all margins. Each of these can be further divided into an array of additional characteristics (Table 3).

Using Occam's Razor, we suggest that a more plausible starting point may be derived from a broader classification of coastal regions based on plate tectonic regimes (Fig. 4), because the processes of crustal plate interactions produce a limited number of possible margin morphologies. These major morphological provinces have been previously defined by

TABLE 3. Physical processes of importance in the coastal margin to the global ocean flux study. (Modified from Knauer 1987).

Processes	Importance
(a) Eastern Boundary Processes	
Wind-driven upwelling	Nutrient supply and coastal productivity
Upwelling filaments	Export to ocean interior
(b) Western Boundary Processes	
Separation of western boundary currents	Export to ocean interior
Shingles, eddies, and meanders	Nutrient supply to continental shelf, particle exchange
Warm core rings	Particle exchange
Western boundary current-induced upwelling	Nutrient supply to shelf and slope waters
(c) Processes Common to All Margins	
Coastline geology	Particle exchange
Canyons	Particle exchange
Topographic waves	Particle deposition
Instability of coastal fronts	Particle exchange
Freshwater inflows	Particle exchange, nutrient supply
(d) Specific Margin Processes	
Downwelling coasts	Particle and water transfer
Ice-controlled coasts	Water exchange, particle input

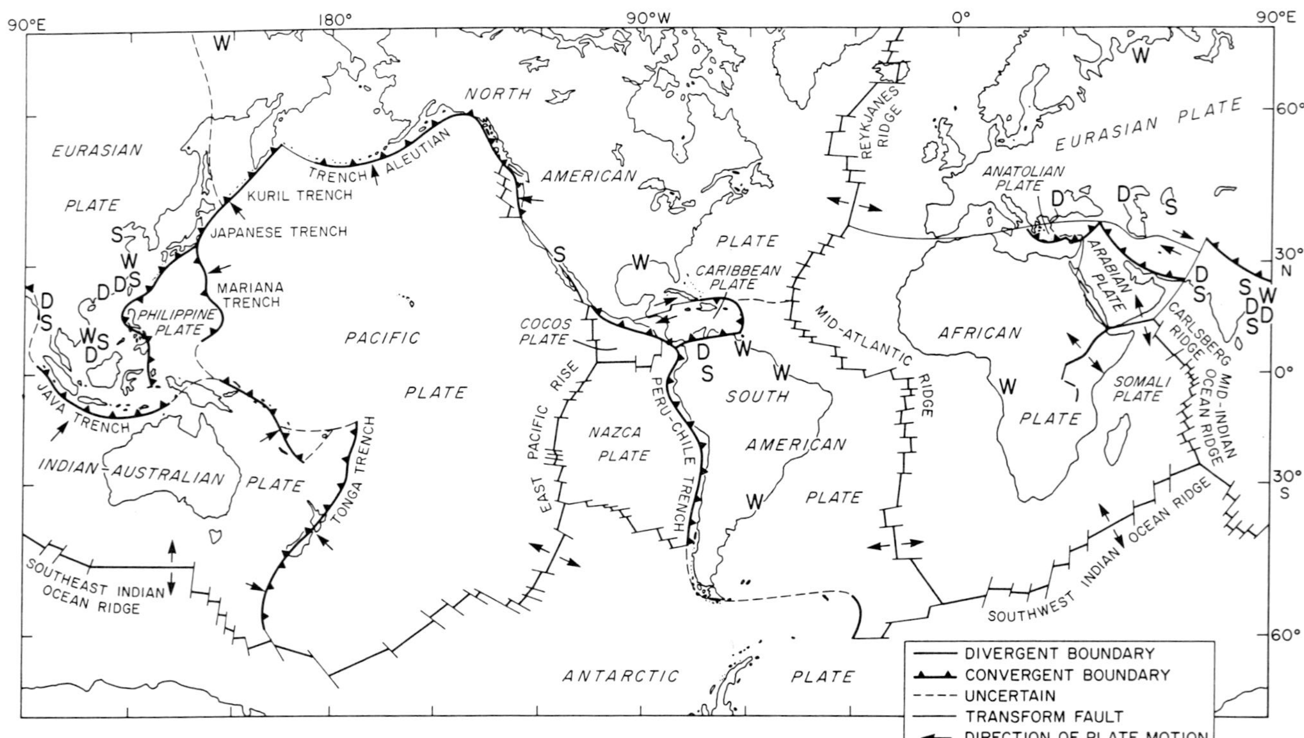

Fig. 4—Global plate tectonic regime map and ten major areas, respectively, of riverine sediment (S), water (W), and dissolved constituent (D) discharges.

Heezen et al. (1959) and are outlined in Table 4. The coastal margin morphological provinces are to a significant extent controlled by their plate tectonic setting. For example, where plates collide at an ocean margin, the collision can produce a marginal trench or trough, a steep slope and a narrow shelf; where a margin is located in passive mid-plate, the profile characteristically includes a broad shelf, gentle slope, and marginal rise or apron that merges with the associated abyssal plains of the deep ocean floor. Thus, to understand on a global scale the role of coastal margins in the biogeochemical cycling of the elements, their interaction with the open ocean, the impact of human activities on these systems, and the feedbacks involved, we suggest that several coastal margin areas be selected for study based initially on plate tectonic and interrelated morphological criteria.

It is implicit that not all areas of the ocean margin can be investigated to the degree required to construct a complete picture of fluxes in the global margin. Accordingly, the most important focus of margin studies are the processes that can be represented subsequently in models. The plate tectonics approach described above lays out one of the basic criteria for selecting suitable areas for intensified process studies. There are some additional criteria that are also relevant to this selection process. These may include overlay of maps of global oceanic production (Fig. 5) and the evaporation–precipitation cycle (Fig. 6) on a global plate tectonic regime map (Fig. 4), as well as overlay of the plate tectonic map by maps or estimates of major river sediment and water discharge rates (see Fig. 4).

TABLE 4. Ocean basin morphological provinces (modified from Knauer 1987).

Order	Feature[a]	
1	**Ocean Basins:** Continents	
2	**Continental Margin:** Mid-ocean ridges, deep ocean floors	
3	**Continental Shelf:** Continental slope, continental rise, marginal trench, marginal plateau, outer rise	
	Associations:	
	Passive (e.g., Atlantic)	*Active (e.g., Pacific)*
	Broad continental shelf	Narrow continental shelf
	Broad complex slope	Narrow slope
	Marginal plateaus	Margin trench/trough
	Continental rise	Outer rise

[a]Note that in each successively smaller order, only the boldface features in the preceding order have been expanded in the table.

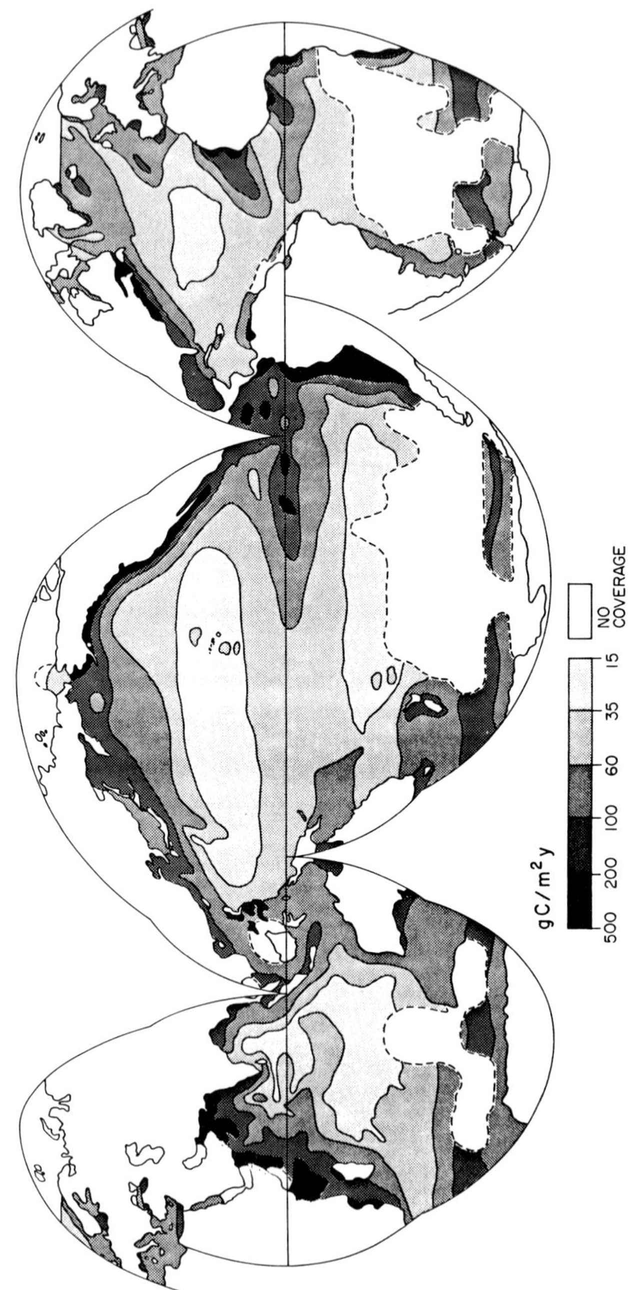

Fig. 5—Map of global total marine productivity. After Berger et al. (1988).

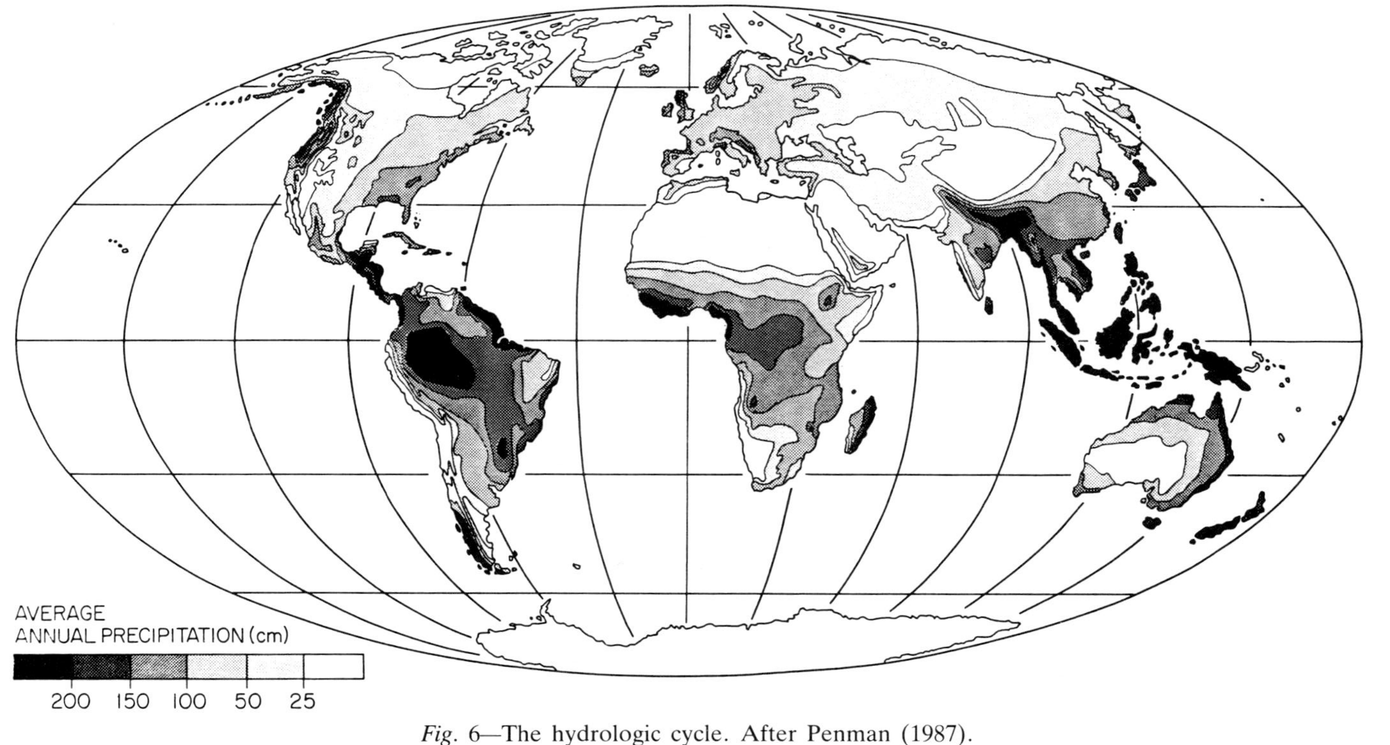

Fig. 6—The hydrologic cycle. After Penman (1987).

This methodology could lead to selection of coastal margin areas based on the criteria of seafloor morphology, organic production, the hydrologic cycle, and the importance of the margin as a depocenter. Furthermore, there needs to exist sufficient understanding of the physical oceanographic conditions, particularly water transport, to permit an adequate experimental design to be formulated. Equally, the area must be suitable for the conduct of adequate physical oceanographic measurements to constrain the transport of material between the shelf and slope regimes. Evaluation of input and export fluxes requires the use of circulation models which already have been developed for several coastal marine systems. A further criterion is that at least one of the selected areas is one in which significant eutrophication is occurring (in the sense that the anthropogenic introduction of nutrients substantially augments the natural influx of nutrients), and/or there is significant transport of organic carbon by rivers into the area. This latter condition could go some way toward satisfying socio-economic interests in margin studies and might enhance the chances of obtaining appropriate funding.

It has been demonstrated with respect to the carbon cycle (see above) that the coastal margins can play a role in sequestering fossil fuel CO_2 owing to processes involving organic and inorganic carbon. These coastal systems also influence the carbon and nutrient chemistry of the open ocean (e.g., Wollast this volume; Agegian 1988). There is a strong need to select coastal margin systems for investigation that are important in a global context. Such selection, when the various studies are synthesized and integrated, will one day enable prediction via model calculations of the role of the coastal margins in the natural global system and as a sink/source and feedback system for materials added worldwide by the activities of humankind.

CONCLUDING STATEMENT

As is the case with many studies of the Earth's biogeochemical systems, the study of ocean margins requires a seemingly paradoxical, dual approach. Throughout the history of development of all natural sciences, it has been necessary to define the scientific questions in a reductionist fashion, bring each question to a point where hypotheses can be formulated and tested via experiments. Many different disciplines may address different aspects of the same system. For example, research on the coastal oceans requires the attention of physical oceanographers, biologists, geologists, chemists, mathematicians, statisticians, and meteorologists. However, for addressing questions of global change, it is not sufficient to stop at this reductionist stage. The second approach, that of integration, involves the reassembly of the results of the reductionist approaches into a single picture of the complex system. The paradox arises because traditional training in reductionist science requires a goal of complete understanding of a finite system. This approach commonly requires consideration of simplified situations, often represented by laboratory experiments and/or mathematical models. The uncertainty that is necessary in attacking the system as a whole may be unacceptable to the "pure" chemist or biologist, etc. What may be needed is a new scientific discipline which considers integration of global geoscience, doing so with the same tools, intensity, personal discipline and quality of logic of reductionist science. James Lovelock has suggested the name "geophysiology," and others have coined the term "Earth system science" or "global geoscience." In any case, it will be useful to conduct research on the coastal oceans as a linked component of the Earth system. The reasons for this approach are clear:

1. The coastal ocean influences the open ocean, the atmosphere, and at least nearby areas of land, if not all the continent.
2. The other components of the Earth system all influence the coastal ocean.

It is, therefore, impossible to study the coastal ocean as a single system with exogenous input conditions; rather, it is necessary to treat it in a global context as one subset of dependent variables, the changes of which have clear importance to human society.

SELECTED BIBLIOGRAPHY

Agegian, C.R., ed. 1988. Biogeochemical Cycling and Fluxes Between the Deep Euphotic Zone and Other Oceanic Realms. Rockville, MD: Natl. Undersea Res. Prog. Res. Rept. 88-1.

Berger, W.H., V.S. Smetacek, and G. Wefer, eds. 1987. Productivity of the Ocean: Present and Past. Dahlem Workshop Report LS44. Chichester: Wiley.

Berger, W.H., K. Fischer, C. Lai, and G. Wu. 1988. Ocean carbon flux: global maps of primary production and export production. In: Biogeochemical Cycling and Fluxes Between the Deep Euphotic Zone and Other Oceanic Realms, ed. C.R. Agegian, pp. 131–176. Rockville, MD: Natl. Undersea Res. Prog. Res. Rept. 88-1.

Bolin, B., and R.B. Cook, eds. 1983., The Major Biogeochemical Cycles and Their Interactions. SCOPE 21. New York: Wiley.

Georgii, H.W., and W. Jaeschke, eds. 1982. Chemistry of the Unpolluted and Polluted Troposphere. NATO ASI Series C, Mathematical and Physical Sciences, no. 96, Dordrecht: Reidel.

Knauer, G.A., ed. 1987. Ocean Margins in Global Ocean Flux Studies. U.S. Global Ocean Flux Studies Planning Report No. 6.

Liss, P.S., and W.G.N. Slinn, ed. 1983. Air-Sea Exchange of Gases and Particles. NATO ASI Series, Series C, Mathematical and Physical Sciences, no. 108. Dordrecht: Reidel.

Meybeck. M. 1984. Les Fleuves et le Cycle Geochimique des Elements. D.S. Thesis, Ecole Normale Superieure, Laboratoire de Geologie, Paris.

Oeschger, H., and C.C. Langway, Jr., eds. 1988. The Environmental Record in Glaciers and Ice Sheets. Dahlem Workshop Report PC8. Chichester: Wiley.

Penman, H.L. 1970. The water cycle. *Sci. Amer.* **223**(3):98–108.

Rowland, F.S., and I.S.A. Isaksen, eds. 1987. The Changing Atmosphere. Dahlem Workshop Report PC7. Chichester: Wiley.

Stumm, W., ed. 1977. Global Chemical Cycles and Their Alterations by Man. Dahlem Workshop Report PC2. Berlin: Abakon.

Wollast, R., and F.T. Mackenzie. 1989. Global biogeochemical cycles and climate. In: Climate and Geo Sciences, ed. A. Berger et al., pp. 453–473. Dordrecht: Reidel.

List of Participants with Fields of Research

BEWERS, J.M.
Bedford Institute of Oceanography
P.O. Box 1006
Dartmouth, Nova Scotia B2Y 4A2
Canada

Marine geochemistry of trace metals and radionuclides

BILLEN, G.F.L.
Université Libre de Bruxelles Groupe de Microbiologie des Milieux Aquatiques
Campus de la Plaine CP 221
Boulevard du Triomphe
1050 Bruxelles, Belgium

Aquatic microbiology

BLACKBURN, T.H.
Dept. of Ecology and Genetics, Building 550
University of Aarhus
Ny Munkegade
8000 Aarhus C, Denmark

Mineralization in marine sediments

BLANTON, J.O.
Skidaway Institute of Oceanography
P.O. Box 13687
Savannah, GA 31416, U.S.A.

Circulation processes on continental shelves; frontal zone dynamics

CHARLSON, R.J.
Dept. of Atmospheric Sciences, AK-40
University of Washington
Seattle, WA 98195, U.S.A.

Atmospheric chemistry

CHURCH, T.M.
College of Marine Studies
University of Delaware
Newark, DE 19716, U.S.A.

Atmospheric and marine chemistry

DUCE, R.A.
Graduate School of Oceanography
University of Rhode Island
Narragansett, RI 02882, U.S.A.

Atmospheric chemistry

EISMA, D.
Netherlands Institute for Sea Research
P.O. Box 59
1790 AB Den Burg, Texel
The Netherlands

Recent marine sediments

FOWLER, S.W.
International Laboratory of Marine Radioactivity, IAEA
19 Avenue des Cestellans
98000 Monaco

Marine biogeochemistry of radionuclides and trace elements

FRASCARI, F.
Istituto per la Geologia marina – CNR
Via Zamboni 65
40127 Bologna, Italy

Dynamics of fine particles and particle-associated pollutants in estuarine and coastal marine environments

GOLDBERG, E.D.
Scripps Institution of Oceanography
A-020
La Jolla, CA 92093, U.S.A.

Scavenging processes in seawater: platinum metals in the marine environment

GORDEEV, V.V.
Institute of Oceanology
USSR Academy of Sciences
Krasikova Str. 23
Moscow 117218, U.S.S.R.

Geochemistry of estuarine zones and sea water, chemistry of hydrothermal solutions on ocean bottom

HAINBUCHER, D.
Institut für Meereskunde
Universität Hamburg
Troplowitzstrasse 7
2000 Hamburg 54, F.R. Germany

Simulation of mesoscale processes in the North Sea

HALIM, Y.
Dept. of Oceanography, Faculty of Science
Moharram Bey
Alexandria, Egypt

Coastal zone ecosystems; eutrophication; dinoflagellates

HOFMANN, E.E.
Dept. of Oceanography
Old Dominion University
Norfolk, VA 23529, U.S.A.

Modeling physical–biological interactions in marine ecosystems and descriptive physical oceanography

HOLLIGAN, P.M.
Plymouth Marine Laboratory
Prospect Place, The Hoe
Plymouth PL1 3DH, U.K.

Marine phytoplankton ecology and biogeochemistry

HSÜ, K.J.
Geologisches Institut
ETH-Zentrum
8092 Zürich, Switzerland

Marine geology, sedimentary processes

JICKELLS, T.D.
School of Environmental Sciences
University of East Anglia
Norwich NR4 7TJ Norfolk, U.K.

Biogeochemical cycling of trace elements in the atmosphere and oceans

KNAUER, G.A.
Center for Marine Science
University of Southern Mississippi
TRL 242
Stennis Space Center, MS 395929, U.S.A.

Biological oceanography, sediment traps and primary productivity

KRAFT, J.C.
Dept. of Geology
University of Delaware
Newark, DE 19716, U.S.A.

Geology of coasts, stratigraphy, sedimentology, and geomorphology. Impact of sea-level rise on peoples' occupancy of the coastal zone

KROM, M.D.
Israel Oceanographic and
Limnological Research
P.O. Box 8030, Tel Shikmona
Haifa 31080, Israel

Marine pollution; chemical oceanography water quality in fishponds

LISS, P.S.
School of Environmental Sciences
University of East Anglia
Norwich NR4 7TJ Norfolk, U.K.

Air-sea gas exchange

MACKENZIE, F.T.
Dept. of Oceanography, MSB 525
University of Hawaii
Honolulu, HI 96822, U.S.A.

Geology–geochemistry

MANTOURA, R.F.C.
Plymouth Marine Laboratory
Prospect Place, The Hoe
Plymouth PL1 3DH, U.K.

Marine organic chemistry

MARTENS, C.S.
Marine Sciences
CB-3300, 12-5 Venable Hall
University of North Carolina
Chapel Hill, NC 27599-3300, U.S.A.

Marine geochemistry and global biogeochemical cycling

MARTIN, J.-M.
Institut de Biogéochimie Marine
Ecole Normale Supérieure
1, rue Maurice Arnoux
92120 Montrouge, France

Environmental chemistry

MCCAVE, I.N.
Dept. of Earth Sciences
University of Cambridge
Downing Street
Cambridge CB2 3EQ, U.K.

Shallow and deep marine sediment transport and deposition; global environmental change record in Pleistocene and Holocene deep-sea cores

MEINCKE, J.
Institut für Meereskunde
Universität Hamburg
Troplowitzstrasse 7
2000 Hamburg 54, F.R. Germany

Physical oceanography—North Atlantic water mass formation and circulation

MILLIMAN, J.D.
Woods Hole Oceanographic
Institution, Woods Hole, MA 02543
U.S.A.

Marine geology/geological oceanography

MOLL, A.
Institut für Meereskunde
Universität Hamburg
Troplowitzstrasse 7
2000 Hamburg 54, F.R. Germany

Mathematical modelling of marine ecosystems – annual cycle of plankton dynamics in the North Sea

MOREL, F.M.M.
Ralph M. Parsons Laboratory,
Building 48-423
Massachusetts Institute of Technology
Cambridge, MA 02139, U.S.A.

Aquatic chemistry

NÖTHIG, E.-M.
Alfred-Wegener-Institut für Polar-und
Meeresforschung
Postfach 12061, am Handelshafen 12
2850 Bremerhaven, F.R. Germany

Plankton ecology of polar oceans and shallow coastal environments

O'KANE, J.P.
Dept. of Civil Engineering, Centre
for Water Resources Research
University College
Earlsfort Terrace
Dublin 2, Ireland

European river ocean system – 2000

QUACK, B.
Institut für Meereskunde an der
Universität Kiel
Düstenbrooker Weg 20
2300 Kiel 1, F.R. Germany

Volatile halogenated organic compounds in and over the sea

SCHAREK, R.
Alfred-Wegener-Institut für Polar-und
Meeresforschung
Postfach 120161, Am Handelshafen 12
2850 Bremerhaven, F.R. Germany

Plankton ecology of polar oceans

SICRE, M.-A.
Laboratoire de Physique et Chimie
Marines
Tour 24-25, 5ème étage
Université Pierre et Marie Curie
4, place Jussieu
75005 Paris, France

Organic geochemistry

SMETACEK, V.S.
Alfred-Wegener-Institut für Polar-und
Meeresforschung
Postfach 120161, Am Handelshafen 12
2850 Bremerhaven, F.R. Germany

Plankton ecology of polar oceans and shallow coastal environments

SPITZY, A.
Max-Planck-Institut für Meteorologie
Bundesstrasse 55
2000 Hamburg 13, F.R. Germany

Carbon cycling in the hydrosphere

STEELE, J.H.
Woods Hole Oceanographic
Institution
Woods Hole, MA 02543, U.S.A.

Comparison of marine and terrestrial ecosystems

SUZUKI, Y.
Meteorological Research Institute
1-1 Nagamine
Tsukuba
Ibaraki 305, Japan

Carbon and nitrogen (DOC and DON) cycle chemical speciation of trace metals and chemical tracer in the ocean

VAULOT, D.
Dept. of Oceanography
University of Hawaii
1000 Pope Road
Honolulu, HI 96822, U.S.A.

Biological oceanography: phytoplankton ecology

WALSH, J.J.
Dept. of Marine Science
University of South Florida
140 Seventh Avenue South
St. Petersburg, FL 33701, U.S.A.

Systems analysis of continental shelves

WHITFIELD, M.
Plymouth Marine Laboratory, The Laboratory
Citadel Hill
Plymouth PL1 2PB, U.K.

Biological involvement in global CO_2 cycle;
particle/water interactions and trace metal cycling in the oceans

WINDOM, H.L.
Skidaway Institute of Oceanography
P.O. Box 13687
Savannah, GA 31416, U.S.A.

Riverine, estuarine and marine trace element geochemistry

WOLLAST, R.F.
Laboratoire d'Océanographie Chimique
Université Libre de Bruxelles
Campus de la Plaine CP 208
Boulevard du Triomphe
1050 Bruxelles, Belgium

Marine geochemistry

Author Index

Subject Index